**Green Energy and Technology**

More information about this series at http://www.springer.com/series/8059

Tomayess Issa · Pedro Isaias
Theodora Issa
Editors

# Sustainability, Green IT and Education Strategies in the Twenty-first Century

Springer

*Editors*
Tomayess Issa
Curtin University
Perth, West Australia
Australia

Theodora Issa
CBS - School of Management
Curtin University
Perth, West Australia
Australia

Pedro Isaias
Institute for Teaching and Learning Innovation
The University of Queensland
Brisbane, St Lucia, Queensland
Australia

ISSN 1865-3529 ISSN 1865-3537 (electronic)
Green Energy and Technology
ISBN 978-3-319-57068-6 ISBN 978-3-319-57070-9 (eBook)
DOI 10.1007/978-3-319-57070-9

Library of Congress Control Number: 2017948384



Printed on acid-free paper

This Springer imprint is published by Springer Nature
The registered company is Springer International Publishing AG
The registered company address is: Gewerbestrasse 11, 6330 Cham, Switzerland

# Foreword

The Curtin University Sustainability Policy (CUSP) Institute has 110 PhDs and 50 Master's students in Sustainability and Climate Policy. We cannot find any other sustainability postgrad institute with so many students, so we must be doing something right. As a Foreword to this fine book I thought I would try and reflect on why we may have hit on a workable formula.

1. I think that we have always been trying to make a difference in the world rather than trying to be a theoretically correct place. When students come to me wanting to do a PhD I always ask them 'who in the world will be able to use your research to make a better world?' If they are so theoretically-minded that they cannot answer that then I don't discourage them but I try to show what other parts of the university would be better for them. Our PhDs always use theory but they are not there to help solve a theoretical issue, but to see how policy in government, business or the broader community can be changed to be more effective in the sustainability agenda.
2. I think we always try to be hopeful. The sustainability agenda can be very despairing as academics have tried to scare people and politicians into action. In my view this doesn't work. People and politicians, in particular, need to see that the problems are real and can be very scary if they continue to worsen but most want to know what they can do about it. Thus we try to work closely with business, governments, NGO's and the broader community, in framing the research and thus ensuring it is acted on. PhDs and Master's students cannot help but be launched into a career of hopeful work when they realize how easy it is to make changes that matter.
3. I think it helps to have a media presence. CUSP academics have always found it easy to be at the forefront of media as our issues are a daily news item. But we work hard at it which means that we cultivate the journalists who we know like to do more in-depth stories. This aligns to a strong web presence and social media linkages that enable our short clips and longer shows to be provided to the broader public (e.g., the ABC Catalyst show early this year was the most popular in downloads of any Catalyst show). We also make our own

films and there are over 100 on the CUSP website simply made by our embedded filmmaker Linda Blagg using a non-corporate approach rather more like the YouTube approach. Linda did her own PhD with us on how a Prezi can be used to tell the sustainability journey of any change-maker (in her case it was the Mayor of Fremantle Brad Pettitt, an associate professor in Sustainability from Murdoch).

4. Such media presence enables people who do not work in sustainability to see that perhaps they could do that kind of work. They come and see whether they can begin their journey by doing research or coursework even though they don't have background or training. Most of our postgrads are in their 30's or 40's and are wanting to change professions. Once they can see a glimpse of how they can build on their background and can even use their business or government job as a case study, then they generally get into it with great gusto.
5. The units we teach are pretty good! They get good feedback anyway. PhDs don't do coursework in Australia but we always encourage our PhDs to sit in on anything that they feel can help. My unit in Leadership for Sustainability has a creative component that is meant to stretch them in how they think and express themselves by doing some artistic project to demonstrate a different approach to solving a sustainability issue. These always start with great suspicion and often fear, but usually end in enormous satisfaction and sometimes tears as we all sense the depth of emotion involved.

So how does this all relate to Green IT? Our fundamental belief is that the fifth wave of innovation, the digital revolution, will be a foundation for the sixth wave of innovation, the sustainability revolution. We must be smart and sustainable. We don't have to teach our postgrads anything about IT, they teach us. But we do need to work with them on how IT can help us become more sustainable. Our definition of this is that sustainability must be simultaneously reducing footprint while improving liveability. And you cannot do this without IT and smart systems thinking.

Two examples from my own research group will illustrate this.

Jemma Green has just completed a PhD on Disruptive Innovation and Low Cost Low Carbon Housing. She examined how it is that Perth suddenly went for rooftop solar and how can it be managed in the grid. Rooftop solar is now the biggest power plant in Western Australia (550 MW) and it took the utilities by surprise. Jemma has created a way of enabling 'citizen utilities' to be formed that can enable two-way flows of electricity and be managed locally to find the most effective markets for the green electrons. This can only be done using blockchain software and once the model was established she set up a firm and has significant global markets from utilities already being established to trial the approach. Not bad for a PhD.

Josh Byrne is a staff member in CUSP and has a significant national media presence. Josh built two houses in an ordinary suburb of Fremantle that were cheap and yet with very low carbon/low water in their footprint. The houses were fully monitored so they could form the basis of research with around 70 different channels of information from various sensors in the house and garden. The rooftop solar was supplemented with a battery system and all the monitoring has been available on his website for anyone to access. Josh has a series of films that have followed his journey and now the first top-line journal articles are appearing that show how his house is fossil fuel-free and largely independent of the main water supply. The role of the grids is also better understood now from this research. This household-scale research has led to a national program monitoring other innovative houses and now innovative precincts (like WGV). Most of all he has shown how a normal Perth family can live well but without a huge footprint, indeed they are a regenerative household.

Such stories are the basis of sustainability, Green IT and education. I hope the other chapters in this book provide much more for our readers of the way we together can create a better world which is after all what education is about.

Curtin University
Western Australia, Australia

Tomayess Issa and Theodora Issa

The University of Queensland
Queensland, Australia

Pedro Isaias

# Preface

In the time of evening instead of all sacrifices give thanks, and Instead of offerings deliver a pure prayer in love, and he who has mouth and word and tongue ought to give thanks for the creatures which are silent.

St. Jacob of Serugh (+521)

ܒܥܕܢܐ ܕܪܡܫܐ ܚܠܦ ܟܠ ܕܒܚܝܢ ܐܘܕܘ ܬܘܕܝܬܐ:
ܘܚܠܦ ܩܘܪܒܢܐ ܨܠܘܬܐ ܕܟܝܬܐ ܒܚܘܒܐ ܩܪܒܘ.
ܘܐܝܢܐ ܕܐܝܬ ܠܗ ܦܘܡܐ ܘܡܠܬܐ ܐܦ ܠܫܢܐ:
ܚܝܒ ܕܢܘܕܐ ܚܠܦ ܒܪܝܬܐ ܫܬܝܩܬܐ.

ܡܪܝ ܝܥܩܘܒ ܕܣܪܘܓ (+521)

With the swift growth in technology usage by individuals and businesses alike, both nationally and internationally, technology becomes a substantial instrument not only for communication, collaboration and cooperation but also for conducting research, entertainment, finance, human resource, marketing and knowledge management. Currently, the recycling of technological devices has become a major dilemma for users and businesses. A modern society should tackle and address this problem in a matter of urgency. The urgency about the matter lies in the lack of ability by the existing infrastructure to handle the increased demands of users and businesses. If the trend continues, the seventh generation [1] will have fewer chances to enjoy an infrastructure similar to the current one.

To tackle and address this problem, it is crucial that the education sector plays an important role in transforming businesses, designers, students' critical thinking and mind-sets towards sustainability and Green IT usage.

This can be achieved by integrating sustainability and Green IT strategies in the education curriculum as well as in businesses and designers' strategies. The education sector should raise and alert users, designers and businesses' moral responsibilities for their contribution to worldwide sustainable development; guiding them to become more sustainable and greenish, sharing with them some strategies, tactics and approaches to change their habits at workforce and study to become greener.

This book aims to address the awareness of sustainability and Green IT in higher education and businesses in developed and developing countries and, furthermore, to raise the new leaders' awareness and mindfulness toward the technology usage problems, as well as to examine the advantages and disadvantages of sustainability and Green IT from the higher education and organizations perspective. Finally, this book will support higher education, business sectors and academic audience (academics, university teachers, researchers and post-graduate students – both Master and Doctorate levels). In addition, it will be beneficial for public and private universities, Information Systems (IS) and sustainability developers and researchers, education managers, professionals related to the Information Society and Information and communication technology (ICT) and education sectors.

The majority of respondents from various countries indicated that sustainability can produce various advantages that include increased productivity, the creation of new jobs and business opportunities, attracting quality employees and satisfying the needs and expectations of customers, stakeholders, the community and society in general. Furthermore, assimilating and integrating sustainability in business strategies and everyday activities will reduce harmful emissions, pollution, health hazards, carbon footprint and the consumption of energy, raw materials and water. However, sustainability can bring some disadvantages from increased marketing failure, scandals by perceived environmental irresponsibility, insurance crises due to environmental disasters, new regulations, security, competition for and cost of raw materials, and increase in fraud including environmental actions and breaches of environmental compliance.

Nowadays, to be sustainable, exercise good stewardship and act sensibly, individuals and organizations must integrate sustainability strategies in their activities. This can be done via various resources and technologies such as Green IT and smart technology in order to conserve resources, energy and raw materials while serving crucial social needs.

There is a rapid increase in the technology used by individuals and businesses alike, both nationwide and worldwide, technological devices now play a significant role in a range of everyday human activities, not only for communication, collaboration and cooperation but also for research, entertainment, finance, human resources, marketing and knowledge management. Currently, the recycling of ICT devices is a major concern for businesses and individuals, since it is not simply a means of addressing environmental damage or a solution to an environmental problem. As a nation, we should tackle this problem as a matter of urgency.

Our planet continues to be damaged by human activities. Hence, there is an urgent need for a major shift in the priorities and mindsets of industries, IT professionals, educators, researchers and technology users since there is no plan B for our planet [1]. Therefore, ideas, opinions and strategies should be developed to manage this problem in both developed and developing countries. Sadly, developed countries are shifting the problem from their own backyard to another area. These countries consider themselves to be sustainable whereas in fact their manufacturing and production activities are being conducted in a developing country.

Developed countries should assist by offering new strategies to the developing countries in order to help them cope with an increasing population and the increased demand for food, health, education, infrastructure and other facilities, especially for the new "Internet" generation. This generation's people want the same commodities and facilities as their counterparts enjoy in developed countries. This will cause a huge problem, not only for the developing countries but globally, since this generation expects to have new devices, including mobiles, computers, laptops and other technological devices, thereby putting huge pressure on the supply of raw materials, especially in developing countries such as India and China.

The urgency of the matter lies in the lack of ability by the existing resources to handle the increased demands of users and businesses. If the trend continues, the seventh generation [3] will be unable to enjoy an infrastructure similar to the current one. Stakeholders at all levels, including universities, must contribute to the transition towards a sustainable world.

The tertiary education sector plays a major role in transforming societies' and students' critical philosophies and world views through their curricula and research. Universities need to raise students' awareness, especially in relation to their moral responsibility to contribute to sustainable development, and guide them to a better, sustainable future. Currently, there is a call for businesses and individuals to integrate sustainability in their business strategies and to include cloud computing technology for example, as a tool for sustainable work especially in the IT departments to cut costs and increase efficiency and productivity.

By adopting sustainable strategies, businesses and individuals who are aware of the carbon footprint and environmental damage can conserve raw materials for the seventh generation, as well as to share their level of commitment to alleviating environmental problems. Therefore, the higher education sector should take responsibility for teaching students about sustainability and raising their awareness, as they will be the leaders of the future whose role will include preserving the current infrastructure, and conserving raw materials and resources for the seventh generation.

In the wake of the global financial crisis in 2007–2008, academics, researchers and universities started to consider how to confront these financial glitches facing organizations of all sizes internationally and nationally. Organizations, and in particular their IT departments, were pushed to handle the pressure successfully. Therefore, to alleviate this problem in organizations and in particular their IT departments now and in the future, universities started to introduce topics related to the concepts of sustainability and Green IT in order to raise awareness of the ways in which ICT harms our planet and to ensure that students acquire the necessary knowledge of these concepts since the majority of the students confirmed that these concepts are new and not part of their vocabulary [2]. In Western Australia, specifically at Curtin University, a postgraduate unit on sustainability and Green IT was developed and delivered to master degree students.

Now, to resolve these problems world-wide, academics need to develop units which focus on sustainability, ethics, moral responsibility, physical and spiritual welfare, and Corporate Social Responsibility (CSR). It is anticipated that such units will shift students' mindsets as the specific activities and assessment tasks will encourage students to adopt sustainable, ethical practices that will benefit them and their communities, and ultimately their country. These students will transfer their knowledge to other generations by education, face-to-face or online via the social networking tools, i.e. Facebook, Twitter, Linkedin, Google+, YouTube, Pinterest and Instagram. The new technology plays a major role in transforming the thinking, attitudes and behaviour of the new generation since they depend on these technologies and we can reshape the new generations' conduct and mindsets via workshops, discussion, debate, communication, cooperation, collaboration and connection.

Time is running out guys, and we need to move quickly as soon as possible, as I mentioned previously that our planet is suffering and deteriorating from our actions and activities, and as human beings, we need to pay attention to our Mother Nature as majority of the damages was carried and done by 'Human Beings'; as a result, we need to take responsibility to tackle this problem as soon as possible to allow the seventh generation to live in similar conditions as us, as there is no plan B for the planet.

We need to integrate and practice sustainability in all our everyday activities, i.e. truth, beauty, moral goodness, economic opportunity, justices, liberty, physical security, cultural creativity, natural environment, high quality of life, good education and political, law, social and technology needs. The practice of sustainability will make our lives healthier, fitter, and stronger for our community and society. Good stewardship and ethical conduct will ultimately be rewarded and celebrated based on our actions locally and globally, and this means living with aspiration, ambition, hope and desire now as well as in future.

In conclusion, this book aims to examine, analyze and present the outcomes from our study on sustainability awareness in developing and developed countries. Currently, sustainability is a buzz word with which people have become familiar via education. The majority of the authors from this book indicated that sustainability should consider and reflect on individuals and organisations duties, in order to change their mind-set, attitudes and activities so as to preserve our planet. The following are comments from students who completed a master degree unit on sustainability and Green IT awareness and responsiveness.

> I would say that completing [ITS65 unit] has given me a new skill altogether, Sustainability Management. This is an area that I have previously professed to have very little knowledge of. However I now feel that I have learnt enough about sustainability to put it into practice in my profession and make a difference to the work that I do.

> I am very glad that I took this unit and I believe it has opened my eyes on new aspects that I did not know before. Furthermore, studying this unit has inspired me to participate in spreading the awareness of Green IT the author's awareness of sustainability was absolutely improved. My mind-set was changed and nowadays, despite of only blindness pursuing the high-end technology product, the author now will pay more attention on the products and manufacturer's sustainable aspects.

> I think [My] University is doing a good job by including this unit in its curricula and being an example to other universities because it is education and learning that play a vital role in changing the attitude and mindsets of the students towards the society and the environment as well. Other higher education institutions should also understand this role and they should take initiatives in transforming their education system and move towards providing sustainable education by introducing sustainability or sustainable development concepts into the curriculum.

> Though the subject on sustainability is offered in very scanty manner in universities, 'it is the subject' everyone must learn and practice in their 'life' to care for the planet and generations to come. This course will definitely impart a good amount of 'concerns' on the sustainability of mother earth by our own actions leading to be 'reasonably good humans' in the process.

> Sustainable and Green IT is a good unit. Which basically focuses on issues and problems technology is causing and how we can stop technology from harming environment. It also mentions that we have to concentrate and work towards sustainable development so that we preserving for future generation.

> University is a place where the students, the future leaders, are nurtured through learning, experiment, and research. So, by introducing sustainability and green IT in the education sector, universities will create awareness and boost students to become socially and environmentally responsible. I also feel that [My] University should make such curriculum available for all students irrespective of what courses they are undertaking.

… Tomayess and Theodora shared with you some of their students' comments which indicated that completing the sustainability unit for the master degree changed their mindset and attitude regarding our planet.

Guys, we need to move rapidly, as I said our planet is in pain and we need to heal this pain quickly, otherwise it will be too late for us and the most important for our next generation, guys there is no plan B for the earth… therefore, take care of it.

Finally, guys we are visitors on this planet and we need to pay attention of our mother nature from economic, environmental and social conditions; once we consider them carefully, we will live in sustainable life, now and in the future. Consequently, the time for action is now and we need to act swiftly otherwise, it will be too late to illuminate the current and future problems.

Curtin University
Western Australia, Australia
Tomayess Issa and Theodora Issa

The University of Queensland
Queensland, Australia
Pedro Isaias

## References

1. Issa T, Isaias P (2015) Sustainable design: HCI, usability and environmental concerns. Springer London
2. Jucker R (2002) Sustainability? Never heard of it!": Some basics we shouldn't ignore when engaging in education for sustainability. Int J Sustain High Educ 3 (1):8–18
3. Newton LH (2003) Ethics and sustainability, sustainable development and the moral life. In: Basic Ethics in Action. Prentice-Hall, Inc., Upper Saddle River, New Jersey, United States

# Acknowledgements

We acknowledge the support of all people who contributed directly or indirectly to this book.

Tomayess and Theodora Issa owe special gratitude first to almighty and their parents: V. Rev. Fr. Boutros Touma Issa; and the Spirit of Bathqyomo Marine Khoury-Issa. Their brother Dr. Touma, his wife Siba, and their daughter Talitha; their sister Tamara, her husband Tony, and their children Tabitha, Antoinette, and Jacob, and finally to their brother Theodore, his wife Mary, and their children Cephas and Mary; for their continuous support, and encouragement since without their help, this work would never have been completed. Tomayess and Theodora would like to acknowledge the support of the School of Information Systems and School of Management, respectively, Curtin University, Australia, for their inspiration and encouragement.

Pedro Isaias would like to acknowledge the other authors of this book for their cooperation and work, as well as for the joint work developed so far that has resulted in various publications. He would also like to acknowledge the support of his family. Without their assistance, this work would not be possible at all.

Curtin University
Western Australia, Australia

Tomayess Issa and Theodora Issa

The University of Queensland
Queensland, Australia

Pedro Isaias

# Contents

**Changing the Students' Mind-set via Sustainability** . . . . . . . . . . . . . . . . 1
Tomayess Issa and Theodora Issa

**Sustainability Perspective and Awareness Amongst Higher Education in Australia** . . . . . . . . . . . . . . . . 17
Tomayess Issa

**Sustainable Development, Ethics, Strategy and International Higher Education: The Case of Australia and France** . . . . . . . . . . . . . . . . 35
Theodora Issa

**From Understanding Net Generation Expectation to Sustainable Student Engagement** . . . . . . . . . . . . . . . . 63
S Zaung Nau

**Understanding "Sustainability" and Attitudes of Students to the Concept of "Sustainable Development" in China and the UK** . . . . . . . . . 77
Austin Williams

**Sustainability Awareness in Thailand** . . . . . . . . . . . . . . . . 103
Nattarinee Phoorisart

**Sustainability Awareness in Vietnam's Higher Education Sector** . . . . . . 149
Anh Tuan Hoang

**The Opportunities and Risks of Sustainability Awareness in Sri Lankan Organizations** . . . . . . . . . . . . . . . . 185
Pathirana Hemage Gihan Dhanushka

**The Advantages and Risks of Sustainability Awareness in the Indian Higher Education Sector** . . . . . . . . . . . . . . . . 215
Nidhin Francis

**Sustainability in Organizations: Bhutan's Perspective** . . . . . . . . . . . . . . 243
Sonam Delma

**Sustainability Awareness in Singapore's Higher Education Sector** . . . . . 283
Nimitra Revi Chandran

**Examining the Opportunities and Risks Associated with Sustainability Awareness in Higher Education in Pakistan** . . . . . . . . . . . . . . . . . . . . . . 307
Muhammad Umar Javed and Umera Imtinan

**Sustainability Awareness in Saudi Arabia** . . . . . . . . . . . . . . . . . . . . . . . . . 339
Abdulaziz Albahlal, Sulaiman Alqahtani and AL-Hassan Al-Muqri

**Sustainability Awareness: Colombia Perspective** . . . . . . . . . . . . . . . . . . . 353
Juan Esteban Otalvaro Restrepo

**Sustainability Awareness in the Brazilian Higher Education** . . . . . . . . . 391
Erika Kuroda

**Sustainability and Green Supply Chain Awareness in Nigerian Organizations** . . . . . . . . . . . . . . . . . . . . . . . . . . . . . . . . . . . . . . . . . . . . . 427
Ogubuike Chukwuemeka Inmpey

**Analysis of University e-Environment in Selected European Countries – IRNet Case Study** . . . . . . . . . . . . . . . . . . . . . . . . . . . . . . . . 473
Peter Svec, Martin Drlik, Tatiana Noskova, Tatiana Pavlova, Olga Yakovleva and Michal Munk

**Design and Evaluation of the Makahiki Open Source Serious Game Framework for Sustainability Education** . . . . . . . . . . . . . . . . . . . . . . . . . . 503
Yongwen Xu, Philip M. Johnson, George E. Lee, Carleton A. Moore and Robert S. Brewer

**System-Events Toolbox Designing a System of Events that Relies on Local Resources to Create Urban Places for Social Cohesion** . . . . . . . . . 525
Davide Fassi and Roberta Motter

**Teaching Assembly for Disassembly: An Under-Graduate Module Experience** . . . . . . . . . . . . . . . . . . . . . . . . . . . . . . . . . . . . . . . . . . . . . . . . . 543
Eleftheria Alexandri

**Designing Participatory Learning Systems** . . . . . . . . . . . . . . . . . . . . . . . . 569
Henriikka Vartiainen

**The Concern about Environmental Responsibility in Public Service Media Management** . . . . . . . . . . . . . . . . . . . . . . . . . . . . . . . . . . . . . . . . . . 585
Tania Fernández-Lombao, Andrea Valencia-Bermúdez and Francisco Campos-Freire

**Index** . . . . . . . . . . . . . . . . . . . . . . . . . . . . . . . . . . . . . . . . . . . . . . . . . . . 601

# Changing the Students' Mind-set via Sustainability

**Tomayess Issa and Theodora Issa**

**Abstract** This chapter discusses and examines the integration of a sustainability unit in the higher education curriculum in Western Australia. The Information Technology Seminar 65 (ITS65) unit was developed subsequent to the repercussions of the global financial crisis in 2007–2008 which saw businesses and users struggling to survive, especially in the area of information technology/systems. The ITS65 unit aims to provide students with an understanding, knowledge and experience of Organizational Sustainable Strategy and Green IT. This chapter will examine whether the completion of the ITS65 unit can transform students' mindset. This chapter provides empirical evidence based on quantitative and qualitative data derived from two sources. The first is the anonymous informal feedback collected during the semester, while the second (the formal) is the other anonymous online system for gathering and reporting students' perceptions of their learning experiences at the university. Evaluations of and attitudes to the ITS65 unit were collected from 89 out from 127 students via the university's 'eVALUate' system. Finally, students' comments indicate that their completion of the ITS65 unit changed their mindset and made them more aware of sustainability and Green IT, and their moral responsibility for the natural environment.

## 1 Introduction

Sustainability has become a very important topic globally, since it concerns global warming and depletion. We need to think seriously about conserving our natural resources and raw materials for the present and future generations. In an effort to tackle this problem, world leaders from 195 countries signed the Paris agreement

T. Issa (✉)
School of Information Systems, Curtin Business School, Curtin University, Perth, Western Australia, Australia
e-mail: Tomayess.Iss@cbs.curtin.edu.au

T. Issa
School of Management, Curtin Business School, Curtin University, Perth, Western Australia, Australia
e-mail: theodora.issa@curtin.edu.au

 
T. Issa et al. (eds.), *Sustainability, Green IT and Education Strategies in the Twenty-first Century*, Green Energy and Technology, DOI 10.1007/978-3-319-57070-9_1

on December 2015 and committed to reducing and preserving the average global temperature so that it will be less than 2 °C above pre-industrial times [23]. By signing this agreement, countries have pledged to make a greater effort to reduce the impact of pollution on the natural environment by developing sustainable strategies to minimize this pollution and greenhouse gas emissions. Countries acknowledged that a sustainable strategy is needed to reduce the pressure on our planet and preserve raw materials and energy for both present and future generations. Now, the way to spread and raise awareness of sustainability among individuals, organizations, communities, societies locally and globally, is by education, following the policies and procedures established by the United Nations' environment projects [22]. This chapter examines the development and integration of a sustainability unit in the higher education curriculum of a university in Western Australia. The ITS65 unit was developed after the global financial crisis in 2007–2008, particularly since businesses and individuals started to use free and different technologies as Information Technology (IT) departments' budget started to shrink and demote based on this crisis. Therefore, businesses approached universities for assistance as they required different technology solutions such as cloud computing, social networking, virtualization and smart technology. These are cheaper and more economical and, most importantly, more sustainable especially in terms of saving resources such as raw materials and energy. In Western Australia, a university developed the ITS65 unit for the IT Master degree to provide students with knowledge, understanding and experience of Organizational Sustainable Strategy and Green IT.

The empirical data presented in this chapter was a combination of quantitative and qualitative approaches for data gathering from 127 students. This empirical data was derived from formal and informal student feedback. Both informal and formal feedback was collected during the semester when students reported their perceptions of their learning experience at the university, including feedback about the unit itself and the lecturer's teaching method. Informal feedback is a teaching and learning innovation requiring students, during the semester, to provide feedback anonymously regarding the unit structure and assessments. This feedback assists the lecturers to improve the delivery of the unit before the end of the semester. The second method is formal feedback, which is collected at the end of the semester through the university's formal feedback process – 'eVALUate' – to gather and report students' anonymous feedback about their learning experience; it gives students the opportunity to provide an evaluation of the unit and the quality of teaching. This study examined whether the students' completion of the ITS65 unit helped to transform their mindset.

The main outcomes indicated that the ITS65 unit materials, activities and assessments changed students' mindset, as they were made aware of issues to which they had been oblivious. Moreover, these issues related to the impact of information technology (IT) tools on environmental issues, and how green IT helps to mitigate the negative impact of this technology. Furthermore, the results indicated that students became more aware of sustainability and green IT and they had begun paying more attention to products and to whether the manufacturers were implementing a sustainability strategy.

The findings from this study will make substantial contributions to the current literature concerning sustainability integrating in the higher education in terms of methodological, theoretical and practical applications. Through education, individuals will become aware of the need for sustainability and their moral obligation to ensure that the needs of the next generation are adequately met.

This chapter examines the reflections and attitudes of students who completed the ITS65 unit for the Master degree in order to determine whether their mindset has been changed regarding sustainability and Green IT adoption now and in future. This chapter is organized as follows: (1) Introduction; (2) What is Sustainability? (3) Sustainability in the Higher Education Curriculum; (4) Methodology and Research Question; (5) Participants; (6) Results; (7) Discussion and Lesson Learned; (8) Conclusion.

## 2 What is Sustainability?

The word 'sustainability' is derived from the Latin *sustinere*, which means to 'support' or 'withstand' [21]. In other words, resources need to be conserved in order to meet current demands as well as those of the future generation. It requires that we maintain and sustain the planet's ecological systems. The three pillars of sustainability are Society, Ecology and Economy otherwise known as the 'triple bottom line' (TBL) of People, Planet, and Profit. TBL is a metric for a corporation's social, environmental, and financial performance and is considered one in the latest series of buzz words to describe business involvement in sustainability. In general, TBL is all about dropping the financial bottom line as a significant indicator of where the organization and individuals stand in the market place and replacing it with a bottom line that properly acknowledges the interplay/relationship of the social, economic and environmental dimensions of our lives [4, 12].

Furthermore, TBL is the notion that businesses have to be concerned with more than the traditional bottom line of profits, but should also be looking at performance in terms of people and the planet. The term is used at the highest level to describe the three main elements of sustainability: environmental, economic and social. The Triple Bottom Line (Economy, Society and Environment) clearly shows the areas where sustainability can be achieved [9]. Finally, the main objectives of TBL are to: eliminate/remove poverty; modify/change unsustainable patterns of production and preserve/save natural resources [4, 12, 25].

The term 'sustainability' was first coined in 1983 by Gro Harlem Brundtland at the World Commission on Environment and Development meetings. The Brundtland report urged businesses and individuals to progress toward economic development in a way that could be sustained without destroying the natural resources or the environment for the next generation [3, 8, 18, 19]. Furthermore, sustainability involves multidisciplinary approaches, political ecology, philosophy, and ethics, and social, economic, and environmental perspectives.

Erek et al. [5, p. 2] defined sustainability as 'a survival assurance, meaning that an economical, natural or social system should be well-looked-after for future generations and, thus, the necessary resources should only be exploited to a degree where it is possible to restore them within a regeneration cycle'. Sustainability involves society (promoting equality/fairness; health; human rights); the economy (people's welfare and equitable/rightful division of resources), and the environment (sustaining nature or natural resources for human and for nonhuman species) [5, 6, 11, 16, 18].

To be sustainable in everyday life is a challenging exercise, as several factors must be considered including local, state, national, and international laws, the development and authorization of development projects, and ethical and cultural perceptions. Therefore, in order to create a sustainable lifestyle, businesses and individuals should consider making several changes in order to establish a sustainable living environment. Changes can be made in the workplace by adopting the latest technologies including Green IT, and adopting strategies that will conserve natural resources for future generations [2, 3, 13, 20, 26].

In 2000, The Millennium Development Goals (MDG) was established by the United Nations (UN), Organization for Economic Cooperation and Development (OECD), the World Bank, and the International Monetary Fund. The eight MDG goals were signed by 189 United Nations member states, and were intended to encourage development by improving social and economic conditions in the world's poorest countries [15]. The eight goals are: Goal 1: Eradicate Extreme Hunger and Poverty; Goal 2: Achieve Universal Primary Education; Goal 3: Promote Gender Equality and Empower Women; Goal 4: Reduce Child Mortality; Goal 5: Improve Maternal Health; Goal 6: Combat HIV/AIDS, Malaria and other diseases; Goal 7: Ensure Environmental Sustainability; Goal 8: Develop a Global Partnership for Development [1, 10, 14]. These particular goals were intended to improve education, health, environment and other concerns in both developed and developing countries to ensure that the basic needs of every human being are catered for especially in regard to sustainability [7].

Sustainability is a complex issue and, to implement it, businesses and individuals require a long-term orientation and commitment to change the way they conduct all their activities. Human beings need to consider their own current needs as well as those of the future generation (known as the seventh generation), and balance their environmental, social and economic activities. Individuals, communities and society need to be made aware of sustainability and how it can improve the quality of life [9, 17, 24]. Patience and commitment are required in order to change the mindset of people to ensure that the future generations will enjoy the natural resources what we are enjoying now.

## 3 Sustainability in the Higher Education Curriculum – ITS65 Unit

The global financial crisis in 2007–2008 prompted businesses and individuals to take the initiative to adopt and implement various technologies in their strategies in order to reduce and minimize the budget of their Information Technology (IT) departments and to raise sustainability awareness among IT managers and employees. Consequently, businesses and individuals suggested that universities offer a

**Table 1** Assessment activities and unit syllabus for ITS65 postgraduate unit

| Unit | Assessments | Unit syllabus |
|---|---|---|
| ITS65 | • Three Journals: 30%<br>• Individual Presentation of an IT Sustainable Strategy & Report Writing: 55%<br>• Wiki for Collaborative Writing: 15% | The ITS65 unit focuses mainly on issues relating to strategic development, IT business, sustainability tools and Green IT and other related issues |

learning unit that would teach students about sustainability and green IT awareness. Such a unit would assist businesses and individuals to understand the impact of IT on the natural environment. Moreover, it would acquaint students with the various technologies such as cloud computing, social networking, virtualization and smart technology, which are cheaper and economical and most importantly, more sustainable especially in terms of conserving resources such as raw materials and energy.

A Western Australian university developed the ITS65 unit for the IT Master degree to provide students with knowledge, understanding, and experience of Organizational Sustainable Strategy and Green IT. As for the learning outcomes, on successful completion of this unit, students: (1) demonstrated awareness of and sensitivity to the importance of sustainable development and business strategies at a time that is crucial for people and the planet in addition to profit; (2) displayed an understanding of the fundamentals of Green IT; (3) applied conceptual tools and frameworks to critically analyze and apply business decision-making practices and policies; (4) translated the theories, concepts and analytical techniques learned into practice. The units' syllabus and assessment tasks were specifically designed with university graduate attributes in mind (Table 1).

The various assessment tasks for the ITS65 unit were designed to change students' mind set in regards to sustainability and Green IT. The three ITS65 assessment tasks are: Journals, Individual Presentations of an IT Sustainable Strategy & Report Writing; and Wiki for Collaborative Writing.

## 3.1 *Reflective Journal Assessment*

This individual assessment is intended to provide students with valuable experience in critically, creatively and reflectively reviewing and recording the main points in material from textbook(s), journal articles and the Internet. In addition, this assessment encourages students to keep up-to-date with their reading and their on-going search of WWW sites for information relevant to the unit. The reflective Journal not only requires a summary of material; it is also intended to indicate students' reactions to and opinions about the material provided by various sources. Each journal must include comments from the textbook(s), e-journals, WWW Internet sites; TV programs and other media regarding Information technology and Green IT. Each week, students are required to visit at least four websites related to the topic, and record the details of each site in their Journal. It is anticipated that this assessment task will develop the following skills: research, writing, technology, information search, critical thinking, reading, and use of Endnote.

### 3.2 *Individual Presentation of an IT Sustainable Strategy & Report Writing*

Students are required to present individually their IT sustainable business strategy as a digital presentation. The presentation (duration 25 min) must be recorded in MP4 format and uploaded with the PowerPoint slides to the blackboard. Following this presentation, students submit a written report based on the digital presentation. This report should be approximately 2,000 to 4,000 words (MAXIMUM) in length, including diagrams, appendices, and references. The report should examine an actual IT department in an Australian or overseas organization. This assessment task is intended to develop the following skills: research, writing, technology, information search, critical thinking, reading, and use of Endnote.

### 3.3 *Wiki Tool*

This is a tool that allows students to add, update, and modify contents on a website; the main purpose of this assessment task is to encourage students' independent learning, collaboration and interaction. Students need to undertake several challenging exercises that require using the Wiki Tool (as individuals or in teams) to analyze and evaluate practical real-case studies, and subsequent uploading the results to the Wiki to be evaluated by the lecturer and student peers. Subsequently, individuals or teams present the results to the class. Students are required to create concept maps based on real case studies or topics that are related to the unit. Wiki is used for this assessment task which is under the blackboard. A mark which contributes to 15% of the total assessment for the unit is given for students' contributions to the weekly group discussions, and for the quality (not quantity) of the discussion questions submitted each week. The lecturer checks the discussion board at least twice a week, on Monday and Thursday afternoons. This assessment tasks is intended to develop the following communication skills: writing, reading, debating, written presentation and oral skills and concept map design. Finally, the inclusion in the Master degree of a sustainability and Green IT unit gives students the opportunity to change their mindset regarding sustainability, and develops their professional and personal skills required for their current studies and for their future in the workforce.

## 4 Methodology and Research Question

This chapter examines the extent to which the ITS65 unit can change the mindset of students in regards to sustainability and sustainable practices. For this purpose, the researchers adopted the informal feedback approach which is a teaching and

**Table 2** Unit evaluation responses (N)-ITS65 2010–2016

| eVALUate year | Responses (N) | Enrolment (N) |
|---|---|---|
| 2010 | 10 | 17 |
| 2011 | 13 | 18 |
| 2012 | 13 | 19 |
| 2013 | 9 | 13 |
| 2014/S1 | 6 | 9 |
| 2014/S2 | 8 | 9 |
| 2015 | 12 | 16 |
| 2016 | 18 | 26 |
| **Total** | **89** | **127** |

learning initiative. This feedback assists the lecturer to improve and enhance the unit before the end of the semester by refining certain aspects of the course to cater for students' needs and improve the teaching and learning methods. The informal feedback questions included: What are the positive aspects of this unit? What do you like about the lecturer? Do you have any suggestions or recommendations for how the learning experience can be improved? (Please provide specific suggestions.) What do you need the lecturer to continue doing, to stop doing, and to start doing?

## 5 Participants

The study participants comprised 127 postgraduate students, mainly from Australia China, Vietnam, Malaysia, India, Indonesia, Iran, Middle East, Pakistan, and Sri Lanka, Russia, South America and South Africa (see Table 2). This cultural mix assisted students to learn from each other by sharing knowledge, skills and cultural perspectives, which led to them developing communication skills, self-esteem and self-confidence. The unit evaluation responses were obtained from 89 students; that is, 70% provided the formal online feedback for the ITS65 unit from 2010 to 2016 via the eVALUate system.

## 6 Results

The study results were derived from students' formal and informal feedback, both of which were collected during the semester to indicate students' perceptions of their university learning experience, and their opinions about the unit itself and the teaching. Table 3 presents the unit evaluation for the ITS65 unit from 2010 to 2016. This unit evaluation was derived from the formal feedback which is required

**Table 3** eVALUate unit evaluation – ITS65 2010–2016

| eVALUate quantitative items | Unit 2010 | University Average 2010 | Unit 2011 | University Average 2011 | Unit 2012 | University Average 2012 | Unit 2013 | University Average 2013 | Unit 2014/S1 | University Average 2014/S1 | Unit 2014/S2 | University Average 2014/S2 | Unit 2015 | University Average 2015 | Unit 2016 | University Average 2016 |
|---|---|---|---|---|---|---|---|---|---|---|---|---|---|---|---|---|
| **The learning outcomes** in this unit are clearly identified. | 80 | 89 | 92 | 89 | 100 | 89 | 100 | 89 | 100 | 88 | 100 | 89 | 100 | 89 | 100 | 89 |
| **The learning experiences** in this unit help me to achieve the learning outcomes. | 90 | 85 | 92 | 85 | 100 | 86 | 100 | 85 | 100 | 84 | 100 | 85 | 100 | 85 | 89 | 85 |
| **The learning resources** in this unit help me to achieve the learning outcomes. | 90 | 85 | 92 | 85 | 100 | 85 | 100 | 85 | 100 | 84 | 100 | 86 | 92 | 85 | 94 | 85 |
| **The assessment tasks** in this unit evaluate my achievement of the learning outcomes. | 90 | 85 | 92 | 85 | 100 | 85 | 89 | 85 | 100 | 84 | 100 | 85 | 100 | 85 | 94 | 85 |
| **Feedback** on my work in this unit helps me to achieve the learning outcomes. | 90 | 79 | 92 | 78 | 100 | 80 | 100 | 79 | 100 | 79 | 100 | 79 | 92 | 80 | 89 | 81 |

(continued)

**Table 3** (continued)

| **eVALUate quantitative items** | **Unit 2010** | **University Average 2010** | **Unit 2011** | **University Average 2011** | **Unit 2012** | **University Average 2012** | **Unit 2013** | **University Average 2013** | **Unit 2014/S1** | **University Average 2014/S1** | **Unit 2014/ S2** | **University Average 2014/S2** | **Unit 2015** | **University Average 2015** | **Unit 2016** | **University Average 2016** |
|---|---|---|---|---|---|---|---|---|---|---|---|---|---|---|---|---|
| **The workload** in this unit is appropriate to the achievement of the learning outcomes. | 90 | 86 | 92 | 85 | 100 | 86 | 89 | 86 | 83 | 85 | 100 | 86 | 83 | 86 | 94 | 86 |
| **The quality of teaching** in this unit helps me to achieve the learning outcomes. | 90 | 84 | 92 | 84 | 100 | 84 | 100 | 84 | 83 | 83 | 100 | 83 | 100 | 83 | 100 | 84 |
| **I am motivated** to achieve the learning outcomes in this unit. | 100 | 85 | 92 | 84 | 100 | 86 | 100 | 85 | 100 | 85 | 100 | 86 | 92 | 86 | 100 | 85 |
| **I make best use** of the learning experiences in this unit. | 100 | 86 | 92 | 86 | 100 | 87 | 100 | 87 | 100 | 86 | 100 | 87 | 100 | 86 | 94 | 87 |
| **Overall**, I am satisfied with this unit. | 90 | 84 | 92 | 84 | 100 | 84 | 100 | 84 | 100 | 83 | 100 | 84 | 100 | 83 | 100 | 84 |

by the university. The results indicated that students enjoyed the unit and were highly satisfied with the learning outcomes, experiences and resources with an average of 96%, 96% and 96% respectively from the eVALUate unit evaluation. This result was supported by comments from the students regarding the learning experience and resources.

> Hopefully [my] University will be able to apply this technique to the rest of the unit in any possible way and can't thank more than enough for providing me such a wonderful learning experience.

> It's been wonderful to have a teacher who is so passionate about the subject matter taught in the unit and so active and innovative in their methods of teaching. The care and support provided throughout the semester really showed me another aspect of learning that I will take with me into future classes and gave me confidence in areas of research and writing that I've not had so far in my postgraduate education.

> ITS65 has put more focus on research and discovery than learning specific tasks. The assessments completed so far complement this idea, requiring students to evaluate and reflect on prescribed publications with the focus on justification of why that conclusion was come to, rather than following a specific set of steps. Other units I have found enjoyable employed a similar tactic of having students 'learn to learn' rather than learning what is in a book.

> University is a place where the students, the future leaders, are nurtured through learning, experiment, and research. So, by introducing sustainability and green IT in the education system, the universities could help in creating awareness and supporting the students to become socially and environmentally responsible. I also feel that [My] University should make such curriculum available for all students irrespective of what courses they are undertaking.

> In my perspective, this is a highly valuable unit as it teaches students about an emerging and important issue. It was useful to learn about what is being done to achieve sustainability in IT and learn how different countries and cultures are adapting towards sustainability. It was also useful to learn about which organisations are raising awareness about the importance of sustainability. This unit prepared me for the future by teaching me about the importance of sustainability and how to achieve sustainability in IT.

Regarding the students' evaluation of the assessment tasks, workload, and quality of teaching the eVALUate average was 96%, 95%, 91% and 96% respectively. This result was supported by comments from the students, both formally and informally.

> The journal assessment on the other hand, is very good to assess students' comprehension and information gathering skill because it teaches students to think critically when presented with information. As I mention before, I enjoyed doing the journal assessments so I might be biased. However, I personally believe that this unit encourages students to read more which is exactly what I need because I am a type of person that dislikes reading but it teaches me that I do not always get what I want and I need to step out of my comfort zone to actually achieve something bigger.

> The wiki exercises also a good assessment that encourages students to communicate to each other. Unlike some other unit that I have taken, there was barely any group discussion that happened in the class. With the allocated team member, most of the group work creates a robust discussion between members that stimulates more debates and more perspective into the discussion.

Regarding the workload and quality of teaching, students made the following observations.

> .....The second part was applying the sustainability module for an exciting business firm, as well as some tips to save money and energy for this organization. This part was not clear in the beginning until I did my presentation and received the feedback beforehand. After that, I understood the idea. Surprisingly, all the class students knew the idea behind Teece's module and they understood every detail about it. It was not easy at the beginning. This point shows what the power of the intensive class could be.

> ....the presentation that we did was really useful and I learned many different things, especially the feedback that we received for each student. So, my point for this idea is adding at least another presentation. I suggest this presentation should be about 'after the company accepts the sustainability plan, what the next phase will be'. This way, students will try to discover the solution and report them to the business people. Also, this is required to write about in the final report.

> There is a lot of interaction in the class which is really good. You speak with a lot of knowledge in class, very energetic and enthusiastic about the course, which makes the class enjoyable to be in. Very hard working and highly organized, when it comes to giving back feedback on assignments, exercises and journal, there are no delays.

> I personally really enjoys and learnt a lot through this unit as it is actively involved student in the modern way where most people nowadays like to read, watch, listen, write and blog on the internet rather than holding a class that only involving passive teaching where lecturer/tutor just passed down information, expecting student to complete their work and ask questions themselves.

> .....Last but not least, the lecturer style of teaching is interesting and inspirational; each of the lectures consists of professional knowledge and group activities. Also, it is very thankful that the lecturer always be patient and happy to assists students in completing their assessments....

> It's been wonderful to have a teacher who is so passionate about the subject matter taught in the unit and so active and innovative in their methods of teaching. The care and support provided throughout the semester really showed me another aspect of learning that I will take with me into future classes and gave me confidence in areas of research and writing that I've not had so far in my postgraduate education.

> I am satisfied with the lecturer specially her proactive approach of teaching. All the best! [Our lecturer] is the best teacher for this unit the topic could have been boring if she could not be teaching us just want to say thank you to such a wonderful teacher.

This positive feedback encouraged the lecturer to include cutting edge information and knowledge about sustainability and green IT, and add the latest, new activities to Wiki to encourage and motivate, inspire and encourage students to interact and engage in the class activities. As a result, students gave outstanding feedback via eVALUate' for the factors of motivation, best use of learning ITS65 and overall satisfaction, with an average of 98%, 98% and 98% respectively. This feedback was supported by the following specific comments from the students.

> This unit was great with such an engaging lecturer! [My lecturer] is very passionate about this subject, she managed to transfer her knowledge to us and engage the whole group. The topics were well presented and explained with a lot of up-to-date information and

> statistics especially in sustainability and Green IT. The assessment tasks were thoroughly explained and an appropriate/constructive feedback was given for making improvements in our work. There was a diverse range of learning resources and tasks, e.g. videos, wikis, group presentations, concept maps, etc. which broadened our experience.

> This unit is a good unit if you want a change of pace in your semester studies. This would teach you a lot of report writing and presentations. It's a recommended unit if you are not confident about your public speaking and seminar skills. The unit as a whole is very friendly and understandable. You can learn a lot about sustainability that you have not considered before. Overall I strongly recommend this unit.

> Overall, I found this unit is very interesting as the topic of sustainability is very current, and the urgency to develop sustainable practices is growing. Sustainability is often in the media through news about climate change; predict an increase in demand through population growth, as well as the overwhelming generation of waste. This unit helped grow on the concept of sustainability by focusing on the ICT components and recommending methods for contributing to the solution.

> Encouraging Class Participation and discussions was very good. I like the ideas that the study groups were done by the lecturer instead of allowing students to form their own group. The course to me was designed to encourage students to learn on our own instead of been thought everything (With excellent support from the lecturer). Help us to be learning how to research for solutions to real world problems. Good Lecturer overall.

> A very relevant unit to raise awareness amongst current and future IT professionals. [My lecturer] is very knowledgeable in her field; she is an excellent lecturer who cares for her students' learning and wellbeing. Overall a very gratifying experience!

> This unit helped me improve myself very much. There were even such things which I had never learnt before. All the journals, assignments as well as wiki helped me enhance my knowledge as well thinking ability. Overall, I am fully satisfied with the unit and the knowledge provided by the lecturer.

These comments confirmed that, according to the students, the ITS65 unit is well-structured, and that the workload, lecturer-to-student feedback, resources and overall learning experience, adequately met the students' need to understand the unit aims, and to improve their professional and personal skills.

## 7 Discussion and Lessons Learned

This chapter presents a case relating to the development, integration and presentation of a Master degree unit about sustainability and green IT at an Australian university. This unit received outstanding feedback from the students and top management at the university's School of Information Systems, confirming the study's aims and objectives, and addressing the questions. The feedback indicated that students who completed this unit experienced an increased awareness of sustainability issues, particularly in terms of the impact of IT on the planet. Furthermore, assessment tasks and activities played a major role in shaping students' attitudes as they gave them the opportunity to think, reflect, search, write, discuss, communicate, collaborate, and present their findings, allowing students to have more control of their own learning.

The formal and informal feedback indicated that students who had completed the ITS65 unit had acquired knowledge about sustainability and green IT, had developed better communication and collaboration with lecturer and peers, and had improved their personal and professional skills.

To ensure the unit's success, the lecturer's role is essential in terms of designing, developing, preparing, and delivering the unit materials and activities, and providing formative and summative feedback to the students to ensure that the unit offers the necessary knowledge and skills. This is a challenging exercise not only in the design and development stages; the lecturer needs to determine whether the whole process is running smoothly according to the specific aims of the unit and the broader aims of the university. The lecturer is responsible for ensuring that students are given every opportunity to acquire knowledge and skills by (the lecturer) providing up-to-date information to the students from journals, books, e-journals and e-books, in order to include the most interesting information about sustainability not only in the unit materials but also in the activities.

This study answered the study questions as completing this unit transform and renovate students' critical thinking and mind-set toward sustainability especially the IT impact on our planet. Furthermore, students who completed ITS65 unit, continue to study with lecturer to complete their dissertation in sustainability and Green IT topic. This action will allow students to apply the theory from ITS65 to obtain new cutting-edge knowledge, and to identify new theoretical and practical significance by completing their dissertation. Furthermore, the lecturer encouraged and assisted her students to publish their dissertation results in a conference/journal paper or book chapter to share it with other researchers and academics worldwide

The study outcomes have made significant contributions to the research and literature in the domain of information technology, predominantly in terms of constructive student outcomes associated with students' knowledge, learning and skills acquisition. Moreover, the lecturer's exceptional teaching method has clearly distinguished her from her university colleagues.

Finally, the inclusion of the ITS65 unit in the IT Master studies was intended to raise students' awareness, knowledge and understanding of sustainability issues, as IT devices from production to recycling have a huge impact on our planet. Business schools and academics should be promulgating the issue of sustainability, as today's students will become future leaders and will be responsible for the future welfare of the planet. IT people should no longer contribute to the problem, but should be part of the solution.

## 8 Conclusion

This chapter examines the process of designing, developing, preparing and delivering a sustainability unit for an IT Master degree in a Western Australia university. This study examines whether students attitudes and opinions will change after their completion of the ITS65 unit. The study outcomes confirmed the study aims and goals, as 70% of the students (who completed the ITS65 unit) confirmed that their mind-set was changed via the unit notes, activities and assessments. Furthermore, students confirmed that the ITS65 unit conveyed to them the concepts of sustainability and green IT and enabled them to assess the damage which is caused by IT from the design to the recycling stages. Finally, the inclusion of this type of unit in undergraduate and postgraduate studies is intended to raise students' awareness of sustainability and encourage them to become more sustainable now and in future. In the future, the researchers will include more students' perspectives and perception regarding this topic in order to strength the research aims and objectives.

## References

1. Acharya T, Daar A, Singer P (2003) Biotechology and the UN's millennium development goals. Nat Biotechnol 21(12):1434–1436
2. Babin R, Nicholson B (2011) How green is my outsourcer? Measuring sustainability in global IT outsourcing. Strategic Outsourcing 4(1):47–66
3. Bonevac D (2010) Is sustainability sustainable? Acad Quest 23(1):84–101
4. Elkington J (2004) Enter the triple bottom line. In: Henriques A, Richardson J (eds) The triple bottom line: does it all add up, Earthscan, London, pp 1–16
5. Erek K, Schmidt N-H, Zarnekow R, Kolbe LM Sustainability in information systems: assortment of current practices in IS Organizations. Americas Conference on Information Systems (AMCIS), San Francisco, 2009. pp 1–9
6. Gomis A, Parra M, Hoffmann M, Mcnulty R (2011) Rethinking the concept of sustainability. Bus Soc Rev 116(2):171–191
7. Issa T (2014) Learning, communication and interaction via Wiki: an Australian Perspective. In: Kaur H, Tao X (eds) ICTs and the millennium development goals. Springer, USA, pp 1–17
8. Issa T, Chang V, Issa T (2010) The impact of cloud computing and organizational sustainability. Paper presented at the Cloud Computing and Virtualization, Singapore, 2010
9. Moscardo G, Lamberton G, Wells G, Fallon W, Lawn P, Rowe A, Humphrey J, Wiesner R, Pettitt B, Don C, Renouf M, Kersham W (2013) Sustainability in Australian business: principles and practice. Wiley-Blackwell, Queensland, Australia
10. Nayyar D (2013) The millennium development goals beyond 2015: old frameworks and new constructs. J Hum Dev Capab 14(3):1–23
11. Nidumolu R, Prahalad CK, Rangaswami MR (2009) Why sustainability is now the key driver of innovation? Harv Bus Rev 87(9):57–64
12. Norman W, MacDonald C (2004) Getting to the bottom of "triple bottom line". Bus Ethics Q 14(02):243–262
13. Philipson G (2013) A framework for green computing. In: International and Interdisciplinary Studies in Green Computing. IGI Global, USA, pp 12–26
14. Potts M, Fotso J-C (2007) Population growth and the millennium development goals. Lancet 369(9559):354–355
15. Saith A (2006) From universal values to millennium development goals: lost in translation. Dev Change 37(6):1167–1199. doi:10.1111/j.1467-7660.2006.00518.x
16. Shaw G, Walters R, Kumar A, Sprigg A (2015) Sustainability in infrastructure asset management. In: Lee W, Choi B, Ma L, Mathew J (eds) Proceedings of the 7th World Congress on Engineering Asset Management (WCEAM 2012), 2015. Lecture Notes in Mechanical Engineering, Springer, Cham, pp 525–534

17. Sikdar SK, Sengupta D, Mukherjee R (2017) Sustainability measurement for technology and business systems: use of currently available tools for quantification. In: Measuring progress towards sustainability. Springer International Publishing, Switzerland, pp 93–127
18. Smith P, Sharicz C (2011) The shift needed for sustainability. Learn Organization 18 (1):73–86
19. Stelzer K (2006) Sustainability = Good Design. Les Ateliters De Lethique 2:1–15
20. Stewart E, Kennedy J (2009) The sustainability potential of cloud computing: smarter design. http://www.environmentalleader.com/2009/07/20/the-sustainability-potential-of-cloud-computing-smarter-design/. Accessed 1 Dec 2009
21. Thiele LP (2016) Sustainability. John Wiley & Sons, Cambridge, England
22. UNEP (2014) Report of the United Nations Environment Programme (UNEP). http://www.un.org/esa/socdev/unpfii/documents/2014/unep.pdf. Accessed 3 Jan 2017
23. United Nations (2015) Historic Paris Agreement on Climate Change. http://newsroom.unfccc.int/unfccc-newsroom/finale-cop21/. Accessed 3 Jan 2017
24. Varadarajan R (2017) Innovating for sustainability: a framework for sustainable innovations and a model of sustainable innovations orientation. J Acad Market Sci 45(1):14–36
25. Wexler MN (2009) Strategic ambiguity in emergent coalitions: the triple bottom line. Corp Commun 14(1):62–77
26. Wiens JA (2013) Is landscape sustainability a useful concept in a changing world? Landsc Ecol 28:1047–1052

# Sustainability Perspective and Awareness Amongst Higher Education in Australia

**Tomayess Issa**

**Abstract** There are positive and negative aspects to the adoption and implementation of sustainability by individuals and businesses. However, if the implementation of sustainability is properly and correctly planned, users will obtain the desired benefits such as financial rewards, brand and reputation enhancement; shareholder satisfaction and human resource efficiency; natural resources conservation; and environmental protection. However, the adoption and implementation of sustainability can have negative consequences in terms of failure and cost. Therefore, to reduce the possibility of these occurring, awareness and training should be made available to individuals and businesses. As discussed in this chapter, the author conducted an online survey with 208 respondents from Australia to examine students' perspective and awareness toward sustainability. The online survey generated a further three positive and two negative aspects associated with students' knowledge and understanding of sustainability. Finally, further research will be conducted in future to examine more diverse groups of stakeholders in order to strengthen the research findings.

## 1 Introduction

Currently, the global warming issue has been a hot topic locally and worldwide because of the rapid changes to the ecosystem. Researchers from numerous disciplines and sectors [1, 9, 15, 25, 39, 40, 46] have recognized that the main reasons for these ecological deviations are our human endeavors which are endangering our environment and planet. Therefore, it is crucial that we give serious consideration to the planet's natural resources, species, water and the most important, air. If the trend continues, the seventh generation [28] will not be able to enjoy resources similar to the current one. Therefore, the world recognizes the fact that the

T. Issa (✉)
School of Information Systems, Curtin Business School, Curtin University, Perth, Western Australia, Australia
e-mail: Tomayess.Iss@cbs.curtin.edu.au



T. Issa et al. (eds.), *Sustainability, Green IT and Education Strategies in the Twenty-first Century*, Green Energy and Technology, DOI 10.1007/978-3-319-57070-9_2

economic element is the one that can have the most impact on society and the environment. Consequently, it is extremely significant to consider the sustainability elements in the economic sector since failure to deliver sustainable outcomes in terms of the environment will produce undesirable consequences.

In the twenty-first century the planet began to change as a result of composite and interconnected developments, and this can present enormous difficulties and problems of inappropriate planet resources by human begins and this can lead to an unsustainable nation. Therefore, sustainability should be integrated in various sectors from education, government, social, economic, industrial and etc. to raise the awareness among users, organizations as well as in the education sector, especially in Australia. All stakeholders should be made aware of the concept of sustainability and of their moral obligation to adopt more sustainable lifestyles which will ultimately benefit the individual, the community and society as a whole.

This chapter presents the Australian attitude toward sustainability awareness, and identifies the positive and negative aspects of sustainability adoption as identified by tertiary students in Australia. This chapter is organized as follows: (1) Introduction; (2) What is Sustainability; (3) Research Method and Questions; (4) Results; (5) Discussion and New Findings; and (6) Conclusion.

## 2 What is Sustainability?

Gro Harlem Brundtland from the World Commission on Environment and Development first coined the word "sustainability" in 1983. Brundtland's report recommended that businesses and individuals undertake their economic development in a way that could be sustained without depleting the natural resources or harming the environment for the next generation [42]. In the twenty-first century, the concept of sustainability has become extremely significant as users and businesses are now required to deliver explanations rather than products, and to identify their markets in terms of more complex consumer demands rather than just merchandise and amenities [23, p. 263].

Erek et al. [10, p. 2] define sustainability as "a survival assurance meaning that an economical, ecological or social system should be preserved for future generations and, thus, necessary resources should only be exploited to a degree where it is possible to restore them within a regeneration cycle". Hence, users and businesses should defend the current resources in line to be used by our next generation in the future. There are a number of advantages associated with the adoption of sustainability. These include: cost reduction, resources preservation, conformity to legislation, improvement of brand reputation, ensuring happier customers and stakeholders, attracting capital investment and capitalizing on new opportunities [29, 34, 38, 40]. By adopting sustainability, a business can distinguish itself from its competitors, both national and international, and the business will become stronger if sustainability is integrated into its business strategy with the full collaboration of the CEO and the top management. In order for sustainability to be integrated successfully in the business strategy, the project manager should gather the necessary information about what is happening in his/her organization at all levels of the business hierarchy.

Once the required information has been collected, it is necessary to secure everyone's cooperation so that all employees and management have the same positive attitude toward sustainability. The project manager must choose the correct moment to introduce the notion of sustainability adoption throughout the organization. The advantages and disadvantages of integrating sustainability in a business strategy should be put to management whose role it is to inform staff of any changes that this requires. Furthermore, in terms of sound business practice, the different attitudes of staff together with their roles, backgrounds, and personalities should be taken into account. Finally, the adoption of sustainability by an organization will be useful when hiring new staff.

According to Moscardo et al. [27], sustainability entails a long-standing alignment and obligation to shift the way businesses conduct their activities in order to balance the needs of the current users and future generations.

Furthermore, there should be recognition that business is part of a complex system that comprises environmental, social, and economic activities. As part of the business strategy, training should be offered so that all personnel are made aware and acquire a better knowledge of sustainability. This learning should not be limited to staff; specific teaching and training should be available to shareholders and the community to make them aware of sustainability, as business needs should be aligned with the needs of the shareholders, society, the economy and the environment. Finally, sustainability is a complex issue and its implementation takes time and patience. However, its successful adoption and implementation as part of a business strategy will enhance the recognition and reputation of an organization both locally and globally.

Based on the current literature review [2, 5, 12, 15, 18, 26, 30, 32, 39, 40, 43] the adoption of sustainability by organizations will bring various advantages in terms of: financial rewards; brand and reputation; human resources; shareholder approval and satisfaction; and environmental preservation.

From the financial perspective, sustainability will reduce energy, water and paper usage, raw materials usage and increase efficiency and cost-effectiveness. Furthermore, brand and reputation will improve corporate social responsibility, reduce risk management, differentiate business, increase green strategy, improve community investments, improve human rights and improve social responsibility and triple bottom line – people, planet and profit. As for the human resources and shareholders, a sustainability strategy will increase productivity, attract quality employees, satisfy customer needs, meet stakeholder expectations, and create new jobs and opportunities. Finally, from the environmental perspective, sustainability will reduce the carbon footprint, harmful emissions, and pollution and health hazards.

On the other hand, sustainability adoption can produce negative consequences, namely failure and cost. Regarding failure, sustainability can increase fraud, litigation and compliance breaches, governance failure and transaction failure due to environmental liabilities, greater security and system failure caused by environmental problems, increase the number of scandals and/or marketing failures as a result of perceived environmental irresponsibility, and finally increase supply chain crises due to suppliers' environmental irresponsibility. Finally, cost can be a major drawback as the integration of sustainability in an organization's business strategy could inflate costs and interest rates, increase insurance crises due to environmental disasters, increase the number of new regulations for environmental protection, and increase competition for and cost of raw materials.

However, despite the disadvantages, organizations and individuals should consider integrating sustainability in their business strategy in order to derive its benefits. Nevertheless, the negative aspects of adoption should be carefully considered so that a risk plan can be put in place before implementation to circumvent or minimize any negative consequences.

## 3 Research Method and Questions

This study addresses a research question, namely: "what are the positive and negative aspects associated toward sustainability awareness in the Australian higher education?" To address the research question, the researcher devised an online survey to assess the negative and positive aspects of sustainability awareness. The online survey was developed based on a review of the current literature.

Online surveys are designed to provide interaction between participant and survey via online mode. The beauty of using online survey is that data downloaded from it are ready for immediate analysis using SPSS or other statistical programs; while the findings can be presented in numerous formats such as tables, figures, calculations of mean and standard deviation based on the researcher's requirements [3, 7, 8, 11, 13, 16, 41].

Via the online survey, participants can provide rich and historical information which the researcher can utilize for the research objectives and aims. An online survey is a "pre-formulated written set of questions to which respondents record their answers, usually within rather closely defined alternatives." In addition, "online surveys are an efficient data collection mechanism when the researcher knows exactly what is required and how to measure the variables of interest" [37, p. 236].

The online survey has many advantages as it is less expensive, provides greater anonymity, is accessible and easy to manage, is less error-prone, and eliminate the paper usage required for the paper and pencil survey, thereby confirming that online surveys are more sustainable compared with the traditional survey methods [20, 33, 34, 35]. However, the online survey has several disadvantages as it is susceptible to technical problems including computer viruses and hacking which can decrease the response rate, according to Fan and Yan [11]. The online survey design offers self-motivated interaction between the respondent and the survey that cannot be achieved via email or paper surveys [6, 7].

The online survey comprised three sections: background, positive and negative aspects. The researcher devised 18 questions for the background section in order to obtain information about each participant's background as well as information regarding the participant's knowledge of sustainability and Green Information Technology. For the "positive" section, the researcher developed 24 statements pertaining to financial, brand and reputation, human resources and shareholders, and environmental issues. The 13 statements for the "negative" section were mainly concerned with failure and cost. Also, participants had the opportunity to offer additional opinions in a comment section in the last section of the survey. A 5-point Likert scale was used in each section of the online survey to "examine how strongly subjects agree or disagree with statements" [37, p. 197]. The 5-point Likert scale ranges from strongly disagree, Disagree, Neutral, and Agree, to Strongly Agree. The online survey contained clear instructions at the top of the page and a progress bar along the bottom to offer feedback to users about their proximity to the finishing point. Furthermore, each page contained only three questions to minimize scrolling, and the concluding page thanked respondents for their participation. The survey was created online using Qualtrics software. Qualtrics website (www.qualtrics.com) distributed the online survey to 209 participants in Australia. Qualtrics is an online survey tool with an outstanding reputation since it is used to develop and summarize the survey results, thereby facilitating the researcher's data collection and data analysis process [24, 31]. The response validity was 99.5% for this study.

## 4 Results

Table 1 shows the number and percentage of online survey participants in terms of gender, age, and qualifications. The survey response rate was 99.5%; 51% are female. The majority of respondents (15%) were aged between 25 and 30 years, while 27% (the highest percentage) of respondents had a bachelor degree.

Table 2 shows the technology used by Australian participants. The online survey results confirmed that 42% of Australian participants are spending up to 5 h per day on the computer for professional work and study; with 45% on the Internet. Furthermore, 68% spend less than an hour on email per day; while 69% spend less than an hour daily on social networking.

Furthermore, the online survey confirmed that 93% of the Australian users are using the Internet to access their email, while 75% use it for banking online and 66% for shopping online (see Table 3).

Furthermore, the online survey identified the devices used to access the Internet. Sixty-eight percent of respondents use laptops, 61% use smartphones and 44% use both PC and desktop (see Table 4).

We noted that Australian users were first introduced to the concepts of sustainability and green information technology via news media, school and Internet with 38%, 33%, and 32% respectively (see Table 5).

**Table 1** Online survey statistics Australia – prepared by Tomayess Issa

| *Number and percentage of online survey* | |
|---|---|
| Questionnaires distributed | 209 |
| Questionnaires returned | 208 |
| Response rate | 99.5% |
| *Gender* | |
| Male respondents | 102 (49%) |
| Female respondents | 106 (51%) |
| *Age* | |
| 17 years and under | 0 (0%) |
| 18–20 | 16 (8%) |
| 21–24 | 19 (9%) |
| 25–30 | 31 (15%) |
| 31–35 | 29 (14%) |
| 36–40 | 26 (12%) |
| 41–45 | 24 (11%) |
| 46–50 | 14 (7%) |
| 51–55 | 19 (9%) |
| 56–60 | 19 (9%) |
| 61–65 | 12 (6%) |
| Over 65 | 0 (0%) |
| *Qualifications* | |
| Primary Education | 4 (2%) |
| Higher Secondary /Pre-University | 37 (18%) |
| Professional Certificate | 26 (13%) |
| Diploma | 30 (15%) |
| Advanced/Higher/Graduate Diploma | 12 (6%) |
| Bachelor's Degree | 56 (27%) |
| Post Graduate Diploma | 12 (6%) |
| Master's Degree | 13 (6%) |
| PhD | 6 (3%) |
| Others | 8 (4%) |

The online survey determined the companies which were associated with Australian users' devices. Apple and Google are the leaders at 33% and 32% respectively (see Table 6).

Furthermore, the survey asked Australian users whether they read the sustainability report of the manufacturer before buying a device. The survey showed that 53% do not read the report; 23% responded "maybe"; and 11% read the report. This indicates that users should act more responsibly when making purchase

**Table 2** Technology used by Australian users – prepared by Tomayess Issa

| Answer | Hours spent on the computer per day | Hours spent on the Internet per day | Hours spent on the email per day | Hours spent on social networking per day |
|---|---|---|---|---|
| | Response % | Response % | Response % | Response % |
| Less than an hour | 22 (11%) | 40 (19%) | 142 (68%) | 144 (69%) |
| Up to five hours | 87 (42%) | 93 (45%) | 61 (29.1%) | 55 (26%) |
| Five to ten hours | 69 (33%) | 47 (23%) | 3 (1.4%) | 3 (1%) |
| Ten to twenty hours | 26 (12%) | 25 (12%) | 3 (1.4%) | 5 (2%) |
| Over twenty hours | 5 (2%) | 3 (1%) | 0 (0%) | 1 (2%) |

**Table 3** Internet usage by Australian users – prepared by Tomayess Issa

| Answer | Response | % |
|---|---|---|
| Email | 194 | 93% |
| Playing games | 82 | 39% |
| Studying | 80 | 38% |
| Working | 96 | 46% |
| Shopping online | 137 | 66% |
| Chatting | 79 | 38% |
| Researching hobbies | 101 | 48% |
| Banking online | 157 | 75% |
| Buying goods or services | 132 | 63% |
| Buying stocks or investing online | 24 | 11% |
| Researching travel information or making reservations | 105 | 50% |
| Others – please specify | 14 | 7% |

**Table 4** Devices used by Australian users – prepared by Tomayess Issa

| Answer | Response | % |
|---|---|---|
| PC | 91 | 44% |
| Desktop | 92 | 44% |
| Laptop | 143 | 68% |
| Netbook | 15 | 7% |
| PDAs | 4 | 2% |
| Workstation | 9 | 4% |
| Tablet | 43 | 21% |
| Smartphone | 128 | 61% |
| Others – please specify | 9 | 4% |

choices, and their awareness of their responsibility to the planet needs to be raised via education and training (see Table 7).

In addition, Australian users change their device after 24–42 months with percentages ranging from 26% to 20% respectively (see Table 8).

**Table 5** First introduced to the concepts of sustainability and green information technology by Australian users – prepared by Tomayess Issa

| Answer | Response | % |
|---|---|---|
| School | 69 | 33% |
| Higher education | 37 | 18% |
| Internet | 67 | 32% |
| Books | 32 | 15% |
| Magazine | 31 | 15% |
| News media | 79 | 38% |
| Conferences | 9 | 4% |
| Others – please specify | 19 | 9% |

**Table 6** Australian users devices – prepared by Tomayess Issa

| Answer | Response | % |
|---|---|---|
| Apple | 67 | 33% |
| Google | 66 | 32% |
| Dell | 22 | 11% |
| IBM | 16 | 8% |
| Others – please specify | 35 | 17% |

**Table 7** Australian users reading the sustainability report of the company before buying a device – prepared by Tomayess Issa

| Answer | Response | % |
|---|---|---|
| Yes | 23 | 11% |
| No | 110 | 53% |
| Maybe | 48 | 23% |
| Not at all | 28 | 13% |

**Table 8** Australian users changing their device – prepared by Tomayess Issa

| Answer | Response | % |
|---|---|---|
| Every six months | 3 | 1% |
| Every 12 months | 16 | 8% |
| Every 18 months | 19 | 9% |
| Every 24 months | 54 | 26% |
| Every 30 months | 15 | 7% |
| Every 36 months | 33 | 16% |
| Every 42 months | 41 | 20% |
| Other – please specify | 27 | 13% |

In response to the survey question "Why do you change your device?" the majority of participants (56%) indicated that they changed because of the functionalities offered by the new device; 53% want to keep up with technology, and 48% want more speed (see Table 9).

Additionally, the online survey sought to determine the Australian users' attitudes to their moral responsibilities toward the planet by asking whether changing

**Table 9** Australian users "Why do you change your device" – prepared by Tomayess Issa

| Answer | Response | % |
|---|---|---|
| Size | 38 | 18% |
| Speed | 100 | 48% |
| Functionality | 116 | 56% |
| Keeping with technology | 110 | 53% |
| Others – please specify | 28 | 13% |

**Table 10** Australian users "changing device frequently will cause damage to our planet" – prepared by Tomayess Issa

| Answer | Response | % |
|---|---|---|
| Yes | 79 | 38% |
| No | 34 | 16% |
| Maybe | 93 | 44% |
| Not at all | 3 | 1% |

**Table 11** Australian users: "Can we change the mindset of designers and users regarding sustainability" – prepared by Tomayess Issa

| Answer | Response | % |
|---|---|---|
| Training | 96 | 46% |
| Education | 148 | 71% |
| Awareness | 140 | 67% |
| Workshop | 45 | 22% |
| Internet | 101 | 49% |
| T.V. | 91 | 44% |
| Social networking | 83 | 40% |
| Others – please specify | 10 | 5% |

their devices frequently will cause damage to our planet. Table 10 shows that 44% responded "Maybe" while 38% showed their awareness that changing devices frequently would cause damage to our planet.

The online survey examined Australian users' recommendations of ways to change the mindset of designers and users regarding sustainability. The survey concluded that via education and awareness (71% and 67% respectively), designers and users could change their mindset and attitude (see Table 11).

A total of 209 valid responses were processed for the subsequent factor analysis. The analysis was conducted separately for the positive and negative aspects with 24 and 13 questions respectively. The Cronbach's Alpha for both negative and positive aspects were 0.966 and 0.959 respectively, indicating an excellent internal consistency of the items in the scale [14]. The Kaiser-Meyer-Olkin measures of sampling adequacy of positive and negative aspects were 0.932 and 0.933 respectively, indicating that a very good sample size was obtained for the purpose of analysis. The Bartlett's test of sphericity was highly significant for both positive and negative

**Table 12** Cronbach alpha, KMO and Barlett's test results – prepared by Tomayess Issa

| Group | Cronbach's alpha | KMO sampling adequacy | Bartlett's test of sphericity |
|---|---|---|---|
| Positive | 0.966 (excellent) | 0.932 (Marvelous) | $\chi^2 = 4142.379$; df = 276 $p < 0.000$ |
| Negative | 0.959 (excellent) | 0.933 (Marvelous) | $\chi^2 = 2522.937$; df = 78 $p < 0.000$ |

**Table 13** Positive – total variance – prepared by Tomayess Issa

| Total variance explained | | | | | | |
|---|---|---|---|---|---|---|
| Component | Initial eigenvalues | | | Rotation sums of squared loadings | | |
| | Total | % of variance | Cumulative % | Total | % of variance | Cumulative % |
| 1 | 13.540 | 56.415 | 56.415 | 6.163 | 25.680 | 25.680 |
| 2 | 1.439 | 5.995 | 62.410 | 5.624 | 23.434 | 49.114 |
| 3 | 1.121 | 4.671 | 67.081 | 4.312 | 17.968 | 67.081 |
| Extraction method: principal component analysis. | | | | | | |

**Table 14** Negative – total variance – prepared by Tomayess Issa

| Total variance explained | | | | | | |
|---|---|---|---|---|---|---|
| Component | Initial eigenvalues | | | Rotation sums of squared loadings | | |
| | Total | % of variance | Cumulative % | Total | % of variance | Cumulative % |
| 1 | 8.776 | 67.507 | 67.507 | 5.713 | 43.947 | 43.947 |
| 2 | 1.046 | 8.048 | 75.555 | 4.109 | 31.608 | 75.555 |
| Extraction method: principal component analysis. | | | | | | |

aspects, indicating that the items of the scale are sufficiently correlated to factors to be found (see Table 12).

Positive aspects have two factors generated where the rotation resulted in a total of 67.081% of the variation from this group. The amount of variances explained by each of these factors is presented below (after rotation attempted); while the Negative aspects have two factors generated where the rotation produced a total of 75.555% of the variation from this group. The amount of variances explained by each of these factors is presented below (after rotation attempted) (see Tables 13 and 14).

To extend the regression coefficients (i.e. slopes), the researcher carried out the factor loadings. The factor loadings of most of the items were adequately high and the one with the cleanest fact structured to be considered as important [4]. Several items under each factor were excluded where the factor loading was below 0.5 based on the rule of thumb of Stevens [44] for a sample size above 100. In addition, Rose et al. [36] suggested that the acceptable factor loading based on sample size between 200 and 249 is 0.40 (see Table 15). Table 15 shows the group pattern matrix for the positive aspects. The pattern matrix revealed three positive aspects namely: (1) Satiate Participants expectancies; (2) 4Rs; and (3) Enhance Accountability for Human Rights (see Table 16).

**Table 15** Rotated component matrix for the positive aspects – prepared by Tomayess Issa

| Rotated component matrix[a] | | | |
|---|---|---|---|
| | Component | | |
| | 1 | 2 | 3 |
| Satisfy customer needs | 0.776 | 0.317 | 0.146 |
| Meet stakeholder expectations | 0.767 | 0.159 | 0.235 |
| Attract new opportunities | 0.751 | 0.335 | 0.121 |
| Attract quality employees | 0.707 | 0.327 | 0.284 |
| Increase productivity | 0.597 | 0.252 | 0.426 |
| Reduce risk management | 0.587 | 0.140 | 0.504 |
| Increase cost-effectiveness | 0.566 | 0.291 | 0.407 |
| Enhance reputation | 0.550 | 0.381 | 0.249 |
| Create new jobs | 0.549 | 0.309 | 0.266 |
| Improve community investments | 0.542 | 0.292 | 0.502 |
| Differentiate businesses | 0.541 | 0.276 | 0.382 |
| Increase efficiency | 0.536 | 0.457 | 0.268 |
| Reduce consumption of raw materials | 0.393 | 0.797 | 0.120 |
| Reduce paper usage | 0.308 | 0.779 | |
| Reduce energy and water usage | 0.398 | 0.764 | 0.153 |
| Reduce emissions | 0.213 | 0.707 | 0.575 |
| Reduce pollution | 0.234 | 0.688 | 0.526 |
| Reduce carbon footprint | 0.191 | 0.638 | 0.546 |
| Reduce health hazards | 0.238 | 0.604 | 0.541 |
| Increase green strategy | 0.419 | 0.597 | 0.323 |
| Improve corporate social responsibility | 0.490 | 0.494 | 0.405 |
| Improve human rights | 0.307 | 0.132 | 0.775 |
| Improve social responsibility investing | 0.329 | 0.275 | 0.755 |
| Increase triple bottom line – People, Planet and Profit | 0.437 | 0.487 | 0.502 |

Extraction method: principal component analysis. Rotation method: Varimax with Kaiser normalization.

[a]Rotation converged in nine iterations.

From Table 16, it can be concluded that sustainability adoption can assist businesses and individuals to improve their efficiency and productivity, and reduce negative impacts on the environment by decreasing the consumption of materials such as paper, and natural resources such as energy and water, and reducing harmful emissions. Moreover, the adoption of sustainable practices can satisfy investor expectations, and improve human rights and social responsibility. Table 16 shows the positive aspects of sustainability.

Furthermore, Table 17 shows the group pattern matrix for the negative aspects. The pattern matrix revealed two negative aspects namely: (1) Increase in Failure, Security and Fraud; and (2) Increase in number of charges and regulations.

**Table 16** The new positive aspects – Australia – prepared by Tomayess Issa

| Component matrix | | |
|---|---|---|
| | Component | *New factor* |
| | 1 | |
| Satisfy customer needs | 0.776 | Satiate participants expectancies |
| Meet stakeholder expectations | 0.767 | |
| Attract new opportunities | 0.751 | |
| Attract quality employees | 0.707 | |
| Component matrix | | |
| | Component | *New factor* |
| | 2 | |
| Reduce consumption of raw materials | 0.797 | 4Rs |
| Reduce paper usage | 0.779 | |
| Reduce energy and water usage | 0.764 | |
| Reduce emissions | 0.707 | |
| Component matrix | | |
| | Component | *New factor* |
| | 3 | Enhance accountability for human rights |
| Improve human rights | 0.775 | |
| Improve social responsibility investing | 0.755 | |

**Table 17** Rotated component matrix for the negative aspects – prepared by Tomayess Issa

| Rotated component matrix[a] | | |
|---|---|---|
| | Component | |
| | 1 | 2 |
| Increase transaction failure due to environmental liabilities | 0.833 | 0.352 |
| Increase marketing failure by perceived environmental irresponsibility | 0.821 | 0.386 |
| Increase governance failure (including environmental actions and environmental compliance failures) | 0.819 | 0.331 |
| Increase litigation and compliance breaches (including environmental actions and environmental compliance failures) | 0.819 | 0.362 |
| Increase scandals by perceived environmental irresponsibility | 0.798 | 0.417 |
| Increase security and systems failures caused by environmental problems | 0.792 | 0.373 |
| Increase fraud (including environmental actions and environmental compliance failures) | 0.768 | 0.306 |
| Increase supply chain crises due to suppliers' environmental problems | 0.690 | 0.543 |
| Increase interest rates | 0.386 | 0.795 |
| Increase competition for and cost of raw materials | 0.325 | 0.790 |
| Increase number of new regulations including environmental | 0.271 | 0.771 |
| Increase insurance crises due to environmental disasters | 0.483 | 0.728 |
| Inflate costs | 0.337 | 0.717 |

Extraction method: principal component analysis. Rotation method: Varimax with Kaiser normalization.

[a]Rotation converged in three iterations.

**Table 18** The new negative aspects – Australia – prepared by Tomayess Issa

| Component matrix | | |
|---|---|---|
| | Component | *New factor* |
| | 1 | |
| Increase transaction failure due to environmental liabilities | 0.833 | Increase of failure, security issues and fraud |
| Increase marketing failure by perceived environmental irresponsibility | 0.821 | |
| Increase governance failure (including environmental actions and environmental compliance failures) | 0.819 | |
| Increase litigation and compliance breaches (including environmental actions and environmental compliance failures) | 0.819 | |
| Increase scandals by perceived environmental irresponsibility | 0.798 | |
| Increase security and systems failures caused by environmental problems | 0.792 | |
| Increase fraud (including environmental actions and environmental compliance failures) | 0.768 | |
| Component matrix | | |
| | Component | *New factor* |
| | 2 | |
| Increase interest rates | 0.795 | Increase in charges and regulations |
| Increase competition for and cost of raw materials | 0.790 | |
| Increase number of new regulations including environmental | 0.771 | |
| Increase insurance crises due to environmental disasters | 0.728 | |
| Inflate costs | 0.717 | |

Table 18 presents the new negative aspects which are presented from the Australian perspective. These two negatives mainly concern increased failure, security and fraud and increased charges and regulations. The Australian students are somewhat concerned about the integration of sustainability in business and individual strategies since it may lead to failure and fraud; moreover, further charges and regulations could be imposed by local, state or federal governments (see Table 18).

Therefore, these negative aspects can be reduced by offering training, workshops, and university courses in order to raise public awareness of sustainability. Universities play a major role in transforming societies' and students' critical philosophies and world views through tertiary education and research. Universities need to raise the students' awareness, especially in relation to their moral responsibility to contribute to sustainable development, and guide them to a better, sustainable future.

The adoption of sustainable strategies by businesses and individuals will reduce the carbon footprint and damage to the natural environment, conserve raw materials for the seventh generation, and ensure their commitment to mitigating environmental problems. Therefore, a major shift in priorities for businesses including Information systems business, educators, researchers and users must be made to adjust and shift their mindsets and level of thinking to recognize IT's impacts, as our planet is now suffering, and we need to tackle the issues of recycling, raw material supply and energy consumption, since there is no plan B for our planet [21].

## 5 Discussion and New Findings

In this chapter, the author examines and evaluates the awareness of tertiary students in Australia regarding the advantages and disadvantages (positives and negatives) of sustainability adoption. For this study, the researcher conducted an online survey of 208 respondents from Australia. The online survey successfully addressed the study's aims and objectives since it generated three new positive and two negative aspects associated with students' knowledge and understanding of sustainability. The new findings from this study are presented in Fig. 1, as the Australian students indicated that integrating and adopting sustainability will Satiate Participants' expectancies; 4Rs; and Enhance Accountability for Human Rights. These new positive aspects indicate that sustainability integration in business or individual strategies will meet investor expectations; satisfy stakeholders' needs, create new opportunities and attract quality employees. Furthermore, it will help to reduce resource consumption, and reduce the four impacts on the environment: harmful emissions, pollution, the carbon footprint, and health hazards.

However, the adoption and implementation of sustainability by individuals and businesses may lead to failures in terms of transactions, marketing, governance,

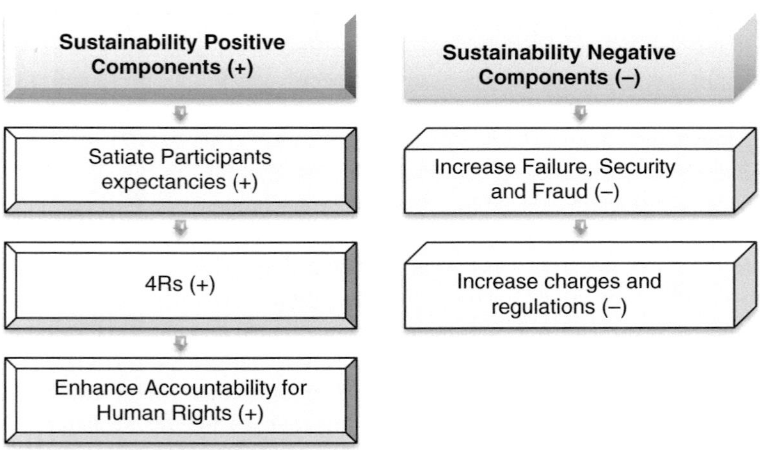


**Fig. 1** Sustainability positive and negative aspects – prepared by Tomayess Issa

litigation, scandals, security and fraud, and environmental problems. Furthermore, users are concerned that the inclusion of sustainability in their plan and policy will increase rates, costs, insurance and new regulations. Therefore, awareness, training, teaching and education are needed in order to minimize these negative aspects with the support of local, state and federal governments and organizations, to change users' mind-set and behaviour by integrating and amalgamating sustainability in their agenda [17, 19, 22, 27, 40, 45].

This change will assist our environment and planet as it is up to us to take care of our planet for the next generation, since there is no plan B for our planet [21]. Users must take responsibility and recognize their impacts on our planet. By integrating sustainability in everyday activities, people's lifestyles can become more sustainable and eco-friendly.

Lastly, all stakeholders including businesses and individuals must contribute to the transition to a sustainable world. Moreover, the universities are playing a historic role in transforming societies' and students' critical philosophy. The urgent call for sustainable development is here, and universities must raise the awareness of students regarding their moral responsibility in contributing to sustainable development, and guide them on the path towards a sustainable future.

Finally, this chapter makes a significant contribution to the current literature, especially regarding the issue of sustainability awareness among tertiary students in Australia and its advantages and disadvantages. With increased awareness of the positive and negative aspects of sustainability adoption, students should be keen to become more sustainable since they are our future leaders. Finally, in future further research will be carried out to examine more diverse groups of stakeholders in order to strengthen the research findings.

## 6 Conclusion

This chapter presents the findings from an online survey conducted with 208 participants, who were invited to offer their opinions on sustainability and its positive and negative aspects. The study generated two negative and three positive aspects of sustainability in an Australian context. The new findings from this study added a new significance to the current literature locally and globally regarding the Australian awareness of and attitude toward sustainability. Stakeholders, namely businesses and individuals, must be responsible for their activities and their impact on the planet since our planet is now suffering as a result of human activities. Therefore, stakeholders should act now to save and protect our planet especially for our next seventh generation, by integrating and adopting sustainability in their strategy to make our life and planet more sustainable, ecologically and environmentally. This study was limited to higher education students in Australia; therefore, in future the researcher will examine different sectors to compare and contrast the stakeholders' awareness, opinions and attitudes to sustainability.

# References


1. Alexander J (2007) Environmental sustainability versus profit maximization: overcoming systemic constraints on implementing normatively preferable alternatives. J Bus Ethics 76 (2):155–162
2. Andreopoulou Z (2012) Green informatics: ICT for Green and sustainability. J Agric Informatics 3(2):1–8
3. Cho Y, Park J, Han SH, Sungjin K (2011) Development of a web-based survey system for evaluating affective satisfaction. Int J Industrial Ergonomics 41:247–254
4. Costello A, Osborne J (2005) Best practices in exploratory factor analysis: four recommendations for getting the most from your analysis. Practical Assess Res Eval 10(7):1–9
5. Datta A, Roy S, Tarafdar M (2010) Adoption of sustainability in IT services: role of IT service providers. In: Americas Conference on Information Systems (AMCIS), Lima, Peru, 2010. pp. 2–10.
6. Dillman D (2007) Mail and Internet surveys: the tailored design method, 2nd edn. Wiley, New York
7. Dillman D, Glenn P, Tortora R, Swift K, Kohrell J, Berck J, Messer B (2009) Response rate and measurement differences in mixed-mode surveys using mail, telephone, interactive voice reponse (IVR) and the Internet. Soc Sci Res 38:1–18
8. Dillman D, Reipus U, Matzat U (2010) Advice in surveying the general public over the Internet. Int J Internet Sci 5(1):1–4
9. Epstein MJ, Roy MJ (2001) Sustainability in action: identifying and measuring the key performance drivers. Long Range Plan 34:585–604
10. Erek K, Schmidt N-H, Zarnekow R, Kolbe LM (2009) Sustainability in information systems: assortment of current practices in IS Organizations. In: Americas Conference on Information Systems (AMCIS), San Francisco, 2009. pp 1–9.
11. Fan W, Yan Z (2010) Factors affecting response rates of the web survey: a systematic review. Comput Hum Behav 26(2):132–139
12. Figueiró PS, Raufflet E (2015) Sustainability in higher education: a systematic review with focus on management education. J Clean Prod 106:22–33. doi:http://dx.doi.org/10.1016/j.jclepro.2015.04.118
13. Fink A (2012) How to conduct surveys: a step-by-step guide. Sage Publications, California, USA
14. Gliem J, Gliem R (2003) Calculating, interpreting, and reporting Cronbach's alpha reliability coefficient for Likert-type scales. In: Midwest Research-to-Practice Conference in Adult, Continuing, and Community Education, Ohio State University, Ohio, 2003, pp 82–88
15. Gomis A, Parra M, Hoffmann M, Mcnulty R (2011) Rethinking the concept of sustainability. Bus Soc Rev 116(2):171–191
16. Gordon J, McNew R (2008) Developing the online survey. Nurs Clin N Am 43(4):605–619
17. Gunn C (2010) Sustainability factors for e-learning initatives. ALT-J Res Learn Technol 18(2): 89–103
18. Humphreys A (2014) How is sustainability structured? The discursive life of environmentalism. J Macromarketing 34(3):265–281
19. Isaias P, Issa T (2013) E-learning and sustainability in higher education: an international case study. Int J Learn High Educ 20(4):77–90
20. Issa T (2013) Online survey: best practice. In: Information systems research and exploring social artifacts: approaches and methodologies. IGI Global, USA, pp 1–19. doi:10.4018/978-1-4666-2491-7.ch001
21. Issa T, Isaias P (2015) Sustainable design: HCI, usability and environmental concerns. Springer-Verlag, London
22. Issa T, Issa T, Chang V (2014) Sustainability and Green IT education practice for incorporating into the Australian Higher Education Curriculum. Int J Sustain Educ 9(2):21–30

23. Jeffers P (2009) Embracing sustainability – information technology and the strategic leveraging of operations in third-party logistics. Int J Operations Prod Manag 30(3):260–287
24. Kocher M (2015) Recipes and research: a survey of cookbook collection users. J Agric Food Inf 16(1):53–59
25. Little G (2009) Third annual Green IT and sustainability survey: US results and services implications. IDC, USA
26. Mebratu D (1998) Sustainability and sustainable development: historical and conceptual review. Environ Impact Assess Rev 18(6):493–520. doi:http://dx.doi.org/10.1016/S0195-9255(98)00019-5
27. Moscardo G, Lamberton G, Wells G, Fallon W, Lawn P, Rowe A, Humphrey J, Wiesner R, Pettitt B, Don C, Renouf M, Kersham W (2013) Sustainability in Australian business: principles and practice. Wiley-Blackwell, Queensland, Australia
28. Newton LH (2003) Ethics and sustainability: sustainable development and the moral life. Basic ethics in action. Pearson Education, USA
29. Nidumolu R, Prahalad CK, Rangaswami MR (2009) Why sustainability is now the key driver of innovation. Harv Bus Rev 87(9):57–64
30. Orsato RJ (2009) Sustainability strategies: when does it pay to be Green? Palgrave Macmillan, New York.
31. Peer E, Paolacci G, Chandler J, Mueller P (2012) Selectively recruiting participants from Amazon Mechanical Turk using Qualtrics. https://ssrn.com/abstract=2100631. Accessed 15 Jan 2017
32. Philipson G (2011) ICT and sustainability. http://www.aiia.com.au/resource/collection/F0FF33B8-BF15-44B3-971E-20E03EC57287/Graeme_Philipson_AIIA_2011-09-29.pdf. Accessed 24 Jul 2012
33. Porter S (2004) Pros and cons of paper and electronic surveys. New Dir Institutional Res 2004:91–97
34. Prahalad CK, Rangaswami MR (2009) Why sustainability is now the key driver of innovation. Har Bus Rev 87(9):56–64
35. Prasad A, Saha S, Misra P, Hooli B, Murakami M (2010) Back to Green. J Green Eng 1(1): 89–110
36. Rose S, Hair N, Clark M (2011) Online customer experience: a review of the business-to-consumer online purchase context. Int J Manag Rev 13:24–39
37. Sekaran U (2003) Research methods for business: "a skill building approach", 4th edn. John Wiley & Sons, MA, USA
38. Sharma A, Lyer G, Mehrotra A, Krishnan R (2010) Sustainability and business-to-business marketing: a framework and implications. Industrial Mark Manag 39:330–341
39. Shaw G, Walters R, Kumar A, Sprigg A (2015) Sustainability in Infrastructure Asset Management. In: Proceedings of the 7th World Congress on Engineering Asset Management (WCEAM 2012), 2015, Springer, pp 525–534.
40. Smith P, Sharicz C (2011) The shift needed for sustainability. Learn Organization 18(1): 73–86
41. Smyth J, Dillman D, Christian L, O'Neill A (2010) Using the Internet to survey small towns and communities: limitations and possibilities in the early 21st century. Am Behav Sci 53:1423–1448
42. Sneddon C, Howarth RB, Norgaard RB (2006) Sustainable development in a post-Brundtland world. Ecol Econ 57(2):253–268. doi:http://dx.doi.org/10.1016/j.ecolecon.2005.04.013
43. Stelzer K (2006) Sustainability = Good Design. Les Ateliters De Lethique 2:1–15
44. Stevens J (1992) Applied multivariate statistics for the social sciences. Erlbaum, Hillsdale
45. Wals AEJ (2014) Sustainability in higher education in the context of the UN DESD: a review of learning and institutionalization processes. J Clean Prod 62:8–15
46. Young W, Tilley F (2006) Can businesses move beyond efficiency? The shift toward effectiveness and equity in the corporate sustainability debate. Bus Strategy Environ 15:402–415

# Sustainable Development, Ethics, Strategy and International Higher Education: The Case of Australia and France

**Theodora Issa**

**Abstract** It is almost one decade to the day when we observed a global financial crisis which hit the markets worldwide, slowed down the global markets and financial systems to nonentity or even adverse growth rates. The negative ramifications had far-reaching effects resulting in a social crisis impacting on millions of people in different parts of the world. There were several calls for governments at all levels, industry, and most importantly to academia to work together with other organizations to create ethically responsible alternatives to address the crisis. Although there can be no turning back, there is a need to move forward; thus, educators in the higher education sector have been motivated to examine options and seek solutions that could mitigate the effects of hard times in future. One of these attempts is discussed in this chapter, which reports on the development and the facilitation of a course (unit) on 'Sustainable Development and Business/ Ethical Strategies' (SDBS/SDES). SDBS/SDES was developed in Australia for delivery to international students from different parts of the world enrolled at a European partner's business school in France. This course (unit), with ongoing updates, was presented six times in France and has enabled students to adopt a new perspective towards change, outlining a new strategy through innovation in an ever-changing, fluctuating, ambiguous, uncertain and contemporary 'risk society' [2]. The topic of sustainable development, intertwined with business and ethical strategies, has gained momentum in various areas including governments, international organizations (e.g. UN PRiME), and business education courses. To understand the multi-disciplinary nature of this vast topic, which in part touches on ethical issues, sustainability and strategic development in contemporary business, the SDBS/ SDES course (unit) was designed to give students the opportunity to engage in class discussions and debates, sharing their unique international experiences. This facilitated the transformation of their mindsets despite their diverse cultural backgrounds. This transformation was measured through the close monitoring of class discussions, and checking ongoing assessments including reflective accounts

T. Issa (✉)
School of Management, Curtin Business School, Curtin University, Perth, Western Australia, Australia
e-mail: theodora.issa@curtin.edu.au



T. Issa et al. (eds.), *Sustainability, Green IT and Education Strategies in the Twenty-first Century*, Green Energy and Technology, DOI 10.1007/978-3-319-57070-9_3

(daily and final), which were designed in consultation with the European partner in France. This paper provides a brief literature review of the concepts under study, followed by an insight into the development and facilitation of the SDBS/ SDES course (unit). It concludes with an analysis of part of the quantitative and qualitative data collected from ($N$ = 138) students on each of the six occasions when the unit was delivered and facilitated in France (February 2010, February 2011, November 2011, November 2012, November 2013 and November 2014). This is coupled with the course (unit) designer/facilitator's personal reflection. Though this chapter has its limitations, it is a step forward in developing an international perspective towards teaching these important issues to Business and Management students at a global level at such crucial times.

## 1 Introduction

Alcaraz and Thiruvattal [1] contend that The United Nations' Principles for Responsible Management Education (PRiME) initiative is call to modify business education in light of changing ideas about corporate citizenship, corporate social responsibility, and sustainability. In an interview, the head of PRiME, Manuel Escudero offered practical suggestions about ways to engage and actively participate in the initiative, persuade skeptics, and benefit from this global project. Alcaraz and Thiruvattal [1] conclude, in this agenda, that experiential learning needs to be at the center of the new learning methodologies in order to create the paradigm shift needed to redefine Business and Management education for the future.

Undoubtedly, this places the onus on the Business and Management academics and more importantly on Universities as a whole including the Business Schools. In this respect, Neu [38] concludes that certain types of Business and Management education are demanded; however, it is education that is practical, concrete, delivered at the worksite, and scheduled around the constraints of production that might make a difference. With regards to the responsibilities of the higher education sector, Issa et al. [23] argued that these are in the hands of academics and business educators who are shaping and influencing tomorrow's leaders and practitioners who use resources that are borrowed from future generations to achieve their ongoing goals. This is a difficult task, and to achieve such a mindset shift that might provide current and future generations with a sustainable future, what is required is the co-operation of and co-ordination at the highest level amongst academics, business school heads, and practitioners, both locally and globally.

There is evidence that steps in the right direction are being taken. Vargas [56] explored the role that education might play in fostering sustainable development. Vargas provides a framework based on the integration of the economic, social, and environmental dimensions. Vargas [56] concludes that educating for a sustainable future requires the inclusion of six features: (1) a holistic, integrated, interdisciplinary approach, (2) teacher training at the pre-service and in-service is necessary, (3) curriculum development needs to integrate not only environmental education, but sustainability concepts anchored on a balanced and interdependent view of the social, economic and environmental dimensions, (4) the educational approach must consider both equality and equity to address the issues of vulnerable peoples since sustainability cannot be attained without consideration of their needs in development discussions, (5) it must be an education that prepares people for active, meaningful participation to prepare citizens for engagement in a democratic process, and (6) educating for sustainability must be respectful of local cultural contexts, particularly in technology transfer from one region to another. In short, the curricula need to be based on the three pillars of sustainable development; (1) social and environmental justice; (2) meaningful political participation; and (3) respect for local and indigenous cultures. By the same token, Young and Tilley [60] articulate the value and importance of moving the sustainable business agenda beyond the notion of eco- and socio-efficiency.

Back in 2011, Davidson et al. [12] described the Culture of Excellence and Ethics approach developed by the Institute for Excellence & Ethics (IEE). They laid out the principles of this framework and the rationale for how this approach achieves sustainability and enduring impact in the development of character, leadership, and culture through a balance of theoretical and practical fidelity with theoretical and practical convenience.

Indeed, here it needs to be acknowledged that changes have been evidenced in higher education [e.g. 21]; there has been an increased emphasis on internships, critical thinking and business ethics that might lead to a transformation that supports greater sustainability. Nonetheless, the role of educational institutions and business educators in fostering sustainable development, through a framework based on the integration of the economic, social, and environmental dimensions [e.g. 50], moral imagination and beyond materialism [8] sustainability [4], values and capitalism [59], should still needs to be enhanced. Korontzis [29] posits that more corporations are now aiming to achieve dual goals: becoming aware of the importance of making a greater contribution to the financial development of their societies while promoting the financial growth of their organizations. Some organizations are doing this through their ongoing work towards establishing an International Standard Organization (ISO) sustainability standard. For example, in Australia, no organization can successfully win a government tender without having this accreditation. It is therefore crucial to teach students that 'sustainable development' and its relationship with 'ethical strategies' is of fundamental importance to the future of individuals, communities, and global societies.

There are evidences from recent research that the consideration of sustainability by businesses will not only enhance reputation, but has a financial return. In this respect, Gomez-Bezares et al. [20] examined the effect of integrating sustainability into corporate strategy on various aspects of shareholder value creation and financial performance in the British capital market. The authors concluded that firms that incorporate sustainability issues into their business operations are better able to leverage their resources toward stronger financial performance and shareholder value creation than other companies.

Despite all the work that has and continues to be done in this regard, still more work is needed to ensure that understanding is achieved at all levels. In their recent publication, Bradley and Ziniel [5] issued a warning acknowledging that one of the least understood aspects of the world-wide 'greening of markets' is the emergence of local 'ethical marketplaces' and the subset of alternative business models described as 'ethical businesses'. Bradley and Ziniel concluded their paper by presenting an example of the way in which radical political change may be part of a wider social movement toward greening markets.

Indeed, enhancing the role of business education makes a difference at all levels including the future government policies, whereas those who would be running the affairs of governments in the future are the students of today at the universities, and any change of their mindset would reflect positively on the policies thus on the society. In this respect, Issa et al. [23] provided empirical evidence of the shift in students' mindsets following a course (unit) about sustainable development. Nonetheless, there continues to be a need for further enhancement of such efforts to achieve a significant transformation of students' mindsets not only at the local but at the global level. This chapter might be a small step towards achieving this ultimate goal, though this work has its own limitations as highlighted towards the end of this chapter.

# 2 Literature Review

## *2.1 Sustainability, Ethics and Global Business*

The issue of sustainability is included and in a very well done manner on the agenda of business and business education. Bradbury [4], Newton [39], and White and Taft [59] argue that sustainability issues are just not 'out there' in the external environment. Discussing the issue of the incorporation of sustainability in business curricula, Bradbury [4] states that students benefit from questioning what it is that sustains their own lives, the lives of those they care about, and the organizations in which they work. If encouraged to do this, students become better systems thinkers as they link their external and internal worlds. Thus, students will increase their knowledge and appreciation of their surroundings. Students will be able to reflect and apply their critical thinking abilities, through their ethical lens, sense making, and moral imagination to evaluate and find solutions to situations. Such an exercise might allow these students to develop strategies that will enhance the sustainability of their businesses, which might ultimately lead to a sustainable community, society and world.

Teaching sustainable business development and business/ethical strategies would be incomplete if it did not include ethics and ethical theories, and their application to business. White and Taft [59, p. 471] argue that whether teaching organizational behaviour, business and society, business ethics, human resource management, or general management, educating management and for that matter educating students about 'business', involves providing them with frameworks to tackle ethical issues and dilemmas. In this regard, White and Taft [59] provide a summary of ethical theories which they divide into two main categories: teleological (consequentialist) which includes utilitarianism, egoism and care; and deontological (categorical imperative or non-consequentialist) which comprises rights, justice (distributive, retributive, compensatory) [43], truth-telling, virtue and ethics of care, all of which are deontological forms of ethical reasoning. These are considered as Western approaches, whereas the Eastern approaches include Confucianism and Buddhism and others [e.g. 16, 31], with a strong return to the Native American ethical perspective [59, 39].

These seem to be very comprehensive, and worth reintroducing to our global and contemporary societies. This is important, as many have been either totally forgotten or have been made ineffective by their strong alliance with the utilitarian or Machiavellian way of thinking which characterize most business leaders especially in the Western World. According to Newton [39] and White and Taft [59] the Native American perspective, for example, covers issues such as moral virtues: generosity, kindness, caring, compassion, understanding, restraint, honesty, mindfulness, non-harming, equanimity, forgiveness, concern for continuity of relationships and strength of community, preference for harmony over truth, peace over justice, and justice through cooperation, connection and interdependence with all living things; consequences to self, others, and community. These features are well beyond the limitations of the most adhered to theory of consequentialism in business [17]. Further, Fisher and Malde [18] argue, when a crisis occurs, the heuristics are not challenged, but are simply put away, and other more appropriate heuristics are put to use until a sense of normality returns, and the use of the old heuristics is resumed. This should not be the case.

Stormer [49] contends that for several decades, business has operated according to the tenets of neoclassical economic theory, where the primary obligation of corporations is to maximize profit for shareholders. However, the larger social mandate for business has changed, represented by the emergence of terminology such as 'sustainable development', 'corporate social responsibility' (CSR) and 'stakeholder groups'. Nevertheless, the theoretical shift implied using such language has not occurred. Issues of sustainable development and CSR continue to be justified in terms of neoclassical economic theory through the rationalization of 'doing well by doing good'. Within this economic paradigm, CSR cannot move beyond enlightened self-interest (acting in socially responsible ways to further one's own ends) because all behaviour must be justified economically. This implies that socially responsible corporate behaviour will simply cease when it becomes uneconomic, regardless of the impact on interrelated systems which in turn will re-impact on the business realm [49, p. 279]. This is unfortunate, and continues to be the case so it might be difficult to move towards a more holistic approach to business globally.

In view of the above, and with the global nature of the contemporary business world, and as highlighted by Jones et al. [27], and Jacobides [25], there seems to be a need for more guidelines at the global level. These guidelines were always in the form of voluntary codes such as the United Nations Declaration of Human Rights, The Coalition for Environmentally Responsible Economies (CERES) principles, The Caux Round Table, known as the CAUX principles, the global Sullivan Principles, amongst others including the corporate codes of ethics. Despite all these ethical theories, guidelines and principles, transgressions are still being committed by businesses with global negative ramifications that in many instances ill-treat and exploit the resources that are at hand, which are as we all know, borrowed from future generations. Despite these guidelines and as contended by Sneddon et al. [46], enhanced levels of ecological degradation, vast inequalities in economic opportunities both within and across societies, and a fractured set of institutional arrangements for global environmental governance all represent seemingly insurmountable obstacles to a move towards sustainability. While these obstacles are significant, we suggest that they might be overcome through a reinvigorated set of notions and practices associated with sustainable development, one that explicitly examines the linkages between sustainability policies and sustainability politics.

In this regard, Jones et al. [27] posit that sustainability has become a key issue for organizations in the twenty-first century as they increasingly acknowledge that their policies and practices have social and/or environmental consequences. Accordingly, many companies are incorporating elements of sustainability into their business practices. Aligned with this trend is the increasing recognition that sustainability-related subjects need to be included in the curricula of business courses and/or units at both undergraduate and postgraduate levels. Having said that, and in a recent publication by McCaffery [33] posits that though there might be some very well-known economic concepts, such as 'moral hazard', which might have been used by the economists to describe actions and judge individuals and governments. McCaffery [33] employs contracts approach to this concept 'moral hazard', using the 'title-transfer' theory of contract to clarify the moral content of moral hazard, thereby increasing its value to scholars in numerous disciplines. McCaffery [33] concludes that this is not only valuable for economic policy discussions because it does not include hidden value judgments. Also, this is important for ethicists because it directly explains a moral dimension of behavior under moral hazard, namely, the violation of property rights, which are in the heart of sustainable development and business/ethical strategies.

Finally, Jacobides [25] contended that it is time to reinvent the way in which companies develop strategy. If we were asked to pick the defining characteristic of today's business environment, we would surely point to its turbulence: unprecedented, unstoppable, and apparently, unlikely to go away. Jacobides [25] goes on to argue that globalization, technological innovation, regulatory restructuring, demographic shifts, and environmental pressures have all conspired to continually redraw the competitive landscape. Consequently, the nature of most industries is changing quickly, and companies are seeing their rivals, suppliers, and customers transform at tremendous speed. Thus, the action(s) currently being taken by some academics and business educators are worthwhile, for example, the understanding of different cultures through an examination of ethical mindsets and its components in 12 countries [22] and the latest publication by Cengage Learning [44] in which the attention is drawn to the Asia-Pacific seeking to draw the attention to seeing the moral dimension in business through an understanding of normative ethical theories, which would lead to taking a moral position through an understanding of the nature of capitalism, corporate social responsibility and corporate governance, the morality and ethics of consumption opening the eye of students that they need to consider the impact of what they consume on their community, society, and the environment, through an understanding of globalization and environmental ethics in business. This publication of Shaw et al. [44] concludes with a look at the organization through the lens of ethical and moral issues, ethics at work leadership and culture and towards an ethical and moral decision-making. Despite all the efforts, this chapter serves as a call to expand these efforts to cover 'business' education at all levels and on a global scale.

## 2.2 Reflection and Critical Reflective Thinking

Mahin [32] argues that business communication uses causal analysis, classification, process analysis, evaluation, as other forms of composition and that the very act of communicating in the social context of a business culture implies an ethical basis, a respect for persons. While previously business students were considered as empty vessels that should be filled by the business educator, more recently, business students have been urged to think in a critical and reflective manner. With regards to teaching business ethics which would allow students to think about the telos of business, Issa and Pick [24] posit that there is a need for research that will provide theoretical underpinnings for the changes that are affecting the curricula in terms of pedagogy and content. Issa and Pick [24] examined a method of teaching business ethics contending that the downfall of several high ranking executives, not only in the business world, but also in governments, attention is directed to the teaching of ethics in business schools. Issa and Pick [24] go on to state that attention is drawn to these schools' purposes, values, methods, research, partnerships and dialogues in terms of their responsibilities regarding management education. In the current academic environment, business ethics are recognized as an important element in curriculum, but these topics have not yet become fully and successfully embedded in the mainstream business education. Issa and Pick [24] conclude that business schools worldwide are adapting their curricula, research, teaching methodologies and institutional strategies to become better aligned with new challenges and opportunities.

Rogers [42] and Thorpe [54] argue that despite the widespread adoption of reflective practices across many fields of study, a critical analysis of the concept of reflection and its application within higher education has been lacking. Tranter and Warn [55] are of the opinion that generic competencies, including critical reflective thinking are currently widely recognized as being essential for an individual's integration into a rapidly changing workplace. To respond proactively to change, one needs to be able to critically consider one's own world view and its relationship to the world views of others. Higher education endeavours to develop such competencies that are essential for the successful adaptation of graduates to the workforce. In management and business education, it seems that these competencies are in demand, and the development of such competencies is given greater priority at the higher education level.

Reflective activities, including critical reflective thinking, play a major role in adult learning, yet it is considered one of the most challenging exercises [13]. While the development of generic skills might prepare students for their next phase in life (i.e. career), nevertheless these students continue to face more challenges as they move from tertiary education into the workforce. Watson [58] states three things are necessary for effective management and business education as a whole: (1) the general intellectual skills produced by a good liberal higher education, (2) the technical knowledge about organizations and the environment in which they have to function, and (3) the specific skills of managing. These can be enhanced by improving the critical reflective thinking ability of students. Indeed, reflection constitutes a major part of the assessment tasks designed for this SDBS/SDES unit that was developed in Australia for delivery and facilitation to students from different parts of the world who are studying in Europe.

## 3 Methodology and Discussions

This research employed a mixed-method approach to allow the generation of a response to the question:

> would a course (unit) such as SDBS/SDES would facilitate the transformation of the mindsets despite their diverse cultural backgrounds.

Denscombe [13] argues that the mixed-method approach has emerged as a 'third paradigm' for social research. It has developed a platform of ideas and practices that are credible and distinctive and that distinguish the approach as a viable alternative to quantitative and qualitative paradigms being used solely. With regards to mixed methods, Johnson and Onwuegbuzie [26] posit that a key feature of this research approach is its methodological pluralism or eclecticism, which frequently results in superior research (compared to mono-method research). This chapter adopts this approach by using a combination of qualitative and quantitative techniques.

Coghlan et al. [10] state that the question regarding methodology should be which approach will collect better data in the particular situation with the available resources. They conclude that, firstly, people will provide you with good data if they trust you and feel that they are part of the process, and have a stake in the results being as accurate and valid as possible. Secondly, all methods of data collection have their inherent biases. The more methods one uses, the more the biases can cancel each other out; hence, the value of mixed methods. Simply, the mixed methods approach and design is proving to be a better alternative to the mono-method approach and design, with support in the literature [e.g. 52, 40, 11, 53, 3]. Furthermore, Tashakkori and Creswell [51] confirm and highlight the benefits of 'the new era of mixed methods'.

This data was collected from 138 International Management Program Masters students over six periods February 2010, February 2011, November 2011, November 2012, November 2013 and November 2014.

Table 1 shows the origin/nationality of these students. The data for this chapter was collected from records provided by the European partner in France together with the qualitative data derived from different sources including daily reflections, feedback on the unit contents and teaching method, and the European partner in France's own evaluation of teaching and material. In total, 861 documents were collected through the administration of the 138 'ice-breaker questionnaires' during the first session of teaching of each of the 6 runs, about 445 daily and final reflections by students over the 6 periods of teaching, and about 276 student evaluations of the teaching and the course (unit), which were presented by the European partner in France and the course (unit) developer/facilitator.

The 'ice-breaker questionnaire' contained 15 questions, 8 of which related to sustainability and the sustainable business development course (unit). The remaining questions were simple questions intended to generate group discussions and break the ice during the first session of each of the runs. For example, one of the questions contained in this 'ice-breaker' asked students to 'name three ways they prefer through which to learn things'.

As for the daily reflection, in this task the students were requested to reflect on what you had learned? 'The Plus', 'The Minus', and 'The Interesting'. In addition, they were asked to indicate what they wish to see 'added', 'stopped', and 'continued' in the way the learning journey was going in this course (unit). This was submitted through e-mail to the lecturer/facilitator in the evening following each of the classes. The lecturer/facilitator would collate and share with the students the next morning the highlights of this reflection.

Reflections continued daily basis, with the ultimate reflection that came in the form of the 'final reflection'. In this reflection, students were asked to reflect on what you had learned? And their opinion was also sought to highlight 'The Plus', 'The Minus', and 'The Interesting' in a holistic way for the whole course (unit). Further explanation was provided to the students that this would be a brief of what they have been doing on daily basis at the end of each class on their daily reflections. In addition, students were requested to recall the 'ice-breaker' that they completed on the first encounter of the teaching period, in which they were asked: 'if they were required to provide data or were involved in forming a strategy for their workplace or their family'. To this another part was added: 'And which frameworks they used for the formation of their strategy then, and now, following this unit how would they improve on the choices that they have taken and reported on earlier, and what sorts of frameworks they would be using following this course.' Whilst students were unable to provide any of the frameworks in the 'ice-breaker' document in response to the question, they were able to come up with a

**Table 1** Student numbers and their countries of origin (obtained from attendance records and ice-breaker document)

| Country, # of students and % of students from each country for the periods of teaching in France | | | | | | | | | | | | | |
|---|---|---|---|---|---|---|---|---|---|---|---|---|---|
| | 01 Feb 2010 | % | 01 Feb 2011 | % | 01 Nov 2011 | % | 01 Nov 2012 | % | 01 Nov 2013 | % | 01 Nov 2014 | % | Total # by country | Total % |
| Argentina | 0 | | 2 | 6.67 | 0 | 0 | 0 | | 0 | | 0 | | 2 | 6.67 |
| Australia | 1 | 4.76 | 0 | | 0 | 0 | 0 | | 0 | | 0 | | 1 | 4.76 |
| Austria | 0 | | 0 | | 2 | 7.7 | 4 | 14.8 | 5 | 22.73 | 5 | 26.32 | 16 | 71.55 |
| Brazil | 0 | | 1 | 3.33 | 1 | 3.8 | 0 | | 0 | | 0 | | 2 | 7.13 |
| Canada | 1 | 4.76 | 2 | 6.67 | 0 | 0 | 1 | 3.7 | 0 | | 0 | | 4 | 15.13 |
| China | | | | | | | | | 1 | 4.55 | 0 | | 1 | 4.55 |
| France | 11 | 52.38 | 4 | 13.33 | 2 | 7.7 | 1 | 3.7 | 4 | 18.18 | 4 | 21.05 | 26 | 116.34 |
| Germany | 0 | | 0 | | 2 | 7.7 | 4 | 14.8 | 0 | | 1 | 5.26 | 7 | 27.76 |
| Hungary | 0 | | 0 | | 2 | 7.7 | 0 | | 0 | | 0 | | 2 | 7.7 |
| India | 2 | 9.52 | 14 | 46.67 | 4 | 15.4 | 4 | 14.8 | 5 | 22.73 | 3 | 15.79 | 32 | 124.91 |
| Italy | 1 | 4.76 | 1 | 3.33 | 7 | 26.9 | 4 | 14.8 | 4 | 18.18 | 2 | 10.53 | 19 | 78.5 |
| Korea | 1 | 4.76 | 1 | 3.33 | 0 | 0 | 0 | | 0 | | 0 | | 2 | 8.09 |
| Morocco | 3 | 14.29 | 1 | 3.33 | 1 | 3.8 | 0 | | 0 | | 0 | | 5 | 21.42 |
| Norway | | | | | | | | | 1 | 4.55 | 0 | | 1 | 4.55 |
| Peru | 0 | | 0 | | 1 | 3.8 | 1 | 3.7 | 0 | | 0 | | 2 | 7.5 |
| Poland | 0 | | 2 | 6.67 | 0 | 0 | 2 | 7.4 | 0 | | 0 | | 4 | 14.07 |
| Singapore | | | | | | | | | 0 | | 1 | 5.26 | | 5.26 |
| South Korea | | | | | | | | | 0 | | 1 | 5.26 | | 5.26 |
| Sri Lanka | 0 | | 0 | | 1 | 3.8 | 0 | | 0 | | 0 | | 1 | 3.8 |
| Taiwan | 1 | 4.76 | 0 | | 2 | 7.9 | 0 | | 0 | | 0 | | 3 | 12.66 |
| Thailand | 0 | | 2 | 6.67 | 0 | 0 | | | 2 | 9.08 | 2 | 10.53 | 6 | 26.28 |
| USA | 0 | | 0 | | 1 | 3.8 | | | 0 | | 1 | 5.26 | 2 | 9.06 |
| | **21** | | **30** | | **26** | | **27** | | **22** | | **20** | | **138** | |

response supported by a framework that they felt would be appropriate to allow them compose this strategy.

The quantitative data consisted of figures and numbers, and was therefore organized and analyzed using Microsoft Excel. However, the qualitative data which was diverse and from different sources as outlined above, needed a more rigorous approach, thus, the contents of the daily reflective journals, the final reflective journal, and the data collected during the first session, were all subjected to content analysis to identify common themes. The qualitative data collected during and after the course (unit) revealed the students' attitudes to the multidisciplinary material covered, and the latest approaches to strategies and sustainable development.

The analysis of the data indicates a distinct shift in students' mindsets; these students consider that sustainability and development/ethical strategies in business are not only important, but need to be incorporated in business strategies as a matter of **urgency**. Because of the careful design of the material and the group work in the class, students appeared to be far more engaged in the course (unit) than it was first anticipated specifically following the skepticism that was witnessed at the beginning of the teaching period every time this course (unit) is run, further enhancing their learning experience. The change in mindset was dramatic, even some of the students stated in their final reflections and in their evaluation of the teaching and material that their dissertation (which they were needed to prepare and present as part of their degree) would be related to the topics discussed during the course (unit), which is considered an achievement.

## *3.1 Teaching Methods*

SDBS/SDES was designed in late 2009 in Australia to be presented for the first time in February 2010 in a European Business School in France, with an emphasis on the role of the lecturer as facilitator, and the extensive participation of students. It was a requirement that assessments needed to be on an individual basis; thus, these assessments were designed to include group work within the class discussions. Thus, class discussions became more animated as students were continuously exposed to the latest issues taking place in the market in relation to the topic(s) under discussion during the time of running the course (unit). For example, during the third run, Rio+2012 was a major issue for discussion. Therefore, the more informed the students became of the issues at hand through the analysis of peer-reviewed journal articles, case studies, video clips and the challenges faced by businesses, countries and continents that are exerting every effort to avert the negative consequences, the more involved they became in class discussions and debates, which significantly enhanced their learning experiences.

Students were not required to have a text book for the SDBS/SDES unit. Instead, they were exposed to different peer-reviewed journal articles, books, books chapters including e-books and extracts, videos, DVDs, and case studies. Therefore, in the syllabus that became available to students 1 month prior to the commencement of the teaching period, students were urged to search the online databases for articles that dealt with the latest issues pertaining to sustainable business and ethical development, talent management, and leaving a legacy and strategy in the corporate world. The students were warned that as the topic was broad, the readings needed to be derived from diverse disciplines. Having said that, and in order to assist the students in their search, they were provided with key areas to search for and topics that would be addressed in the SDBS/SDES course (unit).

These topics and issues included: stewardship, living, environmental ethics as virtues, the responsible person, building competitive advantage through integrated talent management, responsibility for the self, community and nature, strategic thinking, formation, change, and content, sustainable business development, innovation and management thinking, enterprise thinking and the strategic logic of strategic business development, developing business ethics culture in organisations, dynamic capabilities and being responsible for all (community, nature – thinking about the triple bottom line).

The daily and final reflections allowed students to reflect on what they learned throughout the course (unit). Students were specifically warned not to summarize what they have learned, but rather to reflect on what they have learned. The lecturer/facilitator encouraged reflection rather than description without complicating the students' lives in introducing frameworks and models that would assist them in their reflection such as the S.T.A.R.L. and Gibbs [61] reflective circle or The Luft and Harry window 'The Johari Window' [47, 57] but endeavoured to guide them as they go on daily basis on how best to write a reflective account. It was interesting to see the development of the students' reflections throughout the course (unit) on daily basis. Details of these comments will be elaborated upon later in the chapter.

## *3.2 Course (Unit) Development and Discussions*

Involved in teaching sustainable development to engineering students in the UK, Predan et al. [41] consider that sustainable development is a complex concept which concerns a wide range of social, techno-economic and environmental issues. Without addressing all these dimensions, the teaching of sustainable development would not be complete. This is one of the challenges that faced the developer/facilitator of this course (unit). Another challenge was the location of the course (unit) developer/facilitator and the school where the course (unit) was running, since the course (unit) developer/facilitator was in Australia, the school was in Europe and more specifically in France (some 144 km away from the French Capital Paris). Adding to the complexity, the fact that the students were gathered in France from different countries around the world (Table 1). However, technology, although described as one of the major risks of the modern society according to Newton [39] and Donaldson [14], proved to be an invaluable asset during the development of this course (unit). It was through the exchange of e-mails between the developer and the European/French partner's representative that communications were established, and technology enabled contact with students prior to meeting them face to face, which assisted in breaking the ice between all parties concerned. Moreover, Predan et al. [41] highlighted the importance of tackling the teaching of such a course (unit) through the application of a multidisciplinary approach, and this was considered when developing the teaching material and assessments for this course (unit).

The syllabus for this course (unit) was prepared in consultation with the European/French partners, identifying the anticipated outcomes and outlining these in the syllabus as follows: (1) demonstrate awareness of and sensitivity to the importance of sustainable development, talent management and business strategies in such a crucial time for us as humans who are concerned about the next generations and our environment. To try and understand how 'business people and leaders' can assist; (2) apply conceptual tools and frameworks for the critical analysis of business decision-making practices and policies; (3) apply the frameworks, perspectives and skills of critical analysis to contemporary business issues, tackling holistic issues rather than concentrating on economics and finance; and (4) translate the theories, concepts and analytical techniques learned into practice.

Per the students' (anonymous) evaluation of the course (unit), these outcomes were achieved as indicated by the following comments:

> I felt this unit really enhanced my knowledge with respect to the non-economic issues at hand for any business. (Anonymous Evaluation_Student 1)

> I feel that [facilitator] has helped me to understand the more grave issues for a business in today's world and really opened my eyes to the sensitivity of the situation. (Anonymous Evaluation_Student 2)

> I have learned a lot from the theories that were shared in class. (Anonymous Evaluation_Student 3)

In addition to the outcomes, it was evident that it was not only the resources provided, or the discussions facilitated, but also the knowledge and passion of the course (unit) developer, lecturer/facilitator that was under the scrutiny by the students. The students felt that passion and compassion interweaved with multi-disciplinary knowledge in delivering such a complex and holistic topic as sustainable business development and business strategies, is of great importance to the success of this course (unit). This was evident from the following comments received from the students:

> The [facilitator] was very passionate about teaching… communication [developed] and fruitful discussions followed. (Anonymous Evaluation_Student 4)

> The [facilitator] well organized and involves the class in the subject through games and debates as well as extensive case discussions. Learning this subject under [facilitator's] effective teaching was a great experience. (Anonymous Evaluation_Student 5)

> I think the way [facilitator] has [performed] and done the classes is really appreciative – [facilitator's ability] to teach different new concepts [within the limited time]. (Anonymous Evaluation_Student 6)

Thus, it is not only the material, the design or the assessment task design, but rather the developer lecturer/facilitator's passion, holistic understanding and experience of the concepts that were under scrutiny by the students that seemed to have significant importance for the students learning in this course (unit). Indeed, the developer, lecturer/facilitator did not only think globally and teach locally [15], but went a step further by thinking locally and globally and teaching globally with the necessary passion and compassion for the subject.

## 3.3 Course (Unit) Delivery Mode and Discussions

The course (unit) under discussion was delivered over a period of five working days. This included 6 h of seminars per day, except for Thursday, when the session was for only 3 h, giving students the time to work on their 'vision and mission assessment'. During these seminars, the unit developer/lecturer played the role of 'facilitator'. The course (unit) began with an introduction to the issues and challenges that businesses face at the local and global levels, then moved to the introduction of other issues and case studies, derived from the latest literature, books, peer-reviewed journal articles and the news.

This course (unit) was first run in February 2010. This run and through facilitation and group discussions, focused on 'setting up the scene': growth, strategy, values and value systems, environmental ethics: business sustainable development, world risk society, terrorist threat, reflexive modernization, ethics and sustainability, dynamic strategies and strategic management, strategy tools, sustainable development business success, and corporate governance and ethics – impact on strategy. These topics were mainly discussed through brief presentations followed by readings of case studies and peer-reviewed journal articles by group members, who, on several occasions, engaged in debates or further arguments (e.g. [2] and [6]).

When the course (unit) was conducted subsequently, (i.e. February 2011, November 2011, November 2012, November 2013 and November 2014) the material and the follow-on discussions were expanded to incorporate more of the latest sources including some of the latest publications of the European Foundation of Management Development (EFMD). In addition to these sources, and to keep up with the ongoing developments, other sources were introduced. Based on these sources and others, especially with the first day of the second run of this course (unit) in February 2011 coincided the 'St. Valentine's Day', thus, it was felt prudent to relate 'love' to 'business'; thus, the first session following the ice-breaker and introductions session including the examination of the (EFMD) publication on issues facing the responsible manager in the twenty-first century, moved onto 'Love, does it apply to business and global business' [e.g. 6, 34, 28]. Another example of incorporating relevant and up to date material would be what took place in the third run which included a reference to RIO+20, which was held in June 2012, while the fourth run included more recent articles about the economics of well-being [19], runaway capitalism [35], when one business model is not enough [7], and creating sustainable performance [48], the fifth and sixth runs continued on the same trend of including more recent articles such as [30, 36, 37, 45] just to name a few.

During the first seminar of the first run of this course (unit), the lecturer/facilitator sensed the students' suspicion about the inclusion of the topic of 'Love' in a course (unit) that discusses sustainable development and business/ethical strategies; however, despite their scepticism, they appreciated the introduction of 'Love' to the topics discussed in this course (unit):

> 'The way people interpreted love and tried to connect it to organizations was quite interesting too. It was a really light, yet important topic which kept us involved even at the fag-end of the day. (Reflection_Student 1)

Thereafter, and through the facilitation of group discussions, group work, debates and arguments, students could cover other topics which were presented under the different titles. These topics were presented through a short presentation by the lecturer/facilitator, followed by some group work, either reading peer-reviewed journal articles, providing a summary of the main aspects of this article and its relationship with the material or sharing with the students some of the latest case studies that would provide real life examples on how corporations are either adhering to sustainable and ethical strategies or the corporations are well away in their way of thinking of becoming sustainable corporations and what would be the suggestion from the students based on the peer-reviewed journal articles that were shared with the students to improve their way of conducting business for the benefit of the human kind not only now but including the future generation (Table 2).

**Table 2** The topics that were covered in the Sustainable Development and Business/Ethical Strategies course (Unit)

| Topic # | Topic title |
|---|---|
| **Topic 1** | Making the case for sustainable business |
| **Topic 2** | Is our society at risk? – What a question! |
| **Topic 3** | Reflexive modernization – influence on sustainable business development |
| **Topic 4** | Strategic thinking and ethics? – What a match! |
| **Topic 5** | Business processes and global market collapse |
| **Topic 6** | Strategy tools for shifting landscape, values? |
| **Topic 7** | Would they make a difference when it comes to strategy? |
| **Topic 8** | Strategy and society – why the change of heart? |
| **Topic 9** | Sustainable development and business success |
| **Topic 10** | Strategic thinking – corporate governance – a relationship? |
| **Topic 11** | Current status – Where are we heading? What is needed? |
| **Topic 12** | Thinking in a sustainable manner - does it need ethics? |
| **Topic 13** | What is the solution? |
| **Topic 14** | An example from the IT industry strategy |
| **Topic 15** | But how? – Introducing tools and frameworks |

In addition, and to introduce herself and her country, the lecturer/facilitator, shared with students a presentation on 'Australia – Figures and Numbers' which highlighted the sort of data that would be important when making a decision to expand globally. This was very well received by the students as is evident in the following statements:

> Finally, the presentation on Australia and the way you tried to sell the country was not only informative but also very interesting. (Reflection_Student 2).

An extensive amount of material was provided to adequately cover the many different topics, which was appreciated by the students:

> The learning experience of this unit was good, the resources were extensive and induced further thinking and overall I enjoyed this course. (Anonymous Evaluation_Student 7).

> … It was good to move on from just theory to the application… (Anonymous Evaluation_ Student 8)

In fact, students appreciated the efforts exerted, and welcomed the new methods that allowed them to enhance their learning by exposing them to new, diverse and multi-discipline material. Most importantly, the students moved in their learning journey from theory to application, especially when discussions gave them the opportunity to encounter different real-life case studies.

## 3.4 Assessments and Discussions

Students were required to complete three assessment tasks for this course (unit). In agreement with the European/French partners and according to their policy, the assessments needed all to be individual. However, to encourage students, specifically as they are originated in different countries to learn about each other's countries, the class participation (which is the first assessment) had a group component. The second and the third assessments were individual, whereas the students would be providing their vision and mission for a non-governmental organisation (NGO) that they will choose in liaison with the lecturer/facilitator in the second assessment to which feedback will be provided by the lecturer/facilitator, in the third assessment they will provide the strategy for that same NGO showcasing their understanding of the frameworks and models that were provided in the teaching period (Table 3).

Students were encouraged to be innovative and creative in their presentations, for example, recording a video selling their idea to the board of directors of the NGO of their choice in consultation with the lecturer/facilitator as highlighted above.

For the first assessment, and to encourage active participation, students were required to complete an 'ice-breaker questionnaire' consisting of 15 questions, 8 of which were intended to identify the extent to which students knew about sustainable developments and business strategies. The students' responses indicated that the course (unit) was an elective course (unit) for the majority (95%), and in general they had minimal understanding of strategy. Although this exercise took longer than anticipated, students' comments showed an appreciation of this opening activity:

> Today the class started with an ice breaker session wherein we had to answer a questionnaire. After answering these questions when we discussed them in groups, we found that we

**Table 3** Sustainable Development and Business/Ethical Strategies course (Unit) assessments

| First assessment | Second assessment | Third assessment |
|---|---|---|
| Class participation – this entails: | Students' vision and mission to a Non-Governmental Organization (NGO) which entails: | Final Presentation and Report of the strategy they would use to realize their vision and mission which entails: |
| 1. Completion of ice-breaker. | 1. Students to be placed in groups. | 1. A professional power point presentation to be conveyed at the class and assessed by the facilitator with input from other classmates. |
| 2. Face-to-face discussions with the facilitator and other classmates. | 2. Students will choose an NGO in consultation with the facilitator. | 2. A professional final report on the contents of this presentation coming up with strategies to enhance ethical standards in the organization through the sustainable development strategies. |
| 3. Daily Reflections (with very strict guidelines) to be submitted electronically to the facilitator after each session. | 3. Deriving from the unit sources and their own research design a vision and mission statements for this NGO. | |
| 4. Group work on case study and peer-reviewed journal articles and reporting back to the class. | 4. This will form the base of their final Assessment. | |
| 5. Group evaluation (participating in marking groups' presentations). | | |
| 6. Final Reflection. | | |

> had common answers for many general questions. Thus despite our cultural differences we had many things in common. I found this exercise very interesting. (Reflection_Student 3)

Further, to encourage participation in discussions and problem-solving deriving from the literature, students were asked at the beginning of each session to select a green (very well prepared), orange (somewhat prepared) or red (not prepared at all) sheet of paper indicating their preparedness for discussions. On the first day, no sheets were selected, however, on the second day, students selected orange sheets, thereby indicating that they were somewhat prepared to engage in discussions, arguments and debates. It was noticed, however, that students did not select the green sheets throughout the course (unit). This was the case for all the runs of this course (unit). This might be attributed to the fact that students were overwhelmed by the vastness of topics covered in this course (unit) and never felt fully ready to discuss any topic that might face them during the teaching day. Nonetheless, since students were given the chance to read the material prior to their group activities, they were better able to participate in class discussions, arguments and debates.

For the daily reflections, students were urged to reflect upon what they had learned that day in addition to anything that was 'positive or plus', 'negative or minus', 'fascinating or interesting'. Some of the comments included:

> … as the class progressed we discussed the Alphabem of a responsible manager. (Reflection_Student 4)

> This is the first time I am studying sustainable development although I have acquired some knowledge about this topic through interaction with people in the industry. The major gains as far as I was concerned were a few concepts in the ABC manual to sustainable development [alphabem]. I did not really know that concepts such as Article 13 and Agenda 21 existed. I gained a better knowledge about few other terms such as labels and negative growth too. (Reflection_Student 5)

> … Today's session provided [me] with an idea of how to implement ethical philosophy towards building a sustainable business. It also opened my eyes to the notion that we all know but care little about, i.e., while making profits in the short-run, one also has to keep a tab at the sustainability of the business logic and ideals that you are following. (Reflection_Student 6)

Whilst the following extract from a student's reflection is rather detailed, it provides an overview of what was covered in the first two seminars of this course (unit), especially the final part where the student disputes the notion of shareholder and customer value maximization:

> Some words were quite interesting and new like global security…Then we did two case studies. An argument mentioned in that article is that the two goals of shareholder value maximization and customer value maximization cannot be achieved together, though I disagree with it since shareholder value maximization can very well be a by-product of customer value maximization. (Reflection_Student 7)

Moreover, on the first day of the first run, the lecturer/facilitator found the classroom set-up was not as conducive as it should be to encouraging group interactions, and requested that students make suggestions to make the room more discussion-friendly. This attracted some positive comments from students:

> Another interesting [thing] was the sitting arrangement. It is always to have a freedom of choose how to sit – the way you want to and feel relaxed in the classroom. (Reflection_Student 8)

The 6 h of seminars and active workshops took its toll on students. One of the comments in the daily reflection highlighted this:

> I guess the students were looking exhausted and tired because the previous subject was comparatively lighter. But I have no complaints towards that personally. To sum it up, I quite enjoyed the experience today and look forward to tomorrow's session. (Reflection_Student 9)

> Positive…I learned a lot in a short, but also negative, because I even got oral herpes, which I only have in a really stressful situation. (Reflection_Student 10)

For the final reflection, which included the same question that was asked on the first day, another question was added, which reads:

You might recall at the beginning of the week you were asked (in the ice-breaker document) whether you were required to provide data or get involved in coming up with a strategy for your workplace (or might be with family)? And which frameworks you used for your strategy before (if you have done one). Now, following this course (unit) how would you improve on the choices that you have taken and reported on earlier, also what sorts of frameworks you might use to achieve this task?

The students' comments on these issues included:

> The week of sustainable development class was very enlightening. (Reflection_Student 11)

> Teaching process was really interesting as it was always two ways in the form of presentations or debates or group work. It was a place where I learnt a lot in a short period. (Reflection_Student 12)

> Something that is quite remarkable is that I actually now realize the importance of being sustainable in one's approach and not only that but it can actually increase the profits of a company in the long term. I thank our [facilitator] for 5 enlightening days and the class will truly miss you. (Reflection_Student 13)

The above comments clearly demonstrate the transformation of and the shift in students' mindsets; indeed, the word 'enlightenment' reveals how vast the change was. This supports the findings of Issa et al. [23] following a similar course (unit) conducted for Information Systems students in Australia, which was designed at the request of the Head of School then following the successful run of the course (unit) in Europe/France.

For the second assessment task, students were required to produce a 'vision and mission' for an NGO of their choice. There were four aspects to this assessment task; students were to: (1) identify an industry of interest to them liaising with the lecturer/facilitator on the suitability of this choice, (2) identify a major challenge that appears to be affecting this industry, (3) identify a company (preferably a global 'NGO') within that industry that might be faced with this major challenge and, (4) produce a mission statement and a vision that this organization could feasibly implement. The students' final report would allow them to develop and provide a 'sustainable business development or ethical strategy including talent management' that might assist this NGO to survive and serve the 'Seventh Generation', a term coined by Newton [39]. At the start, students were overwhelmed by the demands of the assessment tasks, but they came to understand the complexity of the topics under discussion. In their reflections, students could see the bigger picture and grasp what was required for the assessment tasks.

> Today, I have got some idea about sustainability development in real company. This helped me to doing the assignment easier. (Reflection_Student 14)

The only issue was that students felt that they were treated differently in this course (unit) compared to other courses (units) undertaken at this European French partner's school, especially with the more complex assessment tasks and requirements.

## 4 Lecturer/Facilitator's Personal Reflection

The development and facilitation of this course (unit) had its challenges. Following a telephone discussion with the International Affairs Personnel at the Business School in France, that resulted in a visit by one of the Professors to the course (unit) facilitator/developer at her hotel in Paris in 2009, and following several exchanges of correspondence and telephone calls, the course (unit) was prepared in Australia, and delivered to students from different parts of the world enrolled at a European tertiary institution, more specifically a French Business School, which added to the complexity of the material to be covered. The diverse landscape (European partner, specifically French, Australian developer and facilitator and students from different parts of the world) has provided a complex road map for the course's (unit) development and facilitation. Nonetheless, through technology and perseverance from all stakeholders, it became easier to liaise and communicate throughout.

In this respect, international networks are important, especially with the increasingly diverse mix of students who, no matter where higher education is offered and being undertaken and no matter from where they are originated, continue to seek education in different parts of the world. Australia is considered a melting pot of different nationalities from different parts of the world that are keen to expand their knowledge, even well away from their homeland. From this personal experience, this seems to be also the case in Europe, specifically France. Indeed, higher education in both Australia and Europe specifically France, share the same aim of developing students from different parts of the world to develop the students' abilities to act as responsible and accountable 'global citizens'. Thus, it can be argued here that a course (unit) such as SDBS/SDES would facilitate the transformation of the mindsets despite their diverse cultural backgrounds.

On reflection, this exercise has enhanced the course (unit) developer's knowledge, simultaneously having the honour of sharing knowledge, including the latest ideas and research in relation to sustainable business development and ethical strategies, with students from different parts of the world whose own personal reflections showed an appreciation of this learning experience.

> … So far, I mean, before listening [to] this class, I thought that sustainability related to business is just 'green'. But, now I know what sustainable business is and what corporate social responsibility is. That is not only just about green and environment problem but also business in the community. I learned that the responsible manager considers all stakeholders – employees, customers, government, suppliers, local communities, shareholders, international organizations, non-government organizations, etc. (Reflection_Student 15)

> … I think the group work was also useful and interesting again. It helped us to think, recall all of the memories of the week and try to encase it in practice. Every group did a great job, so we improved during the week; and I believe that now everyone has a clear idea of sustainable development and of ethics. (Reflection_Student 16)

The comment above indicates that networking is a major player in the enhancement of knowledge and business education globally. However, it is not only networking by itself, but a combination of stakeholders' mutual interests that play a role in the success or otherwise of such methods of teaching or such a multidisciplinary course (unit). Figure 1 below identifies these stakeholders.

These stakeholders have different roles to play in the success or otherwise of this teaching model. Nonetheless, the burden is on the developer, lecturer/facilitator who should maintain the global and multidisciplinary knowledge, the passion

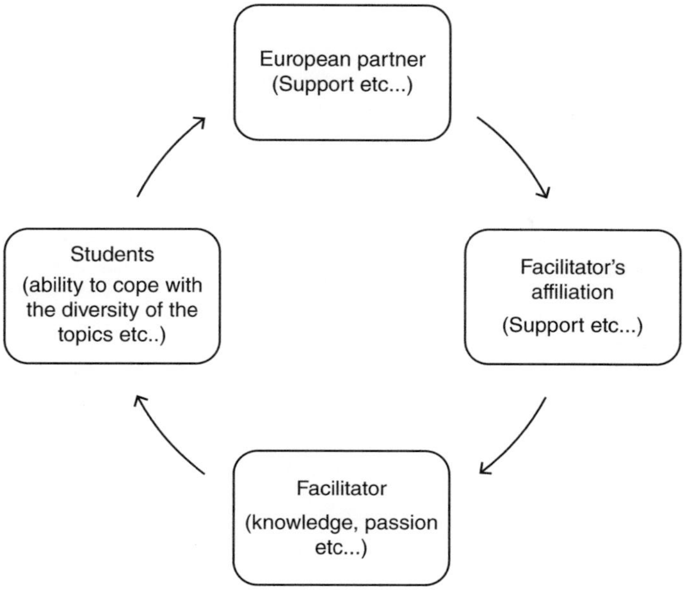


**Fig. 1** Conceptual model of the stakeholders and the looping relationships

and enthusiasm for delivering such a complex multi-faceted course, coupled with the ability to facilitate such a huge multi-faceted and multi-disciplinary topic within a short period for a diverse number of nationalities with different backgrounds in terms of education and experience. Thus, the cultural awareness of the facilitator would be a key to success. In this respect, Issa et al. [23] contributed to the ongoing debate as to whether students' mindsets shift following exposure to new material which changes their mindsets and transforms them into advocates of sustainable development for businesses worldwide. Issa et al. [23] reported their experience of teaching a similar course to Information Systems students in Australia. They concluded that the exposure of students to material pertaining to sustainable development and ethical/business strategies would ultimately change and shift the students' mindsets towards becoming more aware of the responsibilities of businesses towards people, planet and profit.

This has been conceptualized in the following model which might assist others who wish to develop a similar course:

This places great responsibility on the developer, lecturer/facilitator to demonstrate these abilities, and deliver a course (unit) that is valuable and is likely to shift the students' mindsets so that they become more aware and considerate of the triple bottom line of 'people, planet and profit', to balance corporate success and social good. It is a huge challenge that calls for a major effort from the developer, lecturer/facilitator, and great understanding from the different stakeholders.

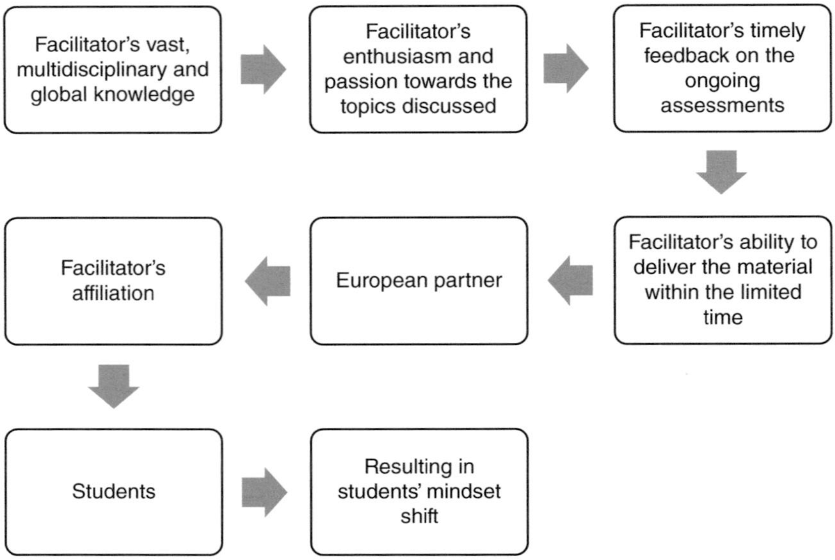


**Fig. 2** Conceptual model of the process and involvement of stakeholders

## *4.1 Limitations and Implications*

Despite the limitation of the small sample size ($N = 138$), this chapter provides empirical evidence that teaching sustainable business development and business/ethical strategies to students, from different cultures and different backgrounds is likely to transform and shift their mindsets towards becoming more aware of the dangers of being fixated only on profit rather than looking at the big picture first. Hopefully, students come to understand that profit can be a by-product, especially when businesses have a long-term rather than short-term vision. This chapter also provides an example of the collaboration that can be achieved between Australia and Europe, mainly France, through an international, higher education network. Despite various inevitable challenges, the patience and enthusiasm of all parties involved ensured that students experienced a valuable learning journey.

Such learning might really assist in enhancing the understanding of the place of moral perceptions in organizations alongside philosophical ethics and casual models of ethical outcomes, which were considered to be lacking according to Chiu and Hackett [9]. This will be achieved when those students who have been engaging in debates and completion of assessments in such a course (unit) will be the managers of the future, thus able to influence the way business is done, leading to a better organization, better community, better society, locally and globally.

It is worthwhile to note here that students' daily comments and final reflection displayed that they were ascertaining the facilitator's level of knowledge reflecting on her teaching methods and interactions with students. In their evaluation of the course (unit), they strongly agreed that the facilitator:

> 'appeared knowledgeable in the subject area' and
>
> 'enthusiastic in teaching such a unit being an effective facilitator'.

These comments indicate that anyone who is keen to teach across borders and across disciplines needs to keep abreast of the developments in diverse disciplines at both local and global levels. This is not an easy topic and it is not an easy task to develop and facilitate such a course (unit). However, if developed properly and with the co-operation of all stakeholders as highlighted in Fig. 2 above, it will benefit students, academics, business educators and practitioners in the long run. This will expand to include the community and society at the global level. In the end, however, such an investment, although huge and immensely demanding, difficult and complex, is worth the effort.

## 5 Conclusion

Despite its limitations, this chapter offers an Australian-European (French) perspective on international networking and co-operation in higher education with regards to teaching students about sustainable development. Students' daily and final reflections, evaluations and facilitator's reflections throughout the time this course was run (i.e. 2011–2014) indicate that theoretical reflection, when students engage in it seriously, is an important instrument that allows students to make better sense of the vast amount of multidisciplinary material presented in a short period (5 days).

The data collected provided evidence that teaching such comprehensive, multidisciplinary courses (units), although complex and difficult, no doubt transforms the students' mindsets and provides them with enlightening insights. This can be possible only if the developer, lecturer/facilitator maintains the necessary up-to-date knowledge combined with the necessary passion and enthusiasm to develop and facilitate such a complex, multidimensional and multidisciplinary course (unit) and be at the receiving end of the total support of the administration and teaching staff at the European (French) Business School, which was the case.

This chapter highlights the importance of such international and academic networks of like-minded academics. It is the strength of such networks that has allowed the author to liaise with the European (French) partner, to develop and present a course (unit) for six times over the period 2011–2014, to students from different parts of the world gathered at a French Business School. This chapter provided a step forward in the process of having it included as a core rather than elective unit for Business and Management students. It is the fact that extra research was conducted whenever the course (unit) was offered each time it ran in France, since further developments were deemed necessary when one considered the ongoing changes in strategy, policy, and any other issues relating to sustainability and business or ethical development. In addition, through their daily reflections, students indicated their expectations of the course and its outcomes; this might be done collaboratively with the European (French) partner well before the Australian facilitator arrives in Europe, to enhance students' learning.

Finally, this chapter provided two conceptual models for stakeholders and their roles in the success or otherwise of the networking and facilitation initiatives that aim at the enhancement of teaching globally, simultaneously shifting students' mindsets so as to enhance their self-awareness, and their awareness in general, of the 'real' role of business that extends beyond making a profit to being a social force for the care of people and environment. These two models could be used for future research in different countries and to address different issues.

Further research is anticipated in this field to provide further evidence that teaching sustainable development and business/ethical strategies to students would provide them with the necessary tools to become 'global citizens' who would care for the triple bottom line 'people, planet and profit', thus allowing businesses to also become 'global citizens' working for the benefit of the world and the environments not only fixated on profits.

***NB:***

(1) ***This chapter is derived and based on the award winning conference paper by the author under the title: 'Sustainable development and business strategies: An international higher education experience.' Proceedings of the 12th International Conference of the Society for Global Business & Economic Development, edited by the Society for Global Business & Economic Development (SGBED), 2213-2222, Singapore (2011). (Recipient of 'Best Paper Award')***

(2) ***Throughout the chapter, the author used 'course' followed by (unit) – as the two different words are being used to describe the same concept – the unit in Australia means a course in France.***

**Acknowledgements** The Author wishes to thank those who made it possible for her to travel to France for six times to teach including the family members led by V. Rev. Fr. Boutros Touma Issa and the Spirit of Bathqyomo Marine Khoury-Issa (2016+) to whom this chapter is presented.

The Author wishes to thank those from the management at the Business Schools in France and Australia for their co-operation throughout the years. Also, to the students who were brilliant in their work and discussions throughout.

## References

1. Alcaraz JM, Thiruvattal E (2010) An interview with Manuel Escudero the United Nations' Principles for Responsible Management Education: a global call for sustainability. Acad Manag Learn Educ 9(3):542–550
2. Beck U (2002) The terrorist threat: World Risk Society Revisited. Theory Cult Soc 19 (4):39–55
3. Bergman MM (ed) (2008) Advances in mixed methods research. Sage Publications, Thousand Oaks

4. Bradbury H (2003) Sustaining inner and outer worlds: a whole-systems approach to developing sustainable business practices in management. J Manag Educ 27(2):172–187
5. Bradley T, Ziniel C (2017) Green Governance? Local politics and ethical businesses in Great Britian. J Bus Ethics 26(1):18–30
6. Caldwell C, Dixon RD (2009) Love, forgiveness, and trust: critical values of the modern leader. J Bus Ethics 93(1):91–101. doi:10.1007/s10551-009-0184-z
7. Casadesus-Masanell R, Tarzijan J (2012) When one business model is not enough. Harv Bus Rev 90(1/2):132–137
8. Cavanagh GF (2010) American business values: a global perspective, 6th edn. Prentice Hall, New Jersey, USA
9. Chiu RB, Hackett RD (2017) The assessment of individual moral goodness. J Bus Ethics 26 (1):31–46. doi:10.1111/beer.12137
10. Coghlan AT, Girault P, Prybylski D (2008) Participatory and mixed-method evaluation of MSM HIV/AIDS Programs in Bangladesh, Nepal and Indonesia. In: Willams B, Sankar M (eds). Evaluation South Asia, vol 35, Asia
11. Credwell JW, Plano Clark VL (eds) (2007) Designing and conducting mixed methods research, 1st edn. Sage Publications, Thousand Oaks
12. Davidson M, Khmelkov V, Baker K (2011) Sustainability and enduring impact: shaping an intentional culture of excellence and ethics. J Character Leadersh Integration 2(1):35–51
13. Denscombe M (2008) Communities of practice: a research paradigm for the mixed methods approach. J Mixed Methods Res 2(3):270–283
14. Donaldson T (2012) Three ethical roots of the economic crisis. J Bus Ethics 100(1):5–8
15. Eck L (2008) Thinking globally, teaching locally: the "nervous conditions" of cross-cultural literacy. Coll Engl 76(6):578–598
16. Elsayed-Elkhouly SM, Buda R (1997) A cross-cultural comparison of value systems of Egyptians, Americans, African and Arab Executives. Int J Commer Manag 7(3/4):102–119
17. Fisher C, Lovell A (2009) Business ethics and values: individual, corporate and international perspectives, 3rd edn. Pearson Education Limited, Essex
18. Fisher C, Malde S (2011) Moral imagination or heuristic toolbox? Events and the risk assessment of structured financial products in the financial bubble. Bus Ethics 20(2):148–158
19. Fox J (2012) The economics of well-being: have we found a better gauge of success than GDP? Harv Bus Rev 90(1/2):79–83
20. Gomez-Bezares F, Przychodzen W, Przychodzen J (2017) Bridging the gap: how sustainable development can help companies create shareholder value and improve financial performance. J Bus Ethics 26(1):1–17. doi:10.1111/beer.12135
21. Hood GK (2006) The notion of spirituality in adult and higher education. High Educ Perspectives 2(1):166–179
22. Issa T (2016) Ethical mindsets: learning from cultural comparisons. In: Sushil, Connell J, Burgess J (eds) Flexible work organizations: the challenges of capacity building in Asia. Springer, India, pp 105–117
23. Issa T, Issa T, Chang V (2011) Would teaching sustainable development business strategies shift students' mindsets? An Australian experience. Int J Environ Cult Econ Soc Sustain 7 (5):257–272
24. Issa T, Pick D (2017) Teaching business ethics post GFC: a corporate social responisbility of universities. In: Jamali DA (ed) Comparative perspectives on global corporate social responsibility. IGI Global, USA, p 290. doi:10.4018/978-1-5225-0720-8.ch015
25. Jacobides MG (2010) Strategy tools for a shifting landscape. Harv Bus Rev 88(1):77–84
26. Johnson BR, Onwuegbuzie AJ (2004) Mixed methods research: a research paradigm whose time has come. Educ Res 33(7):14–26
27. Jones D, Caton B, DeWitt J, Stubbs N, Conner E (2007) Student-to-faculty ratios, teaching loads, and salaries in associate degree nursing programs in the central United States. Teach Learn Nurs 2:17–21
28. Kanter RM (2010) Work pray love. Harv Bus Rev 88(12):38–39
29. Korontzis TC (2013) Corporate social responsibility: impact of corporation on our lives. Eur Sci J 9(22):301–308

30. Lander MW, Kooning L (2013) Boarding the aircraft: trust development amongst negotiators of a complex merger. J Manag Stud 50(1):1–30
31. Lazar I (2005) Spirituality and human ecosystems. In: Zsolnai L (ed) Spirituality and ethics in management. Kluwer Academic Publishers, New York, pp 87–96. Ebook: Springer Science + Business Media, Inc
32. Mahin L (1998) Critical thinking and business ethics. Bus Commun Q 61(3):74–78
33. McCaffrey M (2017) The morals of moral hazard: a contracts approach. J Bus Ethics 26 (1):47–62
34. McDonald G, Pak PC (1996) It's all fair in love, war and business: cognitive philosophies in ethical decision making. J Bus Ethics 15(9):973–996
35. Meyer C, Kirby J (2012) Runaway capitalism. Harv Bus Rev 90(1/2):66–75
36. Mulder LB, Jordan J, Rink F (2015) The effect of specific and general rules on ethical decisions. Organizational Behav Hum Decis Process 126:115–129. doi:http://dx.doi.org/10.1016/j.obhdp.2014.11.002
37. Murphy MC, Dweck CS (2016) Mindsets shape consumer behavior. J Consumer Psychology 26:127–136. doi:http://dx.doi.org/10.1016/j.jcps.2015.06.005
38. Neu D (2008) Market discipline and management education. Organizational Behav Teach Society 32(6):697–715
39. Newton LH (2003) Ethics and sustainability: sustainable development and the moral life. Basic Ethics in Action. Prentice-Hall, Inc., New Jersey, USA
40. Plano-Clark VL, Creswell JW (eds) (2008) The mixed methods reader, vol 1. 1st edn. Sage Publications, Thousand Oaks
41. Predan S, Azapagic A, Clift R (2000) Teaching sustainable development to engineering students. Int J Sustain High Educ 1(3):267–279
42. Rogers RR (2001) Reflection in higher education: a concept analysis. Innovative High Educ 26(1):37–57
43. Sandel MJ (2009) Justice: what's the right thing to do? Farrar, Straus and Giroux, New York
44. Shaw WH, Barry V, Issa T, Catley B, Muntean D (2016) Moral issues in business, 3rd Asia-Pacific Edition. Cengage Learning, Melbourne
45. Slaughter A-M (2013) The Arab spring and climate change. A climate and security correlations series. Center for American Progress – STIMSON – The Center for Climate and Security, Washington
46. Sneddon C, Howarth RB, Norgaard RB (2006) Sustainable development in a post-Brundtland world. Ecol Econ 57:253–268
47. South B (2007) Combining Mandala and the Johari window: an exercise in self-awareness. Teach Learn Nurs 2(1):8–11
48. Speitzer G, Porath C (2012) Creating sustainable performance. Harv Bus Rev 90(1/2):92–99
49. Stormer F (2003) Making the shift: moving from "Ethics Pays" to an inter-systems model of business. J Bus Ethics 44(4):279–289
50. Stubbs W, Cocklin C (2007) Teaching sustainability to business students: shifting mindsets. Int J Sustain High Educ 9(3):206–221
51. Tashakkori A, Creswell JW (2007) The new era of mixed methods. J Mixed Methods Res 1 (1):3–7
52. Tashakkori A, Teddlie C (eds) (2003) Handbook of mixed methods in social and behavioral research. Sage Publications, Thousand Oaks
53. Tashakkori A, Teddlie C (2009) Foundations of mixed methods research: integrating quantitative and qualitative approaches in the social and behavioral sciences. Sage Publications, Thousand Oaks
54. Thorpe M (2000) Encouraging students to reflect as part of the assignment process: student responses and tutor feedback. Active Learn High Educ 1(1):79–92
55. Tranter P, Warn J (2003) Education for flexibility in changing world: a case study of competency development in university graduates. In: HERDSA (ed) 26th HERDSA Annual

Conference, Christchurch, New Zealand, 2003. Higher Education Research and Development Society of Australasia, Inc
56. Vargas CM (2000) Sustainable development education: averting or mitigating cultural collision. Int J Educ Dev 20:377–396
57. Verklan MT (2007) Johari Window: a model for communicating to each other. J Perinat Neonatal Nurs 21(2):173–174
58. Watson SR (2006) 'Management education as if both matter': a response to Jonathan Gosling and Henry Mintzberg. Manag Learn 37(4):429–432
59. White J, Taft S (2004) Frameworks for teaching and learning business ethics within the global context: background of ethical theories. J Manag Educ 28(4):463–477
60. Young W, Tilley F (2006) Can businesses move beyond efficiency? The shift toward effectiveness and equity in the corporate sustainability debate. Bus Strategy Environ 15:402–415
61. Gibbs JP (1998) Amphibian movements in response to forest edges, roads, and streambeds in southern New England. J Wildl Manag 62(2):584–589

# From Understanding Net Generation Expectation to Sustainable Student Engagement

**S Zaung Nau**

**Abstract** While evolving technology brings changes in teaching and learning environment, mega classes with a large number of students from diverse social and educational background convey another challenge in the learning dynamics. It is important to understand the diversity in students' preferred learning styles and expectations of Net generation students. Learning activities should be designed based on the students' preferred learning styles and expectations to enhance the sustainable student engagement which results in improved students' learning experience and achieving the learning outcomes. New learning technology can be deployed to provide the chance to gain better knowledge through experience as well as equal opportunity to participate in classroom discussion regardless of their self-confidence level.

## 1 Introduction

Learning theories have evolved over times and the advances in technology development have changed the complexity and dynamic of the learning process. This chapter reviews how the teaching delivery materials and methods based on the expectations of Net Generation students can facilitate and improve the learning experiences and outcomes in a mega unit accommodating 1500–1700 students per semester. Oblinger and Oblinger [6] state that Net generation students are grown up with high accessibility to the technology and they have significant intuitions to utilize various IT devices and application and to navigate the internet to access required information. They also mention that their ability for visualization, visual-spatial skills, inductive discovery, attentional deployment and fast response time bring significant changes in their learning styles.

S.Z. Nau (✉)
School of Information Systems, Curtin University, Perth, Western Australia, Australia
e-mail: z.nau@curtin.edu.au



T. Issa et al. (eds.), *Sustainability, Green IT and Education Strategies in the Twenty-first Century*, Green Energy and Technology, DOI 10.1007/978-3-319-57070-9_4

## 2 What Is Student Engagement?

According to Reid et al, Student Engagement can be defined as when students transform acquired information into knowledge by engaging in an experience or activity or discussion which facilitate the basis or content of their ensuing learning [10]. Pritchard [9] also recommends that the students' prior knowledge and interest in the subject, the level of difficulty in the subject and the social-cultural context should be considered in planning the lessons. He proposes the guidelines for planning the lessons as:

- To have clear and well-informed learning objectives;
- To build upon the students' prior knowledge and experience;
- To include the social interactions and activities set in appropriate context and
- To guide the students for further learning [9].

## 3 Learning Theories

Constructivist theories have been applied in the field of learning under the broad area of cognitive science since the first half of the twentieth century, [9].

### *3.1 Constructivism*

Constructivism can be defined as learning is the outcome of the mental construction process for transforming new information into individual's knowledge, understanding, and skills [9].

### *3.2 Social Constructivism*

Even though the origin of Constructivist view of learning emerges from the Piaget's cognitive stages of development theory, with the view that the growing child is learning alone. Later, the dimension of the interaction between the learner and others plays important role in the constructive learning theory. Sharing thoughts, ideas, experience, and opinions among peers and teacher in the course of a discussion play important role in the social constructive learning process [9]. Palincsar [7] also suggests that the influence of the social cultural factors such as social structure allowing active or passive compliance, social status, social interaction, individual mental functioning development and social structure of personality should be considered to facilitate the social constructive learning with adult guidance or in collaboration with more capable peers [7].

### 3.3 Cognitive Constructivism

In Piaget's theory of cognitive development, she proposes that experience is required to construct self-knowledge so that human can understand the given information to use it [8].

### 3.4 Connectivism

Nowadays, the exponential growth in accessibility to information has changed the life of knowledge, which becomes phenomenally shorter. Siemens [11] advances the learning theory in the use of technology and networks. He defines the connectivism in learning as "the integration of principles explored by chaos, network, and complexity and self-organization theories. Learning is a process that occurs within nebulous environments of shifting core elements – not entirely under the control of the individual" [11]. He highlights the important role of diversity of opinions and nurturing the connectivity of information sources for the learning process and knowledge building with the continuous learning [11].

He explains that how the learning occurs, its influencing factors, the role of memory, the process of knowledge transfer and types of learning best explained have become more diverse and complex due to significant changes in pace of knowledge growth, development of technology with ubiquity, expectation of Net Generation students and emerging philosophy of knowledge, learning and knowing [12].

### 3.5 Complexity Theory

Later, Morrison [5] introduces the complexity theory as theory of change, evolution, and adaptation in education because technology has dramatically changed the learning and knowledge development process and it becomes more unpredictable with changes in self-learning, roles of feedback for learning, non-learning learning, the connectedness of information sources, teaching pedagogies and various media for teaching and learning deliveries [5].

In addition, Jewitt [4] also suggests applying the multimodal approach by providing all different learning resources in the classroom activities with the technology support, based on the multimodality and activity theories. This multimodal approach has been widely used in designing and developing the learning materials delivered in learning management systems for both face-to-face and online teaching delivery modes.

## 4 Research Questions

The main objective of this research is to examine how it is important to understand Net generation has preferred learning styles and expectations in designing the teaching and learning activities that in turns increase the sustainable students' engagement for better learning experiences. This research is aimed to answer the following research questions;

1. Which learning styles do Net generation students prefer to achieve their learning outcomes?
2. What are the expectations of the Net generation students from their learning activities?
3. Does learning activities based on their preferred learning styles and expectation has impact on their learning outcomes and experiences?

The case study research was conducted to answer the above-mentioned research questions. Details of this case study research will be discussed in the following sections.

## 5 Case Study

In this case-study research, the above-mentioned learning theories were adopted to design the teaching and learning materials, activities and lesson plans for both lectures and tutorials to enhance the student engagement. This case study reflects on the teaching and learning experiences in an introduction to business information systems; which is one of the first-year units at Curtin Business School. This unit has approximately 1500–1700 students enrolled in each semester across all five campuses including Perth, Sydney, Malaysia, Singapore, and Mauritius. The iterative teaching delivery development process as demonstrated in the following diagram is used to continuously improve the teaching deliveries and learning activities.

As the first step of the interactive teaching delivery development process, the students' expectation survey is conducted in the first week of each semester to identify their learning styles and preferences in learning activities. Based on the findings from this survey, the teaching and learning materials are updated and delivered in a multimodal approach to suit with their expectations. At the end of each semester, students' learning experiences and learning outcomes are measured. The findings from these measurements are considered along with reflections from all teaching staff and students' performance on each question in final exam paper to reflect on any improvement for upcoming semesters. Figure 1 shows the iterative process of how the learning activities are improved based on reflection on teaching staff, feedback from students and students' performance on assessments in the unit.

Nearpod application is mainly used as learning and teaching delivery platform to facilitate the classroom activities. According to the social constructivism, social

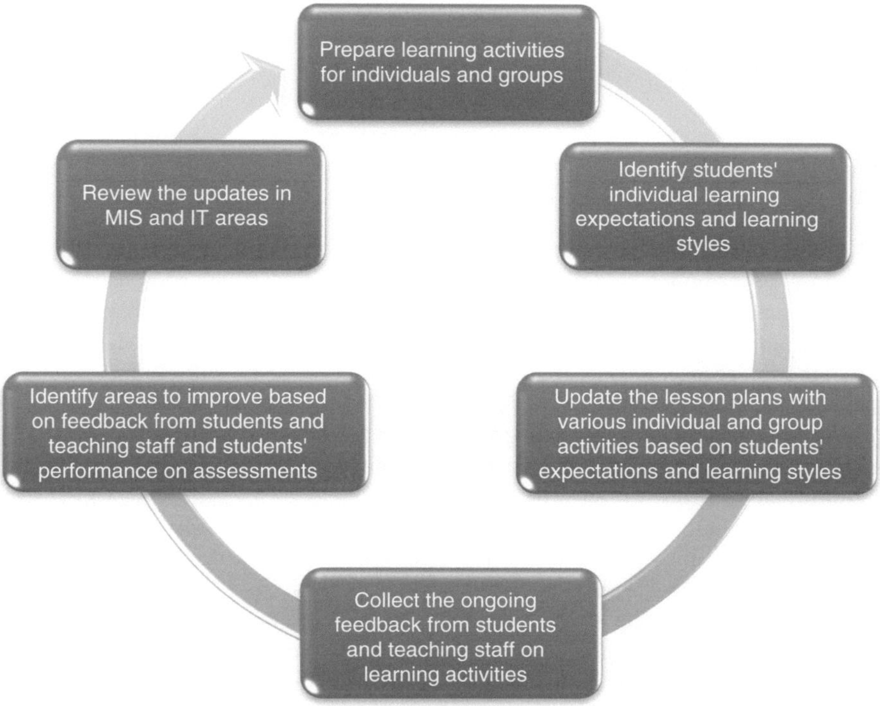


**Fig. 1** Iterative teaching delivery improvement process

interactions among students and between student and teacher are crucial in the social constructive learning process. Nearpod application caters for sharing thoughts, opinions, ideas and experiences individually and as a group. Students can log into Nearpoad session with their name or student IDs (whoever prefer partial anonymous) during their classroom activities to provide their answers, thoughts, and opinions for tutorial discussion questions which are shared later on to learn from each other. This approach enables the students who are shy and need support to gain confidence in classroom participation. It also provides the equal opportunity to all students to participate in tutorial discussions rather than having one or few students dominate in the tutorial discussions.

Followed by the tutorial discussion, case study discussion activity is also included as a group activity to apply the theory and concepts that students shared thoughts and opinions together because students will be able to construct the self-knowledge by experiencing the problem solving with case studies as stated in cognitive constructivism theory. Deploying Nearpod application as learning and teaching delivery platform facilitates the connectivity, sharing diverse opinions and accessibility to technology that play important roles the expectation and learning styles of Net Generation students according to connectivism theory.

As Jewitt [4] suggested the multimodal approach is applied in designing these learning activities. Students need to participate in face to face group discussion as well as an online presentation for their group work.

## 6 Research Method

At the beginning of each semester, the students' expectation survey is conducted to gain a better understanding of their preferred learning styles and expectations. The 328 participants of this study were the students enrolled in the introduction to business information systems unit from all campuses in semester 1, 2016. After removing the cases with missing data, 193 valid respondents with no missing data are included in this analysis.

The link to a set of questionnaire to collect the data for this research was distributed on Blackboard announcement of the unit. In addition, this link is also sent the enrolled students' email. The questionnaire includes 18 questions which can be categorised into students' socio-demographic background, their preferences on the learning activities and concerns for their study. Questionnaire items for these eighteen questions were 5-point Likert scale ranging from 1 (indicating strongly disagree) to 5 (indicating strongly agree). The questionnaire was available online by using the Qualtric (online survey tool) and the collected data was analysed with K-mean cluster analysis by using the SPSS version 21.

Based on the results of this cluster analysis, the learning activities were designed and delivered by using the Nearpod online learning tool which also captured the weekly participation in each classroom. The descriptive analysis will be used in the following section to illustrate the weekly participation in each classroom.

At the end of the semester, the feedback from student was also collected by the university student evaluate systems. The student evaluate report and students' pass rates will also demonstrate the impacts of this iterative teaching delivery improvement process in the following section.

## 7 Results, Discussion and New Findings

In this section, students' preferred learning styles and motivation resulted from the early semester survey will be discussed first. Then, the students' engagement on the weekly learning activities throughout the semester will be explained and their unit evaluations and final exam will be discussed to highlight the learning outcomes of the sustainable student engagement as a result of learning activities based on their expectations. The following survey was conducted in semester 1, 2016 to gain better understanding on the students' preferences in their learning activities and motivation so that classroom activities will be finalised based on these findings. The sociodemographic of the survey participants are as follow:

**Table 1** Sociodemographic of early semester student survey

| | | Frequency | Percent | Valid percent | Cumulative percent |
|---|---|---|---|---|---|
| Gender | Male | 83 | 43.0 | 43.2 | 43.2 |
| | Female | 109 | 56.5 | 56.8 | 100.0 |
| Age | Under 20 | 110 | 57.0 | 57.0 | 57.0 |
| | 20–24 | 50 | 25.9 | 25.9 | 82.9 |
| | 25–29 | 18 | 9.3 | 9.3 | 92.2 |
| | 30–34 | 8 | 4.1 | 4.1 | 96.4 |
| | Above 34 | 7 | 3.6 | 3.6 | 100.0 |
| Field of Study | Information systems | 62 | 32.1 | 32.3 | 32.3 |
| | Marketing | 30 | 15.5 | 15.6 | 47.9 |
| | Management | 21 | 10.9 | 10.9 | 58.9 |
| | Economics and | 27 | 14.0 | 14.1 | 72.9 |
| | Accounting | 39 | 20.2 | 20.3 | 93.2 |
| | Law | 8 | 4.1 | 4.2 | 97.4 |
| | Entrepreneurship | 5 | 2.6 | 2.6 | 100.0 |

As shown in Table 1, the numbers of male and female students in the participants are equally distributed. On the other hand, the majority of participants are the students who are under 20 and it is mainly because the survey was conducted in a first-year common core unit of undergraduate course. Students whose age between 20 and 24 contribute 25% of the participant population and students over 25 years old are accounted for 17% of this survey participants.

Students from seven field of studies participated in this survey and students from information systems study accounted mainly for 32% of total participants, followed by students majoring accounting. Students majoring law and entrepreneurship participated least in this survey.

K-mean cluster analysis is used to classify the students based on their preferences on learning activities and concerns with their studies. Jain [2] mentions that K-means cluster analysis was first published in 1955 and has been widely used since then to identify the patterns and structures in data in exploratory approach. Hartigan and Wong [1] explain that K-means cluster analysis first find the closest and second closest cluster centers for each data point to assign these data points into clusters. Then the cluster center points are recalculated based on the current cluster membership and new cluster/s may be created based on typical convergence criteria, Jain et al. [3].

The result from this cluster analysis identifies three groups of students reflecting on their preferences, motivation, concerns, effort and concerns with learning activities and outcomes of their studies as shown in the Table 2: Cluster analysis on students' preferences in learning activities and their motivation.

**Table 2** Cluster analysis on students' preferences in learning activities and their motivation

| | Cluster | | |
|---|---|---|---|
| | 1 | 2 | 3 |
| I experience interactive group activity during the workshop as motivating. | 1.81 | 4.08 | 3.98 |
| I would like to have individual activity with fewer interactions. | 3.39 | 1.35 | 1.2 |
| Group discussion on case studies followed by challenging questions to other groups. | 2.8 | 4.07 | 4.09 |
| I liked to do better than the other students in the course | 3.17 | 3.8 | 4.08 |
| I was nervous about how I will do on the worksheets and assignments | 3.3 | 4 | 2.89 |
| If I had trouble learning the content, I tried to figure out why. | 3.17 | 3.88 | 3.74 |
| I put enough effort into learning the content and the assignments. | 3.41 | 4.03 | 4.09 |
| I think about how learning this content can help me get a good job | 2.69 | 3.74 | 3.72 |
| I was worried about failing the worksheets and assignments | 2.74 | 4.19 | 2.05 |
| I think about how my grades will affect my overall grade point average | 3.24 | 4.26 | 3.62 |
| The content I learn is more important to me than the grade I receive | 2.41 | 3.18 | 3.5 |
| I was confident I would do well on the worksheets and assignments | 3.39 | 2.53 | 4.08 |
| Understanding the content gives me a sense of accomplishment | 3.2 | 4.12 | 4.15 |
| I liked the assignments that challenge me | 2.8 | 3.64 | 3.75 |

1. The first group of students prefers the individual activities and are not as competitive as other groups. The students in this group also put relatively less effort in their studies and less attempt to overcome the difficulties in learning the content, compared to other groups. In addition, they do not have much expectation for their grades and the learning outcomes of the content due to their less sense of accomplishment. Moreover, the challenging assignments are not their preference.
2. The second group prefers the interactive group activity and is fairly competitive. They also incline to participate the group discussion to solve the case studies which improve their critical thinking. The interesting fact of this group is that they are more grade-oriented and their worry for failing the assessments is significantly higher than the others.
3. The similarity between this last group and the second group is their preference for interactive group activity which involves critical thinking to solve the problems in case studies. But this group is more content oriented with a sense of accomplishment for their understanding of content. Moreover, they have less concern or worry for their grades.

As illustrated in the following Table 3: Number of students distribution in each cluster, 28% of the respondents are the lonely learners who prefer to study by themselves without having interactive group activities. These students have comparatively less expectation or motivation for their study. The second group who are more grade-oriented contributes the highest proportion of the respondents

**Table 3** Number of students distribution in each cluster

| | | | |
|---|---|---|---|
| Cluster | 1 | 54 | 28% |
| | 2 | 74 | 38% |
| | 3 | 65 | 34% |
| Valid | | 193 | |

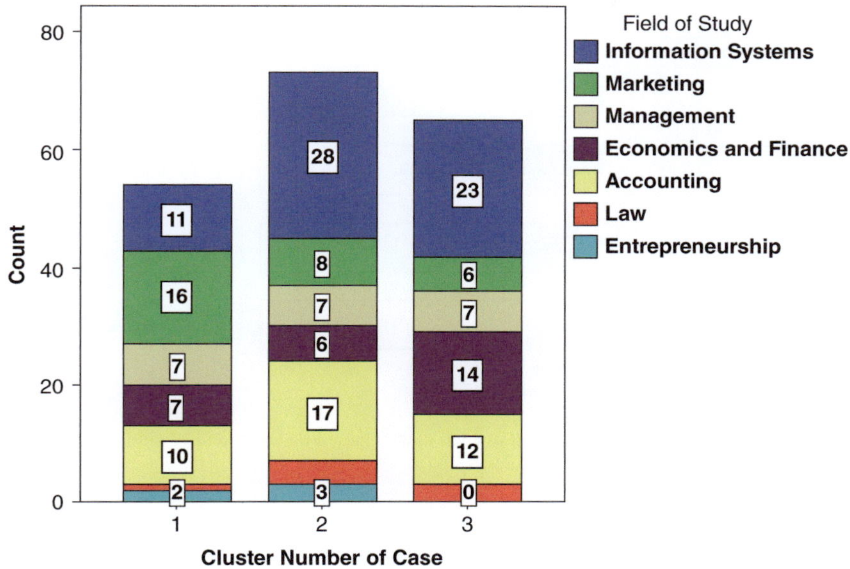


**Fig. 2** Number of students distribution in each cluster based on their field of study

which is 38% and 34% of the respondents are more content oriented. As Pritchard [9] stated in the social constructivist theory, interaction among the learners play an important role for 72% of the respondents in their constructive learning. They prefer sharing ideas, thoughts, and opinions among them by participating the case study activities to develop their critical thinking and problem-solving skills in the social constructive learning process.

Figure 2 illustrates the number of students distributed in each cluster based on their field of study. It is interesting that more than 82% of the information systems students prefer social interactive learning while more than half of the marketing students are lonely or individual learners. The numbers of management students are equally distributed in all clusters and they are accounted to seven students in each cluster. Meanwhile, half of the economics and finance students are more content-oriented learners and enjoy interactive group activities. On the other hand, more accounting students are grade driven for their learning outcomes. This finding indicates that we cannot have pre-assumption on the students' learning preferences based on their field of study. Overall, the number of students is slightly

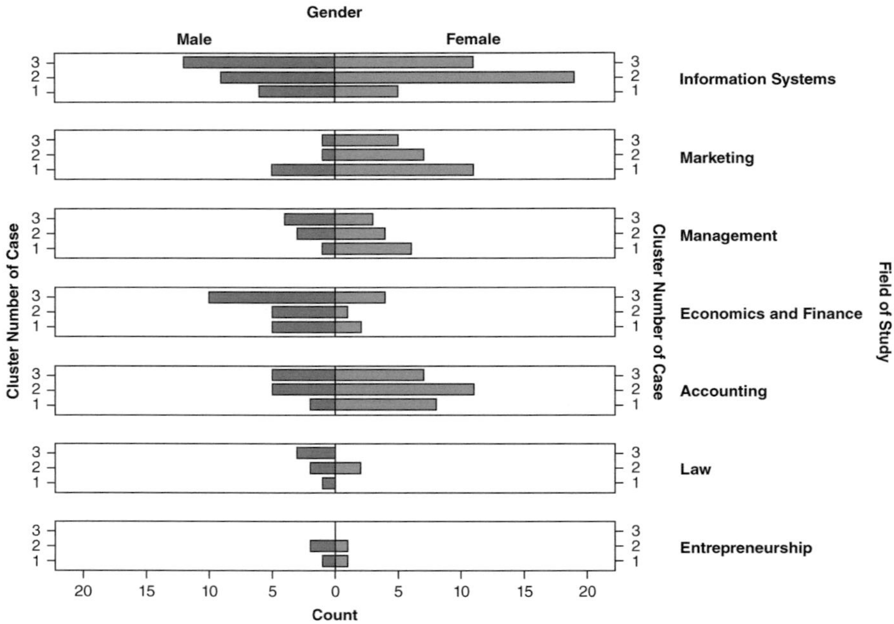


**Fig. 3** Number of students distribution in each cluster based on their field of study and gender

higher in grade-oriented group activity preferred students group than the other two groups. The total percentage of second and third groups is 72% and it can be implied that majority of students prefer to learn with interactive group activities. These findings suggest that it is necessary to include the variety of learning activities by considering the fact that students will have diverse preferred learning styles and expectation on learning outcomes.

The Fig. 3 illustrates that the numbers of grade-oriented students are relatively high for female information systems and accounting student groups while the number of content oriented students are comparatively high in male information systems and economics and finance student groups. The lonely learner group in marketing area is contributed by both male and female students and this group in management area is mainly contributed by the female students. These findings imply that learning style preferences cannot be generalised based on the students' gender and their field of study (Fig. 3).

Based on the learning theories discussed in the previous section, the learning activities are designed to accommodate all students who prefer to learn individually and as a group to enhance the student engagements in the learning activities for sharing ideas, opinions, and thoughts. Student's engagements are recorded in Nearpod application and encourage the students for their great contribution to the classroom activities at the end of each session. These student engagement reports are reviewed to reflect on how to improve learning activities in upcoming

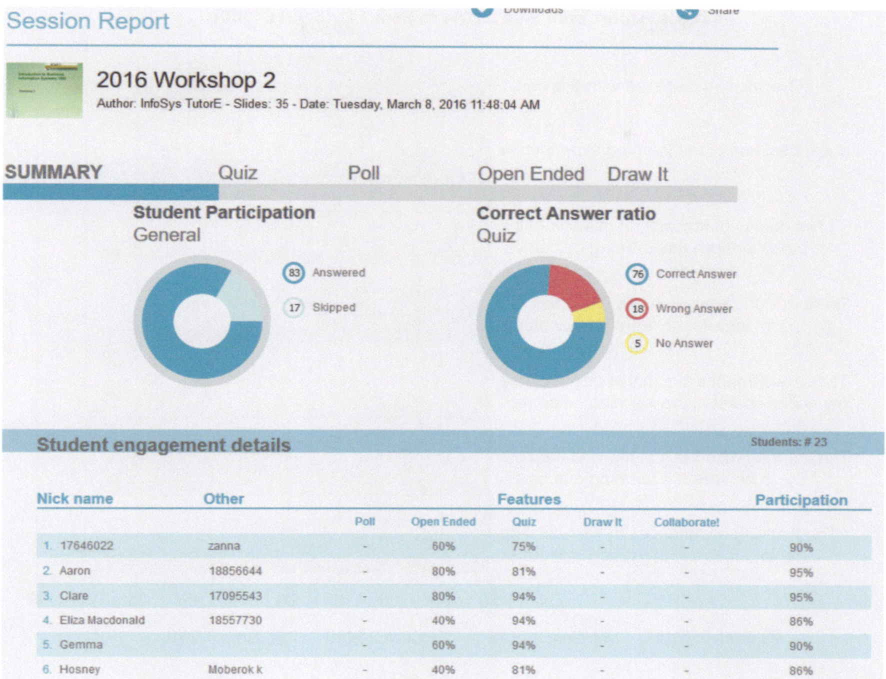


**Fig. 4** Example of student engagement report from each class

weeks. The iterative review and improvement processes are implemented to enhance the student engagement over the semester. As a consequence of sustainable student engagement throughout the semester, the evaluate/feedback from students on their learning experiences and their learning outcome achievements are improved. Figure 4 the "example of student engagement report from each class" is generated from the nearpod application that we used in our class to measure the students' engagement as part of our research.

Interactive teaching delivery development process (as shown in Fig. 1) is implemented in Introduction to Business Information Systems unit since semester 2, 2014. This interactive development process enables us to improve the quality of teaching deliveries over several semesters. As a result, the students' overall unit satisfaction, the teaching quality, learning experience rates have been gradually increased over these semesters as shown in the above Fig. 5. The feedback from students' evaluate indicates that learning resources and learning experiences developed based on their preferred learning style helped them to achieve the learning outcomes which in turn improve the quality of teaching with student-centered approach (Fig. 6).

Moreover, this interactive teaching delivery development process also enables the students to improve their learning outcomes. As shown in the following figure, the fail rates (including failed incomplete for not completing all assessment) and

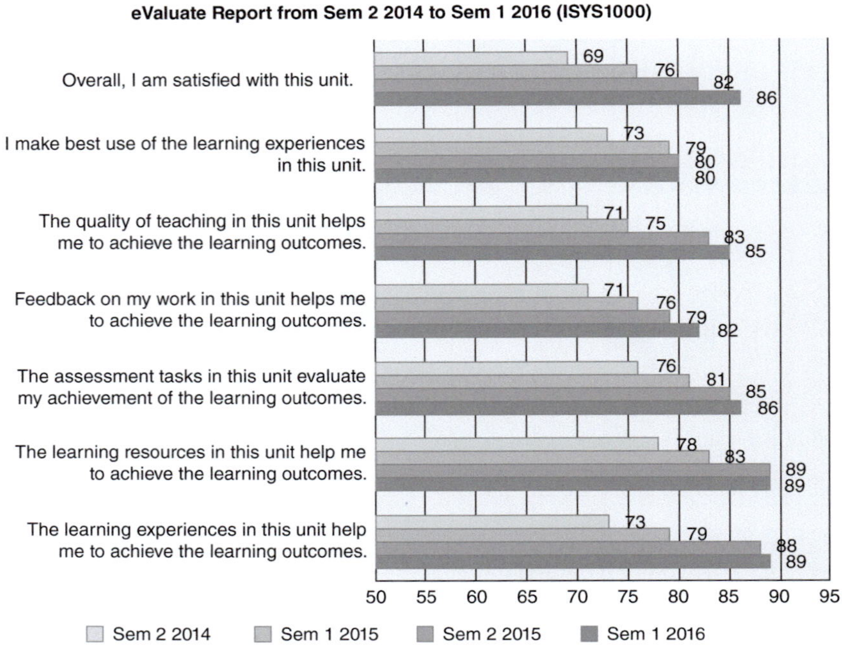


**Fig. 5** Students' evaluate results from semester 2, 2014 to semester 1, 2016

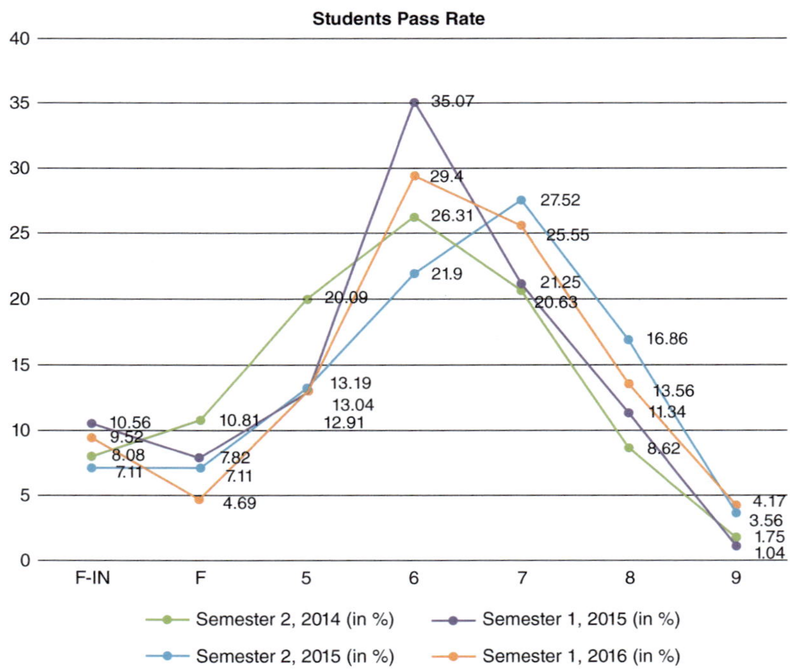


**Fig. 6** Students' pass rates from semester 2, 2014 to semester 1, 2016

just pass rates have been progressively decreased over these four semesters. On the other hand, the percentages of students who achieved grade 7, 8 and 9 (distinction and high distinction) are steadily increased.

## 8 Limitations and Future Research

The participants of this case study research were student who enrolled in the first year common core unit in Bachelor of Commerce undergraduate course. Therefore, the findings from this research cannot be generalised for students from other study levels such as postgraduate or research courses. Moreover, the participants enrolled in this compulsory unit regardless of their interests or study areas. Therefore, their expectations on learning outcomes from this unit could vary from their major units. In addition, it is crucial to keep in mind that the students' expectation and their preferred learning activities will change over time. Moreover, this research has not included the important role of culture factors in students' learning expectation and experience. Diverse culture backgrounds impacts on learning experiences, according to Signorini et al. [13]. They also suggest that we cannot have presumes the cultures are the same as nations.

Therefore, it is necessary to conduct the similar research in major units to gain better understanding on Net generation students' learning expectations on their major study units. Then we could identify the ways or areas- to improve the teaching delivery to maximise their learning outcomes. Additionally, there is also a need to conduct the similar research for postgraduate level and research students because there could be tendency that their learning expectations would vary from the first year unit in undergraduate levels. It is also recommended to include the cultural factors in future research.

## 9 Conclusion

Developing the teaching delivery and designing the learning activities for the mega class to accommodate more than a thousand students with diverse educational, social, cultural backgrounds and experiences are challenging. It is also important to include the variety of learning activities to accommodate all students regardless of their various preferred learning styles. Additionally, it can also be concluded that presumption on students' preferred learning styles cannot be made based on their gender or field of study. It is also required to review the student engagement on current learning activities for iterative improvement. Another important fact to be considered in designing learning activities is to provide the equal opportunity to all students with diverse social and educational backgrounds. As Morrison [5] states in his complexity theory, adopting the evolving technology in learning activities and teaching deliveries is crucial to maintain the sustainable student engagement and to meet the expectation of Net generation students.

## References

1. Hartigan JA, Wong MA (1979) Algorithm AS 136: a k-means clustering algorithm. J Royal Stat Soc Ser C Appl Stat 28(1):100–108
2. Jain AK (2010) Data clustering: 50 years beyond K-means. Pattern Recognit Lett 31 (8):651–666. doi:http://dx.doi.org/10.1016/j.patrec.2009.09.011
3. Jain AK, Murty MN, Flynn PJ (1999) Data clustering: a review. ACM Comput Surv 31 (3):264–323
4. Jewitt C (2006) Technology, literacy and learning: a multimodal approach. Routledge, Abingdon
5. Morrison K (2006) Complexity theory and education. In: APERA Conference, Hong Kong, 2006. pp 28–30
6. Oblinger D, Oblinger J (2005) Is it age or IT: first steps toward understanding the net generation. Educ Net Generation 2(1–2):20
7. Palincsar AS (2005) 12 social constructivist perspectives on teaching and learning. In: Harry D (ed) An introduction to Vygotsky. Routledge, East Sussex, p 285
8. Piaget J (1968) Six psychological studies, vol 462. Vintage Books, New York
9. Pritchard A (2013) Ways of learning: learning theories and learning styles in the classroom, 3rd edn. Taylor and Francis, Hoboken
10. Reid J-A, Forrestal P, Cook J (1989) Small group learning in the classroom, Rev. edn. Primary English Teaching Association, Rozelle
11. Siemens G (2005) Connectivism: a learning theory for the digital age. Int J Instr Technology Distance Learn 2(1):3–10. http://itdl.org/Journal/Jan_05/article01.htm. Accessed 6 May 2017
12. Siemens G (2006) Connectivism: learning theory or pastime of the self-amused. Learning Technologies Centre, Manitoba
13. Signorini P, Wiesemes R, Murphy R (2009) Developing alternative frameworks for exploring intercultural learning: a critique of Hofstede's cultural difference model. Teach High Educ 14(3):253–264

# Understanding "Sustainability" and Attitudes of Students to the Concept of "Sustainable Development" in China and the UK

**Austin Williams**

**Abstract** This chapter is an investigation into the way(s) that words and phrases such as, "sustainability" and "the environment" are used within China and the UK. It is a comparative analysis of responses from 18 to 21 year old students of Architecture in both countries – carried out by questionnaire sampling of students from both regions – and provides an initial assessment of the extent and possible meaning of any key differences in understanding of these terms. The discourse on sustainability is well-established in the UK but is not regularly taught as main-stream in Chinese universities, and yet many tutors, lecturers and practitioners educated outside China who are employed or guest lecture in our particular joint China-British university (and evidence shows that this also happens in other Chinese universities) are bringing a Western educational language of sustainability to Chinese students. In my experience, this has led to some discussion among students and this paper – concerning itself with issues related to sustainability – is the start of a wider assessment of social, cultural, practical and political differences in the class-room: in the West and in China. Drawing on a comparison of student responses from one Chinese university and three academically comparable universities in the UK – including questionnaire responses from 115 students studying in UK and 321 students studying in China (436 respondents in total), I examine their understanding of sustainability that might facilitate the use of these concepts more meaningfully. My findings indicate that there are clear disparities between students in China and UK reflecting a different understanding of expectations, settings and context.

## 1 Introduction

When we talk about sustainability in the west and China, are we speaking the same language? The discourse on sustainability is well-established in the UK but is not regularly taught as mainstream in Chinese universities. This research

A. Williams (✉)
Xi'an Jiaotong-Liverpool University, Suzhou, PR China
e-mail: Austin.Williams@xjtlu.edu.cn

 
T. Issa et al. (eds.), *Sustainability, Green IT and Education Strategies in the Twenty-first Century*, Green Energy and Technology, DOI 10.1007/978-3-319-57070-9_5

examines responses by architecture students to the concept: the subject cohort chosen primarily because the study of architecture routinely examines issues of social, environmental and economic balance (sometimes referenced as the "triple bottom line" of sustainability) as a general requirement of the subject. Energy modelling, material considerations and social impact are well recognized as core ingredients to educating students about how to design a building. This research examines what is understood about sustainability, etc, among certain UK architectural students and compares it with the perceptions and understanding by their peers in China.

Across the world, there is an acknowledged search for clarity within the definitions of "sustainability" and "sustainable development" [19, 34]. The UK government National Planning Policy Framework (NPPF) – the guidance for construction developments in England and Wales for example – defines sustainable development as being the totality of the 59-page NPPF document! Others try to compress the complexity of the term into a sentence. And there is often a tendency in policy circles to recognize difference of interpretation and yet to assume communality of understanding within the discourse. The signatories to the global Universitas 21 Statement on Sustainability, for example, declare that the urgency of the issue requires a "unity of purpose" regardless of one's own predisposition [67].

So even though, in many instances there is an ambiguity to the use and meaning of sustainability, "there appears to be an underlying common understanding" of what we are talking about [25]. Indeed, sustainability's definitional vagueness seems not to have hindered its policy implementation within built environment and education establishments in the West. Similarly, even though Shrader-Fréchette has argued that "(e)cology is not a science enough to provide testable laws" [64] it has done little to prejudice the fact that sustainable development is still seen as an "ecological discourse" seemingly revelling in its flexible interpretation [62]. Similarly, there is a "contested, political and floating nature of the concept" [61], but sustainability somehow still maintains its authority as a guide to morally-correct action. Indeed, I have indicated elsewhere that "sustainability... has unquestionably succeeded in becoming a defining feature of all campus life" [71]. We are coming to the end of the UN Decade of Sustainable Development in Education wherein notions of sustainability, etc, have found themselves as a key feature within the education system.

Unsurprisingly maybe, the use of the word/phrase has currency in China. As far back as 1998, some were noting that "environmental education as an instrument for environmental improvement (has) increasing importance within China's National Environmental Policy" [38] and it is becoming more and more a feature of China's formal education curriculum. Nowadays, Education for Sustainable Development (ESD) has been "recognized as an important area in the new Chinese educational reform" [77]. In the last 10 years, ESD has come under the auspices of "the leadership of the Chinese National Commission for UNESCO" and that ESD now "has become a unique education innovation with increasing recognition" in China [80].

## 2 Self-definition

The words "sustainability" and "sustainable development" are common in the West and East and are regularly used by students and lecturers within the course of conversation or official tuition, but the emergence and prominence of sustainability as an educational and professionalized doctrine in the western teaching and learning establishments over a much longer period means that – unlike China – a generation of UK adults have been well integrated into the mainstream western definitional narrative [29]. China ranks 116th out of 132 countries analyzed by Yale and Columbia Universities' Environmental Performance Index [79] and environmental issues have only entered Chinese educational frameworks in a comparatively recent period [59].

Admittedly China is one of the first countries in the Developing World – as it classifies itself – to strategically introduce sustainable development on a national and regional policy level [49], but Wei Dong Zhou, managing director of The Sustainability Consortium in China has noted that there is a significant challenge for it to apply "sustainability standards developed predominantly in the West in China" [60]. Implicitly for them, there appear to be meaningful differences in understanding about the concepts in the West and China? If so, how can the differences be quantified; and if these differences are overstated, how similar are the similarities?

This chapter is not an investigation into *lingua franca* versus Chinglish, but will try to identify some of the main disparities and similarities in meaning – derived from a range of socio-political, educational and cultural differences. The end then, is to point to potential causes of disparities, but pending further investigation, it is also to let the results of the research enquiry speak for themselves. This chapter seeks to initiate a critical framework for assessing the legitimacy of the meaning of sustainability within the context of UK and China.

## 3 Circular Arguments

The circular sustainability model propounded by William McDonough in his best-selling book "Cradle to Cradle" seems to be a mirror of China's 12th (2011–2015) Five-Year Plan Guiding Principles which say that China "should develop (a) circular economy and low carbon technologies." However, the follow-on sentence in the Chinese document states that "(t)hrough striking a balance between economic development and population growth, sustainable development will be enhanced". Some environmental advocates in the West may find this Malthusianism reference to population restraint appealing [54], but this approach has generally tended to run counter to the West's political distaste for scientifically rational population reduction policies. Even though the population-reduction position has long existed as an undercurrent in the debate in the western world, it has seldom surfaced in popular discourse in post-war Britain. Indeed, moral exhortations against Deng

Xiaoping's 1979 One Child Policy have long been used as Cold War political signifiers in the positional battle between liberal and authoritarian state practices: effectively a moral divider between "us" and "them". This is changing (a number of Western commentators have recently started explicitly referencing the need for population restraint) but the historical context is important.

The aforementioned quotation about sustainable economic development also runs counter to the contemporary Western nervousness about growth, dubbed "growth scepticism" by financial analyst, Daniel Ben-Ami [5]. Accordingly, the discourse in Europe and America tends to focus on "limits to growth" and austerity while the Chinese government explicitly reports that they are in favour of pro-growth environmental measures – to improve the environment "not" at the expense of people's extant living standards.

Recognizing these shifts, one paper in *The International Spectator* suggests that that issues related to the environment should be a little more sensitive to that of China saying that Europe should "resort to a realist but constructive policy" of engagement [28].

# 4 Sustainability

## *4.1 Differing Conceptions of Man and Nature*

The western concept of an environmental "earth-consciousness" [32], was "reinforced and given new persuasive-power" by Apollo 17's famous photograph of the so-called "Blue Marble": earth taken from the moon. For many in the western world, this potent image strongly influenced (and reflected) the environmental discourse of a fragile earth in need of protection [75]. However, that historical moment – that romantic conception of the earth that influenced the promulgation of the Gaia hypothesis in the West [45] – had a different context and connotation for environmental discourse in China. This was, after all, the mid-point of what has subsequently been called China's "decade of chaos" resulting in rural China "left with an environment ravaged by misguided Maoist Era policies of massive relocations, dam building and land transformation for agriculture" [8].

Environmentalism did not come from the same root on these two opposite sides of the planet. While Rachel Carson in the West was defending a representation of nature that was vulnerable to human intervention, saying: "man is a part of nature, and his war against nature is inevitably a war against himself" [11]; Chairman Mao's Great Leap Forward (1960–1962) was popularizing the slogan "Man Must Overcome Nature" [24]. On one hand, in much of the western discourse the environment was proclaimed as somehow sacrosanct with humanity's responsibility to protect and nurture it; while China clearly saw human existence above nature, with a mission to tame it.

Even though the early foundational years of western environmental discourse took place at the period of ping-pong diplomacy and the Shanghai Communiqué (the culmination of a formalized *detente* between President Nixon and Chairman Mao intent on "normalizing relations"), this was still the era mid-point of Mao's Cultural Revolution [15] during which time, tens of thousands of Chinese urban youth were "sent down" to the countryside to learn from a peasant lifestyle. Unsurprisingly, this resulted in a generation who were understandably less enamoured by the sanctity of nature than some of their peers in the West. Also the practical reality of that time was "viewed by Chinese authors as a period of ecological disaster" [18]. A decade later both West and East responded to the early period of reflection about sustainability generated by the Brundtland discourse. But here too, there are clearly differences of approach.

## 4.2 Sustainability in the UK and China

Sustainability has been at the heart of primary and secondary – and tertiary – education in the UK for a long while [71]. Since 1987 western youth have grown up under the shadow of the Brundtland Report, which advocated – in Clause 68 – that " Environmental education should be included in and should run throughout the other disciplines of the formal education curriculum at all levels – to foster a sense of responsibility for the state of the environment and to teach students how to monitor, protect, and improve it." Very quickly, schools in many developed economies (excluding China at that time) enthusiastically pushed the definition of sustainable development almost as a standard text [10]. An Ofsted survey, "Schools and Sustainability: A Climate for Change", for example, states that "(p)rimary schools were more successful than secondary schools in promoting sustainability" but that both were closely involved in pushing the message. Beyond the UK, the UN General Assembly proclaimed 2005 – 2014 to be the UN Decade of Education for Sustainable Development (DESD) across the world with the explicit ambition to "integrate the principles, values and practices of sustainable development into all aspects of education and learning" [65].

## 4.3 Sustainability in China

Admittedly, the debate about sustainability was raised in China in the late 1980s, but one researcher has noted that "the term appears to have been largely developed around agro-ecosystems" [18]. The political and moral discourse was absent as China avoided integration into a global environmental perspective until much later. China signed the Kyoto Protocol only in 2002 and Chinese environmental NGOs (of which there were many at the time) were not present at the United Nations Rio Summit a decade earlier. These were specifically focused on global pressure to consider climate change, rather than sustainability. Indeed, "it was only in Johannesburg in 2002 that Chinese green NGOs made their debut" [13]. But it was only in the 11th Five Year Plan (2006–2010) that Chinese development consciously moved from "Growth Rate" to "Sustainable Development" [68] while the 12th Five Year Plan (2011–2015) "demonstrates a much more robust ambition to make the difficult transition towards a more sustainable model" [27]. It is a contentious point but Chinese novelist and essayist Yu Hua has claimed that, "today more and more people (in China) speak in positive terms about the Cultural Revolution... (reflecting a) dissatisfaction with current realities" [30] and hence there may be a turn towards other values less celebratory of the environment. But in general, "most" Chinese are buying into the environmental discourse and "are skeptical about arguments that the developed world's model should be off-limits to them because the planet's environmental health depends on their taking another path" [3].

So as we are adopt a common frame of reference about sustainability – taking the UK and China as an example – are we speaking the same language?

Even though there had been a 10 year conversation on environmental issues, it was China's 11th Five Year Plan that included environmental issues at its core, but these tended to reflect general social improvements resulting from development *per se*, rather than to "sustainable development" as we have come to appreciate it in the West [73]. The Plan called for technical and infrastructural developments in order to combat environmental degradation, which many in the West discount as "technical fixes" [31]. For example, unlike the sustainability discourse in the UK, which has focused on carbon emissions and on restricting end-user consumption, conceptualized as a policy of *limits*; China's key focus has been on reducing carbon emissions which are recognized as an efficiency *gain*. (The 10th Five-Year Plan prioritized the "synchronization of environmental protection and economic development.") Even the 12th Five Year Plan focused on targets to remediate environmental degradation and to control pollution, while at the same time attempting to boost growth, consumption and efficacy: the latter not traditionally associated with the western model of sustainability.

## 4.4 Development and Pollution

Since 2006, especially since the world's attention was focused on the Beijing Olympics, the issue of pollution has become a significant symbol of China's environmental degradation; what has become known as its "airpocalypse" [17]. The return of pollution after the relaxation of stringent Olympic-period environmental controls, has made the unpleasant and dangerous air quality situation even more evident [33]. However, the issue of pollution in the West, is regularly portrayed as a problem caused *by* humans. Influential commentators like Larsen can state that "(h)umans cause pollution... (p)ollution is caused by the conscious acts of these humans" [37]. However, China tends to view pollution as a problem *for* humans and is therefore prioritizing technical mechanisms to alleviate it [43, 16]. So while the language is the same, it appears that the meanings and practical effects are very different. In very general terms, one (the western discourse) sees humans as a problem, the other (Chinese) sees humans as a solution.

It is a cliché to say that China is cramming 100 years of western development into 20 years, but such reflections are sometimes compelling, not least as a reminder that the West has been through some of these contemporary Chinese problems themselves. In terms of pollution, post-war reforms in healthcare in the UK, for example, helped to foster perceptions of benign healthcare paternalism. After a series of thick, acrid "smogs" in the 1950s, the UK government passed The Clean Air Act in 1956 to make most urban areas zones places where only smokeless fuel could be burnt. In London in 1952, one smog incident claimed the lives of 12,000 people [4], so the legislation was brought in with health and life-saving intentions.

So the early perception of pollution in the West is the same as that understood in China today: that of a large-scale problem arising from rapid development that needs to be technically and socially resolved on the political level. For instance, when 1000 Glaswegians died in smog in 1909, a campaigner from the National Smoke Abatement Society at the time said that it was "a civic crime: the deaths are the result of the actions of the citizens themselves" [48], but ultimately, that was simply a description – it was not a statement of blame. It was not an injunction that they stop lighting fires, given that there was not much of an alternative way for them to keep warm. Similarly, the Clean Air Act introduced cleaner fuels as alternatives for people to keep on keeping warm.

In recent years, there has been a shift in the language and the objectives used. The air pollution discourse in Britain tends to suggest that it is personal consumption that is responsible for the state of the atmosphere; and that morally-culpable individuals should take action to reduce their impact. Thus minimizing personal consumption of fuels, of cars, of energy, etc., has been the driving discourse over the last 15–20 years. So here we witness another difference in approach to pollution East and West – and one that has evolved and changed over time *within* East and West.

In the UK, there is a therapeutic state that seems to be blaming consumers for consuming [40]. So we have a modern western state that seems to advocate personal responsibility to deal with social problems, rather than its post-war incarnation as a welfarist state that tried to provide technical solutions to social problems [20]. China seems to be heading the other way and looking to *increase* consumption, develop its economy and to *reduce* its pollution problems. Even though it can be overstated, there is some evidence that there has been a shift in Chinese heathcare away from personal responsibility [12, 44] as economic reform finally begins to liberate China's health sector with the lure of profitability [22, 36].

In its 10th Five Year Plan (1995–1999), the Chinese Communist Party adopted a new maxim of "sustained, fast, and healthy growth" relying, in part, on the Kuznets curve to introduce better environmental innovation to help the economy keep growing [57]. This is not the same thing as the western discourse on sustainability which has been influenced by The Club of Rome's 1972 maxim, "Limits to Growth".

So, this chapter recognizes that, on many levels, there are differences in theoretical, historical, economic, social and political approaches to the issue of environmental issues and sustainable development [23]. This chapter does not seek to argue whether there is merit in any one position over another, but simply seeks to identify the validity or otherwise of different interpretations and the possible consequences of their propagation.

## 5 Framework and Method

This survey intends to get a reasonably open understanding of some key interpretations of sustainability and sustainable development amongst 18–21 year old students of architecture and the built environment.

Two indications of the central place of "the environment" and "sustainability" within the discipline of architecture and the study of the Built Environment more generally include the growing emergence of "environmental consciousness" as an established discourse within architecture and urbanism [41] and also, the mainstream equating of construction activity with sustainable or unsustainable practices as taught within established architectural education in the West. Successive changes to the Building Regulations in the UK and the launch of initiatives to measure environmental performance such as BREEAM in 1990 in the UK (and LEED in 1998 in the USA) have changed the way architects have specified products and developed detailed designs. Since The Whole Earth Catalogue was first published in 1968 there have been many people writing books about how to design energy efficient buildings.

A number of researchers have examined the subject from an advocate's perspective referring to how we should use "the concept of environmental consciousness to refer to specific psychological factors related to individuals' propensity to engage in pro-environmental behaviours" [56] or those who wish to explore "obstacles... (to) promoting pro-environmental action" [78]. One research paper reflects a more widespread desire to see "the environment" as conveying something innately positive and those with environmentally engaged perspectives as being correct and benign. These are research perspectives that do not seem to countenance the counter perspective: that people may wish to legitimately disengage from pro-environmental behaviour or that the definition might have different meanings from those anticipated and endorsed.

The study has been carried out amongst a specific target group – i.e., architecture & built environment students between 18 to 21 year-old, chosen for the centrality of "environmental thinking" to their discipline [1]. Environmental Science and derivatives thereof (such as Environment & Sustainability modules) are credit-bearing in both UK and China, typically accounting for 10–12.5% of yearly credits in the first years of Built Environment courses.

The respondents comprised 115 students studying in UK and 321 students studying in China. This is 436 respondents in total.

The UK contingent comprised 74 UK nationals and the results do however include the following nationalities: Poland (5), Greece (4), Malaysia (2), Bulgaria (3),

Germany (2), Sweden (2), non-specific Middle East (3) and 1 each from Vietnam, Iran, Russia, Indonesia, India, Spain, Kurdistan, Hungary and Romania. Eleven Chinese students and one Korean student beginning their studies in the UK, and hence registered in the UK contingent, have been excluded for the results to filter for an Asian comparison. Our survey therefore indicates the general results (minus Chinese and Korean) and then identifies trends within the UK national cohort ("the UK cohort").

All students in the Chinese cohort bar 2 (these were excluded from the analysis) had been born in China and few have traveled outside China and this was only for short holidays.

The respondents are representative of the desired catchment representing 18–20 year olds studying architecture including Built Environment modules. All students completed the study and completed the questions fully, although there were a gaps in one or two questions (these are represented as no responses or don't know in the survey results). It is worth noting that in many questions, there was the opportunity for multiple responses and so the numbers may not add up to 100% across the entire cohort.

The survey sample contained 202 males and 243 females reasonably evenly spread. All those students (73%) who responded to the question stated that they received their news from television (27% did not express an opinion). Where gaps in pie charts exist, this is primarily because of no comment by respondents, but also includes some unclear typing/spelling errors.

This research was conducted by on-line questionnaires in autumn, 2013. It required newly arrived first year undergraduate students in Architecture departments to answer anonymously an attitude questionnaire with a requirement that one section be completed before moving on to the next section.

## 6 Advantages and Disadvantages

There are downsides to this methodology in that a clear understanding of the motivations for respondents' choices is not always evident, and also that nuanced responses could not be facilitated. The research method does rely on reasonably adept English language skills (hence the English-language university setting in China). Furthermore, a small number of students spoiled or made errors on their papers that might not have occurred if more direct interviews had taken place. However, the advantages are that the engagement with the project was high borne out by the reasonably high response rate (and a number of faculty in the universities engaged in the survey had been keen to follow up with more comparative surveys) and report (verbally) that the students enjoyed the questionnaire. There were relatively few spoiled papers; the speed of response especially of the Chinese students was impressive, and it could provide a framework for future comparative assessments in the future.

This research was an initial study to lay the foundation for future, more detailed enquiry that will increase the numbers and locations of those surveyed above the scope of this initial proposal (future research and analysis will also address intergenerational comparisons). The scope of this study is limited to catchment groups within one university in China and three in the UK. It is simply to ascertain the level of understanding of students who are alert to the western discourse on sustainability; and those brought up in the Chinese educational model comparing their understanding of sustainability. To this end, results from Chinese students studying in the UK universities (11 students in this sample) and western students studying in China (2 students: 1 from Mauritius and 1 Chinese student born in Germany) were disregarded at this stage of analysis, to keep the data sets clear, although we hope to conduct further research to understand the impact of western university education has on result differentials between those Chinese students completing their studies in UK and those completing in China.

The Chinese university used in this survey (XJTLU) teaches in the English language and was chosen to minimize as much as possible the distortions arising from translation and mis-interpretation of the survey questions which were written in English and were identical for all respondents. All XJTLU students have completed or are completing English language foundation year in China and have a sufficient grasp on the language to be able to complete their initial studies taught fully in the English language. There are undoubtedly many limitations of this form or research, characterized by Harari and Beaty as "traditional questionnaire and survey approaches can easily generate superficial or inaccurate data, particularly when sampling foreign populations" [26], however, this survey recognizes that to translate would introduce translational assumptions that would colour the results in potentially equal and opposite way. For example, Chinese eco-cities regularly use the word *shengtai* (生态) – ecological. The meaning of *shengtai* originates in scientific biology but has come into more common usage over the last 10 years or so to refer to traditional music, dance, fashion, etc. – so-called unspoiled intangible cultural heritage. This meaning features rarely in the local western application of the word "eco" or "ecological".

But actually, the differences in understanding between Chinese and English concepts are a key aspect that is being explored here. We have chosen not to translate the survey into Chinese to avoid distorting, predetermining or directing the students' interpretation. UK academic, Professor Woods has noted that: "where certain clearly defined facts or opinions have been identified by more qualitative methods, a questionnaire can explore how generally these apply" [74]. We note that we have engaged with literate students in tertiary education who are adept at English and occupy a narrow and similar age range.

Another distorting factor that we are alert to is spelled out in Yu Xueying's research showing that the issue of environmental concern is predominantly an urban phenomenon in China [76] and the rural population tend have a different set of values and definitions ascribed to it. This level of filter is not available to us at this stage and this chapter cannot go into this question although the students interviewed are second generation urbanized and from reasonably affluent families; not from immediate peasant or migrant labour family backgrounds.

The research proposition assumed that western students may tend to be more *au fait* with environmental issues and sustainability discourse than the Chinese students as outlined in the general historical incorporation of sustainability ideals into western education outlined above. But we wanted to find out whether the western students would fall into the "limits" and "anti-consumerist" camp (the mainstream "austerity-related discourses" described by Bramall [6], while the Chinese students would be more interested in development and consumption [70]. This chapter also seeks to assess whether the students in China would have a different interpretation of the possibilities arising in the future than might western students.

In order to achieve the purposes of the study the following items were required:

- Students should be able to (attempt to) describe/indicate what they understand by sustainability?
- Students should be able to list, in order of priority, the issues that fall under the rubric of sustainability and how they compare?

A very wide range of multiple choice items were listed for the students to choose from representing positive, negative and neutral conceptions of the issues. We have displayed the statistically meaningful results here. Results eliciting one or two responses in questions not asking for multiple choices, have been omitted for clarity.

This study will address an area of research that is currently under-explored. There are a number of studies asserting what sustainability "is", but very few that explore the received opinion of what people understand by it. The concept of sustainable development has been in the public and political domain for 30 years [10] and yet it is reputed to be incoherent ("indefinable") to many [50]. Based on the future development of China and its increasing contact with the international architectural and sustainability consultancy community, this study is likely to lay the foundations of a wide variety of future opportunities to influence – sensibly and sensitively – this growing discourse.

## 7 Results and Discussion

In this response (Fig. 1) we see that a large majority of students think sustainability is an important issue for their profession, chiming with Williams [72] who suggests that sustainability is the most important change facing the future of the profession. But the fact that almost twice as many UK students as Chinese students said "no" is telling. Twenty percent fewer UK-based students said "yes" than did the Chinese cohort. This seems to refute – almost flip – our assumptions.

Fortunately (in some senses), the results in (Fig. 2) show that the majority of architecture students believe that the profession that they wish to join are helping society. But surprisingly nearly 30% of Chinese students have a negative or undecided view on this topic. In its academic training/educational development

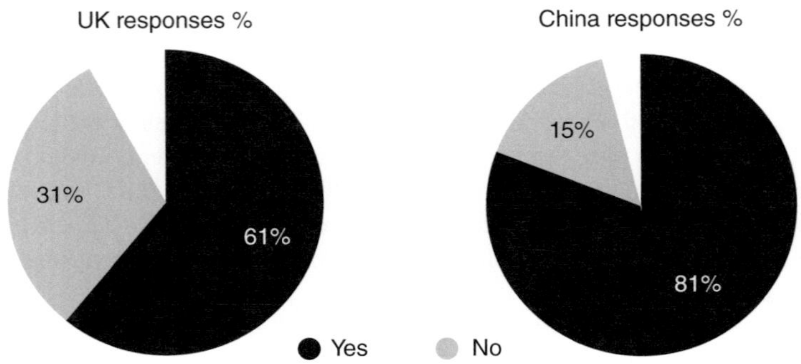


**Fig. 1** Is sustainability the most important issue for architecture?

**Fig. 2** Are architects helping or harming society?

| | helping | harming | Neutral/both |
|---|---|---|---|
| UK % | 71 | 3 | 3 |
| China % | 67 | 15 | 13 |

Chinese architecture tends to defer to technical skills rather than the arts *per se* and those entering architecture schools today, still tend to come from the science streams of secondary school education. There is a clearly expressed aspiration in China for better design over and above the pragmatic response to a client's demands, so dislike/disapproval over the current state of architecture is not necessarily a surprising response.

Unsurprisingly too, the UK response may have to do with confidence – or belief – in "British Architecture in the 21st Century (having) an almost unrivalled reputation around the world for daring, innovation, creativity and flair [53].

It is noteworthy here (Fig. 3) that nearly one-fifth of UK respondents argue that we should build fewer buildings given that the UK has the second lowest construction levels since the war. Construction figures are bolstered somewhat by a few high-cost, high profile projects (the UK government has historically represented 30–40% of construction demand). A 2009 survey states that "Compared to its European counterparts, the UK has suffered from a more pronounced decline in construction activity since the onset of the recession" [63] and at the time of this survey, construction output in the UK was approximately 15% below its pre-crisis peak in 2007. So to be opposed to building more in the UK is an interesting symbol of restraint.

On the other hand, 60% of Chinese respondents preferring fewer buildings is noteworthy. During the course of this survey, China was in the midst of official target plans to construct 32 million new homes within 4 years – a target that it says it has fulfilled. Admittedly, these are not particularly attractive buildings and the desire for fewer may arise from a hope for fewer bad buildings (but this calls

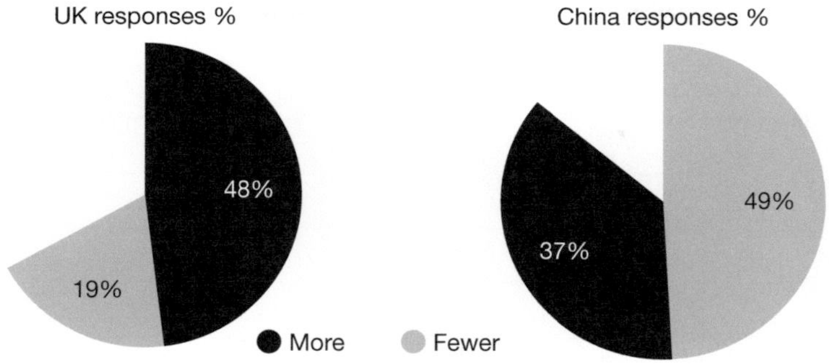


**Fig. 3** Should we build more, or fewer buildings?

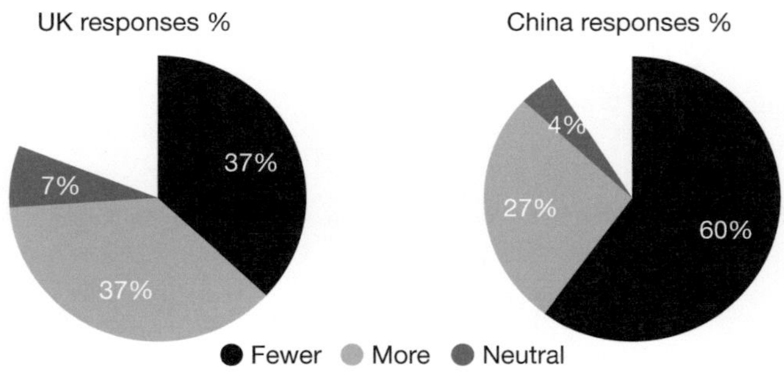


**Fig. 4** Should we build more or fewer cities?

for too much supposition). It is also as likely to be a reflection of the fact that the quantity of buildings being constructed in China allows the respondents the luxury of desiring fewer. After all, 37% of 32 million, for example is still more than 48% of very little. More targeted investigation is needed.

Similar to the Fig. 3, there is a some considerable disquiet within China about the scale and pace of Chinese city-building and its impact on local pre-existing communities, etc. However, it is significant to see (Fig. 4) nearly one-third of UK respondents arguing to build fewer cities since no city has been built in the UK for more than a generation.

It is true that the UK-based cohort could also be commenting on the Chinese urban condition as no parameters were stipulated about where these "more" or "fewer" cities should be.

Even so, it is curious that the UK, a country that has built no city since Milton Keynes in the early 1970s, or Europeans where Berlin is the newest city reunified 25 years ago, should suggest that fewer buildings should be built.

On the bar chart (Fig. 5), we see that 11% of Chinese respondents envision a world of peacefulness (compared to 0% of UK student responses); while 7% of UK students (2% Chinese) see conflict as a determining mood. The highest spike

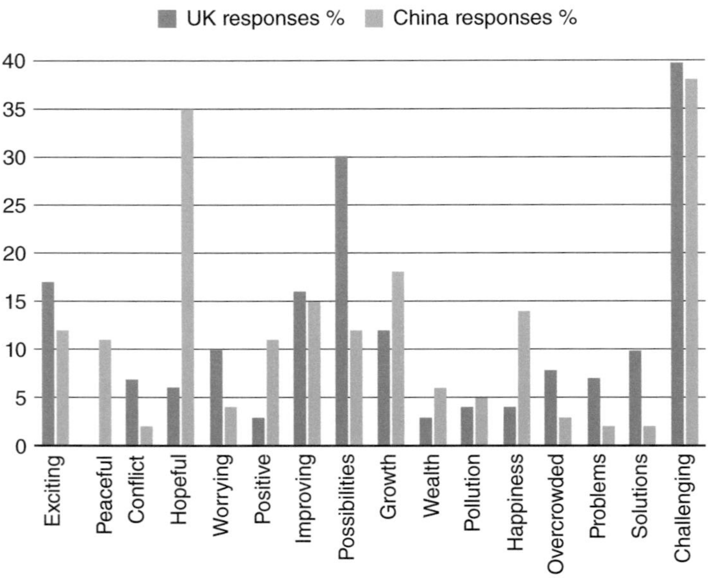


**Fig. 5** Which words would you use to describe the future?

is that the future, it seems, will be "challenging" for both Chinese and western students (38% and 40% respectively), although 10% of UK respondents cite worries compared to 4% of Chinese; and 11% of Chinese see the future positively compared to just 3% of UK students. (Note: students were given a wide range of word choices from which to choose representing an equal distribution of positive and negative values.)

In terms of the "growth" paradigm suggested above, 18% of Chinese compared to 12% of UK anticipate growth (although this is not specifically identified with a positive or a negative attitude to that growth likelihood). Comparatively, 6% predict "wealth" into the future compared to 3% of UK students.

Pollution is only expected by 4% of UK students which may be understandable given the fact of heavy industry decline and the closure of polluting industries, mines etc over recent decades: occurring some time before these respondents were born. From data taken from government sources, it seems that the UK is compliant with its 2010 national emission ceilings for air pollutants [14] and this is much improved in recent decades and thus possibly understandable that pollution might not feature heavily in UK students concerns. China's students however are slightly less concerned – at 5% – notwithstanding the contemporary hype within China about this issue, relating to air and water quality.

Just 2% of Chinese students see problems (unspecified) in the future; and a similar proportion envision solutions. Conversely, 7% of UK students see problems with 10% foreseeing solutions. Two of these responses were from the same individual respondents, i.e., two individuals foreseeing problems and solutions;

the rest allied their "solutions" to their potentially positive responses such as "possibilities", "improvements" or "challenges". However, we are not able to ascertain what these solutions specifically relate to.

From these general responses shown in Fig. 5 there certainly seems to be a certain negative interpretation of the future by the western students over the Chinese, although this is not categorical. However, in terms of the most wide-ranging differences in response categories, we see that 14% of Chinese students equate the future with "happiness" contrasting with just 4% of UK students. Similarly, 35% of Chinese students equate the future with "hopefulness" (possibly they are hopeful) contrasting with just 6% of UK students.

On March 14, 2011, China's 12th Five-Year Plan was announced which has been characterized as a turning point for China's green development "in hope of remedying its domestic environmental crisis" [58]. In this way, "hope" can be a characteristic of a sustainable or environmental mindset, but this survey is simply identifying the general characteristics in terms of ambitions, mood and mentality rather than identifying the specific underlying beliefs. In this way, the huge disparity in "hopefulness" between Chinese and UK students may symbolize a way of seeing the future within which to act: one in which there future is possibly brighter (more hopeful) than that seen by the UK students.

Here (Fig. 6) the high profile issue of sea-level rise are of equal concern to both parties of respondents but not to a great extent: climate change, global warming and greenhouse gas emissions significantly more so but also to an equal degree between cohorts. The disparities are more interesting:

One-fifth of the UK respondents seemed concerned with overpopulation; whereas only one-tenth of Chinese students – those from the most populated country on earth – considered it an important issue.

In UK-based respondents' attitudes, results show that "consumption" and "consumerism" rated 28% and 7% respectively compared to China's 2% and 1% respectively. This is a significant finding reinforcing the concentration within the contemporary critical western sustainability canon about the problems of consumer society. In architectural circles, this attitude has been made plain most recently in Pier Vittorio Aureli's booklet, "Less is Enough: On Architecture and Asceticism" [2] and others who sanctify a more austere, consumer-free life. When the issue of "materialism" is included, Western attitudes – reflected in the UK cohort answers – amount to 40% of all responses concerned with such issues of commercialized lifestyle. This contrasts with just 4% of Chinese respondents (including 1% identifying "materialism" as an issue) identifying commercialism or commercialized lifestyles as an issue. Hardly surprising you might think given that China is the second biggest economy in the world. Indeed, Chinese Premier Li Keqiang announced in 2013 that "the transformation of consumer attitudes and the upgrading of the structure and patterns of consumption... will in turn unleash a huge amount of consumption potential. The per capita income of China's urban residents is currently 3.1 times higher than that of their rural counterparts, and per capita consumption is similarly 3.1 times higher" [42]. Consumerism in China is clearly not frowned upon as it is in environmental discourse in the West.

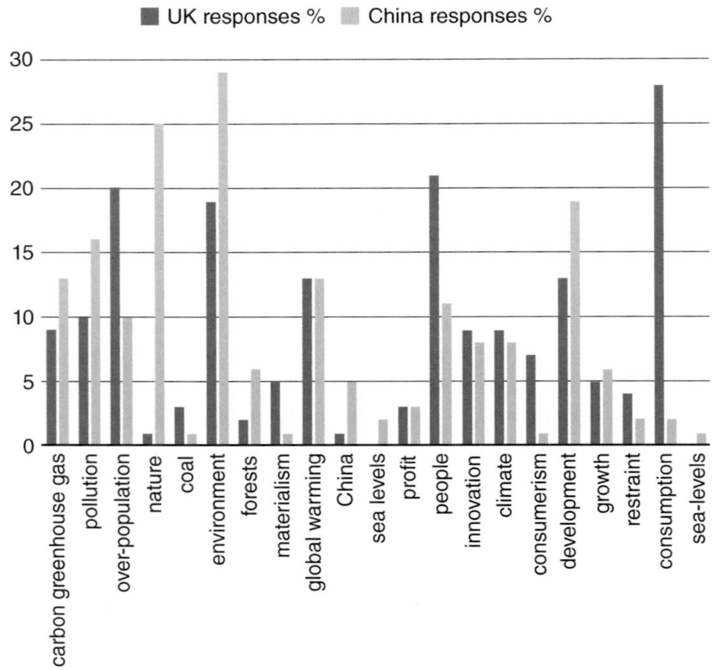


**Fig. 6** What words represent the main concerns of sustainability?

Ten years ago, British Astronomer Royal Sir Martin Rees wrote a book titled "Our Final Hour", subtitled, "A Scientist's Warning: How Terror, Error, and Environmental Disaster Threaten Humankind's Future in This Century – On Earth and Beyond" that posited that "we should forgo the benefits of extravagant energy consumption" [52]. In David Goodman's recent book however, citing government policy, he concludes that in today's China "the new model citizen is someone with high cultural capital, and the economic capacity to consume" [21].

However, we need further research to investigate whether such differences mask a communality? After all, Vijay Vaitheeswaran, the *Economist*'s China business editor says that "increasing consumption per se is not bad. What matters is the quality of the consumption."

In Fig. 6, the issue of "people" gain 21% of western responses compared to 11% of Chinese responses but it is almost impossible to fathom what this might mean. Suffice to say, even the environmental think-tank, The Breakthrough Institute has identified "the rise of Malthusianism in rich countries" [7] as a significant issue and with many leading commentators such as James Lovelock positing that the world is suffering a "plague of people" [46] it is becoming more acceptable to find "people" (their numbers, their consumption habits and the things that they do) to be a problem in Western environmental discourse.

It is unfair to suggest that this is the motivation behind the respondents' choice, but it is a potential influencer. Jeffrey Sach's Reith Lecture pointedly implied the

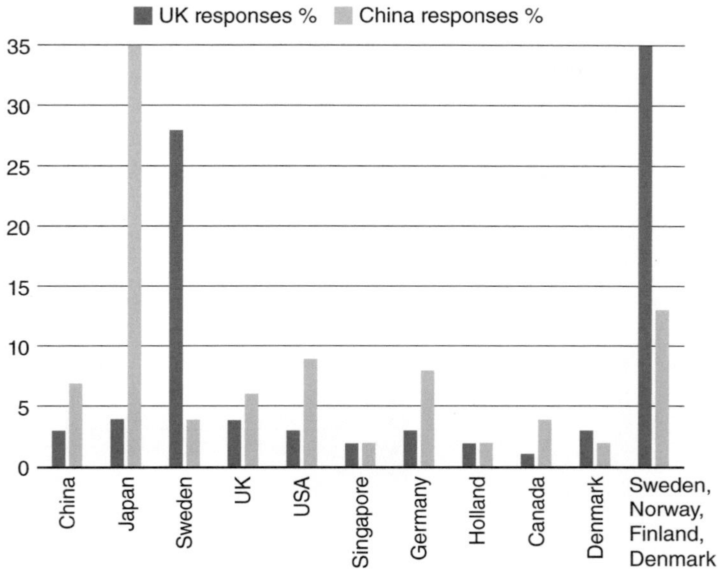


**Fig. 7** Which is the most sustainable country?

negative influences of population levels in China in a more diplomatic manner: "Not only is it the world's most populous country, it is one of the world's most crowded countries, and it is certainly one of the world's most environmentally stressed regions" [55].

Conversely, China has for a long while been retreating from its "One-Child policy" population reduction formulation and hence, maybe, attitudes to population and "people" themselves are becoming a little less fraught in China.

These two questions were free-choice, where the students could write down any answer of their own choosing. Taking these two questions together might be helpful. Firstly, it is worth noting that there were a number of other countries cited by the respondents, especially in answer to the question about which city is the "most sustainable" (Fig. 7). These include Saudi Arabia (1%), New Zealand (1), Poland (1), Brazil (1), Iceland (2), France (2), No-where (2) – but as these are responses made by one or two people they are not statistically significant for our purposes and therefore are not shown in the diagrams. To be clear, many more countries were included in the overall list of responses that were deemed to be the "most sustainable", (Fig. 7) than in the list of those "least sustainable" countries, (Fig. 8) but this long list included choices by just one person or so, so the above results are the main results excluding these small numbers, for the sake of clarity. The UK was deemed to be the least sustainable by one person in the UK; and by no-one in China (Fig. 8).

In UK responses, Sweden tops the chart as the most sustainable country (28%) and grouped together with other Scandinavian and Nordic countries amounts to (42%). In China, the picture is less pronounced with 4% choosing Sweden and

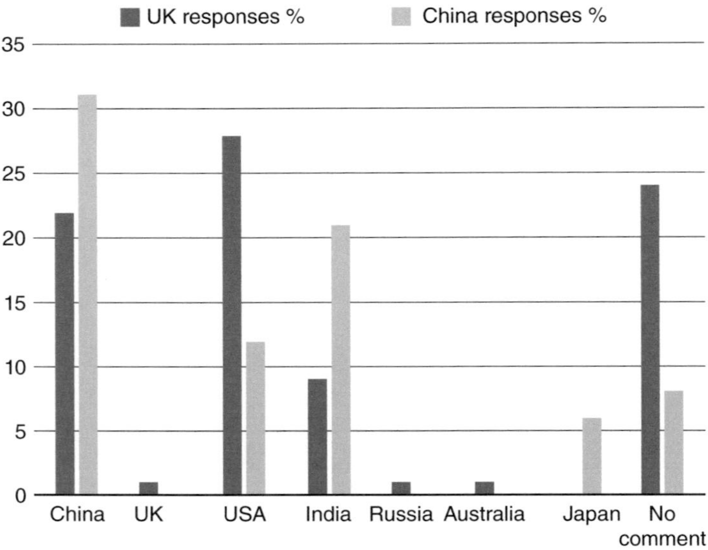


**Fig. 8** Which is the least sustainable country?

13% citing (Sweden, Norway, Finland, Denmark) as the most sustainable country/countries respectively.

Four percent of UK respondents and 6% of Chinese respondents consider the UK to be the "most sustainable" country, while just 1% of UK respondents (but no Chinese respondents) consider the UK to be the "least sustainable" country.

Conversely, 22% of UK respondents and 31% of Chinese respondents believe China to be the least sustainable, while 3% of UK respondents and 7% of Chinese respondents believe China to be the most sustainable.

The Chinese cohort also had a significant percentage of the "most sustainable country" responses allocated to USA (9%) and Australia (6%), (Fig. 7) countries that didn't register with the UK cohort at all. India is voted "least sustainable" by the Chinese cohort (Fig. 8).

Perhaps most surprising that the number one "most sustainable country" for the Chinese cohort was Japan, with 41% of the vote. Japan obtained 4% of the votes cast by the UK cohort. The West may be more prone to see Japan through the prism of natural disasters, Fukushima nuclear melt-down and non-eco-friendly whaling, but these are lesser issues in China (although anti-whaling sentiment is growing). Recent World Economic Forum Global Sustainable Competitiveness Indices that measure sustainability as the productivity level of an economy with reference to environmental stewardship and social sustainability, placed Japan at Number 12 in a list of 176 countries: (Japan rose to Number 6 position in 2014). Japan can also claim to be sustainable itself due to its Basic Environment Law that enshrines, in many ways, future generations' environmental rights. In a recent

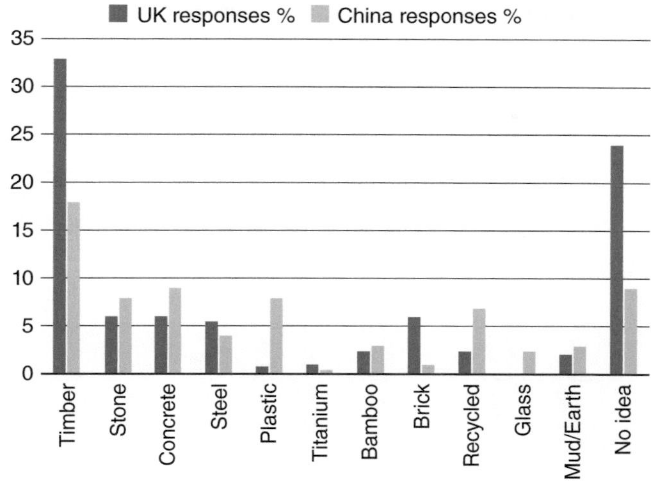


**Fig. 9** What is the most sustainable material?

comparative analysis of attitudes to sustainability in Sweden and Japan, Professor Terue Ohashi from Reitaku University in Japan has analyzed words and phrases such as "society without social gap or anxiety" that have appeared in Japan but not in Sweden, and "democracy", "equality" and "education" that have appeared in Sweden but not in Japan [51]. Maybe this East Asian interpretation gives us a clue towards the Chinese cohorts' thinking.

The percentage of "no comment" responses or those failing to present an opinion about the "least sustainable" country (24% in both cohorts) is quite significant.

While 32% of UK respondents cited timber as the most sustainable material, (Fig. 9) 30% of Chinese students said the opposite. Only 18% of the Chinese students said that timber (and that includes references to "timber", "wood", "softwood" and "trees") was the most sustainable material with a significant proportion (11%) of the UK cohort saying that it was the least sustainable. It is important to note that two of UK students made the qualification of "timber from sustainable sources", in their citation of timber as the "most sustainable material".

In (Fig. 10), 29% of UK students cited concrete as the least sustainable material as did 17% of Chinese students, with only 9% and 6% of UK and Chinese students respectively identifying concrete as the most sustainable material.

Eighteen percent of Chinese students' responses cited "plastic" as the least sustainable with 8% stating that it was the most sustainable.

Of all UK respondents, 24% said that they didn't know (or failed to enter an opinion) which was the most sustainable material and 26% similarly didn't know, or would not venture an opinion on the least sustainable materials.

The low numbers indicate that there is not a consensus (or clear knowledge) of what materials are the most or least sustainable. However the inversion of results for the sustainability of timber was unexpected.

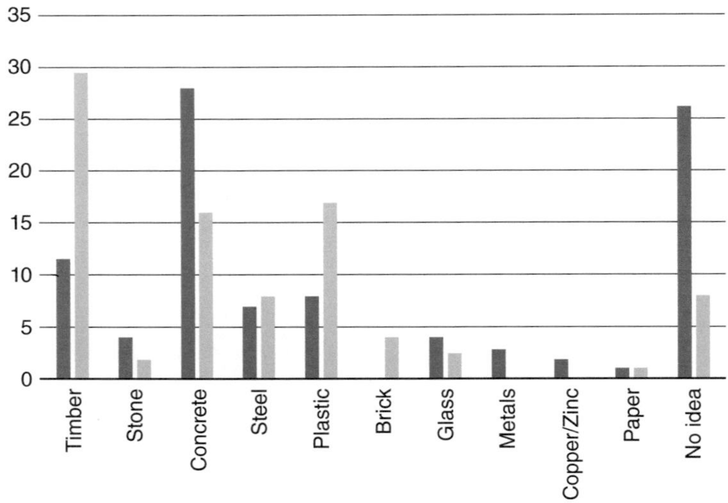


**Fig. 10** What is the least sustainable material?

## 8 Conclusion

This chapter has presented the initial framework of a model for assessing the legitimacy of the meaning of sustainability within the context of UK and China. There are limitations to this initial research that can be developed and improved. However, these indicative results should have wider implications – in terms of pedagogy, policy and practice – and it could be developed to understand other perceptual differences beyond the two countries. Further exploration can also be developed on this chapter's research question (related to understanding sustainability) amongst *different* demographics within these two countries; as well as exploring perceptual difference within and between more targeted regions within these countries.

More investigation will be needed to translate these results and to drill down further into their interpretation. As we stated at the beginning of this report, the UK (and also Europe and America) tends to focus on carbon emissions and the need to save energy, but "for most Chinese these are intangible issues" [9]. Even at the state and local level, sustainability has a variety of interpretations and China tends to see sustainability as a pragmatic rather than a moral objective.

The model could be developed and refined in various ways. The use of research categories to frame the responses and then open-ended questions to provide scope for wider comparison can be developed further to draw out some of the general reasoning behind choices. For the purposes of this attitudinal survey, the primarily intention has been to explore the initial reaction to trigger words in the sustainability lexicon. The questions need to be honed and also a secondary level of inquiry about why respondents choose their answers would add a new depth and clarity to future research. At the moment, the initial research question was premised on a suspicion of difference of opinion between UK and Chinese students on the understanding of sustainability and is borne out somewhat – indicating that greater caution might be appropriate in the assumption that there is a communality of approach, ambition or even material choices related to this subject. However, there needs to be a more detailed investigation, that should address several secondary issues arising, that might aid clarity, including for example, the strong identification among UK students in Scandinavian sustainability (Fig. 7) which could be the subject of a similar comparative survey between those two countries; and China's interest – culturally as well as environmentally – in "nature" expressed in Fig. 6. *Inter alia*, both express a "difference" that needs further clarification. The perception of humanity's relationship to nature is clearly different in these two countries and requires more analysis.

For comparative purposes, for example, we have used a range of diagrammatic representations although some of these issues lend themselves to a graphic form along the lines of the SPeAR diagram sustainability assessment tool [47]. This is the next phase of research.

At present, in this specific field there are no academic comparative assessments between West and East, and it is worth assessing to see whether the discourse – which is assumed to be global by its very nature and origin – has deficiencies or discrepancies. Just as there is no consensus on how to measure democracy [35] so this research chapter intends to improve clarity in the understanding of the different interpretations: the better to choose an appropriate model for the given setting.

In similar examples of this kind of research such as Ref. [51] which examined the interpretations of an ideal society (linked to sustainability) in Sweden and Japan, Ohashi found that when asked to describe their ideal city, while both countries cite equally the importance of "environmental consciousness", the Swedish considered "democracy", "equality" and "education" to be a fundamental part of a good society, while Japan primarily cited "social cohesiveness". As an expression of cultural difference, this is useful.

Even though many scholars admit that we do not all agree on the meaning of the term sustainability [39] there are still some fundamental assumptions made by the use of the term. Moreover, this chapter should result in a challenge to some of the concepts and precepts of environmental thinking in the West compared to the East. This should be an important modifier in the practical "assessment for sustainability" industry that proactively aims to ensure that "plans and activities make an optimal contribution to sustainable development" [69]. With a global agreement in September 2015 on universal Sustainable Development Goals, including, for example, the following: "Ensure sustainable consumption and production patterns" [66] it would be worthwhile having a sense of what these demands mean in a global context. This chapter should go some way to opening up a more critical investigation of the real, harmonized, global meaning – if any exists – of "sustainability" and "sustainable development".

## References

1. Altomonte S (2010) Environmental education for sustainable architecture, review of European studies. Can Cent Sci Educ 1(2):12–21
2. Aureli PV (2013) Less is enough: on architecture and asceticism. Strelka, Moscow
3. Bauer JR (2006) Forging environmentalism: justice, livelihood, and contested environments. ME Sharpe, New York, p 28
4. Bell ML, Davis DL (2001) Reassessment of the lethal London fog of 1952: novel indicators of acute and chronic consequences of acute exposure to air pollution. Environ Health Perspectives 109(Suppl 3):389–394
5. Ben-Ami D (2010) Ferraris for all: in defence of economic progress. Policy, London, p xi
6. Bramall R (2011) Dig for victory! Anti-consumerism, austerity and new historical subjectivities. Critical Psychol 4(1):68–86
7. Breakthrough Institute (2011) Dialogue 3: modernizing environmentalism and conservation. In: Modernizing liberalism: breakthrough dialogue 2011. Sausalito, California. 16 Jun 2011
8. Brown C et al (2002) The pressures of industry on the Chinese environment: a tale of two South China Cities. NSF/REU Program-Society and Environment in South China. p 2
9. Brubaker R (2012) China and sustainability: connecting the dots between economy and ecology, *The Guardian*, 10 Sep 2012
10. Brundtland GH et al (1987) Report of the World Commission on Environment and Development: Our Common Future. General Assembly. World Commission on Environment and Development. Annex A/42/427
11. Carson R (1963) The silent spring of Rachel Carson. CBS Reports, CBS (Columbia Broadcasting Service), Season 4, Episode 14, New York, USA. TV documentary
12. Chan CK et al (2008) Social policy in China: development and well-being. Policy, Bristol, p 126
13. Chan G et al (2011) China engages global governance: a new world order in the making? Routledge, London, p 228
14. DEFRA (2015) 2010 to 2015 government policy: environmental quality. Department for Environment, Food & Rural Affairs. London
15. Domes J, Näth ML (1977) China after the cultural revolution: politics between two party Congresses. J Asian Stud 37(03):530–553. May 1978
16. Donald JW, Davis CW (2009) Carbon dioxide: harmless, ubiquitous, and certainly not a pollutant under a liability policy's absolute pollution exclusio. Seton Hall Law Rev 39 (10):107–127
17. Economist (2015) Airpocalypse, http://www.economist.com/blogs/graphicdetail/2015/02/india-v-china/. Accessed 18 Feb 2016
18. Edmonds R (2012) Patterns of China's lost harmony: a survey of the country's environmental degradation and protection. Routledge, London, p 18
19. Faber N et al (2005) The sustainability of "sustainability" – a study into the conceptual foundations of the notion of "sustainability". J Environ Assess Policy Manag 7:1–33
20. Furedi F (2003) Therapy culture: cultivating vulnerability in an uncertain age. Routledge, London, p 256
21. Goodman DSG (2014) Class in contemporary China. Polity, London, p 27
22. Goodman DSG (2008) The new rich in China future rulers, present lives. Routledge, London, p xii
23. Gray RH (2010) Is accounting for sustainability actually accounting for sustainability…and how would we know? An exploration of narratives of organisations and the planet. Account Organizations Soc 35(1):47–62
24. Grumbine RE (2010) Where the dragon meets the angry river: nature and power in the People's Republic of China. Island, Washington, p 33
25. Haas T, Olsson K (2014) Emergent urbanism: urban planning & design in times of structural and systemic change. Ashgate Publishing, Farnham, p 110

26. Harari O, Beaty D (1990) On the folly of relying solely on a questionnaire methodology in cross-cultural research. J Manag Issues 2(3):267–281
27. Hilton I (2011) China's Green Revolution: Energy, Environment and the 12th Five-Year Plan, Available via chinadialogue reports, www.chinadialogue.net/reports. Accessed 23 Feb 2015
28. Holslag J (2010) China's scepticism of clean energy champion Europe. Int Spectator Italian J Int Aff 45(1):75–89
29. Hornsby A (2012) The good analyst: impact measurement and analysis in the social-purpose universe. Investing for Good, London
30. Hua Y (2014) China waits for an apology, *New York Times*, 9 Apr 2014
31. Huesemann M, Huesemann J (2011) Techno-Fix: why technology won't save us or the environment. New Society Publishers, British Columbia, Canada
32. Jasanoff S (2001) Image and imagination: the formation of global enviromental consciousness. In: Edwards P (ed) Changing the atmosphere. MIT, Cambridge, USA
33. Jing L (2013) Pollution-free days of Beijing Olympics now just a happy memory, *South China Morning Post*, 10 Aug 2013
34. Johnston P et al (2005) Reclaiming the definition of sustainability. Environ Sci Pollut Res Intl J 14:60–66
35. Kekic L (2007) The world in 2007: democracy index. The Economist Intelligence Unit
36. KPMG Advisory. (2011). China's 12th Five-Year Plan: Healthcare sector. http://www.kpmg.com/cn/en/IssuesAndInsights/ArticlesPublications/Documents/China-12th-FiveYear-Plan-Healthcare-201105-3.pdf. Accessed 9 Sept 2015
37. Larson J (1993) Elegant technology. J Econ Issues 27:421–442
38. Lee JCK, Tilbury D (1998) Changing environments: the challenge for environmental education in China. Geography 83:227–236
39. Lee K, Holland A, McNeill D (2000) Global sustainable development in the twenty-first century. Edinburgh University, Edinburgh, pp 227–236
40. Lehoux P (2006) The problem of health technology: policy implications for modern health care systems. Taylor & Francis, London, p xxix
41. Lewis P (2011) From ego to eco: architecture's new 'ecosophy'. In: Williams A, Dounas T (eds) Masterplanning the future: modernism, east, west and across the world. TRG Publications, London, pp 137–143
42. Li K (2012) Promoting coordinated urbanization – an important strategic choice for achieving modernization. http://static.squarespace.com/static/541ff5f5e4b02b7c37f31ed6/5429942de4b0455e57050a1b/542994cde4b0455e570531a1/1412011213445/Li-Keqiang-China-urbanization-speech.pdf. Accessed 9 Sep 2012
43. Liu JQ (2011) Reining in China's energy targets. In: China's Green Revolution: Energy, Environment and the 12th Five-Year Plan. *China Dialogue*. p 24
44. Liu XZ, Wang JL (1991) An introduction to China's health care system. J Public Health Policy 12(1):104–116
45. Lovelock J, Margulis L (1974) Atmospheric homeostasis by and for the biosphere: the Gaia hypothesis. Tellus 26:2–10
46. Lovelock J (2007) The revenge of Gaia: Earth's climate in crisis and the fate of humanity. Basic Books, New York, p 3
47. McGregor A, Roberts C (2003) Using the SPeAR assessment tool in sustainable master planning. In: US Green Building Conference, Pittsburgh, PA, USA
48. Moseley S (2002) The big smoke: fifty years after the 1952 London smog. Brunei Gallery, School of Oriental and African Studies, University of London. 9–10 Dec 2002
49. NDRC (2012) The People's Republic of China's report on sustainable development. National Development and Reform Commission, Beijing, China
50. Newton J, Freyfogle ET (2005) Sustainability: a dissent. Conservation Biology 19:23–32
51. Ohashi T (2009) Democracy as the indicator of sustainability – from the research of 'Questionnaire on Ideal Society Part II' in Japan and Sweden. Reitaku Int J Econ Stud 17:9

52. Rees M (2004) Our final hour: a scientist's warning: how terror, error, and environmental disaster threaten humankind's future in this century – on earth and beyond. Basic Books, New York, p 116
53. RIBA (2014) The Brits who built the modern world. Royal Institute of British Architects, London, p 4
54. Robertson T (2012) The Malthusian Moment: global population growth and the birth of American environmentalism. Rutgers University, New Jersey, USA, p 8
55. Sachs J (2007) Lecture 2: Survival in the Anthropocene. Available via Reith Lectures BBC. http://www.bbc.co.uk/radio4/reith2007/lecture2.shtml. Accessed 18 Apr 2007
56. Sánchez MJ, Lafuente R (2010) Defining and measuring environmental consciousness. Revista Internacional de Sociología (RIS) 68:731–755
57. Schwartz J (2004) Environmental NGOs in China: roles and limits. Pac Aff 77:28–49
58. Schwerin E (2012) China and a sustainable future: green growth and sustainable public procurement policies and strategies. In: International Public Procurement Conference Conference Proceedings, Part V: Sustainable and Green Procurement. p 3534
59. Shihping K-H, Wang Y-L (2013) A comparative study of sustainability management education in China and the USA. Environ Educ Res 19:64–80
60. Singh A (2013) The sustainability consortium extends its reach to China. *GreenBiz*. 20 Sep 2013
61. Tregidga H (2007) Power and politics of organisational sustainable development: an analysis of organisational reporting discourse (Thesis, Doctor of Philosophy). University of Otago. http://hdl.handle.net/10523/516. Accessed 20 Aug 2015
62. Tregidga H, Milne MJ and Kearins K (2011) Sustainable development as a floating signifier: recognising space for resistance. Launceston, Australia: The 10th Australasian Conference on Social and Environmental Accounting Research, 5–7 Dec 2011
63. UK Contractors Group (2012) Construction in the UK Economy: the benefits of investment, UK Contractors Group. p 3
64. Ulanowics RE (2000) Toward the measurement of ecological integrity. In: Pimentel D, Westra L, Noss F (eds) Ecological integrity: integrating environment, conservation, and health. Island, Washington, p 99
65. UN (2015) Decade of education for sustainable development, mission. http://www.unesco.org/new/en/education/themes/leading-the-international-agenda/education-for-sustainable-development/mission/ Accessed 21 Jul 2015
66. UN (2015) Transforming our world: the 2030 Agenda For Global Action – final draft of the outcome document for the UN Summit to adopt the Post-2015 Development Agenda. Goal 12. All Permanent Representatives and Permanent Observers to the United Nations. 8 Jul 2015
67. Universitas 21 (2009) Statement on sustainability. The Network for Internationsl Higher Education
68. van Someren TCR, van Someren-Wang, S (2012) Green China: sustainable growth in east and west. Springer Science & Business Media, Berlin
69. Verheem R (2002) Recommendations for sustainability assessment in the Netherlands. In: Environmental impact assessment in the Netherlands: views from the Commission for EIA in 2002. Commission for Environmental Impact Assessment, the Netherlands. p 9
70. Whalley J, Xin X (2006) China's Fdi and non-Fdi economies and the sustainability of future high Chinese growth. Working Paper. National Bureau of Economic Research. p 18
71. Williams A (2010) Corroding the curriculum: sustainability v. education. Acad Quest 23:70–83
72. Williams D (2007) Sustainable design: ecology, architecture, and planning. Wiley, London
73. Wong S-W, Tang B (2005) Challenges to the sustainability of development zone: a case study of Guangzhou Development District, China. The Hong Kong Polytechnic University, Hong Kong
74. Woods P (1996) Qualitative research. In: University of Plymouth Faculty of Education (ed) Study Guide for E835, Educational Research in Action. Open University, London
75. Wuebbles DJ (2012) Celebrating the 'Blue Marble'. Eos 93:509–511
76. Xueying Y (2014) Is environment 'a city thing' in China? Rural–urban differences in environmental attitudes. J Environ Psychol 38:39–48

77. Yang G, Lam C-C, Wong N-Y (2010) Developing an instrument for identifying secondary teachers' beliefs about education for sustainable development in China. J Environ Educ 41:195–207
78. Zelezny LC, Schultz PW (2000) Promoting environmentalism. J Soc Issues 56:365–371
79. Zhang J (2012) Delivering environmentally sustainable economic growth: the case of China. Asia Society, University of California. p 5
80. Zhang T (2010) From environment to sustainable development: China. Int Rev Educ 56:13

# Sustainability Awareness in Thailand

**Nattarinee Phoorisart**

**Abstract** As we know, the world is rapidly changing. At the present time, many people and institutions all over the world are becoming increasingly concerned about the negative impacts of environmental and climate changes because people still want to fulfill their needs but neglect to consider our planet's non-infinite supply of natural resources. For this reason, the sustainability concept has been established for preventing, protecting, restoring, and maintaining the natural resources. Sustainability has to do with meeting the needs of the current generation while safeguarding the needs of future generations. It also aims to increase the economic growth and wellbeing of the society. Consequently, this study will examine the resource users' reactions and attitudes to sustainability awareness in Thailand because Thailand is the one of developing countries that will suffer the impacts of environmental and climate changes. Thai people should be educated in order to increase their awareness of the sustainability concept. The survey consisted of questions under three sections intended to elicit information about the respondents themselves, their usage of the Internet and communication devices, and their awareness of the opportunities and risks associated with sustainability. The survey results (115 valid responses) showed that Thai people think the sustainability concept can expand their job opportunities and reduce pollution and gas emissions such as $CO_2$, $CH_4$, and $N_2O$. On the other hand, some Thai people believe that if the organizations do not implement sustainable development, this can result in fraud, transaction failure, government failure and threats to the environment. Finally, this research has several limitations. Firstly, it took a long time to receive responses from potential participants since some did not appreciate the importance of the study, or had limited time to respond outside of office hours. Secondly, the target group was restricted to Thai workers who had different roles within their organizations. Lastly, there was the issue of incomplete questionnaires because some respondents did not complete the last section of the survey or there was missing data elsewhere in the survey.

N. Phoorisart (✉)
School of Information Systems, Curtin University, Perth, Western Australia, Australia
e-mail: n.phoorisart@postgrad.curtin.edu.au



T. Issa et al. (eds.), *Sustainability, Green IT and Education Strategies in the Twenty-first Century*, Green Energy and Technology, DOI 10.1007/978-3-319-57070-9_6

# 1 Introduction

## *1.1 Background of Chapter Topic*

These days, people are encountering a rapidly changing world in terms of technology, the environment, culture, lifestyles, etc. Although the technological revolution in particular has many benefits, it also has negative implications for individuals, societies, and the planet. People are generally pleased when positive changes mean that they can better fulfill their needs. However, the negative impact of change has also been recognized and acknowledged. Subsequently, individuals, organizations and governments have initiated campaigns to mitigate the negative impact of change in order to find ways to sustain our world [52]. Some of the negative impacts of change that threaten the world's sustainability include environmental deterioration, pollution, the proliferation of electronic waste, and so on.

As a result of concern about the planet and its resources, the sustainability concept has been established and many countries have become more aware of sustainability and have attempted to implement guidelines and recommendations by establishing sustainability policies. Our world has a limited amount of natural resources, thus the notion of sustainability is related to the consumption of natural resources to meet the needs of the present generation while safeguarding enough resources to meet the needs of future generations who should not bear the brunt of negative impacts. Also, it is related to preventing wastage, and protecting, maintaining, and restoring the natural resources for present and future use [52].

Nowadays, the sustainable development is very popular in many countries. It is not only about the environmental issues, but is also applied to many other fields such as business operations, communities, and the economic sphere. In other words, the sustainability is about balancing environment, society, and economy. It is related to the well-being and health of people, and includes energy and water consumption [1, 9].

Thailand is one of the developing countries that contribute to sustainable development through the business operations, environmental policies, quality of life, liability, human rights, and so on. Nevertheless, it still has several problems resulting from irresponsible actions by individuals or organizations that have little understanding of or concern for sustainable development. Consequently, the objective of this study is to examine the attitudes to and awareness of sustainability of Thai users who utilize the resources and technology and have various roles in different organizations. Moreover, the study will explore the attitudes of Thai users regarding the opportunities and risks associated with the adoption of sustainability by businesses in Thailand.

More importantly, this study will attempt to define the concept of sustainability, describe the goals of sustainable development, and examine the current progress toward sustainability that involves many institutions all over the world. Also, this study attempts to increase the awareness of Thai users and encourage other readers to be more concerned about this planet's limited resources, and take action accordingly to conserve as many as possible.

## 2 Background of Research

### 2.1 Introduction

The first part of this chapter talks about the fundamental of sustainability as a brief introduction. It covers the meaning of sustainability and three main dimensions of sustainability (environmental, economic, and social). After that, it outlines the history of sustainability, explaining when and how it began and by whom it was initiated. Moreover, this section explains the classification of sustainability goals, showing the different targets in each goal. Seventeen goals will be explained in this section. The triple bottom line is a concept that is related to sustainable development, so it is discussed in this chapter together with Corporate Social Responsibility (CSR), which has been adopted by many organizations at the present time.

The second part of this chapter discusses the implications of sustainability for the business and education sectors. Then, the current level of progress of sustainability will be discussed. There are discusses all conferences of climate change that was occurred in the past.

The third part of this chapter looks at sustainable development in developing countries and sustainability in Thailand in particular. Further, it includes a brief history of Thailand up to and including the present. Finally, it discusses the advantages of sustainability based on the survey results, and sustainability in the future.

### 2.2 What Is Sustainability?

Sustainability is a notion that commenced with the Brundtland Report of 1987 and Sustainable Development is the pathway of sustainability. The 1987 Brundtland Report of The World Commission on Environment and Development (WCED) defined sustainable development as: "Sustainable Development is development that meets the needs of the present without compromising the ability of future generations to meet their own needs" [50] p. 286. However, the meaning of sustainability is complex and broad. Many authors have tried to define this concept [52] said that it is the relationship between humankind and nature. There are three scopes of sustainability, which are environmental, economic, and social. Also, sustainability is concerned with the livelihood of future populations given the planet's limited natural resources. The world needs replaceable capitals that should be enough for the present generation and the next generation. Therefore, we should create a reserve supply whenever we expend natural resources [52]. Moreover, one can say that sustainability means the conservation of limited resources, and the safeguarding of the economy and society in the long term. This idea is crucial for individuals and businesses if they are to prepare for a sustainable future [39]. Additionally, Beilharz and Hogan [12] defined sustainability as the handling of natural resources by means of ecological restoration, pollution management, and environmental impact assessment. Nowadays, sustainability is a concept that has been adopted by governments and organizations in many countries. Sustainability not only relates to safeguarding the environment but also relates to: (1) liveliness and ecological stewardship; (2) organizational stewardship and human resources; and (3) achieving operational and commercial goals [71]. Regardless of the various definitions of sustainability, all contain the three main dimensions discussed in the following sections.

### 2.2.1 Environmental Sustainability

The first dimension concerns the environmental issues and solutions. There are several environmental issues in our world that should be considered including climate change, quantity of water resources, land degradation and pollution [28]. For this reason, people have attempted to sustain our planet, livelihood, and preserve the limited natural resources through using all resources at a sustainable rate. In other words, current resources should not be wasted. Numerous studies have been conducted on environmental sustainability issues such as ecologically-friendly practices, environmental crises, environmental aesthetics, to name a few [35].

### 2.2.2 Economic Sustainability

The second dimension relates to how the economy can be sustained worldwide. An organization should be concerned not only with profit, but also the influence of resources extraction and production processes [21]. Moreover, this second dimension is also related to increasing production efficiency and replenishing the resources that have been exploited. Economic sustainability should take into account production cost, transportation cost, purchase price, taxes, insurances, and so forth [60].

### 2.2.3 Social Sustainability

Society is the third dimension of sustainable development. Important social issues in our world are fair wages, human health, working conditions, human rights, safety, and knowledge and education [60]. There have been many efforts to find ways to meet people's basic needs, increase well-being and welfare standards, ensure social justice, provide education to all, increase the worth for living, etc. [108]. In conclusion, the combination of three dimensions will be [19]:

Environmental + Economic Sustainability = Viable
Environmental + Social Sustainability = Bearable
Economic + Social Sustainability = Equitable

## 2.3 History of Sustainability

The notion of sustainability first emerged in 1713. Since then, concern about sustainability has increased worldwide with various institutions and sectors contributing to the history of sustainability. The term sustainability applied in the context of the forestry industry, first appeared in 1713 in *Sylvicultura Oeconomica* by Hans Carl von Carlowitz. Known as the father of sustainability, he was the first to publish a work on the issue of environmental destruction and the protection of natural resources for the future. This word *Nachhaltigkeit* means sustainability in the long-term [39, 52]. In 1798, Thomas Malthus promulgated his principles about farming, so sustainability became an important point of learning for economists [52]. In 1931, Harold Hotelling, an American economist, developed the concept of using non-renewable supply – a concept that is still applied today [52]. From that time, sustainability became more recognized among economists.

In 1972, the United Nations Conference on the Human Environment discussed the need for a common, consolidated approach to environment protection. Twenty-six principles were established, intended as a guide for people and nations world-wide as a means of preserving and enhancing natural and human environment [80]. In 1980, the World Conservation Strategy (WCS) was announced by the International Union for the Conservation of Natural Resources (IUCN). It said that sustainability accomplishment should begin with the fulfilment of basic needs and the alleviation of deficiency and suffering of population [80]. In 1982, The World Charter stated that “mankind is a part of nature and life depends on the uninterrupted functioning of natural systems” [80].

In 1987, the World Commission on Environment and Development (WCED) presented the correlation between the economics and the environment. The report merged social, economic, cultural and environmental issues and global solution together [80]. The United Nations’ Bruntland Report (1987) defined sustainable development as “Development that meets the needs of the present generation without compromising the ability of future generations to meet their own needs” [39, 52, 80]. Since 1987, sustainability has been extensively adopted by many countries because governments and organizations have realized that the environment, economy, and society are important for individuals and our world. Problems have arisen because natural resources have been continuously used for production purposes. In 1989, Dr. Karl-Henrik Robert established the essential elements needed for social sustainability and expressed the four circumstances for sustainability, which are the basis for an organization at the present time. The four circumstances are to concentrate with the withdrawal of natural resources, the material invention from society, ruin by physical methods, and the ability to meet people’s needs (Heiberg 2012). In 1992, the first United Nations Conference on Environment and Development (UNCED) was held in Rio de Janeiro and approved a plan for environment and development in the twenty-first Century [80]. Nowadays, there are numerous theories, regulations, campaigns, technologies and innovations that relate to the environmental, economic, and social pillars of sustainable development.

## 2.4 Sustainable Development Goals

There are 17 main goals of sustainable development [80] as follows:

Goal 1 ***No Poverty:*** There is poverty in many countries all over the world. In Thailand, poverty has been significantly reduced from 58% in 1990 to approximately 12.60% in 2015 because the of the improvement of the country's infrastructure and price support programs [8]. There are about 39 countries in Asia that have significant levels of poverty data, including Bangladesh (31.5%), Mongolia (27.4%), Myanmar (25.6%), Nepal (25.2%), India (21.9%), Cambodia (18.9%), Thailand (12.6%), Vietnam (9.8%), and Taipei (1.6%) [8]. The main causes of poverty are a country's history, war and political instability, national debt, discrimination and social inequality, and susceptibility to natural disasters [114]. This goal is to eradicate poverty for all societies universally and ensure that people have equal rights in their society such as rights to use services, rights to economic resources, rights to own property or natural resources, etc. [80].

Goal 2 ***Zero Hunger:*** Several factors may cause people to go hungry; they are: (1) The poverty trap where people cannot provide food for themselves and their families because of lack of land, lack of education, and lack of money to buy seeds and tools for farming [116]. (2) Lack of investment in agriculture because there are inadequate transport routes and the cost of transport to other parts of the country is prohibitive [116]. (3) Climate change as a result of human activities and natural disasters. (4) War interrupts agri-business and food production. As a result, people become homeless. (5) Prices of food products are always uncertain [116]. This goal has been established in order to support agricultural production and rural infrastructure, and to increase agricultural investment and sustain food production systems [80].

Goal 3 ***Good Health and Well-Being:*** This goal has been established in order to improve the living standards of people so that they can live happily in their country. Strategies that can help to achieve this goal include: provide education, develop infrastructure, increase and improve health-care services, increase availability family planning advice, decrease mortality rate, and develop the medical profession [80].

Goal 4 ***Quality Education:*** This goal is intended to ensure "inclusive and quality education for all and promote lifelong learning". The UN sees education, from pre-primary to university level – as the means by which poverty can be alleviated and employment can be increased [80]. Moreover, to ensure that there is equality (a human rights issue) at all levels of education, enough teachers should be provided for all students, and standard quality study courses should be offered [80].

Goal 5 ***Gender Equality:*** The aim of this goal is to safeguard women, girls, and children who have suffered from discrimination and abuse by males within or outside the family. Numerous laws have been established for protecting females from violence. Gender equality also impacts on social development because it is related to basic human rights, fairness, and well-being [115]. This goal attempts to eliminate the barriers of gender inequality between men and women in educational, economic, political, social, and behavioural spheres, and facilitates empowerment [115].

Goal 6 ***Clean Water and Sanitation****:* This goal aims to provide drinking water for all people, reduce the amount of rubbish deposited in water, eliminate water pollution, increase recycling and reuse of water, and improve the quality of water [80]. Also, there are policies to increase the efficiency of water use, prevent water scarcity, and manage water resources and sanitation [80].

Goal 7 ***Affordable and Clean Energy****:* It is anticipated that this goal will help to provide energy sources that are inexpensive, consistent, and modern for citizens [80]. The improvement of energy efficiency is also required because energy is very important for individuals, households, businesses, manufacturing, agriculture, and other sectors of the economy. Many new technologies have been designed to meet daily needs as well as support renewable energy [80].

Goal 8 ***Decent Work and Economic Growth****:* This goal supports the sustainability of economic growth that should increase productivity and be sufficient for collective demand in the long-term [70]. Also, this goal supports employment, technological improvement, and innovations that are appropriate for several types of jobs and different sizes of business such as micro-sized, small-sized, and medium-sized [80]. Moreover, it encourages women and men to be engaged in decent and productive work, and receive equal payment [80]. Finally, the important thing is to protect the labor rights and provide a safe workplace for all employees [80].

Goal 9 ***Industry, Innovation, and Infrastructure****:* Some countries have inadequate infrastructure and undeveloped technology, which in turn impact on many industries and prevent economic growth. For this reason, the first aim is to improve quality, recover and repair a tool, and sustainable infrastructure for increasing the facility, economic growth, and welfare of all citizens [80]. For instance, an efficient IT infrastructure can support business; a communication infrastructure enables the sharing of knowledge and education; transport infrastructure facilitates movement around a country; an energy infrastructure benefits all citizens, and so forth. The second aim is to encourage the development of technology in countries and the use of technology and innovation in different types of industries [80].

Goal 10 ***Reduce Inequalities****:* Inequalities of gender, education, human rights, income, social, and origin exist within and across countries. Despite laws protecting human rights, inequality is still widespread. The purpose of this goal is to eliminate inequality for all people and ensure equal opportunities for everyone in all spheres of life [80].

Goal 11 ***Sustainable Cities and Communities****:* The aim of this goal is to reduce the ecological footprint and carbon emissions in the city as well as improve people's quality of life [61]. It is intended to ensure that everyone has a good community, decent accommodation for the family, good transport systems, and willing to help each other [61]. Moreover, this goal attempts to protect the world's cultural and natural resources including decreases the amount people deaths in all countries [80]. Also, as well as sustaining the economic, social, and environmental aspects of a country, this goal aims to ensure that people feel safe in their environments, whether these be rural or urban.

Goal 12 ***Responsible Consumption and Production:*** The resources of the planet are limited, so people need to consume the resources efficiently and act responsibly to conserve what we have. Also, sustainable management is required for manufacturing processes because the recycling of natural extraction will assist to sustain the limited resources [80]. Furthermore, the everyday consumption of products and services generates a great deal of waste. For this reason, waste management is a good way of exercising responsibility by means of activities such as prevention, reduction, recycling, and reuse of waste products [80]. Finally, governments should conduct campaigns in order to increase the public's awareness of sustainable development [80].

Goal 13 ***Climate Action:*** As a result of natural disasters and human activities, climate change has become an important problem for all countries in the world and will affect the whole world in the long run. Therefore, this goal supports actions that will address the climate change problem and encourages governments, organizations and individuals to find ways of planning, preventing and reducing the climate change [80].

Goal 14 ***Life Below Water:*** The impact of water pollution and a solution for this are issues that people need to consider. Consequently, this goal is related to preventing and reducing water pollution of all kinds, including overfishing in the sea and the rivers [80]. There are many factors that increase water pollution including industrial waste, sewage water, mining activities, marine dumping, oil outflow, burning of fossil fuels, global warming and animal waste [111].

Goal 15 ***Life on Land:*** Many countries have problems of land and forest degradation as well as other environmental issues resulting from land disturbance, overpopulation, landfills, deforestation, natural disasters, and so on [65]. These problems impact on human health, the economy, the tourism industry, and ozone layer depletion. For this reason, this goal provides the sustainable management for solving the problems such as preservation and recovery of soil. Also, the United Nation has established guidelines and regulations for protecting wildlife and preventing the trading of species at risk of extinction [80].

Goal 16 ***Peace, Justice, and Strong Institutions:*** This goal emphasizes the reduction of violence and death rates, and also advocates justice for all because people everywhere crave peace [80]. Moreover, corruption is one problem that is evident in society through people's personal experiences or via the media. Corruption can take the form of abuse of authority, exploitation, extortion, fraud, etc., especially in some countries. Therefore, the reduction of corruption and bribery in all forms is supported by this goal [80]. For instance, in Thailand, there is corruption in some institutions; thus, the government has established and promulgated a 'No Cheating' policy, which has been incorporated in education as to increase children's moral development and their knowledge and awareness of corruption and its effects.

Goal 17 ***Partnerships for the Goals:*** This goal has five aspects: (1) Finance: to assist developing countries improve domestic resource deployment, debt relief, and debt restructuring [80]. (2) Technology: using technology for sharing knowledge and information about sustainable development [80]. (3) Capacity-Building: is about the implementation of effective capacity building in developing countries to support national policies [80]. (4) Trade: increase the amount of export and import in developing countries [80]. (5) Systemic issues: this topic is related to enhancing policy coherence and global partnership for sustainable development [80].

## 2.5 Triple Bottom Line – People, Planet, and Profit

From the very beginning, the Bottom Line has been used by business as a means of measuring economic performance because investors and businessmen are mainly interested in profit and good returns on investments. All businesses, in order to maintain viability, focus on administrative accounting, financial statements, and net profit [75]. Subsequently, the notion of Corporate Social Responsibility (CSR) emerged that changed the ways in which businesses were assessed. CSR means that organizations need to consider the impact of their production and service activities, and modify their processes in order to be more responsible in terms of the three pillars of sustainability; hence, the introduction of the Triple Bottom Line concept developed by Elkington in 1998 [75]. The Triple Bottom Line is intended to balance economic, environmental, and social concerns arising from the process of production and service provision [36]. Therefore, the Triple Bottom Line is used as a guideline for business management that includes assessing and reporting business performance in terms of economic, social, and environmental factors both internally and externally [36]. As shown in Fig. 1, the Triple Bottom Line has been popular on the sustainability development agenda of corporations and organizations. It consists of three pillars as follows:

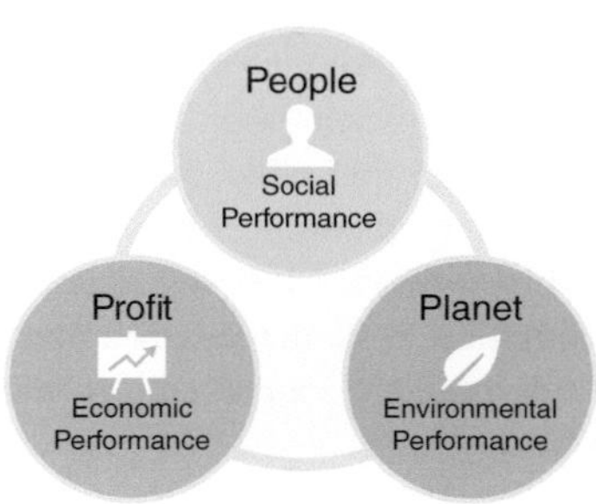


**Fig. 1** Triple Bottom Line – People, Planet, Profit. *Source* Boonlert [14]

The first pillar is **People**, which relates to Social Performance (see Fig. 1). It covers business operations as well as assisting people in the society. In other words, it is a relationship of constituencies that consist of workers, suppliers, customers, stakeholders, and society [13]. All employees have an undeniable right to fair treatment; hence, all decisions by management should take into account human rights of everyone within an organization. Also, an organization has to look after the well-being of all employees and ensure that they are happy and feel safe in the workplace and with fellow-employees [13]. Subsequently, this will attract quality employees to an organization. For suppliers, an organization should follow the Supplier Relationship Management (SRM) principles for managing and improving the interactions and transactions between an organization and its suppliers [69]. Moreover, customer satisfaction is the one of many factors that affects the viability of a business because the word-of-mouth assessments can almost make or break a business. Therefore, an organization needs to concentrate on its customers and satisfy their needs as well as attracting future customers [13]. Most customers desire, for example, high product qualities, information about new and future products, and friendliness and fairness from staff that are willing its local community and support local people. This can be done by providing better customer service, improving the reliability of services and in the case of industries, reducing pollution resulting from the manufacturing process. Such activities will enhance the organization's social performance in the long-term period. There are the international standards that international organizations have been following.

For instance, the International Organization for Standardization or ISO26000, provides the standard, concept, or guidance for social responsibility to listen to their customers [44]. It is not so much about requirements – it is more about the willingness of organizations to adopt and comply with the guidelines. Also, it supports the good intentions and positive actions of organizations in terms of the society; moreover, it increases the awareness of social responsibility [44]. There are seven main principles: (1) Principle of legal compliance; (2) Principle of respect for authoritative inter-government agreements or internationally recognized instruments; (3) Principle of recognition of stakeholders and concerns; (4) Principle of accountability; (5) Principle of transparency; (6) Principle of respect of fundamental human rights; and (7) Principle of respect for diversity [66]. The second pillar is **Planet**, which concerns Environmental Performance (see Fig. 1). The design and process of production and service can have a negative impact on the environment such as global warming, water pollution, climate change, deforestation, loss of biodiversity, etc. Therefore, this pillar covers everything regarding the safeguarding of the ecology, including planning, preservation, prevention, reduction, recycling, reuse, and compensation. Most organizations within developed and developing countries have complied with the environmental standards that have been established. However, each country can establish the different environment policies for using in its own country.

The International Organization for Standardization or ISO14000 series, concerns environmental management [107]. The structure of this standard includes

Environmental Management System (EMS), Environmental Auditing and Related Environmental Investigations (EA), Environmental Labeling (EL), Environmental Performance Evaluation (EPE), Life Cycle Assessment (LCA), and Term and Definitions (T&D) [49]. Moreover, the steps that should be implemented on an on-going basis in compliance with ISO14000 are: environmental policy, planning, implementation, checking and corrective action, and management review. This international standard assists businesses and organizations to manage the environmental issues (from eco design to finished products), create a positive work environment, achieve cost saving in the long-term, improve waste and energy management, attract new opportunities, develop positive corporate image, enhance business reputation, and so on [45]. The last pillar, **Profit**, is related to Economic Performance (see Fig. 1). The compensation is rather easy to measure because the return shows in accounting transactions such as annual reports, balance sheets, retained earnings, financial statements, cash flow, etc. because the Accounting Standard and Regular Financial Reporting have supported this pillar. This pillar is concerned with increasing the company's profitability while taking into account the social benefits and cost, and environment impact [66]. The adoption by businesses of the Triple Bottom Line has several benefits. It: (1) reveals the work performance of all sections within an organization; (2) facilitates risk assessment and risk management; (3) satisfies the stakeholders' needs; (4) is appropriate for sustainable development; (5) promotes a sustainable economy; (6) develops a sustainable economy [66]; (7) increases revenue and market share; (8) reduces energy expenses; (9) reduces wastage and associated loss; and (10) increases employee productivity [112].

## 2.6 *Corporate Social Responsibility (CSR)*

Corporate Social Responsibility (CSR) is strongly related to Sustainable Development for business. Business operations can have both positive and negative effects on people, society and the environment. Accordingly, Corporate Social Responsibility (CSR) has been established as a guideline that expects organizations to be socially responsible [62]. All corporations are concerned about profit but they have to be more concerned about short-term and long-term social impacts [10]. The following quotation succinctly expresses the importance of CSR: "Business has become, in the last half century, the most powerful institution on the planet. The dominant institution in any society needs to take responsibility for the whole .... Every decision that is made, every action that is taken, must be viewed in light of that kind of responsibility" [62] p. 60. The words 'Corporate' 'Social' and 'Responsibility' can be explained as follows:

'Corporate' refers to any large company that is seeking to benefit from several activities. 'Social' intends as a group of people who live in the same society. They may know or may not know each other before but they can make a relationship and help each other.

'Responsibility' is the admission of both good and bad results from all actions that are taken, including taking all responsibilities for developing, and sustaining business [24].

### 2.6.1 Roles of Corporate Social Responsibility

All people within the organization should be part of Corporate Social Responsibility (CSR). Personnel at the different levels of an organization have different roles in terms of Corporate Social Responsibility (CSR). These are as follows [64]:

1. The responsibilities to shareholders or investors are: respect for stakeholders, disclose all necessary information to shareholders and investors, and maintain the confidentiality of internal information.
2. The responsibilities to directors and managers are support for good governance, develop managing system, provide financial report accurately, and disclose all financial report.
3. The responsibilities to employees are: pay fair wages, offer welfare according to labour protection law, look after employees' health and security in the workplace, develop employees' skills, and provide staff training.
4. The responsibilities to consumers are: the provision of products and services, making available to customers/ consumers information about the company and its products, concern for the customers' health, and responding appropriately to all complaints.
5. The responsibilities to business partners are: offer a fair contract and reject all forms of corruption and bribery.
6. The responsibilities to community and environment are: support the local workers, establish a policy of environmental protection, and learn about the local culture.
7. The responsibilities to business' competitors are: promote and maintain fair competition.

However, due to the diversity of cultures and traditions in different countries, Corporate Social Responsibility (CSR) takes different forms depend on the country [62].

### 2.6.2 The Benefits of Corporate Social Responsibility

Business activities impact on society and the environment. Therefore, organizations need to take responsibility for all the impacts of their activities. Corporate Social Responsibility (CSR) contributes to the prosperity of society, so it assists business in several ways. CSR can improve profitability, promote long-term profits, increase financial performance, increase the return of investments and raise sales volume. This is because many investors and shareholders believe that a company that is concerned about social and environmental impacts will be looked upon more favourably in the marketplace [33].

Moreover, CSR enhances community relations because it is proactively attempting to benefit society, thereby establishing positive relationships between individuals, companies, and government [33]. CSR improves the business ethics within the company, and among business partners and competitors. The good public image of a company is necessary for its viability; thus the adoption of Corporate Social Responsibility (CSR) can maintain a favourable public image and enhance the company's reputation [33]. A good public image helps to attract more consumers and quality employees, and provides more access to the money market and stock exchange. Furthermore, it helps to create a better environment and improve the welfare of the community [33]. Finally, a company that has adopted CSR will see positive returns, both tangible and intangible, from all company stakeholders.

### 2.6.3 Divisions of Corporate Social Responsibility (CSR)

There are three main divisions of Corporate Social Responsibility: CSR-after-process, CSR-in-process, and CSR-as-process [77].

- **CSR-after-process** concerns the social and environmental activities that are not directly related to the organization or business process. Examples include reforestation, scholarships for students, education funds, aid to victims, to name a few [76].
- **CSR-in-process** relates to the social and environmental business activities that have an effect on the stakeholders and environment of the organization [7]. These concern employees' welfare, production processes that have no negative impact on the environment, the responsibilities to customers, and so on. In other words, CSR-in-process encompasses all business operations that are indicative of responsibility to society and the environment. There are many means of providing CSR-in-process and preventing any problems. For example [41]:
  - Reinforcing the knowledge and understanding of CSR-in-process;
  - Promoting the value of CSR-in-process to managers, employees, and stakeholders to increase their awareness when working;
  - Establishing policy to support CSR-in-process;
  - Communicating with all levels of employees to secure and maintain their cooperation;
  - Acknowledging the participation of employees when they comply with CSR-in-process policy in all working processes;
  - Motivating employees through a reward system;
  - Assessing the performance of CSR-in-process.
- **CSR-as-process** refers to the organizations that are established specifically in order to assist the community and environment. These are not-for-profit organizations or foundations [76].

### 2.6.4 Types of Corporate Social Responsibility (CSR)

The notion of "doing well by doing good" increases the awareness of social responsibility [74]. The activities of Corporate Social Responsibility (CSR) have been categorized as follows:

**Cause Promotions:** The first activity involves providing capital, materials, or other resources for expanding awareness and concern about social issues. This activity supports fundraising, collaboration, and recruitment of volunteers to help tackle social issues. An organization can launch and manage a project itself, or cooperate with partners, or become a sponsor [51].

**Cause-Related Marketing:** The second activity is related to marketing activities and supporting community projects or charities. For example, an organization may offer to donate to a cause a certain percentage from the sale of a product or service. This activity can help to achieve economic, social and business goals. Moreover, it gives customers the opportunity to support the cause or charity by purchasing a product or service [74].

**Corporate Social Marketing:** This activity began with the social marketing concept introduced by Kotler and Zaltman in 1971 [43]. The main purpose of corporate social marketing is to change the behaviour of people and show how organizations can affect customer behaviours. It supports the development and improvement of public health, community connection, injury prevention, and environmental preservation. The credibility of corporate social marketing depends on: (1) qualities of the organization; (2) qualities of the corporate social marketing creativities; and (3) qualities of the cause [43].

**Corporate Philanthropy:** Social problems are directly addressed by this activity, in the form of donations of money or goods, or cash grants for education. Most corporate philanthropy projects have been in response to external needs and requests. The projects are not related to business objectives, but are intended to assist people in rural areas for example, or small schools that lack the necessary resources for teaching and learning [51].

**Community Volunteering:** This is an activity whereby organizations promote and encourage all employees and business partners to devote their time and effort to help the local community [51]. It is a proactive way of helping to eliminate or alleviate one or more social problems and shows that an organization is concerned about those problems. An organization can conduct this activity on its own or collaborate with other non-profit organizations [51].

**Socially Responsible Business Practices:** An organization manages and conducts all business activities carefully, in order to prevent and avoid causing social and/or environmental problems [51]. Also, an organization can work collaboratively with employees and business partners to address social problems through a business campaign. This can improve community welfare [51]. This can be done, for instance, by designing and manufacturing eco-friendly products, and packaging products in recyclable materials in order to reduce waste [51].

**Developing and Delivering Affordable Products and Services:** The last activity concerns the production and selling of low-priced goods and services which enable people with low income to purchase them [51]. This gives companies the opportunity to access a new market while improving people's quality of life.

### 2.6.5 Corporate Social Responsibility (CSR) vs Sustainable Development (SD)

The principle of Sustainable Development (SD) was established because there has been a steady degradation of the environment for many decades. An increasingly urgent need to protect the environment has led individuals, groups and governments in many countries to establish institutions and guidelines for environmental responsibility [77]. In 1992, the United Nations' "Earth Summit" conference on environment and development reached important conclusions regarding Sustainable Development (SD), and called for the integration of economic development, social development, and environmental development [84].

Corporate Social Responsibility (CSR) also has three aspects related to Sustainable Development (SD): (1) an organization has to be responsible for the business execution or activity that affects society and environment; (2) sustainable business can adopt different methods for sustainable development; and (3) Corporate Social Responsibility (CSR) can be a tool by means of which organizations can increase business sustainability and support sustainable development. Therefore, Corporate Social Responsibility (CSR) has gradually led to achieving Sustainable Development [117].

## *2.7 What Does Sustainability Mean for Business/Education?*

As we know, our world is constantly changing and new businesses and industries are always emerging in response to people's needs and wants, and their desire to improve their quality of life. Therefore, businesses have an important influence on people's behaviour and decision-making. However, business operations contribute to: (1) negative environmental impact such as deforestation, climate change, scarcity of natural resources, the extinction of wildlife, etc.; and (2) social issues such as gender inequality, lack of education, poverty, discrimination, inequitable job opportunities, etc. For this reason, investors and shareholders have become increasingly concerned about social and environmental problems, and business sustainability. Nowadays, businesses can be sustained only with new thinking, new behaviors, and new balances of environmental, economic, and social factors [32]. Some consumers desire a change, but they cannot recognize an environmental, social, or economic problem if it does not directly affect their lives and their families. Therefore, Sustainable Development is an opportunity businesses to gain a competitive advantage by bringing about change through management, and enhancing the awareness of employees and customers regarding sustainability [32]. Moreover, it is about green business encouraging the design and manufacture of eco-friendly products, which benefits the community and increases profits.

The advantages of adopting Sustainable Development should be enough to motivate management and business leaders. There are three interconnected dimensions to Sustainable Development when adopted by a business or organization. Firstly, there is the creation of a sustainable business that achieves profitable growth, increases investors' returns, maintains good relationships with business partners, is seen as a good employer, encourages creativity and innovation, develops a sustainable business model, develops efficient processes, and increases its competitive advantage [32]. Secondly, it is about living within environmental limits, which is related to replenishment of resources, development of logistics and transportation, end-of-life discarding of products, improvement of waste management, reduction of harmful emissions and pollution, sustaining diversity, attention to building design, and renewable energy [32]. Lastly, it is about ensuring a fair society, which is related to locality and communities, education and development, relieving poverty, improving human rights, ensuring safety and security, ethics and governance, privacy and equality, and health and wellbeing [32]. Nevertheless, Sustainable Development and Sustainable Business begin with education because education is crucial for all aspects of daily life and the development of oneself, family relationships, workplace environment, social environment, community, country, and our world. Education prevents social awkwardness, dispels superstitions, helps people to keep up with technology, introduces people to the wider world, provides skills to improve productivity, imparts ethical values, and contributes to a country's prosperity [109].

Furthermore, the main purpose of Sustainable Education is to develop a deeper understanding of sustainability and to support student learning [11]. Even the smallest, positive actions can assist to sustain the world; thus, children should be educated about the importance of the environment, develop skills to resolve environmental issues, increase their awareness of and concern about environment, improve behaviours and practices that help to protect the environment, and understand the notion of Sustainable Development [11].

## 2.8 *Current Progress Level of Sustainability*

The United Nations is the main organization that has addressed the issue of sustainable development in the past, and continues to do so. Since 1995 annual conferences have been held in locations all over the world and the issue of climate change has been on the agenda every year. The United Nations Climate Change Conferences held in the framework of the United Nations Framework Convention on Climate Change [83] or (22 to date) have encouraged UNFCCC Parties to sign an agreement acknowledging the impact of climate change. The main purpose of this agreement is to prevent dangerous human interference with the climate system [102] (Table 1).

## *2.9 Sustainability in Developing Countries*

A developing country is a country that is less developed or underdeveloped when compared to other countries. It usually depends on assistance from developed countries, and adopts the management ideas of developed countries [63]. In all, 139 countries are listed as developing countries, some of which include Afghanistan, Bhutan, China, Ghana, India, Myanmar, Nepal, Thailand, Vietnam, etc. [106]. The distinction between developed countries and developing countries can be measured by the gap between rich and poor in terms of the income and wealth of the country's population [63]. Moreover, developing countries mostly need to develop their commercial transport infrastructure, their education standards, technology, communication, well-being of its inhabitants, etc. because the lack of these things hinders the country's development [63]. Developing countries are characterized by the following problems: (1) common poverty such as lack of food, incapacity, lack of basic infrastructures, illness, and poor health, (2) low labour productivity, (3) short life expectancy, (4) low standard of education, (5) poor health care, (6) unemployment, and (7) overpopulation [72]. Consequently, the sustainability is very critical in developing countries because it can help develop countries reduce these problems [2]. It is a valuable concept for rehabilitating cities, relieving poverty, and protecting the natural environment because the main purpose of Sustainable Development is to balance environmental, social, and economic factors in the present and future [4, 20]. The United Nations Development Programme (UNDP), United Nations Environment Programme (UNEP), and United Nations Human Settlements Programme (UNHABITATE) were established to help developing countries achieve Sustainable Development [4]. However, Sustainable Development is still difficult to achieve in some developing countries because of the lack of human capital and strategies to support imported technology [4, 3].

## *2.10 Sustainability in Thailand*

### 2.10.1 History of Thailand

Thailand was known as 'Siam' from 1855 to 1949 [59]. In July of 1948, by popular vote, the name 'Siam' was changed to 'Thailand' and this became official on May 11, 1949 [16]. The meaning of Thailand is 'land of the free' and the people have

**Table 1** The Conference of the Parties about climate change

| Name | Date | Places | Main Point |
|---|---|---|---|
| COP1 (The Berlin Mandate) | March 28–April 7, 1995 | Berlin | At that time, the politicians in Berlin were concerned about unemployment and the reduction of economic growth more than the impact of climate change. Therefore, despite negotiations, the proposal was not accepted and the disappointing outcome attracted negative media coverage [5]. |
| COP2 | July 8–19, 1996 | Geneva, Switzerland | Attended by over 1,500 representatives of governments, intergovernmental organizations, and NGOs. The Parties accepted the latest (1995) scientific findings of the Intergovernmental Panel on Climate Change (IPCC) findings on climate change proffered in its second assessment (1995); advocated flexibility rather uniform policy for all countries, and suggested mid-term targets that were binding on all Parties. This conference discussed the progress of the Berlin Mandate regarding the development and transfer of the technology, and established the Green Environment Fund. Geneva noted 'The Geneva Declaration' but the agreement was not adopted [27] |
| COP3 (The Kyoto Protocol on Climate Change) | December 1–10,1997 | Tokyo, Japan | The Kyoto Protocol and its legislative instrument, the Merlin Mandate. The Kyoto Protocol was an agreement whereby Annex 1 countries pledged to reduce greenhouse emissions by 6–8%. Also, it required the United States to reduce its greenhouse gas emissions by around 7% [85]. |
| COP4 | November 2–14, 1998 | Buenos Aires, Argentina | It was concerned with the adoption of the Buenos Aires Plan of Action that specified the financial machinery, the development of technology, the execution of activities under the Kyoto Protocol, and the work schedule of the Kyoto Protocol [87]. |
| COP5 | October 25–November 5, 1999 | Bonn, Germany | It discussed about the progress of Buenos Aires Plan of Action, the calculation of GHG Inventories within Annex I group, the development and transfer technology, the reinforcement of capability in developing countries, and the greenhouse gas emission from the ocean liner [86]. |
| COP6 | November 13–15, 2000 | The Hague, Netherlands | The main discussion was still related to Buenos Aires Plan of Action because there was the argument about the proposal of United States. Further, there was the first report from countries of non-Annex I [88]. |

(continued)

**Table 1** (continued)

| Name | Date | Places | Main Point |
|---|---|---|---|
| COP6 | July 17–17, 2001 | Bonn, Germany | It was the continuous discussion of the reinforcement of capability in developing countries, the transfer of technology, the financial support, and the financial support under Kyoto Protocol [89]. |
| COP7 | October 29–November 10, 2001 | Marrakech, Morocco | The Marrakech Ministerial Declaration was established and set up (1) The Least Developed Countries Fund (LDC), (2) Special Climate Change Fund (SCCF), (3) Adaptation Fund (AF) [90]. |
| COP8 | October 23–November 1, 2002 | New Delhi, India | There was Delhi Ministerial Declaration about Climate Change and Sustainable Development. Furthermore, it prepared and reviewed the information about Emission Reduction Units, Certified Emission Reductions (CERs), Assigned Amount Units and Removal Units, and National Registries [91]. |
| COP9 | December 1–12, 2003 | Milan, Italy | The main discussion was the format and method for land use, land-use change, and forestry [92]. |
| COP10 | December 6–18, 2004 | Buenos Aires, Argentina | It discussed about Buenos Aires program on adaption and response measure, the support of capability in developing countries, the model of CDM, and the global observing system [93]. |
| COP11 | November 28–December 10, 2005 | Montreal, Canada | The main point of this conference were the direction for running the Least Developed Countries Fund and the guidance for financial machinery [94]. |
| COP12 | November 28–December, 2006 | Nairobi, Kenya | The main point of this conference was the review of financial machinery, the guidance of Global Environment Facility, development and transfer technology, the reduction of greenhouse gas emission from the forest destruction in developing countries [95]. |
| COP13 | December 3–15, 2007 | Bali, Indonesia | Bali Action Plan was established. It is the long-term cooperation plan that common but differentiated responsibilities and respective capability. Further, it discussed the development and transfer technology under SBSTA and SBI [96]. |
| COP14 | December 1–12, 2008 | Poznan, Poland | This conference discussed the quantity of emission reduction and progressing the Bali Action Plan [97]. |
| COP15 | December 7–18, 2009 | Copenhagen, Denmark | The important negotiation was Copenhagen Accord, which was unofficial document. Also, it discussed the goal of the reduction of greenhouse gas emission of the countries in Annex I [98]. |

(continued)

**Table 1** (continued)

| Name | Date | Places | Main Point |
|---|---|---|---|
| COP16 | November 29–December 10, 2010 | Cancun, Mexico | Cancun Agreement was signed and Cancun Adaption Framework was also established. Additionally, it considered the assessment of the Special Climate Change Fund [99]. |
| COP17 | November 28–December 11, 2011 | Durban, South Africa | Ad Hoc Working Group on the Durban Platform for Enhanced Action was established. Green Climate Fund was also launched in this conference [100]. |
| COP18 | November 26–December 8, 2012 | Doha, Qatar | There was seven institutions in this conference, which are (1) Conference of the Parties to the UNFCCC (COP), (2) Meeting of the Parties serving as Conference of the Parties to the Kyoto Protocol (CMP), (3) Subsidiary Body for the implementation [26], (4) Subsidiary Body for scientific and Technological Advices (SBSTA), (5) Ad Hoc Working Group on the Durban Platform for Enhanced Action (ADP), (6) Ad Hoc working Group on Long-term Cooperative Action under the Convention (AWG-LCA), (7) Ad Hoc Working Group on Further Commitments for Annex I Parties under the Kyoto Protocol (AWG-KP) [101]. |
| COP19 | November 11–23, 2013 | Warsaw, Poland | There was the important decisions about the advice of the clean development, report of the green climate fund, review the financial machinery, and so on [103]. |
| COP20 | December 1–14, 2014 | Lima, Peru | The most important decision was about Lima Call for Climate Action. Also, it discussed the long-term climate finance, National Adaption Plans, and so on [104]. |
| COP21 | November 30–December 13, 2015 | Paris, France | This was the most important for conference of the parties because it was the first success of The United Nations Framework Convention on Climate Change [83]. All the nations (195 countries) over the world achieved the agreement of climate change together. The important objective is the average temperature must not increase more than 2 degree Celsius.<br>Moreover, there was the adoption of Paris agreement, which has six main purposes: (1) the reduction of greenhouse gas emission, (2) the adapting to negative impact, (3) the financial support, (4) the development and transfer technology, (5) the increasing of capability, and (6) the clearness [105]. |
| COP22 | November 7–18, 2016 | Marrakech, Morocco | There are no details about this conference at the present [83]. |

made it knows as the 'land of smiles'. Thailand has had four kingdoms representing four main periods of history. These are:

1. The kingdom of Sukhothai, which means 'The Dawn of Happiness' [18, 6] was the golden era of Thai culture that developed from the middle of thirteenth to fourteenth century [18]. The historical milestone of the Sukhothai era was the creation of the Thai alphabet by '*King Ramkhamhaeng The Great*'. This Thai alphabet has 44 consonants, 32 vowels, and 5 tonal sounds [6]. Moreover, this is considered as the era when there were fish in the waters and rice in the fields as well as numerous Buddhist monasteries throughout the land.
2. The kingdom of Ayutthaya, which was the island city of the fourteenth to eighteenth centuries was encircled by three rivers – the Chao Phraya, the Pasak, and the Lopburi, and included rice walkways. Ayutthaya was in an ideal location as it was safeguarded from periodic flooding and attacks from enemy warships [82]. Furthermore, It was the hub of economic and religious activities in Asia and became a centre for trade between the East and the West [82]. Foreign traders included the Chinese, Vietnamese, Indians, Japanese, and Persians. Also, Europeans arrived in Ayutthaya Kingdom as religious missionaries or to establish commercial connections. Europeans included the Portuguese in 1511, the Dutch in 1605, the English in 1612, the French in 1662, and the Greeks in 1675 [54]. Finally, the society of Ayutthaya was divided into three social classes: the common people and the slaves, the nobles, and the princes.
3. The Kingdom of Thonburi was established by '*King Taksin The Great*' and remained as the capital for only 15 years [48]. During this period, King Taksin The Great also continued the relationship with the foreigners such as the Chinese and Dutch [22]. Also, people benefitted financially from the junk trade of the Chinese because this was the era of Chinese commerce that extended to Thonburi [17].
4. The Kingdom of Rattanakosin was established in 1782 and still exists today [59]. The capital of the Rattanasin period was '*Siam*', built by '*King Buddha Yot Fa Chulalok*' or '*Rama I*' [59]. In 1809–1824 was the reign of '*Rama II*' or '*King Buddha Loet Lad*' who dedicated himself to encouraging Thai literature and wrote a new version of the Indian epic called Ramakien [59]. '*Rama III*' or '*King Nang Klao*' (1824–1851) further developed relationships with foreigners and improved the local farming production [59]. '*Rama IV*' or '*King Mongkut*' (1851–1868) established laws to improve the rights of women and children; he also opened new trade and transportation routes [59]. '*Rama V*' or '*King Chulalongkorn*' (1868–1910) restructured Thai society and economy. He emancipated the slaves and declared Thailand as an independent kingdom [59]. '*Rama VI*' or '*King Vajiravudh*' (1910–1925) took steps to westernize Thailand and began compulsory primary school education [59]. '*Rama VII*' or '*King Prachadhipok*' (1925–1935) reformed the government, changing it from an absolute monarchy to a democratic system [54]. The reign of '*Rama VIII*' or '*King Ananda Mahidol'* lasted from 1934 until 1946 which brings us to the current period under the reign of '*Rama IX*' or '*King Bhumipol Aduldej'* [54].

### 2.10.2 Thailand at the Present

Today, the Thai government and many institutions are concerned about sustainability. Various industries and business sectors have committed to sustainable development and Corporate Social Responsibility (CSR). At the twenty-first Conference of the Parties in 2015, all nations including Thailand, agreed on climate change issues and recommendations [81].

### 2.10.3 The Importance to Thailand of the Climate Change Agreement

It is the new agreement that covers many climate actions more than the Kyoto Protocol. This agreement will commit to all members and begin around 2020. The main purposes of this agreement are to reduce the amount of greenhouse gas emissions, to facilitate adaption to climate change impacts, to extend financial support to developing countries, to develop and transfer technology to developing countries, to reinforce the capabilities of developing countries, the honesty in each operation process, etc. The achievement of these is the responsibility of several government sectors in Thailand such as the Department of Industrial Works, that is responsible for the Greenhouse Gas Inventory and Change of Land Use Pattern, the Department of Environmental Quality Promotion (EDQP), the Ministry of Natural Resources and Environment, and so on [57]. Climate change is an important issue in the world that affects the developing countries such as Thailand. Thailand has to encounter and deal with changes in weather patterns, including the fluctuation of seasons, natural disasters, the degradation of natural resources, and the new diseases. Furthermore, since Thailand is an agrarian country, climate change poses a dangerous threat to sustainable development and its goals of economic growth, the elimination of poverty, the balance of ecology, and so on.

Consequently, several institutions in Thailand are establishing goals and strategies for preventing, reducing and dealing with the climate change impact. For example:

- In order to deal with the impact of climate change, the relevant government departments are managing: (1) the water usage, flood, and drought, (2) agriculture and food stability, (3) tourism industry, (4) the public health, (5) natural resources, (6) the habitation and human security [58].
- The reduction of greenhouse gas emissions: the appropriate government departments have goals to reduce the carbon dioxide (CO2) emissions by 31.19 million tons per year, and to reduce greenhouse gas emission by 79.92 million tons per year [29].
- In order to deal with climate change, the relevant institutions are: (1) developing the information, research, and technology; and (2) increasing the public's awareness of climate change [58].

### 2.11 *Advantages (Based on the Survey)*

The survey results showed that participants perceived the following to be the advantages of adopting sustainability. It will:

- Increase the awareness of Thai people regarding sustainability;
- Increase the Thai people's understanding of the sustainability concept;
- Encourage Thai people to be concerned about the impact of environmental issues and climate change;
- Extent the job opportunities for employees;
- Obtain new knowledge about the Conference of the Parties;
- Pay attention to the interrelationship of the environment, economy, and society.

Know the most important sources of knowledge that Thai people obtain and share their knowledge as well as affect their daily life.

## 3 Research Methods and Questions

### 3.1 *Research Question*

The primary research question for this study is:

*What are the factors that are influencing the attitude of Thailand's higher education sector towards sustainability in terms of its opportunities and challenges?* The primary research question will guide the research including the analysis of results and the conclusion drawn from these results. This study focuses on the attitude of Thai people regarding sustainability awareness, the opportunities of sustainability, and the risks associated with the adoption of sustainability.

### 3.2 *Research Methods*

This research uses the quantitative method, which presents the result in numerical and statistical formats. It is a method that focuses on finding the facts and reaching a quantitative conclusion [67]. The conclusion must be supported by numerical evidence. For data collection, statistic calculation procedures such as questionnaire, survey, experiment, and observation are used [68]. Furthermore, the quantitative method is related to the deductive approach. It is about the presentation of mathematical model and the measurement of many variables. In this research, the main purpose is to explain, compare, and measure some objects such as opportunities, risks, satisfaction, or the reaction of participants to something [47].

## 3.3 Research Design

### 3.3.1 Unit of Analysis

The unit of analysis is the 'unit', the 'who', or the 'what' that the researcher desires to study and analyse such as individuals, groups of people, objects (book, photos, newspapers), areas, social interactions, villages, districts, or countries [78]. The unit of analysis for this research focuses on a group of people in Thailand in order to ascertain their level of awareness of sustainability. Additionally, this research uses the five-point Likert scale for observing the participants' attitudes towards sustainability in Thailand. The five points in the Likert scale are: strongly disagree, disagree, neutral, agree, and strongly agree. This survey also sought information about the attitude of Thai people towards the adoption of sustainability by businesses. The online survey comprised 20 questions divided into three sections: demographics, positive, and negative attitudes towards sustainability.

### 3.3.2 Target Population and Sample

The population covers the gathering of all elements for observation. The target population may be a group of people, objects, areas, etc. that the researcher wants to examine in order to draw conclusions from the results obtained [56]. The target sample comprises the participants. For this research, the target population and sample was a group of Thai people who are working in various organizations where they hold different positions such as receptionist, factory manager, accounting staff, human resource personnel, programmer, auditor, and so on.

### 3.3.3 Data Collection

In this section, the process of data collection for this research is discussed. Data for this research was collected by means of Qualtrics, free online software. It provides a survey platform for the researcher. The online survey was distributed to 217 participants via the following methods.

1. Mobile applications such as Line, WhatsApp, Messenger, and Text Messages
2. Social networks such as Facebook and Twitter
3. Email
4. In cases where participants wanted more information about the topic, a face-to-face explanation was easier as a means of increasing their understanding. Therefore, information regarding sustainability was printed out as a document for use in face-to-face meetings (see Fig. 2).

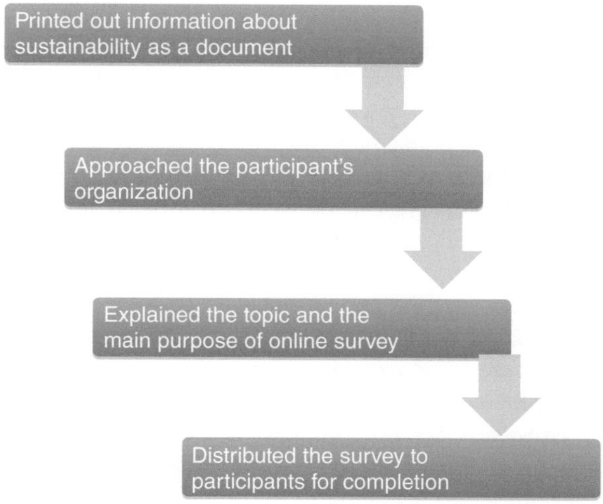


**Fig. 2** The survey distribution by face-to-face meeting (prepare by author)

### 3.3.4 Data Analysis

Data analysis is used to evaluate all the data that the researcher has collected. This allows the results from the online survey to be understood and, later, interpreted [110]. The summarization of data depends on whether the data is quantitative or qualitative. Also, the data should be presented in a format that is easy to understand and interpret. However, several problems can arise during the data collection process, including missing data or non-compliance. These problems can affect the process of data analysis because researchers need to decide whether or not to include such data in the analysis [110]. This research is related to quantitative data; thus, IBM's Statistical Package for the Social Sciences or SPSS version 22.0 is used for the analysis of the collected data. The output is analysed and presented in statistical and numerical format.

### 3.3.5 Research Instrument

The research instrument is the tool that is necessary for conducting the research. The quality of the collected data depends on the quality of the instrument. A research instrument should have the following characteristics: validity, reliability, discrimination, and efficiency. The research instrument assists researchers to gather and evaluate the data, including carrying out the statistical analysis [53]. There are many tools for different research purposes, but three well-known ones are: (1) Test, (2) Questionnaire, and (3) Interview. For this research, the online survey has been used as the research instrument for collecting and analysing the data because it is appropriate for quantitative research.

Currently, online surveys are increasingly being used as the researchers' primary data collection tool because of their several advantages. The online surveys have many design options, such as colour, graphic, and interactive features, to suit different kinds of research questions. Also, the online surveys can provide prompt feedback from the participants as soon as they complete the online survey, so researchers do not need to ask or wait for the return of feedback [23]. Moreover, the online surveys can be easily distributed to participants because there are the links for email, text messages, social networking, application, or post on the website. It can encourage the participants to complete the whole survey because they need to answer the current question in order to continue to the next question. The online surveys are cost-saving and time-saving [46, 73]. Additionally, the online surveys are anonymous for all participants, so they are comfortable about disclosing their genuine attitudes via their responses to the survey questions [38].

However, the more graphic or interactive features can affect the time taken to download an online survey [23]. There is the problem that some people do not want to access the Internet in order to complete an online survey because they think it wastes time [25]. Also, some people may not appreciate the importance of an online survey so they do not read it all carefully or they ignore it [38].

#### 3.3.6 Reliability and Validity

The concept of research reliability is related to the research instrument that should be reliable, responsible, predictable, stable, truthful, and correct. If the research instrument is of high quality, the researcher will obtain effective measurements [53]. Therefore, the reliability of research should begin with a good instrument. For instance, this research used the Qualtrics survey instrument, which is popular among researchers worldwide. Three types of validity can be used in quantitative research:

1. Content validity, which is the systematic investigation of the survey. It should cover everything that researchers need to observe for the purposes of their research. Content validity is also known as logical validity, course validity, and textbook validity [55].
2. Predictive validity, concerns the ability of an instrument to predict the outcomes [79].
3. Construct validity, is the improvement of relevant knowledge and measurement of the research construction [53].

## 4 Results

### *4.1 Survey Results*

Survey results were collected from a sample size of 217 respondents. Survey links were distributed to 217 respondents and 186 survey feedbacks were returned but 115 are valid. However, some questions were omitted so there was the problem of missing data (Table 2).

**Table 2** The genders of online survey respondents (prepared by author)

| Gender | Respondents % |
|---|---|
| Males Respondents | 95 (51%) |
| Female Respondents | 91 (49%) |

**Table 3** Range of ages of online survey respondents (prepared by author)

| Range of ages | Respondents % |
|---|---|
| 17 years and under | 0 (0%) |
| 18–20 | 3 (2%) |
| 21–24 | 17 (9%) |
| 25–30 | 84 (43%) |
| 31–35 | 42 (22%) |
| 36–40 | 18 (9%) |
| 41–45 | 9 (5%) |
| 46–50 | 6 (3%) |
| 51–55 | 12 (6%) |
| 56–60 | 0 (0%) |
| 61–65 | 2 (1%) |
| Over 65 | 1 (1%) |

Gender representation was almost equal with 51% being male (Table 3).

Most (43%) of the research participants fell within the 25–30 years age group. No respondents were in the 17 years and under group. The sample target was working people with the different positions as follows: Admin, Receptionist, General Manager, Factory Manager, Business Owner, Accounting Staff, Managing Director, Teacher, Human Resource Officer, Purchasing Manager, Engineering Manager, Deputy Director, Customer Services Officer, Reservation Officer, Sales, Corporate Affair Assistant, Accounting Manager, Programmer, Doctor, Marketing Officer, Overseas Education Consultant, Technician, Engineer, Data Analytics Manager, Lawyer, Cabin Crew, IT Manager, Sales Engineer, Business Development Manager, Audit Officer, Project Manager, Client Service Manager, and Sales Supervisor (Table 4).

The majority (16%) of respondents came from the fields of Accounting, Science and Engineering. Thirteen percent – 'others' – were from Chemical Science, Tourism, Architecture, Arts (Thai language), and English for Communication, Medicine, Educational Research, Journalism and Mass Communication, Auditing, Computer Business, Oil and Gas, Sociology, and Film Production(Table 5).

From the results, the respondents' highest level of education are 60%, 21%, and 10% for Bachelor's Degree, Master's Degree, and Diploma respectively (Tables 6 and 7).

From the results produced when respondents could choose from more than one option, we can see that Thai users are using the Internet for working (77%), email

**Table 4** The main fields of study (prepared by author)

| The main fields of study | Respondents % |
|---|---|
| Accounting | 30 (16%) |
| Business Law | 4 (2%) |
| Economics and Finance | 9 (5%) |
| Information Systems | 2 (1%) |
| Information Technology | 19 (10%) |
| Computer Science | 9 (5%) |
| Management | 22 (12%) |
| Marketing | 27 (15%) |
| Health Sciences | 0 (0%) |
| Humanities | 8 (4%) |
| Science and Engineering | 29 (16%) |
| Art and Design | 2 (1%) |
| Others | 23 (13%) |

**Table 5** The respondents' highest level of education (prepared by author)

| Student status | Respondents % |
|---|---|
| Primary Education | 2 (1%) |
| Higher Secondary/ Pre-University | 6 (3%) |
| Professional Certificate | 3 (2%) |
| Diploma | 18 (10%) |
| Advanced/ Higher/ Graduate Diploma | 4 (2%) |
| Bachelor's Degree | 109 (60%) |
| Post Graduate Diploma | 0 (0%) |
| Master's Degree | 39 (21%) |
| PhD | 1 (1%) |
| Others | 0 (0%) |

**Table 6** The computer and Internet usage per day (prepared by author)

| Answers | How many hours do you spend on the computer per day? | How many hours do you spend on the Internet per day? | How many hours do you spend on the Internet for social networking (i.e. Facebook, etc.) per day? | How many hours do you spend on the Internet for email per day? |
|---|---|---|---|---|
| | Respondents % | Respondents % | Respondents % | Respondents % |
| Less than an hour | 21 (12%) | 21 (11%) | 49 (27%) | 90 (49%) |
| Up to 5 h | 43 (24%) | 69 (38%) | 90 (49%) | 52 (29%) |
| 5–10 h | 77 (43%) | 59 (32%) | 25 (14%) | 35 (19%) |
| 10–20 h | 37 (20%) | 31 (17%) | 14 (8%) | 4 (2%) |
| Over 20 h | 3 (2%) | 3 (2%) | 4 (2%) | 1 (1%) |

**Table 7** Internet usage by Thai users (prepared by author)

| Answers | Respondents % |
|---|---|
| Email | 136 (75%) |
| Playing games | 59 (32%) |
| Studying | 78 (43%) |
| Working | 141 (77%) |
| Shopping Online | 74 (41%) |
| Chatting | 114 (63%) |
| Researching hobbies | 38 (21%) |
| Banking online | 69 (38%) |
| Buying goods or services | 51 (28%) |
| Buying stocks or investing online | 24 (13%) |
| Researching travel information or making reservations | 58 (32%) |
| Others | 3 (2%) |

**Table 8** Devices used by Thai users (prepared by author)

| Devices | Respondents % |
|---|---|
| PC | 90 (49%) |
| Desktop | 28 (15%) |
| Laptop | 78 (43%) |
| Netbook | 22 (12%) |
| PDAs | 0 (0%) |
| Workstation | 5 (3%) |
| Tablet | 38 (21%) |
| Smartphone | 161 (88%) |
| Others | 2 (1%) |

(75%), and chatting (63%). Therefore, it may be concluded that the Internet is an important device for the Thais' working life (Table 8).

This question allowed the respondents to choose more than one option. The results show that most Thai users (88%) are accessing the Internet via their Smartphones (Table 9).

This results show that Thai users obtain information about the concept of Sustainability and Green Information Technology from the Internet (69%), and Magazines and News media (22%). Moreover, government agencies and friends are two other channels through which they can obtain information about sustainability. However, several respondents stated that they are unfamiliar with the concept of Sustainability and Green Information Technology (Table 10).

Most Thai users are using devices from Apple, Google, and others. It is 46%, 32%, and 11% respectively. For the 'other', the respondents answered that they are using devices from Samsung, Acer, ASUS, HP, iMobile, and Dell (Table 11).

**Table 9** Means by which respondents were first introduced to the concepts of sustainability and green information technology (prepared by author)

| Answers | Respondents % |
|---|---|
| School | 38 (21%) |
| Higher education | 35 (19%) |
| Internet | 124 (69%) |
| Books | 35 (19%) |
| Magazines | 39 (22%) |
| News media | 39 (22%) |
| Conferences | 7 (4%) |
| Others | 12 (7%) |

**Table 10** Thai users' devices (prepared by author)

| Company's name | Respondents % |
|---|---|
| Apple | 83 (46%) |
| Google | 57 (32%) |
| Dell | 16 (9%) |
| IBM | 4 (2%) |
| Other | 20 (11%) |

**Table 11** Thai users reading the company's sustainability report before purchasing a device (prepared by author)

| Answers | Respondents % |
|---|---|
| Yes | 50 (28%) |
| No | 70 (39%) |
| Maybe | 45 (25%) |
| Not at all | 15 (8%) |

**Table 12** Frequency with which Thai users change their devices (prepared by author)

| Answers | Respondents % |
|---|---|
| Every 6 months | 5 (3%) |
| Every 12 months | 25 (14%) |
| Every 18 months | 14 (8%) |
| Every 24 months | 52 (29%) |
| Every 30 months | 7 (4%) |
| Every 36 months | 41 (23%) |
| Every 42 months | 18 (10%) |
| Others | 18 (10%) |

Survey results show that most Thai users do not read the company's sustainability report prior to purchasing a device (Table 12).

Many Thai users change their devices every 2 or 3 years but some Thai users change their devices only when they are damaged (Table 13).

**Table 13** The reasons that Thai users give for changing their devices (prepared by author)

| The reason | Respondents % |
|---|---|
| Size | 33 (19%) |
| Speed | 73 (41%) |
| Functionality | 102 (58%) |
| Keeping with technology | 68 (39%) |
| Others | 22 (13%) |

**Table 14** Thai users who think that changing devices frequently will cause damage to our planet (prepared by author)

| Answers | Respondents % |
|---|---|
| Yes | 109 (62%) |
| No | 18 (10%) |
| Maybe | 44 (25%) |
| Not at all | 5 (3%) |

**Table 15** Thai users: can we change the mind-set of designers and users regarding sustainability? (Prepared by author)

| Answers | Respondents % |
|---|---|
| Training | 50 (29%) |
| Education | 68 (39%) |
| Awareness | 64 (37%) |
| Workshop | 32 (18%) |
| Internet | 89 (51%) |
| T.V. | 37 (21%) |
| Social Networking | 88 (50%) |
| Others | 3 (2%) |

The survey results show that Thai users change their devices for the following main reasons: 58% cite functionality, 41% are concerned about speed, and 39% want to keep up with technology (Table 14).

These survey results indicate the Thai users' awareness that changing devices frequently will cause damage to our planet. Sixty-two percent said 'yes', while 25% said 'maybe' (Table 15).

The survey results suggest that Thai users believe that the Internet and Social Networking (51% and 50% respectively) are capable of changing people's mind-set.

## 4.2 *Process for Carrying Out Factor Analysis*

There are five steps for carrying out factor analysis as follows:

**Step 1:** Is the data suitable for factor analysis?

The first step is to determine whether the sample size is adequate for factor analysis. This can be done by applying the Kaiser-Meyer-Olkin (KMO) and Bartlett's Test [113] and examining the results.

**Step 2:** How will the factors be extracted?

Through rotation, a group of items are reduced to a simple factor. There are several methods of factor extraction: Principal Component Analysis (PCA), Principal Axis Factoring (PAF), Maximum Likelihood, Unweighted Least Squares, Generalized Least Squares, Alpha Factoring, and Image Factoring [113].

**Step 3:** What criteria will assist in determining factor extraction?

The application of criteria will help to determine the number of factors that should be retained after the extraction. The retained factors should have an eigenvalue of more than 1 and 70–80% of the total variance [30].

**Step 4:** Selection of rotational method

The factor rotation method is chosen in this step. There are two main methods with subgroups as follows [15]:

1. Orthogonal Rotation: the subgroups are Quartimax Rotation, Varimax Rotation, and Equamax Rotation.
2. Oblique Rotation: the subgroups are Quartimin Rotation, Covarimin Rotation, and Oblimin Rotation.

**Step 5:** Interpretation and labelling

In this step, factors are named and the word(s) used to label each factor depend on the researcher's interpretation. The name of factor can be conclude all variables in a short word [113].

## 4.3 *Factor Analysis Reporting*

This section presents the quantitative research results and research findings regarding the Thai users' attitudes towards and awareness of the issue of sustainability, including their opinions about the opportunities and risks associated with sustainability adoption.

### 4.3.1 Type of Factor Analysis

Exploratory factor analysis was used for this research because it aimed to observe and explore the results from a group of variables, define the factors, and interpret the results of variables.

### 4.3.2 Measures of Sample Adequacy

Sample adequacy can be measured using the result of Kaiser-Meyer-Olkin (KMO) and Bartlett's Test table. Kaiser-Meyer-Olkin (KMO) is used to verify that the respondents' data used for factor analysis is appropriate [113]. The result of KMO should be greater than 0.50 or close to 1.0. If it is less than 0.50, it will be inadequate for factor analysis [42]. Bartlett's Test of sphericity is related to the significance of the research and the value should be less than 0.05 (sig $< 0.05$) [42].

### 4.3.3 Type of Factors

All factor analysis results for opportunities and risks were extracted from 'Principal Axis Factoring method' or 'Common Factor Analysis method', which examines the smallest number of factors that can account for the common variance of a set of variables [40]. It is one of several methods used by researchers.

### 4.3.4 Factor Rotation

For this research, Varimax is chosen for factor rotation for the results of opportunities and risks.

### 4.3.5 Result of the Reliability of Data

The reliability of data presents as the table of Cronbach's Alpha. The main purpose is to analysis reliability and internal consistency of the online survey questions that involve the use of Likert-Type Scale. Also, Cronbach's Alpha is the popular test for researcher to examine the level of validity and accuracy in the online survey questions [37].

George and Mallery (2003, p. 231) [34] provide the following rules of thumb:

- > 0.9 = Excellent
- > 0.8 = Good
- > 0.7 = Acceptable
- > 0.6 = Questionable
- > 0.5 = Poor
- < 0.5 = Unacceptable

For this research, there are the examinations of sustainability opportunities and sustainability risks. The results of data reliability (Tables 16 and 17) are 0.933 for opportunities and 0.911 for risks, which are the range of excellent reliability.

**Table 16** Cronbach's Alpha (opportunities) (prepared by author)

| Reliability statistics | |
|---|---|
| Cronbach's Alpha | N of Items |
| 0.933 | 24 |

**Table 17** Cronbach's Alpha (risks) (prepared by author)

| Reliability statistics | |
|---|---|
| Cronbach's Alpha | N of Items |
| 0.911 | 13 |

**Table 18** KMO and Bartlett's Test for the opportunities (prepared by author)

| KMO and Bartlett's Test | | |
|---|---|---|
| Kaiser-Meyer-Olkin Measure of Sampling Adequacy | | 0.880 |
| Bartlett's Test of Sphericity | Approx. Chi-Square | 1525.619 |
| | df | 276 |
| | Sig. | 0.000 |

### 4.3.6 Factor Analysis Result for Opportunities

Values for KMO [31]:

Values between 0.5 and 0.7 = Mediocre
Values between 0.7 and 0.8 = Good
Values between 0.8 and 0.9 = Great
Values above 0.9 = Superb

As can be seen from Table 18, result of the KMO and Bartlett's test produced a Kaiser-Meyer-Olkin (KMO) value of 0.880. This means that the factor analysis is useful and the sample size of respondents is adequate because it is greater than 0.50. For Bartlett's Test of Sphericity, the significant result is 0.000. Therefore, the correlation between variables is significant and adequate because it meets the concept of Sig < 0.05.

The Rotated Factor Matrix produced seven variables for factor 1 and five variables for factor 2. The three highest values for factor 1 are 0.720, 0.664, and 0.654. Therefore, it may be concluded that Thai users believe that the adoption of sustainability by an organization can increase productivity, create new opportunities, and satisfy customer needs. Moreover, the three highest values for factor 2 are 0.842, 0.838, and 0.788. This indicates that Thai users believe that the adoption of sustainability by an organization can help to reduce emissions, pollution, health hazards, carbon footprint, and the consumption of energy and water (Table 19).

### 4.3.7 Naming Factor for Opportunities

Factor 1 comprises *Increased productivity, Attracting new opportunities, Satisfying customer needs, Attracting quality employees, Improving community investment, Meeting stakeholder expectations, and Creation of new jobs*. All variables have been named as Organizational Occasions because these variables are related to productivity, employees, customers, jobs, stakeholders, and investment (Table 20).

Factor 2 comprised *reduced emissions, reduced pollution, reduced health hazards, reduced carbon footprint, and reduced energy and water usage*. All variables have been named 5Rs that are related to the reduction of five problems when adopting the sustainability (Table 21).

**Table 19** Factor matrix for the opportunities (prepared by author)

| Rotated factor matrix[a] | | |
|---|---|---|
| | Factor | |
| | 1 | 2 |
| Increased productivity | 0.720 | |
| Attracting new opportunities | 0.664 | |
| Satisfying customer needs | 0.654 | |
| Attracting quality employees | 0.645 | 0.335 |
| Improving community investment | 0.641 | |
| Meeting stakeholder expectations | 0.623 | |
| Creation of new jobs | 0.600 | |
| Reduced risk management | 0.596 | |
| Improved human rights | 0.580 | |
| Increased triple bottom line – People, Planet and Profit | 0.544 | |
| Differentiates the business | 0.511 | |
| Enhanced reputation | 0.492 | |
| Reduced emissions | | 0.842 |
| Reduced pollution | | 0.838 |
| Reduced health hazards | | 0.788 |
| Reduced carbon footprint | | 0.781 |
| Reduced energy and water usage | | 0.678 |
| Reduced paper usage | 0.306 | 0.524 |
| Improved corporate social responsibility | 0.469 | 0.520 |
| Increased green strategy | 0.483 | 0.504 |
| Increased efficiency | | 0.469 |
| Improved social responsibility investing | 0.328 | 0.446 |
| Increased cost-effectiveness | 0.368 | 0.443 |
| Reduced consumption of raw materials | 0.367 | 0.370 |

Extraction Method: Principal Axis Factoring. Rotation Method: Varimax with Kaiser Normalization
[a]Rotation converged in three iterations.

**Table 20** Factor number 1 for opportunities (prepared by author)

| Variables | Factor Name |
|---|---|
| Increased productivity | Organizational Occasions |
| Attracting new opportunities | |
| Satisfying customer needs | |
| Attracting quality employees | |
| Improving community investment | |
| Meeting stakeholder expectations | |
| Creation of new jobs | |

**Table 21** Factor number 2 for opportunities (prepared by author)

| Variables | Factor Name |
|---|---|
| Reduced emissions | 5Rs |
| Reduced pollution | |
| Reduced health hazards | |
| Reduced carbon footprint | |
| Reduced energy and water usage | |

**Table 22** KMO and Bartlett's Test for the risks (prepared by author)

| KMO and Bartlett's Test | | |
|---|---|---|
| Kaiser-Meyer-Olkin Measure of Sampling Adequacy | | 0.854 |
| Bartlett's Test of Sphericity | Approx. Chi-Square | 785.219 |
| | df | 78 |
| | Sig. | 0.000 |

### 4.3.8 Factor Analysis for Risks

Values for KMO [31]:

Values between 0.5 and 0.7 = Mediocre
Values between 0.7 and 0.8 = Good
Values between 0.8 and 0.9 = Great
Values above 0.9 = Superb

As a result of KMO and Bartlett's Test (Table 22), the value of Kaiser-Meyer-Olkin is 0.854, which means the sample size of respondents is adequate because it is greater than 0.50. For Bartlett's Test of Sphericity, the significant value is equal 0.000. Therefore, the factor analysis produced results that are useful, significant, and sufficient.

The Rotated Factor Matrix (Table 23) yielded five variables for factor 1 and four variables for factor 2. The three highest values for factor 1 are 0.833, 0.777, and 0.725, suggesting that environmental irresponsibility can affect the loves of Thai users. Furthermore, Thai users also believe that failure to comply with environmental guidelines and regulations can lead to failures in governance, litigation and transactions.

### 4.3.9 Naming Factors for Risks

Factor 1 consists of five variables: *Increased supply chain crises due to suppliers' environmental problems, Increased marketing failure by perceived environmental*

**Table 23** Factor matrix for the risks (prepared by author)

| Rotated factor matrix[a] | | |
|---|---|---|
| | Factor | |
| | 1 | 2 |
| Increased supply chain crises due to suppliers' environmental problems | 0.833 | |
| Increased marketing failure by perceived environmental irresponsibility | 0.777 | |
| Increased scandals by perceived environmental irresponsibility | 0.725 | |
| Increased insurance crises due to environmental disasters | 0.687 | 0.303 |
| Increased number of new regulations including environmental | 0.656 | |
| Increased security and systems failures caused by environmental problems | 0.580 | 0.419 |
| Increased competition for and cost of raw materials | 0.567 | |
| Increased Interest rates | 0.547 | 0.339 |
| Cost inflation | 0.491 | |
| Increased governance failure (Including environmental actions and environmental compliance failures) | | 0.888 |
| Increased litigation and compliance breaches (Including environmental actions and environmental compliance failures) | | 0.671 |
| Increased transaction failure due to environmental liabilities | 0.520 | 0.647 |
| Increased fraud (Including environmental actions and environmental compliance failures) | | 0.628 |

Extraction Method: Principal Axis Factoring. Rotation Method: Varimax with Kaiser Normalization.
[a]Rotation converged in three iterations.

*irresponsibility, Increased scandals by perceived environmental irresponsibility, Increased insurance crises due to environmental disasters*, and *Increased number of new regulations including environmental*. These variables have been named as Environmental Hazards because they are related to the environmental issues (Table 24).

Factor 2 has four variables: *Increased governance failure, Increased litigation and compliance breaches, Increased transaction failure*, and *Increased fraud*. It has been named as 4Is (Table 25).

**Table 24** Factor number 1 for risks (prepared by author)

| Variables | Factor Name |
|---|---|
| Increased supply chain crises due to suppliers' environmental problems | Environmental Hazards |
| Increased marketing failure by perceived environmental irresponsibility | |
| Increased scandals by perceived environmental irresponsibility | |
| Increased insurance crises due to environmental disasters | |
| Increased number of new regulations including environmental | |

**Table 25** Factor number 2 for risks (prepared by author)

| Variables | Factor Name |
|---|---|
| Increased governance failure (including environmental actions and environmental compliance failures) | 4Is |
| Increased litigation and compliance breaches (including environmental actions and environmental compliance failures) | |
| Increased transaction failures due to environmental liabilities | |
| Increased fraud (including environmental actions and environmental compliance failures) | |

## 5 Discussions and New Significance

This section discusses the findings of this research. The survey results were analysed using SPSS for Windows and the results were presented in mathematical format as follows. The participants in this research comprised 51% Males and 49% of Females with an average age of 25–30 years. No respondents were under 17 years of age as the research targeted people in employment. Thai people spend 5–10 hours on the computer per day and use the Internet for working, email, and chatting (77%, 75%, and 63% respectively). Moreover, they are often access the Internet via the smartphone because it is very convenient and can be used everywhere. The findings showed that the Internet is the most important channel though which Thai people obtain information about the concept of sustainability and green information technology. Furthermore, the results showed that most Thai people do not read the sustainability report before buying a device. Thai people also believe that our planet would be damaged if people often change their devices. The respondents think the factors that can change the mind-set of designers and users are the Internet (51%), Social Networking (50%), education (39%) and awareness (37%). For the sampling adequacy, the KMO and Bartlett's

Test were applied to the survey results for both opportunities and risks. The values were all acceptable with 0.880 for the sustainability opportunities and 0.854 for the sustainability risks, indicating the adequacy and usefulness of the sampling.

The following were named as the factors:

- Opportunities
  - Organizational Occasions that came from *Increase productivity, Attract new opportunities, Satisfy customer needs, Attract quality employees, Improve stakeholder expectations, and Create new jobs.*
  - 5Rs that came from *Reduce emissions, Reduce pollution, Reduce health hazards, Reduce carbon footprint, and Reduce energy and water usage.*
- Risks
  - Environmental Hazards that came from *Increase supply chain crises due to suppliers' environmental problems, Increase marketing failure by perceived environmental irresponsibility, Increase scandals by perceived environmental irresponsibility, Increase insurance crises due to environmental disasters, Increase number of new regulations including environmental.*
  - 4Is that came from *Increase governance failure (Including environmental actions and environmental compliance failures), Increase litigation and compliance breaches (Including environmental actions and environmental compliance failures), Increase transaction failure due to environmental liabilities, Increase fraud (Including environmental actions and environmental compliance failures).*

## 6 Limitations

This research uses the quantitative research method that obtains feedback from participants through the online survey. The online survey is a research instrument used by many researchers although it has several limitations.

### *6.1 Time*

The conduct of this research through the survey online took quite a long time because there were some participants who did not understand the research topic and did not see the importance of completing the online survey. Therefore, it took a longer time to obtain the feedback.

### *6.2 Sampling*

The sample for this research consisted of Thai users who are using the resources and technology as well as are working in different positions in several organizations. Some of participants had a bit of time after they finish their office hour.

### 6.3 *Incomplete Surveys*

Several respondents were unable to see all parts of the online survey when they answered it on their smartphone. Therefore, the respondents did not answer some questions and some sections. As a result, the researcher tried to advise the respondents to complete the online survey on a personal computer because this would enable them to see all the questions.

## 7 Conclusion

Sustainability is becoming important for many countries at the present time. Sustainable development helps to sustain our world because the world is rapidly changing and people constantly want to satisfy their needs and desires. For this reason, many problems occur within countries and eventually become world problems. People are using new technology, new inventions, and other instruments to exploit natural resources to satisfy their needs. Problems such as deforestation waste generation and excessive consumption of resources all affect the balance of sustainable development, which has three pillars: environmental, economic, and social.

Thailand is one of the developing countries that have begun to be concerned about sustainability because of the negative impacts of environmental issues and climate change. Further, several organizations have adopted sustainable development, triple bottom line, and corporate social responsibility (CSR). For this reason, this research aims to observe the user's attitudes towards sustainability, including its associated advantages and risks when businesses in Thailand adopt sustainability practices. The research question of this study is: *What are the factors that are influencing the attitude of Thailand's higher education sector towards sustainability in terms of its opportunities and challenges?* To answer the research question, an online survey was designed and distributed to participants. The results of the data analysis provided new findings and allowed a conclusion to be drawn. Sustainable Development is a concept that has been adopted by many governments, businesses and industries. Seventeen sustainable development goals have been established by the United Nations. These are: (1) no poverty, (2) zero hunger, (3) good health and well-being, (4) quality education, (5) gender equality, (6) clean water and sanitation, (7) affordable and clean energy, (8) decent work and economic growth, (9) industry, innovation and infrastructure, (10) reduced inequalities, (11) sustainable cities and communities, (12) responsible consumption and production, (13) climate action, (14) life below water, (15) life on land, (16) peace, justice and strong institutions, (17) partnership for the achievement of goals. Additionally, the first Conference of the Parties or United Nations Framework Convention on Climate Change [83] took place in 1995. It focused on the problem of climate change in several countries but did not achieve any real success until 2015 when all participating nations reached an agreement on climate change action. Thailand is also contributing to the Paris Agreement and providing several model schemes that will be used within the country to address the issue of climate change.

## References


1. Adams WM (2013) Sustainable development? In: Johnston RJ, et al (ed) Geographies of global change: remapping the world 2002. Routledge, London and New York. pp 412–426
2. Adenle AA, Haslam GE, Lee L (2013) Global assessment of research and development for algae biofuel production and its potential role for sustainable development in developing countries. Energy Policy 61:182–195. doi:http://dx.doi.org/10.1016/j.enpol.2013.05.088
3. Alkire S, Santos ME (2011) Acute multidimensional poverty: a new index for developing countries. In: Proceedings of the German Development Economics Conference, Berlin
4. Amoateng P (2015) Sustainable development in developing countries: ramifications of urbanisation and poverty. http://www.openpop.org/?p=1054. Accessed 1 June 2017
5. Arts B, Rüdig W (1995) Negotiating the 'Berlin mandate': reflections on the first 'conference of the parties' to the UN framework convention on climate change. Environ Politics 4(3):481–487. doi:10.1080/09644019508414218
6. Asia News Monitor (2012) Thailand: King Ramkhamhaeng the Great of the Sukhothai Kingdom. *Asia News Monitor*, 15 Jan 2012
7. Asia News Monitor (2014) Thailand: SEC supports banks to adopt CSR in process and anti-graft practices. *Asia News Monitor*, 02 Sep 2014
8. Asian Development Bank (2015) Poverty in Thailand. http://www.adb.org/countries/thailand/poverty. Accessed 18 June 2016
9. Atkinson G, Dietz S, Neumayer E (2008) Handbook of sustainable development. Edward Elgar Publishing, UK
10. Attig N, El Ghoul S, Guedhami O, Suh J (2013) Corporate social responsibility and credit ratings. J Bus Ethics 117(4):679–694. doi:10.1007/s10551-013-1714-2
11. Australian Association of Environmental Educators (2014) Welcome to environmental and sustainability education. http://www.aaee.org.au/. Accessed 1 June 2017
12. Beilharz P, Hogan T (2006) Sustainability. In: Sociology: place, time & division. Oxford University Press, South Melbourne, Australia. pp 414-420
13. Bingham JB, Dyer Jr WG, Smith I, Adams GL (2011) A stakeholder identity orientation approach to corporate social performance in family firms. J Bus Ethics 99(4):565–585. doi:10.1007/s10551-010-0669-9
14. Boonlert K (2016) Sustainable business with triple bottom line. https://nairienroo.wordpress.com/2016/02/06/. Accessed 1 June 2017
15. Brown JD (2009) Choosing the right type of rotation in PCA and EFA. Statistics Corner 13(3):20–25
16. Cavendish R (1999) Siam officially renamed Thailand. History Today 49(5)
17. Chartsiri J (2009) The emergence and development of the Thonburi Kingdom in the context of the Chinese Era. 1727–1782. PhD Dissertation, Chulalongkorn University, Bangkok
18. Chen Liang CDs (2002) Journey into golden Thai history. *China Daily*, 01 Jun 2002, p 5
19. Circular Ecology (2015) Sustainability and sustainable development – what is sustainability and what is sustainable development? Circular Ecology. http://www.circularecology.com/sustainability-and-sustainable-development.html#.WSxt8YlHIU. Accessed 18 June 2016
20. Cobbinah PB, Black R, Thwaites R (2011) Reflections on six decades of the concept of development: evaluation and future research. J Sustain Dev Africa 13(7):134–149

21. Coca-Prados J, Gutiérrez-Cervelló G (2013) Economic sustainability and environmental protection in mediterranean countries through clean manufacturing methods. Environ Sci. Springer, The Netherlands. doi:10.1007/978-94-007-5079-1
22. Cotterell A (2011) Asia: a concise history. John Wiley and Sons, Singapore. doi:10.1002/9781119199601
23. Couper MP, Traugott MW, Lamias MJ (2001) Web survey design and administration. Public Opin Q 62(2):230–253
24. Crowther D, Aras G (2008) Corporate social responsibility. David Crowther, Guler Aras and Ventus Publishing ApS, UK
25. Dillman D, Rejpus U, Matzat U (2010) Advice in surveying the general public over the Internet. Int J Internet Sci 5(1):1–4
26. Djamasbi S, Siegel M, Tullis T (2010) Generation Y, web design, and eye tracking. Int J Hum Comput Stud 68(5):307–323
27. Dowdeswell E, Bolin B, Ashry ME, Desai N, Chimutengwende C, Obasi GOP (1996) News from the Framework Convention on Climate Change COP-2. http://www.austlii.edu.au/au/journals/MelbJIL/2002/7.html. Accessed 1 June 2017
28. Ekins P (2011) Environmental sustainability: from environmental valuation to the sustainability gap. Prog Phys Geogr 35(5):629–651. doi:http://dx.doi.org/10.1177/0309133311423186
29. Electricity Generating Authority of Thailand (2015) The outcome of COP 21 and Thailand. http://www.egat.co.th/en/index.php. Accessed 18 June 2016
30. Field A (2000) Discovering statistics using SPSS for Windows. Sage, London
31. Field A (2009) Discovering statistics using SPSS. Sage, London
32. Fisk P (2010) People planet profit: how to embrace sustainability for innovation and business growth. Kogan Page Publishers, UK
33. Forte A (2013) Corporate social responsibility in the United States and Europe: how important is it? The future of corporate social responsibility. Int Bus Econ Res J 12(7):815
34. George D, Mallery P (2003) SPSS for windows step by step: a simple guide and reference, 11.0 update, 4th edn. Pearson Education, Boston, p 231
35. Giblett R, Lester L (2008) Environmental sustainability. Media Cult Stud 22(2):167–170
36. Gimenez C, Sierra V, Rodon J (2012) Sustainable operations: their impact on the triple bottom line. Int J Prod Econ 140(1):149–159. doi:http://dx.doi.org/10.1016/j.ijpe.2012.01.035
37. Gliem JA, Gliem RR (2003) Calculating, interpreting, and reporting Cronbach's alpha reliability coefficient for Likert-type scales. Paper presented at the 2003 Midwest Research to Practice Conference in Adult, Continuing and Community Education. Columbus
38. Gordon JS, McNew R (2008) Developing the online survey. Nurs Clin N Am 43(4): 605–619. doi:http://dx.doi.org/10.1016/j.cnur.2008.06.011
39. Heiberg R (2012) What is sustainability? The Post Carbon Reader Series: Foundation Concepts. doi:978-0-9709500-6-2. https://assets-production-webvanta-com.s3-us-west-2.amazonaws.com/000000/03/97/original/Reader/PCReader-Heinberg-Foundation.pdf. Accessed 1 June 2017
40. Hill BD (2011) The sequential Kaiser-Meyer-Olkin procedure as an alternative for determining the number of factors in common-factor analysis: A Monte Carlo simulation. PhD, Oklahoma State University, Ann Arbor
41. iBiz (2014) CSR-in-process. ASTV. http://www.manager.co.th/iBizChannel/ViewNews.aspx?NewsID=9570000118255. Accessed 1 June 2017
42. IBM Knowledge Center (2011) Factor Anaysis. http://www.ibm.com/support/knowledgecenter/SSLVMB_20.0.0/com.ibm.spss.statistics.cs/fac_telco_kmo_01.htm. Accessed 18 June 2016
43. Inoue Y, Kent A (2014) A conceptual framework for understanding the effects of corporate social marketing on consumer behavior. J Bus Ethics 121(4):621–633. http://10.1007/s10551-013-1742-y. Accessed 18 June 2016
44. ISO (2010) ISO 26000:2010 Guidance on social responsibility. http://www.iso.org/iso/catalogue_detail?csnumber=42546. Accessed 1 June 2017
45. ISO Central Secretariat (2015) ISO 14001 key benefits. http://www.iso.org/iso/iso_14001_-_key_benefits.pdf. Accessed 18 June 2016

46. Issa T, Isaias P, Vijayarajan A (2014) Risks and opportunities of integrating social networking sites in the Australian and Portuguese workplace/ communities: an exploratory study. Int J Web Based Communities 10(4):466–489
47. Jha NK (2008) Research methodology, 1st edn. Abhishek Publications, Chandigarh
48. John Pike (2013) Thailand – Thonburi Period (1767-1782). http://www.globalsecurity.org/military/world/thailand/history-thonburi.htm. Accessed 18 June 2016
49. Casto KM, Ellisen EP, Trnovec T, Kross BC, Ginter E (1996) ISO 14000: Origin, Structure, and Potential Barriers to Implementation. Int J Occup Environ Health 2(2):99–124
50. Kohn MM, Chang E, Karduck AP, Televski A (2010) SMART CAMP: environmental sustainability through intelligent automation technologies. 24th IEEE International Conference on Advanced Information Networking and Applications Workshops, WAINA 2010 IEEE Computer Society, pp 286-292. doi:10.1109/WAINA.2010.97
51. Kotler P, Lee N (2011) Corporate social responsibility: Doing the most good for your company and your cause. Reviewed Work: Academy of Management Perspectives 20(2) (May, 2006):90–93
52. Kuhlman T, Farrington J (2010) What is sustainability. Sustainability 2:3436–3448. doi:10.3390/su2113436
53. Kumar R (2011) Research methodology: a step-by-stepguide for beginners. Sage, London
54. Lambert T (2016) A short history of Thailand. http://www.localhistories.org/thailand.html. Accessed 18 June 2016
55. Lyman HB (1963) Test scores and what they mean. Prentice-Hall, Inc., New Jersey
56. McLeod SA (2014) Sampling methods. http://www.simplypsychology.org/sampling.html. Accessed 18 June 2016
57. Ministry of Natural Resources and Environment (2015) The knowledge of United Nations Framework Convention on Climate Change. http://webeng.mnre.go.th/. Accessed 18 June 2016
58. Ministry of Natural Resources and Environment (2015) The model scheme for climate change. http://webeng.mnre.go.th/. Accessed 18 June 2016
59. Mishra P (2010) History of Thailand. ABC-CLIO, Santa Barbara
60. Neugebauer S, Martinez-Blanco J, Scheumann R, Finkbeiner M (2015) Enhancing the practical implementation of life cycle sustainability assessment – proposal of a tiered approach. J Clean Prod 102:165–176. doi:http://dx.doi.org/10.1016/j.jclepro.2015.04.053
61. Newman P (2010) Sustainable Cities. In: Davis M, Lyons M (eds.) More than Luck. Ideas Australia needs now. Centre for Policy Development, Sydney, NSW, pp 149–158
62. Post J, Lawrence A, Weber J (2002) Business and society: corporate strategy, public policy, ethics, 10th edn. McGraw-Hill, New York
63. Punnett BJ (2013) Management: a developing country perspective. Taylor and Francis, Florence
64. Rangan K, Chase LA, Karim S (2012) Why every company needs a CSR strategy and how to build it. Harvard Business School, Cambridge
65. Rinkesh (2015) What is environmental degradation? http://www.conserve-energy-future.com/causes-and-effects-of-environmental-degradation.php. Accessed 18 June 2016
66. Sarunee A (2009) Triple Bottom Line. *Fringer*, 25 Feb 2009. http://www.fringer.org. Accessed 18 June 2016
67. Saunders M, Lewis P, Thornhill A (2012) Research methods for business students, 6th edn. The Pitman Publishing, England
68. Saunders M, Tosey P (2013) The layers of research design. Rapport: The Magazine for NLP Professionals 14.4 (2012):58–59
69. Schuh C, Strohmer MF, Easton S, Hales M, Triplat A (2014) Suppliers relationship management. Apress Publisher, New York. doi:10.1007/978-1-4302-6260-2
70. Setterfield M (2010) Handbook of alternative theories of economic growth. Edward Elgar, UK
71. Sheehan B (2009) What is sustainability? Control Eng 56(9):12
72. Siyavula (2009) Characteristics of developing countries. https://cnx.org/contents/4fntVCbD@1/Characteristicsof-developing-. Accessed 1 June 2017

73. Smyth J, Dillman D, Christian L, O'Neill A (2010) Using the Internet to survey small towns and communities: limitations and possibilities in the early 21st Century. Am Behav Sci 53:1423–1448
74. Steckstor D (2012) The effects of cause-related marketing on customers' attitudes and buying behavior. Springer Gabler Verlag, Germany. doi:10.1007/978-3-8349-7078-7
75. Sumeteeprasit J (2014) Triple Bottom Lines. GotoKnow. https://www.gotoknow.org/posts/558991. Accessed 1 June 2017
76. Thai CSR Network (2008) CSR Catagories. Thai CSR Network. http://www.thaicsr.com Accessed 18 June 2016
77. Thai CSR Network (2012) What is CSR? Thai CSR Network. http://www.thaicsr.com Accessed 18 June 2016
78. Trochim WMK (2006) Unit of analysis. http://www.socialresearchmethods.net/kb/unitanal.php. Accessed 18 June 2016
79. Trochim WMK (1999) Research methods knowledge base [on-line textbook]. 2nd edn.
80. UN (2011) The history of sustainable development in the united nations. The United Nations. http://www.uncsd2012.org/history.html. Accessed 18 June 2016
81. UN News Centre (2015) UN chief hails new climate change agreement as "monumental triumph". http://www.un.org/sustainabledevelopment/blog/2015/12/un-chief-hails-new-climate-change-agreement-as-monumental-triumph/. Accessed 18 June 2016
82. UNESCO (1991) Historic city of Ayutthaya. UNESCO. http://whc.unesco.org/en/list/576
83. UNFCCC (2016) Event: UNFCCC COP 22. http://climate-l.iisd.org/events/unfccc-cop-22. Accessed 18 June 2016
84. United Nations (1997) UN Conference on Environment and Development (1992). http://www.un.org/geninfo/bp/enviro.html. Accessed 18 June 2016
85. United Nations (1998) Report of the Conference of the Parties on its third session, held at Kyoto from 1 to 11 December 1997. Paper presented at the COP3, Kyoto, Japan
86. United Nations (1999) Report of the Conference of the Parties on its fifth session, held at Bonn from 25 October to 5 November 1999. Paper presented at the COP 5, Bonn, Germany
87. United Nations (1999) Report of the Conference of the Parties on its fourth session, held at Buenos Aires from 2 to 14 November 1998. Paper presented at the COP 4, Buenos Aires, Argentina
88. United Nations (2001) Report of the Conference of the Parties on the first part of its sixth session, held at the Hague from 13 to 25 November 2000. Paper presented at the COP 6, Hague, the Netherlands
89. United Nations (2001) Report of the Conference of the Parties on the second pert of its sixth session, held at Bonn from 16 to 27 July 2001. Paper presented at the COP 6, Bonn, Germany
90. United Nations (2002) Report of the Conference of the Parties on its seventh session, held at Marrakesh from 29 October to 10 November 2001. Paper presented at the COP 7, Marrakesh, Morocco
91. United Nations (2003) Report of the Conference of the Parties on its eighth session, held at New Delhi from 23 October to 1 November 2002. Paper presented at the COP 8, New Delhi, India
92. United Nations (2004) Report of the Conference of the Parties on its ninth session, held at Milan from 1 to 12 December 2003. Paper presented at the COP 9, Milan, Italy
93. United Nations (2005) Report of the Conference of the Parties on its tenth session, held at Buenos Aires from 6 to 18 December 2004. Paper presented at the Cop 10, Buenos Aires, Argentina
94. United Nations (2006) Report of the Conference of the Parties on its eleventh session, held at Montreal from 28 November to 10 Decenber 2005. Paper presented at the COP 11, Montreal, Canada
95. United Nations (2007) Report of the Conference of the Parties on its twelfth session, held at Nairobi from 6 to 17 November 2006. Paper presented at the COP 11, Nairobi, Kenya

96. United Nations (2008) Report of the Conference of the Parties on its thirteenth session, held in Bali from 3 to 15 December 2007. Paper presented at the COP 13, Bali, Indonesia
97. United Nations (2009) Report of the Conference of the Parties on its fourteenth session, held in Poznan from 1 to 12 December 2008. Paper presented at the COP 14, Poznan, Poland
98. United Nations (2010) Report of the Conference of the Parties on its fifteenth session, held in Copenhagen from 7 to 19 December 2009. Paper presented at the COP 15, Copenhagen, Demark
99. United Nations (2011) Report of the Conference pf the Parties on its sixteenth session, held in Cancun from 29 November to 10 Decem 2010. Paper presented at the COP 16, Cancun, Mexico
100. United Nations (2012) Report of the Conference of the Parties on its seventeenth session, held in Durban from 28 November to 11 December 2011. Paper presented at the COP 17, Durban, South Africa
101. United Nations (2013) Report of the Conference of the Parties on its eighteenth session, held in Doha from 26 November to 8 December 2012. Paper presented at the COP 18, Doha, Qatar
102. United Nations (2014) First steps to a safer future: introducing the United Nations Framework Convention on Climate Change. Unite Nations
103. United Nations (2014) Report of the Conference of the Parties on its nineteenth session, held in Warsaw from 11 to 23 November 2013. Paper presented at the COP 19, Warsaw, Poland
104. United Nations (2015) Report of the Conference of the Parties on its twentieth session, held in Lima from 1 to 14 December 2014. Paper presented at the COP 20, Lima, Peru
105. United Nations (2016) Report of the Conference of the Parties on its twenty-first session, held in Paris from 30 November to 13 December 2015. Paper presented at the COP 21, Paris, France
106. United Nations Statistics Division (2013) List of Developing Countries. http://www.iugg2015prague.com/list-of-developing-countries.htm. Accessed 18 Jun 2016
107. Valdez HEaARC (2002) ISO14000 Standards and the US Construction Industry. Environ Pract 4(4):210–219
108. Vallance S, Perkins HC, Dixon JE (2011) What is social sustainability? A clarification of concepts. Geoforum 42(3):342–348. doi:http://dx.doi.org/10.1016/j.geoforum.2011.01.002
109. Vidyarthi K (2015) Top 15 reasons why education is extremely important. https://listsurge.com/top-15-reasons-education-important/. Accessed 1 June 2017
110. Vogt WP, Vogt ER, Gardner DC, Haeffele LM (2014) Selecting the right analyses for your data: quantitative, qualitative, and mixed methods. Guilford, New York
111. Water Wise (2016) Cause of pollution. http://www.waterwise.co.za/site/water/environment/causes-of-water-pollution.html. Accessed 18 June 2016
112. Willard B (2012) New sustainability advantage: seven business case benefits of a triple bottom line, 10th edn. New Society, Gabriola Island
113. Williams B, Onsman A, Brown T (2010) Exploratory factor analysis: a five-step guide for novices. J Emerg Prim Health Care 8(3):1–13
114. Williams D (2013) What are causes of global poverty? http://borgenproject.org/what-causes-global-poverty/. Accessed 25 June 2013
115. World Bank (2011) World Development Report 2012: Gender equality and development. World Bank Publications, Herndon
116. World Food Programme (2016) What causes hunger? World Food Programme. http://www.wfp.org/hunger/causes. Accessed 18 June 2016
117. Youssoufou Hamadou D (2014) CSR and sustainable development. Cadernos de Estudos Africanos 28:141–162

# Sustainability Awareness in Vietnam's Higher Education Sector

**Anh Tuan Hoang**

**Abstract** Sustainability is becoming more imperative all over the world, with many countries signing environmental agreement to prevent the issue from becoming worse. In November 2016, a number of international conferences on sustainability have been held in Marrakech, Morocco such as the 22nd session of the Conference of the Parties (COP 22), 12th session of the Conference of the Parties to the Kyoto Protocol (CMP 12), and the first session of the Conference of the Parties serving as the meeting of the Parties to the Paris Agreement (CMA 1). The success of these conferences confirmed the commitment of Paris Agreement implementation and continued cooperation of many countries on the climate change issue. Vietnam is a developing country that has taken part in the implementation of sustainable development; however, there are still environmental concerns due to unawareness of the issue and the irresponsible behaviours of organizations and individuals. This study aims to examine the advantages and disadvantages associated with sustainability awareness in Vietnam's higher education sector. Via an online questionnaire, 281 participants were surveyed. Results indicate that the three advantages of sustainability adoption are: improved efficiency and reduction in resource usage, fulfilment of investor expectations and differentiating achievements, and reduction of environmental impacts. The two disadvantages are the possibility that there will be breaches of regulations and compliance; and an increase in the number of scandals, crises and failures due to environmental problems, irresponsibility and disasters. Educators in Vietnam should consider the sustainability issue in their courses in order to raise the awareness of students. Additionally, to ensure the success of sustainability implementation, businesses should ensure that all staff at every level of the organisation is clear about sustainability and sustainability practices.

A.T. Hoang (✉)
School of Information Systems, Curtin University, Perth, Western Australia, Australia
e-mail: tuaninbox@gmail.com



T. Issa et al. (eds.), *Sustainability, Green IT and Education Strategies in the Twenty-first Century*, Green Energy and Technology, DOI 10.1007/978-3-319-57070-9_7

# 1 Introduction

In this era of rapidly developing information technology, changes are apparent in all areas of life, including communication, entertainment, education, business and society. Technologies, as a catalyst, has transformed the way people get connected and enjoy life by providing new communication channels and fulfilling people's needs with advanced facilities, products and services; however, it also causes negative impacts on the planet in regards to demand of resources [32]. Climate change and its consequences, which are natural disasters such as flood, earthquake and hurricane, are causing huge damage and loss of life that human beings have to face every year. Governments, organizations and individuals have recognized and started to take actions to reduce the severity of these issues.

In an attempt to prevent resource depletion and save the planet, the concept of sustainability has been introduced and embraced by many countries and international organizations. Based on this, several policies have been formulated together with guidelines and recommendations to provide a direction for governments and organizations in their pursuit of sustainable development [16]. The aim of sustainability is to ensure that resources meet the needs of current development but will also sufficient to meet the needs of the future. The concept of sustainability has become popular all over the world, especially in the business environment because this is where most resources are used for manufacturing and it is the core of development. Businesses are required to not only ensure their economic profits, but also maintain the balance of environmental and societal values of the community in which they operate [42]. As a developing country, Vietnam is adopting sustainability management via its strategic policies for the economy, education, society and so on. Nonetheless, there are still several issues in the management of sustainability due to the lack of awareness and irresponsibility of individuals and organizations. Consequently, this study seeks to discover the extent, to which Vietnamese people are aware of sustainability, especially students in the higher education sector. Furthermore, the study will examine Vietnamese perceptions of the advantages and disadvantages of sustainability to the business environment especially from the higher education perspective.

# 2 Sustainability

This section comprises three sections. The first is introductory and discusses definitions of sustainability, traces the history of the sustainability concept, and describes the various types of sustainability: social, environmental and economic. Next, the meanings of sustainability in terms of education, and the status of sustainability, are explored. This section explains why people and businesses need to be aware of sustainability and how governments have agreed and supported sustainability adoption. It also explains the current situation regarding sustainability adoption in different regions, especially developing countries. The third section examines the policies for sustainability in Vietnam and discusses the advantages and disadvantages associated with the adoption of sustainability. Some of the advantages are resource reduction, corporate image enhancement, improved performance, less environmental impact; and some of the risks are an increase of crises and fraud, cost of investment, and inflation.

Sustainability is usually discussed in different contexts as a global phenomenon in the current age that has emerged in response to the detrimental effects of climate change and resource scarcity. It is considered to be the ability of nature and humankind to endure or continue to exist and evolve collaboratively. Sustainability is often referred to as the management of resource usage, which requires wise consumption, and effective management to ensure adequate conservation of resources for future use. Generally, it is recognized as a vibrant and mutually dependent existence of all elements including nature and all other entities on earth [38]. Specifically, it is associated with human activities to maintain and control the well-being of mother earth. Hence, sustainability is a concept that is connected with human usage of resources in a way that supports its continuous availability [49]. Put simply, it means using resources efficiently and preserving them for the well-being of the next generation [28].

Originally, the concept of sustainability was related to forestry, and meant collecting no more than the forest could reproduce. The German term "Nachhaltigkeit", which means sustainability, was first mentioned in the seventeenth century following a concern about the depletion of natural resources [32]. In the 1970s, became a concept related to environmental, economic and societal business performance in the modern world. This happened during a period of global resource scarcity which stemmed from overpopulation, food shortage, environmental issues and global warming [47]. The concept achieved global recognition after the report published by the World Commission on Environment and Development in 1987 which defined sustainable development as "development that meets the needs of current generations without compromising the ability of future generations to meet their own needs" [WCED 1987, 45 quoted in 13]. Since then, this terminology has been fully adopted by businesses for accounting and management purposes, and acknowledged as the three pillars of social (people), environmental (planet) and economic (profit) presentation. As a combination of three elements, sustainability is accomplished only when all of them are achieved; in other words, sustainability ensures the growth of society with cost efficiency and without environmental degradation.

Traditionally, businesses refer to the bottom line in financial statements as a measurement of business health, which is either the profit or loss. With the advent of the Triple Bottom Line (TBL) in the mid-1990s, John Elkington [44] introduced a new framework to gauge performance which went beyond the conventional way of financial evaluation. It does not only evaluate performance based on financial values such as profits or return on investment, but also considers environmental and societal accomplishments. For example, in regard to supply chain management, the process of selecting qualified suppliers requires a manufacturer to consider criteria such as product and service quality and delivery, responsibility to community and employee, and technological solutions for pollution control. These three factors pertain to business operations (profit), social responsibility (people) and environmental concerns (planet), respectively [40].

The goal of sustainability is to preserve resources, ensure environment protection and ensure social equity. Businesses are considered sustainable when on the one hand their economic advantages, which are the first pillar – profit or economic factor of the triple bottom line (3BL), are attained through their operational activities; but on the other hand the environmental and social elements, the second and third pillars, must be met. As described, triple bottom line is a means of assessing business performance in terms of the three factors of profit, people and planet; therefore, businesses must consider all of these factors in their strategic plan as well as management activities. Although the most important thing for businesses is maximizing profit by optimizing manufacturing cost and operation expenditure, their activities cannot make any positive impact on environment and society if they are not implementing an efficient waste processing system or if they are using child labour. As shown in Fig. 1, the three dimensions of sustainability are of equal significance, and sustainability is achieved only when each pillar is considered without sacrificing either of the other two.

Goodland [22] stated that economic sustainability is the preservation of capital and the consumption of added value without using it. As the ultimate goal of any business is to maximise profits, the business must continuously increase its productivity to obtain the best return on investment and meet stakeholders' expectations. Economic sustainability is the traditional measure used to assess the health of a business. It is usually achieved by implementing strategic business schemes and wise financial investment such as the application of new and green technologies, optimizing operating processes, innovating to create new products and services.

Environment is the surroundings and their conditions in which people, animals and plants are living and interacting; therefore, environmental sustainability refers to the protection of all things related to its long-term existence and involves the preservation of wildlife, protection of land and water resources, and reduction of air and noise pollution. In other words, it refers to the state when human being use natural resources within their capacity and give them the opportunity to recover

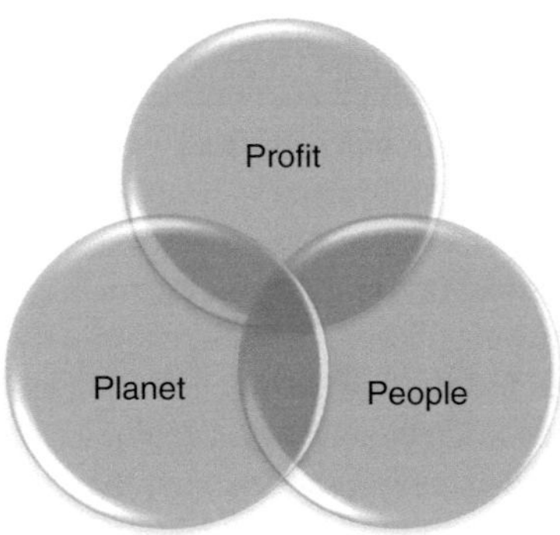


**Fig. 1** The three pillars of sustainability – prepared by the author (Adopted from Slaper and Hall [44])

and last forever [44]. With regards to business, environmental performance is measured by assessing the impacts of business activities on the environment and includes the level of fuel and electricity consumption, waste management.

Social sustainability is defined as the management of human and social aspects which ensure fairness and apply management best practices in terms of human resources and communities [18]. Pertaining to this definition, social sustainability is the approach to ensure long-term societal development like promoting community, ensuring equality and eliminating poverty. Sustainability creates a common value in business, and is represented by the triad of profit, planet and people. Triple bottom line is a business success factor in the long term because of various decisive factors such as environmental regulations, stakeholders and customer expectations. Porter and Kramer [2011 quoted in 26] described the shared value as the practices used to maximize business ability and performance, at the same time improving its operating environment economically and socially. Using best practices of creating shared value, businesses concentrate on boosting their profits, enhancing their people value and eliminating practices that are detrimental to the planet. By making these their strategic goal, businesses retain their leading positions in the market. Velde et al. [53] reported that socially responsible investors' portfolios achieved higher performance than conventional investment methods. More critically positive outcomes were produced, when looking at some aspects of sustainable investment, indicating the added shared value created by corporate social responsibility. Although each pillar is considered as an individual measurable criterion, the three pillars are interactive and overlap. Doane and MacGillivray [12] posit that the three factors of sustainability are inter-connected. Good financial performance can help business survive in the short term; nevertheless, it does not ensure long-term success in terms of economic, social or environmental aspects. Assuming the right prediction of sustainability, just concentrating on economic performance might be an issue in the long period. As a consequence, each of the three factors of sustainability should be given equal importance.

With the current level of development and the speed of globalization, more and more resources are required for manufacturing, transportation, consumption and so on. Additionally, the rapidly increasing world population has created a huge demand for food, water and other necessaries of life. According to United Nation prospect report, the world population was 7.3 billion in 2015 and at the current rate, this number will reach 9.7 billion in 2050 [50]. Importantly, the world population has increased by 1 billion since 2003 and 2 billion since 1990. At the present rate of population growth, there will be a strong demand for food, transportation, energy and so forth. As Evans [16] stated in the report on globalization and scarcity, "the issues could emerge as an important catalyst for collective international action to tackle global challenges – in the process helping to ensure that a globalization that is already efficient also becomes more sustainable, equitable and resilient". As being sustainable is crucial for long-term development, people should be aware of sustainability, particularly the youngsters because they are the future caretakers of the country. As discussed earlier, the triple bottom line is a novel model for measuring business performance in which three dimensions of economic, environmental and social sustainability are assessed against business achievements. Students need to be given the appropriate skills to be prepared for their future career path because "environmental education should be an integral part of the educational process, aimed at practical problems of an interdisciplinary character that build a sense of values, and contribute to public well-being" [Stapp et al. 1998 quoted in 20]. Moreover, as declared in United Nations development plan for the period from 2005 to 2014 [5], it is urgent to transfer the necessary analytical skills and applicable perspective to students and to give them the opportunity to apply these workplace skills at school. Additionally, Ayson [5] maintains that businesses want to employ new graduates with knowledge about sustainability practices. Students' learning experiences are changing from passive to proactive in a more collaborative and engaging environment. This transformation is aligned with the shift in the curriculum and assessment approach, which now has a strong focus on both hard and soft skills such as communication, interpersonal skills, and critical thinking and sustainability approach.

To solve the problem of climate change, governments have signed many agreements to commit to reducing their greenhouse gas emissions in order to decrease the impact of climate change and promote sustainable development. Therefore, an environmental treaty was signed under the United Nations Framework Convention on Climate Change (UNFCCC) organized in Rio de Janeiro, Brazil in 1992, which then become effective on 21 March 1994. The goal of the UNFCCC agreement is to prevent further detriment to the climate caused by greenhouse gases resulting from human activities [51]. The expectation of the treaty was that by the year 2000 the countries listed in Annex I, which is called Annex I countries, would cut their greenhouse gas emissions to the level of 1990 because they are the most significant countries contributing to the current level of emissions; however, the Convention only provided recommendations without any binding rules.

The Kyoto Protocol, which is linked to UNFCCC, was adopted in 1997 but did not enter into force until 2005 because of a complex approval process. This Protocol is a commitment of developed countries to limit greenhouse gas emissions based on the UNFCCC, which was only for encouragement purposes. The Protocol has two commitment phases: the first 5-year period from 2008 to 2012 and the second one from 2013 to 2020 which was adopted in the Doha Amendment in 2012 [52]. The latest Conference of Parties 21 in Paris is the most successful treaty to date because 27 countries out of 197 made a commitment to take accelerated actions to combat climate change. It was an agreement to limit the earth warming level up to 1.5 °C above the level of the pre-industrial period by 2010 [3].

There is no clear definition about what constitutes a developing country since it depends on a particular context. For example, the World Trade Organization (WTO) does not define developed or developing countries; its members declare themselves to be developed or developing countries [57]. However, the World Bank currently defines developing countries as the ones with a gross national income per capita less than US$12,475, and this definition is generally used by most governments and organizations [56]. Based on this definition, this section reviews the current sustainability situation of some countries; it examines the sustainability policy and strategy in China, and the awareness of people in Brazil. As the second largest economy in the world, China depends mostly on fossil fuel (60%), particularly coal, for its energy needs. Complying with the Paris Climate Change Agreement, China is investing significantly in smart solutions and technology to cut down the level of greenhouse gas emissions. For example, it is committed to shrinking the number of fossil-fuel power plants to no more than 50%. Additionally, the imposition of strict regulations on local firms and the application of advanced technology with appropriate metrics in monitoring and controlling pollution level indicates China's commitment to reduce the detriment of climate change [48].

According to a study by Barbosa and Veloso [6] on the daily consumption behaviour of Brazilians, the daily routine practices of a large number of Brazilians are environmentally unfriendly because of cultural influence and sustainability unawareness. For instance, participants tend to leave fans and air conditioners on even when they are not in the room; the reason given was to cool the room down. In regards to awareness, less than 10% of surveyed participants are very familiar with the terms "sustainable", "environmental friendly", "carbon emissions" and "transgenic". In the social class context, 43% of participants who are middle to lower class know nothing about the concept while the remaining respondents partly understand concepts such as "preserving the planet" (61%), "garbage recycling" (27%) and "reducing energy consumption" (19%) [6]. Vietnam is a developing country in Southeast Asia. It was reunified in 1975 after a thousand years under imperial China, nearly 100 years under French colonial rule and 25 years of civil war. Vietnam was devastated due to long periods of chaos; however, it has been reformed through many renovation policies, which have helped the country to develop and achieve strong growth.

Since 2004, the Vietnamese government has adopted and approved many policies and action plans to promote and orientate the country towards sustainable development. For example, Vietnam's prime minister adopted the "Strategic Orientation for Sustainable Development in Vietnam" in 2004, the "Strategy for sustainable development in Viet Nam for the period of 2011-2020" in 2012, the "The National Action Plan on Sustainable Development in Viet Nam for 2011-2020" in 2013 [35]. The purpose of these strategies and policies was to shape Vietnam so that it was aligned with the three dimensions of sustainability of economy, social and environment. The strategies were intended to shape Vietnam's economy and: ensure the basic development of the country regarding food security, sustainable development in agricultural and rural areas; promote sustainable local and regional development nationwide; and maintain green economic growth with the adoption of clean and renewable energy. Additionally, the government encourages and then enforces sustainable production and consumption so that people awareness can be raised and gradually becomes an inherent attitude and behaviour in the long term. In regards to societal sustainability, Vietnam's government tries to improve social equity by implementing social security policies to enhance the quality of life, reduce poverty, close the gap between the poor and the rich, create more sustainable work with equal employment opportunities, and provide qualified health and medical services. Environmentally, policies are intended to enforce legislation compliance to prevent damage to ecosystem by protecting and ensuring sustainable water sources and land, developing new forest with sustainable use, reducing air and noise pollution, and assuring effective hazardous waste management [35].

With directional policies and schemes, Vietnam has achieved salient accomplishments in all facets of sustainable development. Economically, Vietnam has achieved remarkable outcome in regards to growth and poverty reduction. It has shown huge achievements in different sectors, including food and aquaculture production, industrial and services. For example, it has become the second largest rice exporter from being a country with a scarcity of food; furthermore, industry and services have accounted for 80% of GDP instead of 40% agriculture dependence 20 years earlier. In 2010, Vietnam's GDP was triple that of 2000, which was estimated at over US$100 billion. Compared to 2000, its GDP per capita was threefold in 2011. These achievements have helped Vietnam move from being a low income to a middle-income country. Socially, results from the attempts to overcome poverty, improve public health services, education and training, and increase the employment rate are quite promising. For example, poverty rate has plunged nearly 20–9.45%, which is equivalent to around 2% per year, for a period of 8 years from 2002 to 2010. Gender equality has improved with a large number of women being nominated for positions in the National Assembly and state administrative offices. Environmentally, many actions have been taken to prevent pollution and conserve the biodiversity of Vietnam, with heartening results. For instance, 80% of medical waste is properly collected and disposed of in compliance with environmental requirements. The percentage of rural population with access to a clean water source has doubled to over 80% in 2010 from 40% in 2003 [4].

Although the level of economic development has tremendously changed people's living standards and encouraged advancement in different areas, it has placed a heavy burden on sustainability. The core point of sustainability is ensuring economic growth without destroying the environment and harmfully affecting societal balance; hence, sustainability will fail if any one of the three sustainability pillars comprises the success of the other two pillars. During the process of economic transformation, Vietnam shifted from an agricultural economy to an industry and services economy, which gave rise to greater energy consumption. It is crucial to plan and implement appropriate policy to not only control the rate of economic growth but also shape it to comply with sustainability standards. According to Duc Luong [14], energy production in Vietnam from 2001 to 2011 increased significantly from over 40,000 kilotons of oil equivalent (kTOE) to nearly 70,000 kTOE; similarly, the amount of energy consumption also escalated from nearly 30,000 kTOE to about 50,000 kTOE in the same period.

It is noteworthy that the source of energy during this period was mostly fossil fuel such as coal and peat. According to Duc Luong 2015, the production of different types of energy from 2001 to 2011, including coal, peat, crude oil, biofuel, waste, natural gas and hydro. With the current increasing rate of growth in population and the economy, there are many challenges facing the Vietnam government in implementing energy efficiency and conservation programs because of the lack of information and expertise in resource management. For instance, information and data for energy efficient improvements such as end-user usage statistics for different sectors and analysis data for supply and demand assessment are insufficient to provide an insight for management purposes [14]. The government's strategy for maintaining strong economic growth and the current population growth in the future will unavoidably result in a high demand for energy. The energy demand during the 20 years from 2010 to 2030 has tripled [17].

In addition to energy consumption, transportation issues such as congestion and environmental pollution are becoming increasingly serious in Vietnam. However, the option to commute sustainably is not a choice for a large majority of Vietnamese people due to the limitation of public transportation and poor infrastructures. As Le and Trinh [34] discussed, the Vietnamese people's perception of negative impacts of the most popular means of transport, which is the motorcycle, is understood; their willingness to switch to public transport modes such as buses still depends on the quality of the transport infrastructure. Regarding consumption behaviour, economic growth leads to an increase in people's incomes and improved lifestyle, with the latter changing from subsistence to consumption. People are no longer satisfied with just meeting their basic needs, especially when a large percentage of the population, which is middle class, have better access to new products and services. As stated in the consumption study of De Koning et al. [8], there is a huge intention-action gap in the sustainable behaviour of middle-class people. For example, people usually use energy-efficient products to save their own money, not for sustainability reasons. Additionally, the current situation regarding food hygiene and transport are very hard to comply with because of the lack of available information and facilities; as a result, governments need to take responsibility for providing more opportunities for sustainability and promoting sustainable lifestyles.

Sustainable development plays a critical role in Vietnam's growth, especially in relation to businesses. It provides a lot of advantages which includes gaining reputation, meeting stakeholders' expectation, and financial improvements and so on. Nevertheless, it has some shortcomings such as increased cost of material and technologies, higher risk of fraud and governance failure. Figure 2 shows the advantages and disadvantages of being sustainable.

## *2.1 Sustainability Advantages*

### 2.1.1 Financial

Environmental issues are not the only concern of sustainable development; economic efficiency is of the utmost importance. The perceived economic advantages are the decisive elements that encourage business to become more sustainable. Economic measures that help businesses to become sustainable include reducing resource usage such as energy, water and paper; decreasing raw material consumption; and increasing operational efficiency.

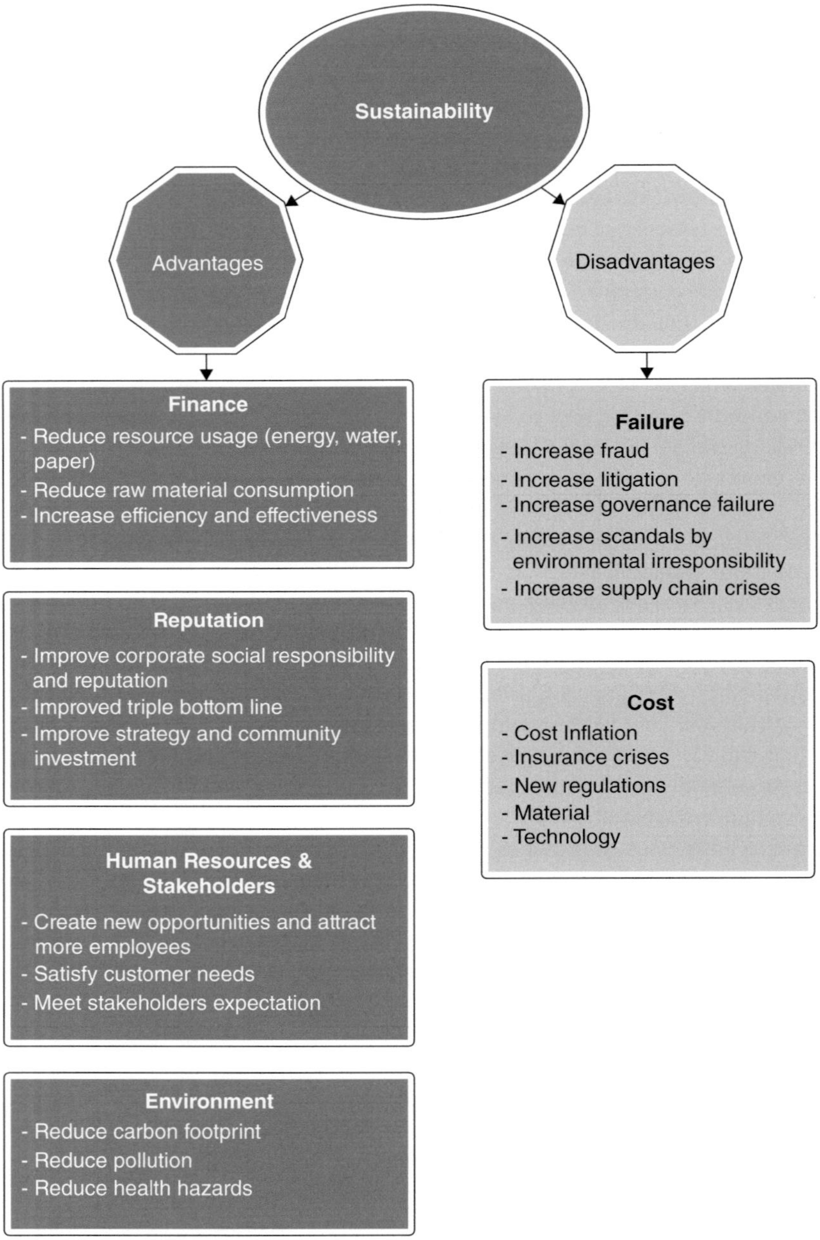


**Fig. 2** Advantages and disadvantages of Adopting Sustainability – prepared by author

Reducing Resource Usage

Effective operation is a criterion of all businesses because it helps reduce cost and maximize advantages. There are many ways to address this such as reducing the amount of resources that the company uses every day, including energy, water, and paper. In fact, implementing this practice provides a twofold benefit because businesses not only save monthly costs, but also achieve regarding credibility regarding the preservation of natural resources.

Decreasing Raw Material Consumption

Preserving materials is another approach to help business increase profits, reduce costs and attain sustainability because it minimizes input expenses, which lead to manufacturing cost reduction, preventing resource depletion and boosting revenue. Allwood [2004 quoted in 10] states the application of technology to achieve sustainability manufacturing contributes to the effective use of materials without carbon emissions, non-renewable materials usage or waste generation.

Increasing Efficiency

Efficiency is an on-going process in all businesses to minimize expenses and maximize revenue. Efficiency increase is implemented in all aspects of business management, including resource usage efficiency and business process optimization. It is one factor that encourages businesses to adopt green information technology such as virtualization and cloud computing to fully utilize energy and reduce their infrastructure investment. Green technologies help organizations reduce operational costs, monitor and control their carbon footprint, thereby contributing to their overall resource and waste management strategy [11].

### 2.1.2 Reputation

In regards to reputation, sustainability factors include enhanced corporate social responsibility, business uniqueness, increased triple bottom line, improved strategy and investment.

Enhanced Corporate Social Responsibility

Jamali [27] suggested four dimensions of corporate social responsibility; these entail responsibility in terms of the economy, ethics, compliance with legal requirements, and philanthropy. Enhanced corporate social responsibility means increasing its responsibilities to the communities where it operates and its employees' welfare in addition to its commitments to stakeholders. The adoption of sustainability helps businesses improve their social responsibility status in different areas such as workplace environment, customer and supply chain relationship, and corporate and community connection.

Improved Triple Bottom Line

Triple bottom line refers to the framework which businesses use for their decision-making process based on economic (profit), environmental (planet) and social (people) considerations [55]. There is always a compromise between these three factors when businesses are improving their sustainability; therefore, improved triple bottom line differentiates businesses among other corporations in the field.

Improve Community Investment

When businesses become more sustainable and acquire social sustainability status, they gain more trust not only from customers, but also from investors who commit capital to the business. Since the triple bottom line is becoming a criterion for evaluating a company, investors usually prefer corporations with greater sustainability achievements [46].

### 2.1.3 Human Resources and Stakeholders

Attract New Employees

Job seekers tend to gravitate towards companies that have a clear human resource strategy and constructive culture, in addition to a good sustainability status [17]. Eweje [17] gives an example of a job offer refusal from an applicant because the sustainability report of the company was not as good as the applicant's current one. Hence, businesses will be able to attract more talent if sustainability issues are on their priorities list.

Meet stakeholder expectations

Stakeholders are people who are interested in the performance of business and include owners, employees, suppliers, customers; moreover, they have a significant influence on businesses' activities. Nowadays, the business environment is changing from one where there is single decision-making by the CEO to one where stakeholders have an input because the efficiency of business operations depends on inter-related parties. Thus, adopting sustainability practices help business build trust in stakeholders and attract more support from them

### 2.1.4 Environmental

Reduce Carbon Footprint

The amount of carbon dioxide emissions produced during the life cycle of a product is known as the carbon footprint [Wiedmann and Minx 2008, quoted in 33]. These emissions cause the greenhouse gas effect and contribute to the increase of global warming. Sustainability practices will help reduce the amount of carbon emissions; therefore, it will help prevent the impacts of climate change.

Reduce Pollution

Pollution such as air, water, noise, and light pollution causes numerous environmental disasters related to global warming; such disasters include species depletion, to name just one harmful consequence [2]. With the adoption of sustainability practices, business will help lessen the impacts of climate change caused by pollution issues.

Reduce Health Hazards

In addition to other environmental controls, employees' health and safety are of the utmost significance because they concern not only immediate working forces, but also other involved parties. Consequently, risk prevention and management such as safety procedures for the work environment should be carefully considered to prevent catastrophes, especially in hazardous industries such as construction and mining [30].

## 2.2 *Sustainability Disadvantages*

Despite the advantages of sustainability, several disadvantages and possible risks need to be carefully considered and prepared for if sustainability is to be successfully adopted by business. Two of these are cost and failure. Cost is a major consideration when businesses adopt sustainability because it directly relates to the triple bottom line while, failures are the potential incidents which might prevent sustainability from being implemented, or the violation of established sustainability standards.

### 2.2.1 Cost of Inflation

The inflation of cost is considered to be a significant hindrance to sustainability adoption in business due to the additional expenses incurred during the design and manufacture of products and services. For example, the eco-labelled standards and requirements usually mean an increase in product price that must be borne by the customer; however, not all customers are willing to pay more [59]. Therefore, complying with environmental standards such as eco-labelling is a trend and requirement of industries, but it presents a barrier to sustainability implementation.

### 2.2.2 New Regulation

Sustainability legislation is the effort of governments to enforce sustainability implementation and compliance; but this might be a hindrance to businesses when they have to invest many resources for the purpose of compliance. As a consequence, they may lose their competitive advantages in the market if there are no equivalent regulations for the less developed countries [Jaffe, Peterson, Portney, and Stavins, 1995, quoted in 9]. However, businesses could achieve a higher competitive position in a long run when they have to invest, innovate and optimize their processes to comply with the legislation, which in turn helps them become more sustainable and developed [Porter and Linde, 1995, quoted in 9].

### 2.2.3 Fraud Allegation

According to Petraşcu and Tieanu [37], fraud has several meanings such as the engagement in illegal activities with the intention of cheating, or the action of providing wrong or misleading information in order to gain advantages. Given the number of fraud cases discovered recently in large, well-known companies, fraud is considered a significant threat to businesses in regards to cost, reputation, and litigation. The greater the number of environmental regulations when governments try to enforce and commit to sustainability strategy, the more probability there is of fraudulent behaviour.

### 2.2.4 Increase Supply Chain Crises

Sustainability is not just an internal process; it involves many parties including suppliers and partners. Seuring [43] defines sustainable supply chain management as the cooperation among partners in the supply chain so that the three pillars of sustainable development – the economy, environment and society – are maintained. It also involves the control of materials and the stream of information. The purpose is, on the one hand, to satisfy social and environmental standards; on the other hand, the economic factor must still be considered in order to meet competitive expectations. It is a much more complex process to adopt sustainability in supply chain management; hence, crises can arise there is poor management.

### 2.2.5 Technology and Material Costs

Many authors [15] mention that investment costs incurred by a sustainability strategy can be a hindrance. It is always a dilemma in sustainability projects because adopting sustainability usually requires a change in the business processes and application of technology. These projects always involve investment in equipment, training and operations; therefore, before they can derive any advantages from sustainable improvement, businesses have to spend money on improvement projects.

### 2.2.6 Sustainability and the Future

From the time sustainability was first mentioned in 1987 and then raised as a common concern of all countries, many conferences have organized and agreements signed; however, there has not been much progress and governments are still discussing the best actions and promises for a low carbon future. Pandey [36] pointed out that to date there has been no successful means of tackling the issue of climate change; additionally, no substantial progress in climate change cooperation has been made, indicating that developing countries are postponing their decisions and not making commitments before they see practical actions from committed countries, especially industrialized ones. Adams [1] states the holistic and striking concept of sustainability is respectful and helps to bring everyone together, but it does not assist in achieving any goals.

It will take time for any country to achieve sustainability; therefore, governments need to start implementing their commitment. As reported in the IUCN Renowned Thinkers Meeting, most governments have a department that is in charge of environment issues. To some extent, organizations and businesses have also responded to sustainability challenges, but more effort needs to be made to balance the three pillars of sustainable development [1]. Sustainability is achieved when economic, social and environmental factors contribute equally to businesses' success; nonetheless, the economic factor is the most overarching element and overwhelms the social and environmental ones. The tipping point is that environmental issues are becoming extremely urgent, which requires changes to improve this situation. It is imperative to transform these factors sustainably so that all elements are equally considered and measured to ensure the harmony between people and planet.

## 3 Research Method and Research Question

This study aims to examine the following research question "To what extent are Vietnamese tertiary students aware of the advantages and disadvantages associated with the adoption of sustainability by business?" *T*o address the research question the deductive approach was followed in the analysis of secondary data, which begins with the researcher reviewing the literature relevant to sustainability, and inductive approach is used when examining the data from questionnaires distributed to respondents. The survey research was employed in this research to obtain information from respondents from Vietnam. Survey research is commonly linked with the deductive research approach and is widely used in business and management research. The sample is adequate, the findings are considered as representative of the whole population without the need for a massive amount of data collected from the entire population [41]. Surveys are appropriate for explanatory or descriptive research that entails massive data from participants and answers questions such as what, where, who, how many, and how much. Standardized questions are mostly appropriate in this approach since it allows all respondents to interpret them in the same way. Furthermore, surveys allow researchers to collect quantitative data that can then be analysed using statistical software. Patterns and relationships among variables can be easily extracted using these kinds of software [39, 41].

As survey research is one of many different options, the advantages and disadvantages should be considered. Some advantages of survey research are: data can be collected from a large number of people; there is quicker access to results and data analysis with the help of statistical software; there is a variety of means of distribution, especially for hard-to-reach respondents; and it can collect a broad range of data pertaining to values, points of view and behaviours [58]. For example, surveys can be distributed online so that respondents can complete them using their devices such as mobile phones, tablets or computers, and can be done at the respondents' convenience. Moreover, this enables researchers to obtain responses easily and quickly. However, there are also many disadvantages to survey research such as the time required to distribute questionnaires and receive responses, the limit to the number of questions, data quality, and the level of knowledge of participants regarding the subject of the research interest. For instance, data quality can be affected when respondents lose interest in answering questions or they provide incorrect responses. In some types of surveys such as semi-structured or in-depth interviews, the knowledge of researchers is very important when preparing and organizing questions and the context of the interview [41, 58]. Moreover, the wording of questions must be clear and unambiguous so that respondents can interpret them correctly.

As stated previously, there are many ways to conduct survey research and each one has its own pros and cons. This study sought to discover the extent to which tertiary students in Vietnam are aware of sustainability and its associated advantages and disadvantages; therefore, the web-based survey was considered the most appropriate way to distribute the questionnaires to as many people as possible. Smyth et al. [45] stated the other advantages of the web-based survey as opposed to the postal survey are the convenience, cost saving for researchers and a wider reach to a large population. The web-based survey has numerous advantages over other methods such as the paper survey which requires a lot of effort in distributing and collecting responses. Additionally, the cost and time associated with the web-based survey are reduced because researchers can prepare the survey and upload it to a social network such as Facebook or Twitter, and send it through email or instant messaging services like Viber, WhatsApp.

The survey questionnaire was created using Qualtrics software provided by Curtin University. It contained 20 questions, 5 of which were intended to elicit demographic information from participants; the remaining questions concerned sustainability awareness. The online tool, Qualtrics, gives researchers informative and quick reporting. For example, survey results can be used directly on IBM's Statistical Package for the Social Sciences (SPSS – Version 24) statistical software to analyse data and obtain trending information in real time without additional data input as in a paper-based survey. Furthermore, the software provides an interactive and intuitive interface with ease of access from different types of devices, which assists the researcher to reach a greater number of participants. The survey link was distributed via many channels including Facebook, email, WhatsApp, and Viber, over a 1-month period.

## 4 Results

Data were collected using Qualtrics research software, which generated an online survey web link. The link was distributed to participants via several communication channels such as Facebook social network, email, Viber and Google's hangout instant messaging services. The survey has three sections comprising demographic information, technology usage patterns, and awareness. The survey also explained the concepts of sustainability so that respondents could understand clearly the aspects being discussed.

The number of collected responses was 281; however, some questions were unanswered which made the data invalid and inappropriate for analysis. Therefore, data collected from 128 out of 281 respondents were used for analysis by SPSS statistical software.

Demographic information involves respondents' age, gender and education level. Figure 3 shows the percentage of male and female respondents, which are 58% and 42% respectively. Table 1 illustrates the age distribution of participants. The majority of respondents are aged between 21 and 24 years, which accounts for 47.51%; the percentage of respondents in the age ranges of 25–30, 18–20 and 31–35 are 21.07%, 17.24% and 8.05%, respectively. The remaining age groups account for a very small percentage of less than 2%.

**Fig. 3** Gender distribution – prepared by author

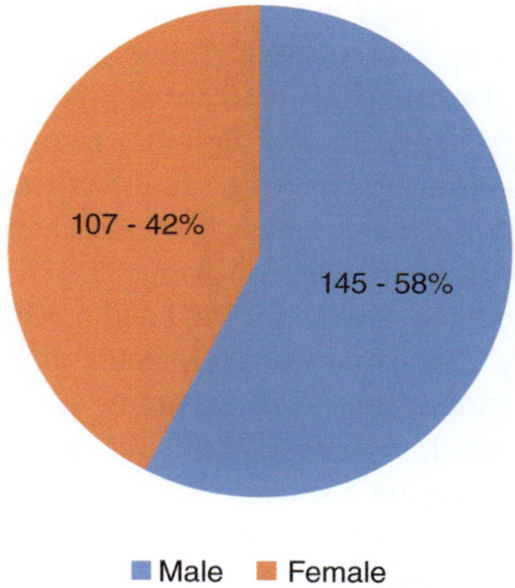

**Table 1** Age of respondents – prepared by author

| Age range | Percentage | Responses |
|---|---|---|
| 17 years and under | 1.53% | 4 |
| 18–20 | 17.24% | 45 |
| 21–24 | 47.51% | 124 |
| 25–30 | 21.07% | 55 |
| 31–35 | 8.05% | 21 |
| 36–40 | 1.53% | 4 |
| 41–45 | 1.15% | 3 |
| 46–50 | 0.38% | 1 |
| 51–55 | 0.38% | 1 |
| 56–60 | 0.38% | 1 |
| 61–65 | 0.77% | 2 |
| Over 65 | 0.00% | 0 |

**Table 2** Education level of participants – prepared by author

| Student status | Percentage | Responses |
|---|---|---|
| Primary Education | 8.30% | 19 |
| Higher Education/Pre-University | 23.14% | 53 |
| Professional Certificate | 8.30% | 19 |
| Diploma | 9.17% | 21 |
| Advanced/Higher/Graduate Diploma | 3.93% | 9 |
| Bachelor's Degree | 29.26% | 67 |
| Post Graduate Diploma | 5.24% | 12 |
| Master's Degree | 6.99% | 16 |
| PhD | 0.87% | 2 |
| Others | 4.80% | 11 |

Table 2 illustrates the education level of participants. From the table, it is easily seen that bachelor degree has the highest percentage of 29.26%. Next is 23.14% of participants who are at higher-secondary or pre-university levels. 9.17% of participants have a diploma; respondents with a primary education or professional certificate have the same percentage of 8.30%. The number of participants in the post-graduate and "Others" categories are the lowest at 3.93% for advanced or higher graduate diploma, 5.24% for post-graduate diploma, 6.99% for master's degree, 0.87% for PhD and 4.80% for the "Others" category.

The survey also collected information about the pattern of technology usage of participants such as hours spent on computer, Internet, social networking and email. Figures 4, 5, 6 and 7 show these statistics.

Figure 4 shows a significant percentage (42.17%) of respondents use a computer for up to 5 h a day; 31.30% spend 5–10 h; 16.09% spend 10–20 h, 9.13% spend less than an hour; and only 1.3% spend over 20 h on a computer. Regarding

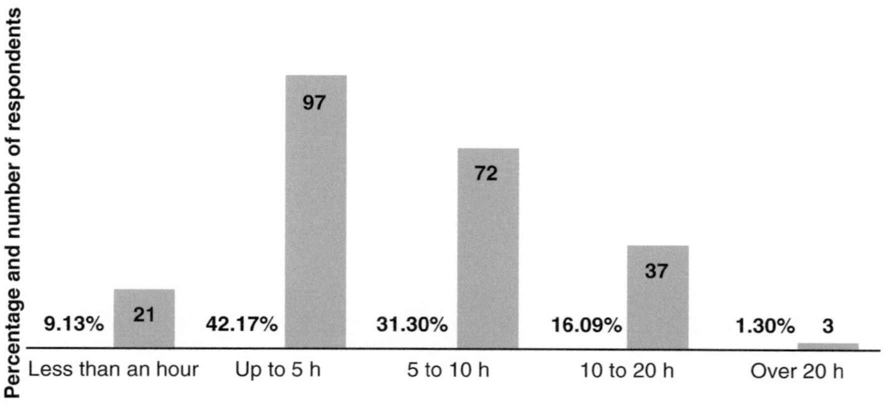


**Fig. 4** Hours spent on computer – prepared by author

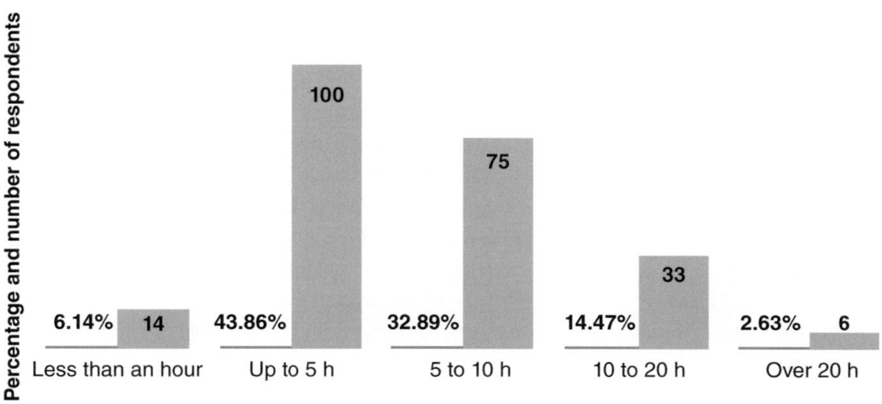


**Fig. 5** Hours spent on Internet – prepared by author

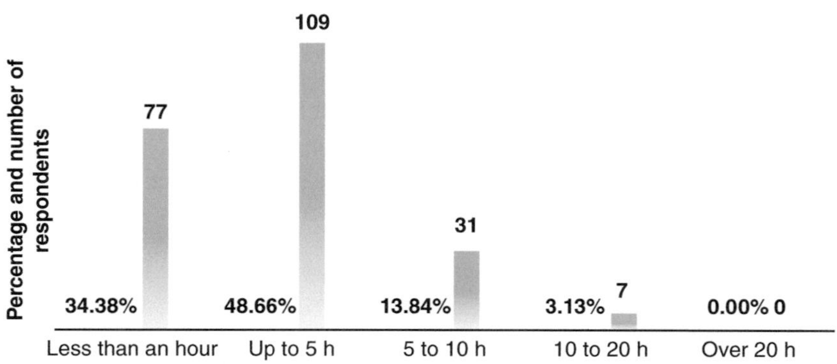


**Fig. 6** Hours spent on social networking – prepared by author

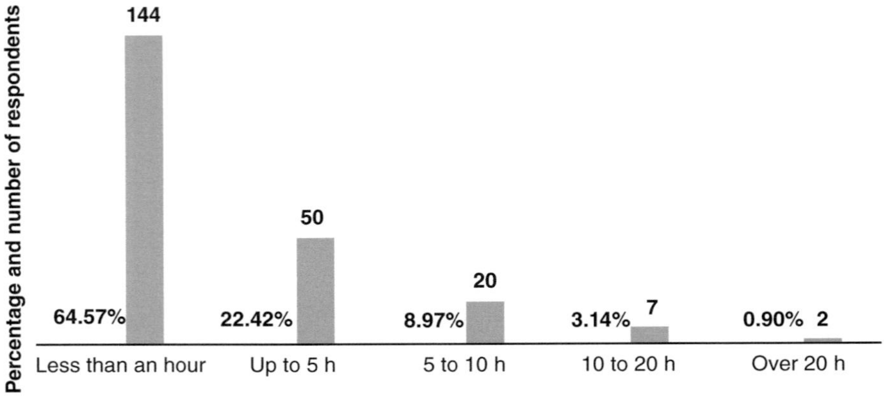


**Fig. 7** Hours spent for email on internet – prepared by author

the number of hours that respondents spend accessing the internet Fig. 5, survey results shows 43.85% spend up to 5 h a day; 32.89% spend 5–10 h a day; 14.47% spend 10–20 h a day; 6.14% spend less than an hour a day; and only 2.63% spend over 20 h a day. Survey statistics show that nearly half of the respondents access social networking sites for up to 5 h; 34.38% spend less than an hour a day; 13.84% spend 5–10 h a day; 3.13% spend 10–20 h a day and none of them spend more than 20 h a day on social networking site (see Fig. 6). More than half of the respondents use the internet for email each day; around one-fifth of them access email up to 5 h a day; 8.97% spend 5–10 h a day; 3.14% spend 10–20 h a day; and a very small portion of them use email more than 20 h a day (see Fig. 7). For those questions where participants could choose more than one answer, results (see Fig. 8) indicate that respondents mostly the use internet for work (73.78%), studying (77.33%) and communicating (72.44% for chatting and 74.67% for email). This result confirms the importance of the internet to Vietnamese people. Additionally, Fig. 9 shows that most respondents use laptops (79.82%) and smartphones (82.96%) to access the internet.

One can see that the majority (71.50%) of respondents were first introduced to the concept of sustainability through the internet; while school in general, tertiary education institutions, books, news media and magazines have much lower percentages with 28.04%, 22.43%, 20.56%, 20.09% and 19.16% respectively (Fig. 10).

Nearly half of respondents (41.86%) read the company's sustainability report before they buy new equipment; while 19.53% of respondents do not read reports and 29.77% is not sure whether they read the report (see Fig. 11).

The percentages of respondents who change their technology equipment every 1, 2 and 3 years are 13.55%, 17.76% and 14.02% respectively; while 5.61% of participants change devices every 6 months (see Fig. 12).

A large number of people changes devices for reasons of speed, which is 61.73%. Functionality accounts for 46.43% of the reason why people change devices and 40.31% want to keep up with technology (see Fig. 13).

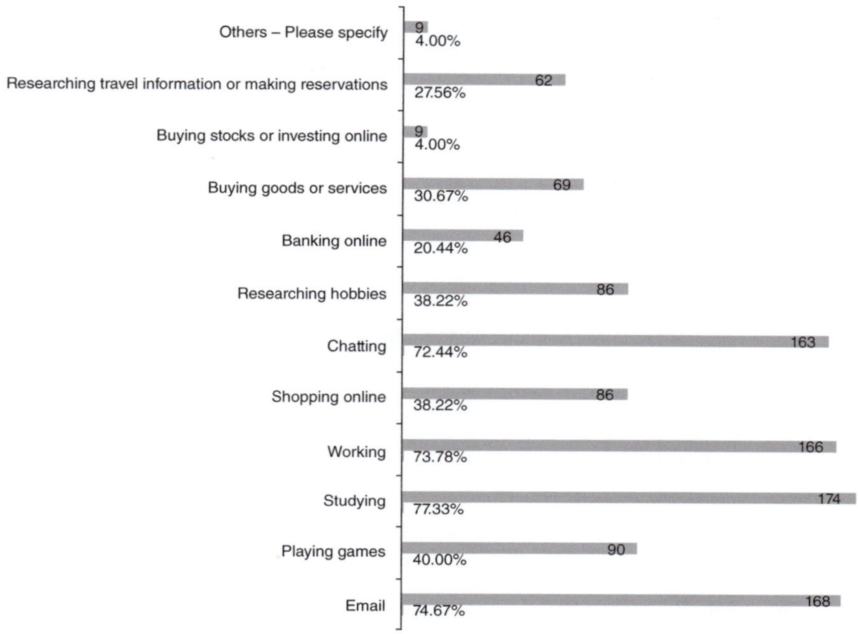


**Fig. 8** Purpose of Internet access by respondents – prepared by author

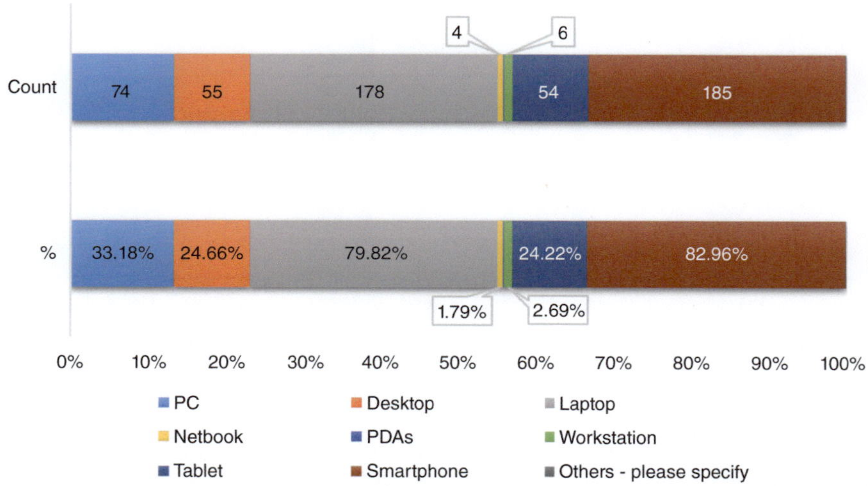


**Fig. 9** Devices respondents use to access Internet – prepared by author

When being asked whether the regular changing of devices damages the environment, 45.18% of respondents answered in the affirmative, while 14.72% disagreed. 38.07% were unsure (see Fig. 14).

Finally, over 50% of respondents believe that education and information from the internet can change the mind-set of users and product designers regarding

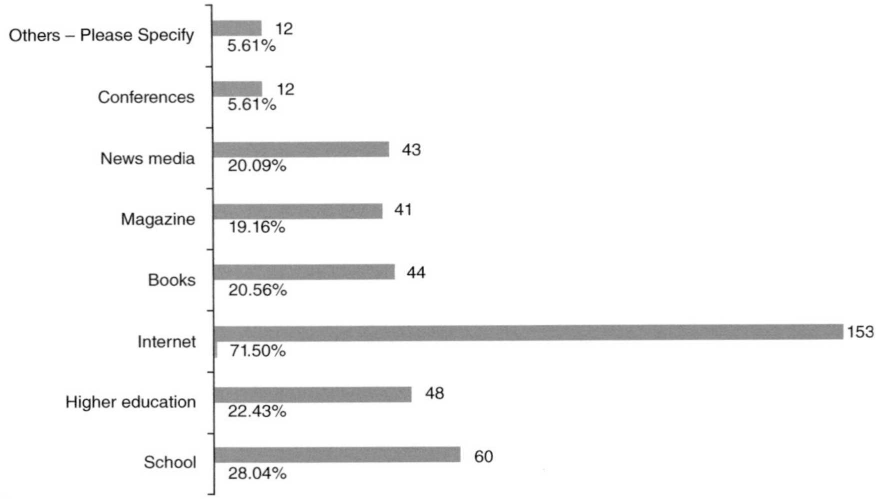


**Fig. 10** Means of first introduction to sustainability – prepared by author

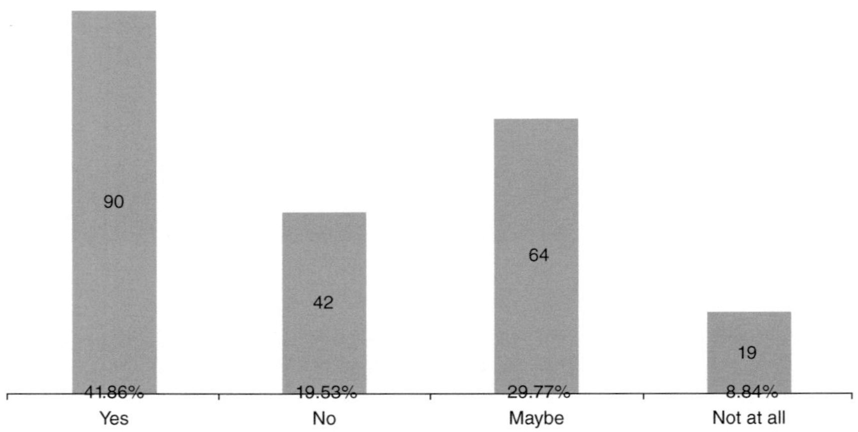


**Fig. 11** Reading sustainability report before buying new equipment – prepared by author

sustainability. The percentage for social networking, training, awareness, TV and workshop are 52.06%, 45.88%, 35.05%, 32.99% and 30.41% respectively, while only 2.06% think other means such as events and exhibitions can really change mind-sets of users and designers (see Fig. 15).

As mentioned above, data were collected from respondents in Vietnam aged from 17 to 65 years, especially in the higher education level, in order to study their awareness of sustainability. The study demonstrates the current level of understanding about the advantages and disadvantages of sustainability.

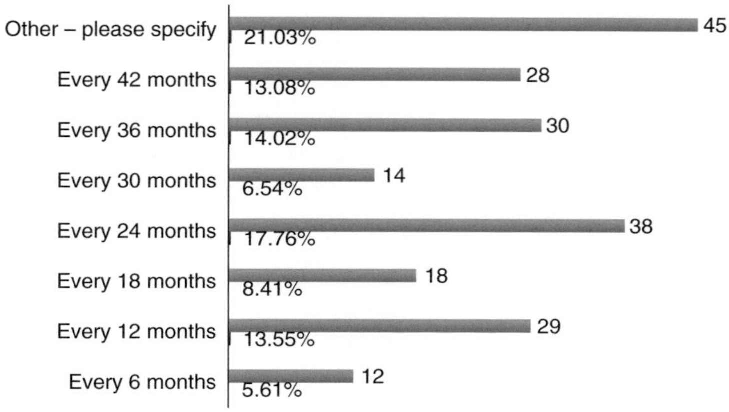


**Fig. 12** Device change frequency – prepared by author

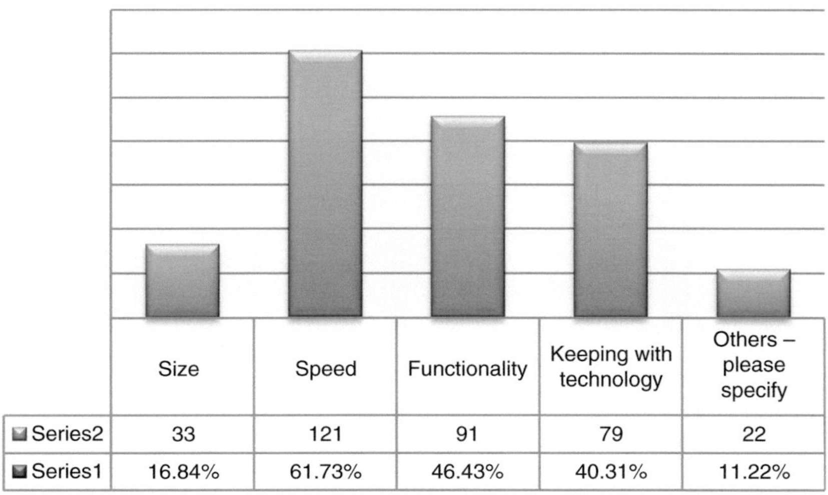


**Fig. 13** Reasons for changing device

Using IBM SPSS Statistics software version 24, the reliability of surveyed data was verified based on the Cronbach's Alpha test which is the test of internal consistency of collected data and required for Likert scale survey [21]. Twenty-four items about advantages of sustainability in business are calculated, result is shown in Table 3 below:

Similarly, 13 items about disadvantages of sustainability in business are calculated and its result is shown in Table 4.

According to George and Mallery [19], the result of the Cronbach's Alpha test are categorised as excellent (>0.9), good (>0.8), acceptable (>0.7), questionable (>0.6), poor (0.5), and unacceptable (<0.5). Based on the calculated value of

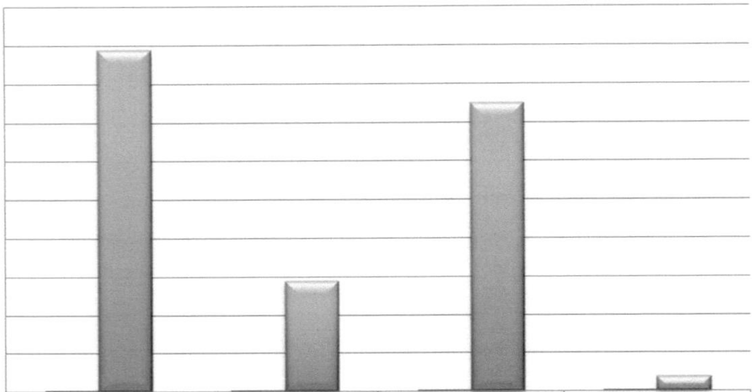

| | Yes | No | Maybe | Not at all |
|---|---|---|---|---|
| Series1 | 45.18% | 14.72% | 38.07% | 2.03% |
| Series2 | 89 | 29 | 75 | 4 |

**Fig. 14** Changing device can damage environment

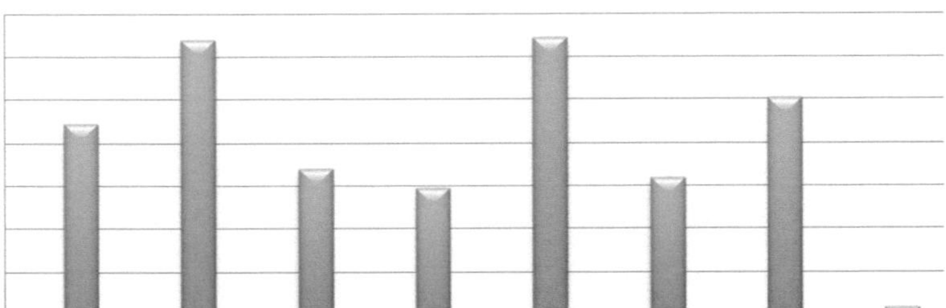

| | Training | Education | Awareness | Workshop | Internet | T.V. | Social Networking | Others – please specify |
|---|---|---|---|---|---|---|---|---|
| Series1 | 45.88% | 65.98% | 35.05% | 30.41% | 66.49% | 32.99% | 52.06% | 2.06% |
| Series2 | 89 | 128 | 68 | 59 | 129 | 64 | 101 | 4 |

**Fig. 15** Changing mind-set of users and designers

**Table 3** Cronbach's Alpha (advantages) – prepared by author

| Reliability statistics | |
|---|---|
| Cronbach's Alpha | No. of items |
| 0.914 | 24 |

**Table 4** Cronbach's Alpha (disadvantages) – prepared by author

| Reliability statistics | |
|---|---|
| Cronbach's Alpha | N of items |
| 0.930 | 13 |

0.914 for positive items and 0.930 for negative items, the consistency and reliability of the survey data was confirmed. Therefore, it was suitable for analysis to answer the research questions. Factor analysis is a statistical technique usually used in social science studies to deal with a complex data set by finding the interrelationship of a set of variables, which are used to measure a common thing. Factor analysis reduces the number of variables by extracting common information from a large number of items to create fewer representative ones [23, 54].

Exploratory Factor Analysis (EFA) and Confirmatory Factor Analysis (CFA) are the two main categories of factor analysis. EFA, as its name implies, is experimental in nature and helps researchers to build a new theory from initial constructs by exploring both the advantages and disadvantages. CFA, on the other hand, verifies the proposed hypothesis which is based on prior understanding of different factors [54]. In general, factor analysis reduces the number of observed factors to unobserved ones which explain the correlations among them to validate a proposition [29]. Principal Component Analysis is another method, and resembles exploratory factor analysis. The primary difference between the two is that exploratory factor analysis assumes that only a smaller number of unobserved variables have originated from observed ones, while principal component analysis is merely calculating a smaller group of variables to represent all observed variables [24].

The first step in factor analysis is to calculate the values using KMO and Bartlett's Test of Sphericity. The results of these tests are shown in Table 5:

Field [2000 quoted in 31] states that the Kaiser-Meyer-Olkin Measure of sampling adequacy (KMO-Test) can be used to check the adequacy of the sample size. If the value is greater than 0.5, the sample size is considered adequate. KMO Test shown in the table above is 0.878 which is greater than 0.5; this means that the sample size can be used for analysis. Additionally, Bartlett's Test of Sphericity verifies the relationships of variances in a correlation matrix. When the value of significance is less than 0.05, the data are considered relevant in factor analysis [25, Field 2000 quoted in 31]. The Sig. value of the Bartlett's Test above is 0.000. The results of these two tests indicate the appropriateness of factor analysis for this study.

Secondly, a table of communalities for advantages is evaluated which shows the relationship between a variable and all other variables. This means it expresses the percentage to which a variance can be explained by factors. Only items with a value above 0.5 should be considered for further analysis [7]. Table 6 shows 12 item with values less than 0.5; hence, they should be removed from the list.

Thirdly, an explanation of total variance is shown in Table 7. This table shows all the variances are displayed as 13 factors with 3 factors having Eigenvalues

**Table 5** KMO and Bartlett's Test for advantages – prepared by author

| | | |
|---|---|---|
| Kaiser-Meyer-Olkin Measure of Sampling Adequacy | | 0.878 |
| Bartlett's Test of Sphericity | Approx. Chi-Square | 1123.820 |
| | df | 78 |
| | Sig. | 0.000 |

**Table 6** Communalities for advantages

| Communalities | | |
|---|---|---|
| | Initial | Extraction |
| Reduce energy and water usage | 1.000 | 0.554 |
| Reduce paper usage | 1.000 | 0.493 |
| Reduce consumption of raw materials | 1.000 | 0.458 |
| Increase efficiency | 1.000 | 0.442 |
| Increase cost-effectiveness | 1.000 | 0.285 |
| Enhance reputation | 1.000 | 0.457 |
| Improve corporate social responsibility | 1.000 | 0.464 |
| Reduce risk management | 1.000 | 0.197 |
| Differentiate businesses | 1.000 | 0.433 |
| Increase green strategy | 1.000 | 0.425 |
| Improve community investments | 1.000 | 0.469 |
| Improve human rights | 1.000 | 0.474 |
| Improve social responsibility investing | 1.000 | 0.524 |
| Increase triple bottom line – People, Planet and Profit | 1.000 | 0.465 |
| Increase productivity | 1.000 | 0.587 |
| Attract quality employees | 1.000 | 0.579 |
| Satisfy customer needs | 1.000 | 0.670 |
| Meet stakeholder expectations | 1.000 | 0.579 |
| Attract new opportunities | 1.000 | 0.578 |
| Create new jobs | 1.000 | 0.525 |
| Reduce carbon footprint | 1.000 | 0.729 |
| Reduce emissions | 1.000 | 0.821 |
| Reduce pollution | 1.000 | 0.759 |
| Reduce health hazards | 1.000 | 0.643 |

Extraction method: Principal Component Analysis.

greater than 1.0, which indicates that they are usable. Furthermore, these three factors are considered explainable for advantages because they add up to a cumulative percentage of 72.381% of all factors [24].

The fourth step is the evaluation of the rotated component matrix as shown in Table 8.

Table 8 above shows the rotated component matrix for the study results. The first factor represents six variances with values from 0.606 to 0.751; this implies the advantages of sustainability are an increase in business productivity and efficiency, and a reduction of resource consumption; therefore, the triple bottom line is increased and quality employees are attracted to the company. The second factor comprises four elements with values ranging from 0.658 to 0.896. They explain

**Table 7** Variance for advantages

| Total variance explained | | | | | | | | | |
|---|---|---|---|---|---|---|---|---|---|
| Component | Initial eigenvalues | | | Extraction sums of squared loadings | | | Rotation sums of squared loadings | | |
| | Total | % of variance | Cumulative % | Total | % of variance | Cumulative % | Total | % of variance | Cumulative % |
| 1 | 7.095 | 54.578 | 54.578 | 7.095 | 54.578 | 54.578 | 3.812 | 29.327 | 29.327 |
| 2 | 1.290 | 9.922 | 64.500 | 1.290 | 9.922 | 64.500 | 3.079 | 23.683 | 53.010 |
| 3 | 1.025 | 7.881 | 72.381 | 1.025 | 7.881 | 72.381 | 2.518 | 19.371 | 72.381 |

Extraction method: Principal Component Analysis.

**Table 8** Rotated component matrix for advantages – prepared by author

| Rotated component matrix[a] | | | |
|---|---|---|---|
| | Component | | |
| | 1 | 2 | 3 |
| Increase productivity | 0.751 | | 0.138 |
| Reduce energy and water usage | 0.690 | 0.271 | |
| Attract quality employees | 0.688 | | 0.325 |
| Reduce paper usage | 0.670 | 0.206 | |
| Increase triple bottom line – People, Planet and Profit | 0.632 | 0.214 | 0.139 |
| Increase efficiency | 0.606 | 0.192 | 0.196 |
| Satisfy customer needs | 0.581 | | 0.572 |
| Reduce consumption of raw materials | 0.551 | 0.390 | |
| Improve social responsibility investing | 0.528 | 0.361 | 0.339 |
| Increase cost-effectiveness | 0.457 | | 0.274 |
| Reduce emissions | | 0.896 | 0.104 |
| Reduce pollution | 0.277 | 0.823 | |
| Reduce carbon footprint | | 0.808 | 0.273 |
| Reduce health hazards | 0.415 | 0.658 | 0.195 |
| Improve corporate social responsibility | 0.336 | 0.579 | 0.128 |
| Increase green strategy | | 0.502 | 0.414 |
| Meet stakeholder expectations | 0.338 | −0.102 | 0.674 |
| Attract new opportunities | 0.148 | 0.333 | 0.667 |
| Differentiate businesses | | 0.119 | 0.644 |
| Create new jobs | | 0.331 | 0.639 |
| Improve community investments | 0.163 | 0.248 | 0.617 |
| Enhance reputation | 0.320 | 0.326 | 0.498 |
| Improve human rights | 0.454 | 0.218 | 0.470 |
| Reduce risk management | | | 0.433 |

Extraction method: Principal Component Analysis. Rotation method: Varimax with Kaiser Normalization.

[a]Rotation converged in seven iterations.

**Table 9** Three advantages factors – prepared by author

| | Factor name |
|---|---|
| Increase productivity | Efficiency and Resource Usage Reduction |
| Reduce energy and water usage | |
| Attract quality employees | |
| Reduce paper usage | |
| Increase triple bottom line – People, Planet and Profit | |
| Increase efficiency | |
| Reduce emissions | 4Rs |
| Reduce pollution | |
| Reduce carbon footprint | |
| Reduce health hazards | |
| Meet stakeholder expectations | Fulfil investor prospects and gain differentiated achievements |
| Attract new opportunities | |
| Differentiate businesses | |
| Create new jobs | |
| Improve community investments | |

the four advantages of reduction in emissions, pollution, and carbon footprint and health hazards. Finally, the last factor of variances with values from 0.617 to 0.674 show that a company's values are meeting stakeholder expectations, improving community investment, creating new jobs and opportunities, and therefore differentiate businesses from others when sustainability is adopted.

The final step is to label the correlated variances. Table 9 below shows the grouping of factors according to their values:

From Table 9, it can be said that the adoption of sustainability can help businesses to improve their efficiency and reduce resource consumption; reduce four impacts on the environment such as emissions, pollution, carbon footprint and health hazards; and meet investor expectations and differentiate business. Figure 16 shows the advantages of sustainability.

Similar to advantages, variances for sustainable disadvantages are analysed and tested based on KMO and Bartlett's Test. Table 10 shows KMO and Bartlett's Test for disadvantages.

Table 10 shows that the value for the KMO Test is 0.878 and significance of Bartlett's Test is 0.000 which indicates the reliability and relevance of factor analysis.

Secondly, the table of commonalities for disadvantages is evaluated which shows the relationship between a variable and all other variables. This means that it expresses the percentage to which a variance can be explained by factors. Only items with a value above 0.5 should be considered for further analysis [7]. Communalities for disadvantages are different from those for advantages in that no variance is below 0.5 (Table 11).

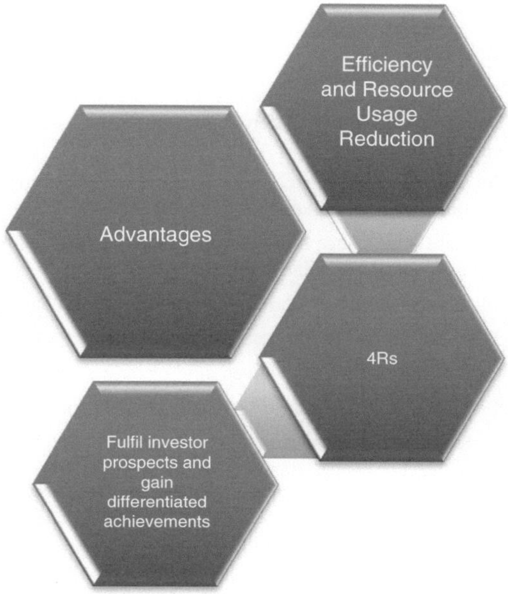


**Fig. 16** Advantages of sustainability

**Table 10** KMO and Bartlett's Test for disadvantages

| | | |
|---|---|---|
| Kaiser-Meyer-Olkin Measure of Sampling Adequacy | | 0.878 |
| Bartlett's Test of Sphericity | Approx. Chi-Square | 1123.820 |
| | df | 78 |
| | Sig. | 0.000 |

Thirdly, the total variance values are shown in Table 12. This table shows all the variances are displayed as 13 factors with 3 factors having Eigenvalues greater than 1.0, which indicates that they are usable. However, only two factors are calculated for the disadvantages of sustainability. These two factors are considered explainable for advantages because they add up to a cumulative percentage of 64.5% of all factors [24].

The fourth step is to evaluate the rotated component matrix as shown in Table 13.

Table 13 demonstrates two factors for disadvantages with variance values ranging from 0.608 to 0.888 and from 0.668 to 0.839, respectively. Factor one involves an increase in supply chain crises due to suppliers' environmental problems, an increase of marketing failure and scandals as a result of perceived environmental irresponsibility, increase of fraud, increase of transaction failures due to environmental liabilities, increase of security and systems failures caused by environmental problems, increase of interest rates and cost inflation and

**Table 11** Commonalities for disadvantages

| Commonalities | Initial | Extraction |
|---|---|---|
| Increase fraud (including environmental actions and environmental compliance failures) | 1.000 | 0.543 |
| Increase litigation and compliance breaches (including environmental actions and environmental compliance failures) | 1.000 | 0.658 |
| Increase governance failure (including environmental actions and environmental compliance failures) | 1.000 | 0.554 |
| Increase transaction failure due to environmental liabilities | 1.000 | 0.623 |
| Increase security and systems failures caused by environmental problems | 1.000 | 0.596 |
| Increase scandals by perceived environmental irresponsibility | 1.000 | 0.679 |
| Increase marketing failure by perceived environmental irresponsibility | 1.000 | 0.725 |
| Increase supply chain crises due to suppliers' environmental problems | 1.000 | 0.793 |
| Inflate costs | 1.000 | 0.657 |
| Increase interest rates | 1.000 | 0.521 |
| Increase insurance crises due to environmental disasters | 1.000 | 0.650 |
| Increase number of new regulations including environmental | 1.000 | 0.705 |
| Increase competition for and cost of raw materials | 1.000 | 0.681 |

Extraction method: Principal Component Analysis.

**Table 12** Variance for disadvantages

Total variance explained

| Component | Initial eigenvalues | | | Extraction sums of squared loadings | | | Rotation sums of squared loadings | | |
|---|---|---|---|---|---|---|---|---|---|
| | Total | % of variance | Cumulative % | Total | % of variance | Cumulative % | Total | % of variance | Cumulative % |
| 1 | 7.095 | 54.578 | 54.578 | 7.095 | 54.578 | 54.578 | 5.374 | 41.338 | 41.338 |
| 2 | 1.290 | 9.922 | 64.500 | 1.290 | 9.922 | 64.500 | 3.011 | 23.162 | 64.500 |

Extraction method: Principal Component Analysis.

increase of insurance crises due to environmental disasters. Factor two represents variances such as an increased number of new regulations including environmental ones, increased competition for and cost of raw materials, and increased litigation and compliance breaches.

Finally, the Grouping and Naming of correlated variances produces the Table 14:

The disadvantages of sustainability can be summarized as two factors (see (Fig. 17). The first one is the increase of scandals, crises and failures due to environmental problems, irresponsibility and disasters. The second one concerns a greater number of regulations and compliance breaches.

**Table 13** Rotated component matrix for disadvantages – prepared by author

| Rotated component matrix[a] | | |
|---|---|---|
| | Component | |
| | 1 | 2 |
| Increase supply chain crises due to suppliers' environmental problems | 0.888 | |
| Increase marketing failure by perceived environmental irresponsibility | 0.830 | 0.190 |
| Increase scandals by perceived environmental irresponsibility | 0.766 | 0.305 |
| Inflate costs | 0.745 | 0.318 |
| Increase fraud (including environmental actions and environmental compliance failures) | 0.706 | 0.211 |
| Increase interest rates | 0.697 | 0.187 |
| Increase security and systems failures caused by environmental problems | 0.653 | 0.412 |
| Increase insurance crises due to environmental disasters | 0.639 | 0.493 |
| Increase transaction failure due to environmental liabilities | 0.608 | 0.503 |
| Increase governance failure (including environmental actions and environmental compliance failures) | 0.548 | 0.504 |
| Increase number of new regulations including environmental | | 0.839 |
| Increase competition for and cost of raw materials | 0.230 | 0.793 |
| Increase litigation and compliance breaches (including environmental actions and environmental compliance failures) | 0.460 | 0.668 |

Extraction method: Principal Component Analysis. Rotation method: Varimax with Kaiser Normalization.
[a]Rotation converged in three iterations.

**Table 14** Two disadvantages factors – prepared by author

| | Factor name |
|---|---|
| Increase supply chain crises due to suppliers' environmental problems | Increase of scandals, crises and failures due to environmental problems, irresponsibility and disasters |
| Increase marketing failure by perceived environmental irresponsibility | |
| Increase scandals by perceived environmental irresponsibility | |
| Inflate costs | |
| Increase fraud (including environmental actions and environmental compliance failures) | |
| Increase interest rates | |
| Increase security and systems failures caused by environmental problems | |
| Increase insurance crises due to environmental disasters | |
| Increase transaction failure due to environmental liabilities | |
| Increase number of new regulations including environmental | More regulations and compliance breaches |
| Increase competition for and cost of raw materials | |
| Increase litigation and compliance breaches (including environmental actions and environmental compliance failures) | |

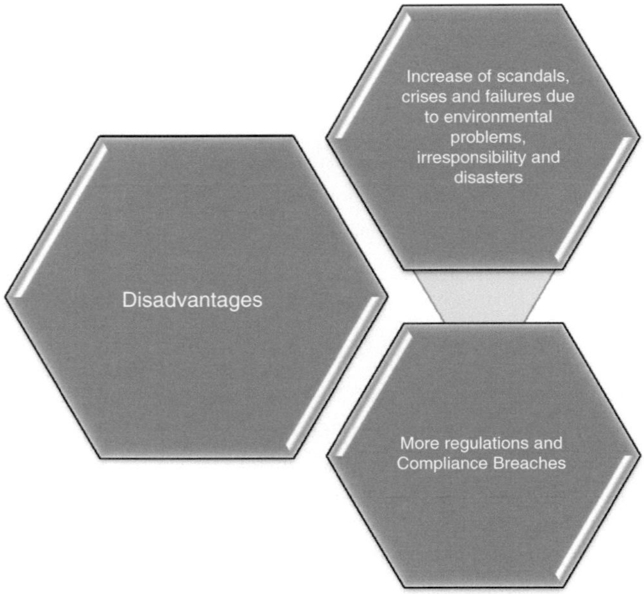


**Fig. 17** Disadvantages of sustainability

## 5 Discussion

Climate change, resource scarcity, natural disasters such as flooding, hurricanes, air and water pollution and so on are the consequences of anthropogenic exploitation of resources needed for development. These issues have negatively impacted on the three pillars of sustainable development, which concern people, profit and planet. Sustainability is an urgent issue in the current world; therefore, many countries are cooperating to minimise the negative impacts of economic growth and development. Vietnam is a developing country in Southeast Asia that has several concerns and activities regarding sustainable development. Several organizations are implementing sustainability strategies in order to comply with legislation; nonetheless, there are still many environmental issues due to the unawareness and irresponsibility of individuals and business. Hence, this research study aims at discovering the level of awareness of students in the higher education sector, and their perception of the advantages and disadvantages of sustainability adoption. This is important because they will be part of the business environment in the future, and will be responsible for implementing sustainability. An online survey was distributed to acquire responses from higher education students in Vietnam. Data analysis shows respondents' perspective on sustainability and elaborates on the advantages and disadvantages to businesses that consider adopting sustainability. This information will help educators to prepare a suitable curriculum to raise student awareness and provide business with an insight into the advantages and disadvantages of being sustainable.

Throughout the findings of this study were thoroughly discussed. The factor analysis sections provide the answer to the research question: *to what extent are Vietnamese tertiary students aware of the advantages and disadvantages associated with the adoption of sustainability by business?* Data analysis of survey results was used to generate three factors for the advantages of sustainability awareness in Higher Education concerning business environment, such as efficiency and resource reduction, the 4Rs and fulfilment of investor prospects and achieving business differentiation. In addition, the two factors for disadvantages of sustainability awareness generated were the increase in the number of scandals, crises and failures due to environmental problems, irresponsibility and disasters; and an increase in the number of regulations and compliance breaches. Factors for advantages will raise awareness of sustainability in tertiary education students, who will be the future workers, business owners and so on. With an awareness of sustainability, students can capitalize on their knowledge and embrace sustainability best practices to increase triple bottom line performance. At the same time, understanding the factors generated for disadvantages helps prevent the negative impacts of sustainability adoption in future business environments. Finally, this research was limited to examine the Vietnam's Higher Education Sector in relation to sustainability; further research will be carried out to examine other sectors in the future.

## 6 Conclusion

This chapter aims to examine the Vietnam's Higher Education Sector awareness of sustainability. An online survey was conducted to examine this perspective. The online survey outcomes have shown the extent of students' awareness of sustainability and their understanding of the advantages and disadvantages of sustainability in business. Result shows that resource reduction, performance increase and meeting stakeholder expectations are the advantages, but there are still risks of failure because of compliance breaches and increased chance of scandals and crises occurring when implementing sustainability. Finally, based on the study outcomes, it suggests initiatives that educators can take to improve students' awareness and prepare them for the future labour market. Lastly, these approaches should be adopted by businesses to ensure their successful implementation of sustainability.

## References

1. Adams WM (2006) The future of sustainability – rethinking environment and development in the twenty-first century (trans: Geography Do). Report of the IUCN Renowned Thinkers Meeting. The World Conservation Union. http://cmsdata.iucn.org/downloads/iucn_future_of_sustanability.pdf. Accessed 1 June 17

2. Amran A, Periasamy V, Zulkafli AH (2014) Determinants of climate change disclosure by developed and emerging countries in Asia Pacific. Sustain Dev 22(3):188–204. doi:10.1002/sd.539
3. Annesi-Maesano I (2016) United Nations climate change conferences: COP21 a lost opportunity for asthma and allergies and preparing for COP22. J Allergy Clin Immun 138 (1):57–58. doi:10.1016/j.jaci.2016.02.043
4. Anonymous (2012) Implementation of sustainable development – National Report. Paper presented at the United Nations Conference on Sustainable Development (RIO+20). https://sustainabledevelopment.un.org/index.php?page=view&type=400&nr=852&menu=1515. Accessed 1 June 2017
5. Ayson S (2011) Chapter 13 sustainable management education: a graduate skills approach. In: Gabriel Eweje, Martin Perry (eds) Business and sustainability: concepts, strategies and changes (critical studies on corporate responsibility, governance and sustainability, Vol 3), Emerald Group Publishing Limited. pp 293–308. doi:10.1108/S2043-9059(2011)0000003021
6. Barbosa L, Veloso L (2014) Consumption, domestic life and sustainability in Brazil. J Clean Prod 63:166–172. doi:10.1016/j.jclepro.2013.09.020
7. Chetty P, Datt S (2015) Interpretation of factor analysis using SPSS. https://www.project-guru.in/publications/interpretation-of-factor-analysis-using-spss/. Accessed 10 Nov 2016
8. De Koning JIJC, Crul MRM, Wever R, Brezet JC (2015) Sustainable consumption in Vietnam: an explorative study among the urban middle class. Int J Consumer Stud 39(6): 608–618. doi:10.1111/ijcs.12235
9. Debnath SC (2015) Environmental regulations become restriction or a cause for innovation – a case study of Toyota Prius and Nissan Leaf. Procedia Soc Behav Sci 195:324–333. doi:10.1016/j.sbspro.2015.06.360
10. Dehghanian F (2012) Supply chain sustainability and raw material management: concepts and processes. IGI Global, Hershey
11. Deng Q, Ji S (2015) Organizational green IT adoption: concept and evidence. Sustainability 7(12):16737–16755
12. Doane D, MacGillivray A (2001) Economic sustainability – the business of staying business. New Economics Foundation. http://isites.harvard.edu/fs/docs/icb.topic140232.files/RD_economic_sustain.pdf. Accessed 1 June 2017
13. Drexhage J, Murphy D (2010) Sustainable development: from Brundtland to Rio 2012. Paper presented at the High Level Panel on Global Sustainability. http://www.surdurulebilirkalkinma.gov.tr/wpcontent/uploads/2016/06/Background_on_Sustainable_Development.pdf. Accessed 1 June 2017
14. Duc Luong N (2015) A critical review on energy efficiency and conservation policies and programs in Vietnam. Renew Sustain Energy Rev 52:623–634. doi:10.1016/j.rser.2015.07.161
15. Engert S, Rauter R, Baumgartner RJ (2016) Exploring the integration of corporate sustainability into strategic management: a literature review. J Clean Prod 112 (Part 4):2833–2850. doi:10.1016/j.jclepro.2015.08.031
16. Evans A (2010) Globalization and scarcity – multilateralism for a world with limits. Center on International Cooperation. http://cic.nyu.edu/sites/default/files/evans_multilateral_scarcity.pdf. Accessed date 1 June 2017
17. Eweje G (2011) Business & sustainability: concepts, strategies and changes. Critical studies on corporate responsibility, governance and sustainability, vol 3. Emerald Group Publishing Limited, Bradford
18. Fahian Anisul H, Mark S, Marta Z (2014) Social sustainability in developing country suppliers – an exploratory study in the ready made garments industry of Bangladesh. Int J Oper Prod Manag 34(5):610–638. doi:10.1108/IJOPM-10-2012-0467
19. George D, Mallery P (2003) SPSS for windows step by step: a simple guide and reference, vol 11. 4th edn. Allyn & Bacon http://wps.ablongman.com/wps/media/objects/385/394732/george4answers.pdf. Accessed 1 June 2017
20. Giap BN (2015) Education for environment: a useful concept. J Viet Env 6(3):188–190. doi:10.13141/jve.vol6.no3.pp188-190

21. Gliem JA, Gliem RR (2003) Calculating, interpreting, and reporting Cronbach's alpha reliability coefficient for Likert-type scales. In: Paper presented at the Research to Practice Conference in Adult, Continuing, and Community Education, Columbus, OH
22. Goodland R (2002) Sustainability: human, social, economic and environmental. Encycl Global Environ Change 5:481–491
23. Gorsuch RL (2013) Factor analysis. 2nd edn. Taylor and Francis, Hoboken
24. IBM Exploratory Factor Analysis with categorical variables (n.d.). http://www-01.ibm.com/support/docview.wss?uid=swg21477550. Accessed 1 June 2017
25. IBM (2016) KMO and Bartlett's test. http://www.ibm.com/support/knowledgecenter/SSLVMB_20.0.0/com.ibm.spss.statistics.cs/fac_telco_kmo_01.htm. Accessed 09 Oct 2016
26. Isaksson RB, Garvare R, Johnson M (2015) The crippled bottom line – measuring and managing sustainability. Int J Prod Perform Manag 64(3):334–355. doi:10.1108/IJPPM-09-2014-0139
27. Jamali, D (2014) CSR in developing countries through an institutional lens. In: G Eweje (ed) Corporate social responsibility and sustainability. Emerging trends in developing economies (critical studies on corporate responsibility, governance and sustainability, vol 8), Emerald Group Publishing Limited, pp 21–44. doi:10.1108/S2043-905920140000008005
28. Kennedy D (2007) Sustainability. Science 315(5812):573–573
29. Khan AA, Iliescu DD, Shah RJ, Sneath CE, Hutchinson AA (2015) Principal component and factor analysis to study variations in the aging lumbar spine. IEEE J Biomed Health Inform 19(2):745–751. doi:10.1109/JBHI.2014.2328433
30. Kim K, Cho Y, Zhang S (2016) Integrating work sequences and temporary structures into safety planning: automated scaffolding-related safety hazard identification and prevention in BIM. Autom Constr 70:128–142. doi:10.1016/j.autcon.2016.06.012
31. Kootstra GJ (2004) Exploratory factor analysis: theory and application. http://www.let.rug.nl/~nerbonne/teach/rema-stats-meth-seminar/Factor-Analysis-Kootstra-04.PDF. Accessed 1 June 2017. p 2006
32. Kuhlman T, Farrington J (2010) What is sustainability? Sustainability 2(11):3436
33. Lazarevic D, Martin M (2016) Life cycle assessments, carbon footprints and carbon visions: analysing environmental systems analyses of transportation biofuels in Sweden. J Clean Prod 137:249–257. doi:10.1016/j.jclepro.2016.07.075
34. Le TPL, Trinh TA (2016) Encouraging public transport use to reduce traffic congestion and air pollutant: a case study of Ho Chi Minh City, Vietnam. Procedia Eng 142:236–243. doi:10.1016/j.proeng.2016.02.037
35. Nguyen T (2015) Mainstreaming sustainable development at national level Vietnam experiences, opportunities and challenges. http://www.unosd.org/content/documents/1397DAY%201%20Vietnam.pdf. Accessed 1 June 2017
36. Pandey CL (2014) The limits of climate change agreements: from past to present. Int J Clim Change Strategies Manag 6(4):376–390. doi:10.1108/IJCCSM-03-2013-0026
37. Petraşcu D, Tieanu A (2014) The role of internal audit in fraud prevention and detection. Procedia Econ Finance 16:489–497. doi:10.1016/S2212-5671(14)00829-6
38. Robertson M (2014) Sustainability principles and practice. Taylor & Francis Group, London
39. Robson C (2011) Real world research: a resource for users of social research methods in applied settings, 3rd edn. Wiley, Chichester
40. Sarkis J, Dhavale D (2015) Supplier selection for sustainable operations: a triple-bottom-line approach using a Bayesian framework. Int J Prod Econ 166:177
41. Saunders M, Lewis P, Thornhill A (2012) Research methods for business students, 6th edn. Pearson, Harlow
42. Schüz M (2012) Sustainable corporate responsibility – the foundation of successful business in the new millennium. Cent Eur Bus Rev 1(2):7–15
43. Seuring S (2008) Sustainability and supply chain management – an introduction to the special issue. J Clean Prod 16(15):1545–1551
44. Slaper T, Hall T (2011) The triple bottom line: what is it and how does it work? Indiana Bus Rev 86(1):4–8

45. Smyth JD, Dillman DA, Christian LM, O'Neill AC (2010) Using the internet to survey small towns and communities: limitations and possibilities in the early 21st century. ABS 53(9): 1423–1448. doi:10.1177/0002764210361695
46. Steurer R, Langer M, Konrad A, Martinuzzi A (2005) Corporations, stakeholders and sustainable development I: a theoretical exploration of business–society relations. J Bus Eth 61(3): 263–281. doi:10.1007/s10551-005-7054-0
47. Stoycheva S The history of sustainability: a critical assessment of metrics and their changes. Paper presented at the 4th International Conference on Management, Leadership and Governance, ICMLG 2016 Russia Academic Conferences International Limited, pp 475–480
48. Tambo E, Duo-quan W, Zhou X-N (2016) Tackling air pollution and extreme climate changes in China: implementing the Paris climate change agreement. Environ Int. 95:152–156. doi:10.1016/j.envint.2016.04.010
49. Todorov V, Marinova D (2011) Modelling sustainability. Mathematics and Computers in Simul 81(7):1397–1408
50. United Nation (2015) World population prospects (trans: Affairs DoEaS). https://esa.un.org/unpd/wpp/. Accessed 1 June 2017
51. United Nation (2016) First steps to a safer future: introducing the United Nations framework convention on climate change. http://unfccc.int/essential_background/convention/items/6036.php. Accessed 05 Sep 2016
52. United Nation (2016) Kyoto protocol. http://unfccc.int/kyoto_protocol/items/2830.php. Accessed 06 Sep 2016
53. Velde EVd, Vermeir W, Corten F (2005) Corporate social responsibility and financial performance. Corp Gov Int J Bus Soc 5(3):129–138. doi:10.1108/14720700510604760
54. Williams B, Onsman A, Brown T (2010) Exploratory factor analysis: a five-step guide for novices. J Emerg Prim Health Care 8(3):1–13
55. Winkler R, Deller S, Marcouiller D (2015) Recreational housing and community development: a triple bottom line approach. Growth Change 46(3):481–500. doi:10.1111/grow.12100
56. World Bank (2016) World Bank country and lending groups. https://datahelpdesk.worldbank.org/knowledgebase/articles/906519-world-bank-country-and-lending-groups. Accessed 05 Sep 2016
57. World Trade Organization (2016) https://www.wto.org/english/tratop_e/devel_e/d1who_e.htm. Accessed 05 Sep 2016
58. Wyse SE (2012) Advantages and disadvantages of surveys. https://www.snapsurveys.com/blog/advantagesdisadvantages-surveys/. Accessed 1 June 2017
59. Yenipazarli A (2015) The economics of eco-labeling: standards, costs and prices. Int J Prod Econ 170(Part A):275–286. doi:10.1016/j.ijpe.2015.09.032

# The Opportunities and Risks of Sustainability Awareness in Sri Lankan Organizations

**Pathirana Hemage Gihan Dhanushka**

**Abstract** In recent years, the issue of global warming has been a hot topic because of rapid changes in the environment. The researchers in various sectors have identified that the main causes of these rapid environmental changes are our human activities. In comparison with historical civilization extinction, that clearly indicates the upcoming danger towards the modern society. Therefore, it is highly important to pay more attention to the limitations of our natural resources. The extinction of habitats and ancient human civilizations has gradually begun years ago. Many natural species have been completely eradicated, due to unhealthy ecological systems. The modern world is making a consolidated effort to protect earth's natural resources. They emphasize that the extraction of natural resources must be limited because it is vital that resources exist for future generations. The formula now being used to tackle sustainability issues consists of Economy, Environment and Society. Those three elements are interrelated. The change in one element does have a direct impact on other elements. Therefore, the world understands the fact that the economic element is the one that can have a greater influence on society and the environment too. Therefore, it is highly important to address the sustainability elements in the economic sector because failure to deliver sustainable outcomes to the environment will result in severe negative consequences. A better understanding of Sri Lanka's sustainability awareness at the organizational level is very timely, based on two factors: firstly, the country is one of the developing countries, and secondly, it is engaged in international business. Therefore, the awareness of Sri Lanka's sustainability from the perspective of organizations is factually analyzed in order to determine whether there is an awareness of the opportunities and risks associated with sustainability. To understand the level of awareness, the researcher has designed and distributed an online survey. As a result, 178 completed survey answers were able to be considered from a total of 312 participants.

P.H.G. Dhanushka (✉)
School of Information Systems, Curtin University, Perth, Western Australia, Australia
e-mail: scrumy10@yahoo.com

 
T. Issa et al. (eds.), *Sustainability, Green IT and Education Strategies in the Twenty-first Century*, Green Energy and Technology, DOI 10.1007/978-3-319-57070-9_8

# 1 Introduction

Tackling global warming and ensuring ecological balance are highly important factors in protecting and maintaining the balance of our Earth. The current global warming has to a point where natural species are being gradually made extinct. If we do not take any initiatives to stop the current bad cycle, the extinction of human being is not a farfetched proposition because all living creatures and natural elements are interdependent. Upon disturbance of any element, rapid fall-out will result. Therefore, it is much wiser to pay more attention to nature and natural resources whilst we go about our daily activities. In the contemporary world, the daily activities have become economic activities, such as becoming a stakeholder of a money-based society. For example, the majority of people are engaged in some sort of business activity, either online or not online. It is well understood that business activities can cause disruptions to the natural world. To gain further understanding of current level of awareness of sustainability in Sri Lanka, as this chapter aims to identify the level of awareness of sustainability and what are the underlying gaps within Sri Lanka. To evaluate the process, countries' organizations have been chosen because the organizations are the combinations of mass education network integrate on achieving many tasks in the economy. Therefore it is the most appropriate venue to evaluate responsible individuals' level of sustainability awareness and make changes to the future through their actions. Following, this chapter will be gradually introduce background of research, research methods and questions used to gather information from its workforce, results from the gathered data sets, discussion of new significant if any arises through the survey results, what are the limitations in the country in understanding sustainability component and finally the conclusion of the whole research.

# 2 What Is Sustainability?

"Sustainable" can be considered as the ability to maintain something over time [15]. The core concept of sustainability is not to extract resources from elements in Mother Nature excessively [20]. Sustainability puts more emphasis on saving natural elements for future generations and also to maintain all-balanced ecological systems on the planet. Further, the main aim is to achieve sustainability by assisting nature to maintain balance in its ecological system, whilst obtaining economic benefits from human activities. The main reason that we have to become more sustainable is because of the scarcity of resources [15]. The literature further discusses the complications that the world can face in future, upon failure to maintain a balance in nature. Therefore, Organizations like the United Nations have stepped up to address any issues causing imbalance in nature and to minimize the impact from human activities. The notion of sustainability is mainly focused on environment and its health, as well as society because social activities do have a greater impact on nature, with the ever-increasing human population and the human need factor travelling parallel with increase of human growth. The final element is the economy because due to the rapid growth of human activities, human needs reach extreme levels in economic terms, due to variance in ethical and social influences. In the end, this everlasting cycle of activities within the constant parent variable of environment, is directly influenced by those mentioned elements [11].

If insufficient sustainability measures to address environmental issues from human activities, there will be severe consequences to face for everyone on the planet at present and in the future. Further, Huesemann [16] states that over-use of natural resources will causes imbalance in nature and its activities. Alternatives must be considered so that maximum value can be achieved with minimal consumption of resources in an effort to be sustainable. Currently, global warming is the most concerning issue in the world and is a result of the misuse and over-use of natural resources. Devastating consequences like pollution, over population and the increase in ill-health have all been documented and proven factually in the literature [2].

Therefore, it is highly important to focus on the subject of sustainability because after taking into consideration the scarcity of natural resources and changes in nature. Hence, this research was carried out by targeting awareness and sustainability education. The main reason for focusing on this domain was, as mentioned in the above paragraphs, because organizations play a key role in society. The economic element is one of the three pillars of the sustainability domain (Fig. 1).

Figure 1 shows that economic variables and values change as social attributes change due to individual behavioral changes. As such, for education, cultural, political and religious reasons, society does value changes constantly over time. Economic values and activities parallel the changes in society. Since the environment is the parent of all those activities, these social activities and economic activities are clearly reflecting in the environment. If the environment at risk, it indicates the alarming situations in the social and economic elements. Further this research, under the literature review, the importance of awareness in sustainability, human activities and environment will be understood. Upon failure, what are the

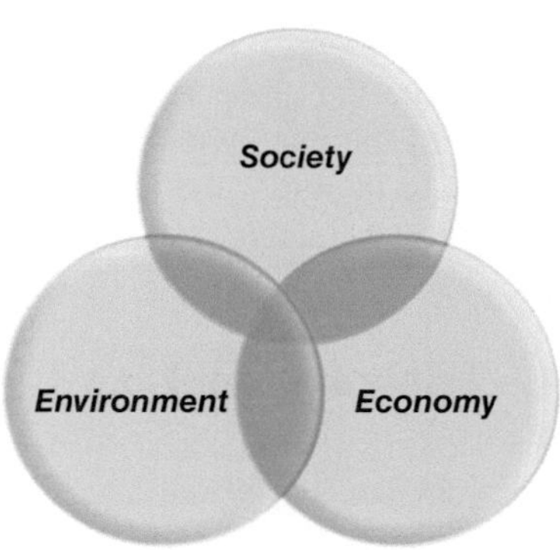


**Fig. 1** Three-Pillar structure of sustainability [37]. Prepared by Gihan Dhanushka

consequences from the point of view of sustainability that the world can face in the near future? To address all those issues further down in this chapter they will be discussed in detail.

## 3 History of Sustainability and Consequences of Being Unsustainable

According to McLamb [24], sustainability concerns arose soon after the Industrial Revolution in the nineteenth century in Europe. According to the literature, there was much environmental pollution that caused severe disruptions to Mother Nature. McLamb [24] discusses the evolution of sustainability and the human growth factor, and also the impact on nature. In 1700, in Great Britain they began to replace human labor with machinery to bring a new level of efficiency to human activities. Although it was the beginning of a new and more efficient world, there were many negative complications attached to this revolutionary wave such as pollution and over-use of natural resources, to name just two. Vitousek et al. [35] confirmed that the Industrial Revolution caused severe damage to nature. The excessive misuse and abuse of natural resources has led to a massive depletion of natural resources, the extinction of species, and pollution. McLamb [24] discusses in detail the population movement that occurred as a result of industrialization and how it affected nature. For example, after manual labor force was replaced by machinery, there was a massive migration to the cities which then became over-populated. As a negative effect of the move, the authorities had to clear a great deal of land to meet the demand for living space as people moved from one location to another. Extensive man-made constructions caused severe deterioration in the natural elements as such water, soil and natural habitats. Over time, the negative effects have accrued, resulting in disruptions to the natural environment, evident in things such as weather changes and global warming. Back in 1800, alarmingly pollution, health issues and other natural disasters were occurring all around the world. The researches had found the core of this issue was imbalance of population and over-use of fossil fuels. Thereafter, the amount of toxic elements released into the natural environment has created massive pollution across the globe [14]. Further, Deutz et al. [7] have demonstrated the complexity of the issue with a broadly analyzed extinct civilizations table. Deutz et al. wanted to highlight the fact that the extinction of civilization can result from the failure to take actions to restore the balance in environment (Table 1).

## 4 Triple Bottom Line

According to Slaper and Hall [33], the triple bottom line is an accounting framework, which includes three dimension of performance: Social, Economic and Environmental. This triple bottom line is commonly known as the three Ps: People,

**Table 1** Environmental factors in civilization collapse [7] – prepared by Gihan Dhanushka

| | |
|---|---|
| Sumerian Civilization | Poorly designed irrigation system in farming industry caused rising water table to come closer to the surface, leaving a salt concentration after the water evaporated. Then the salty environment affected the crops and gradually as the crops were destroyed so was the civilization. |
| Garamantian Empire (Sahara) 500 BC–300 AD | Underground irrigation canals being used to feed up to 3000 miles with massive irrigation network and over-exploitation of ground water caused collapse of this civilization. |
| Roman Civilization | Due to invention of iron, to build the iron they needed more coal; to produce coal, they cleared massive stretches of land, leaving many lands abandoned. With time, those abandoned lands became breathing grounds for malaria and eventually this weakened the nation and resulted in defeat. |
| Mayan Civilization | Loss of soil fertility, massive deforestation and civil clashes caused this civilization to become extinct. |
| Ankor Wat, Khmer Empire, Cambodia | Deforestation caused constant flooding and washed-off sand filled their canal water systems. |

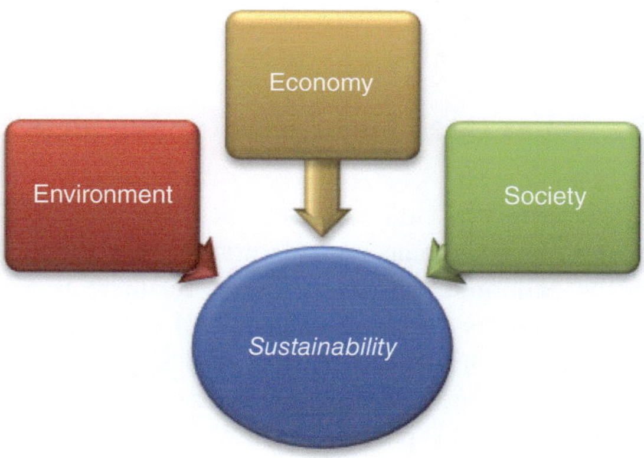


**Fig. 2** Sustainability development structure – prepared by Gihan Dhanushka

Planet and Profit. These three components of sustainability development were first drafted and discussed at the 2005 World Summit on Social Development [34]. The main objectives of the agenda/effort of drafting a blueprint for sustainability development are to eradicate poverty, change unsustainable patterns of production, and conserve natural resources. The current sustainability development framework has been drafted to address any remerging sustainability issues regarding these three elements. Due to various social and economic factors, human nature and behavioral changes do have direct impacts on nature. Sustainability is the median space in the sustainable development structure as shown in Fig. 2.

According to Slaper and Hall [33], sustainability cannot be measured by dollar value because it depends on other variables such as economic and social elements.

**Table 2** Triple Bottom Line Index elements – prepared by Gihan Dhanushka

| Economic Measures | Environmental Measures |
| --- | --- |
| • Personnel income<br>• Job growth<br>• Level of sustainability awareness<br>• Percentage of firms in each sector | • Sulphur dioxide concentration<br>• Selected priority pollutants<br>• Electricity consumption<br>• Hazardous waste management |
| Social Measures | |
| • Unemployment rate<br>• Female labor participation rate<br>• Average commute time<br>• Median house hold income | |

But there are many elements that can be taken into consideration when deciding the success of each and every element as shown in Table 2.

By considering the elements outlined above (Table 2), the level of sustainability can be measured at a certain level. When choosing vendors or goods and services, if those elements are considered by a third party, this would help to encourage sustainable awareness within society and also to push the society more towards sustainable measures to be followed.

## 5 What Does This Mean for Business/Education?

Ecosystems are vital to society because without the basic requirements of water, air and land, humans cannot survive. Hence, without human beings, there would be no economic activities. Therefore, there is a very strong and interdependent relationship between ecosystem and economy [28]. Philipsen [28] further discusses the relationship between the ecosystem and negative impacts caused by human to nature and the importance of having a more sustainable way of living by adopting solar-powered installations such as solar-powered stoves and heaters. He further stresses the importance of having political involvement in achieving sustainability because political involvement is essential to regulate new sustainable measures introduced into society. The same work also discusses the level of use of natural resources and the time frame of the beginning of the Industrial Revolution. It is vital that we become sustainable as the natural environment depends on humans to maintain its ecological systems. Therefore, many of the world's nations have identified the key underlying issue regarding this subject, and that it that environmental changes are largely due to economic factors. According to Simon and Eitzen [32], businesses and business deals in the western world have been conducted using corrupt and illegal business practices. For example, using prostitutes as a reward for corrupted politicians and other fringe benefits establishing more toxic and harmful production in Vietnam by American military personnel [32].

# 6 Impact on Environment of Failure to Follow Sustainable Standards

Although it improved society in many ways, the Industrial Revolution depleted natural resources and created pollution. The Revolution was not restricted to one industry in terms of economic development. It ranged from the farming industry to every industry in the world. For example, people were using machinery to generate mass production. Toxic fertilizers and deforestation to grow crops gradually caused natural disasters such as droughts and the imbalance of resources below and above the ground. Further, according to Morgan [25], other severe consequences that the world has to face is drought and dust storms. Mass deforestation further exposes top layer of soil to sun, directly causing severe imbalance in the soil. The result is the creation of desert and drought due to the evaporation of the underground water system. Farmer [10] further explains that pollution, the use of chemical fertilizers and exploitation of the land has left unhealthy soil for the current generation.

# 7 Concept of Sustainability in Business

The Green concept should have been adopted by every industry in the world according to policies and procedures established by the United Nations Environment Protection Committee. Every organization in any industry must adhere to corporation rules and regulations. If there is failure to do so, there will be tougher penalties to enforce the sustainability measures [1]. Hence, the economic environment has become much more sustainable and controlled by sustainability regulations because every industry has now begun to follow sustainability practices. Suppliers of goods and services are also required to follow sustainable ways of conducting business. Upon failure to behave within sustainability measures, they will become isolated in the economic world and profits will decrease.

## *7.1 Education*

The importance of awareness of sustainability and its effects on the environment is well understood. It has been clearly proven, that since the Industrial Revolution, level of environmental pollution has become alarmingly high due to the continued consumption or destruction of natural resources [23]. The present generation is inheriting an unbalanced ecological system. The greater challenge, which is global warming, is yet to be overcome [36]. Previously it has been clearly demonstrated that with the growth in the world's population, there has emerged a gradual awareness of the importance of sustainability. This generation has understood the core issue and has undertaken to address the correct segment in the Triple Bottom Line framework, particularly the economic element.

## 8 Sustainability in Developing Countries

After the Second World War, the rise of separate nations has established new modes of living. After massive destructions and environmental and societal collapses, most nations had to rebuild their societies and environment from scratch [27]. That is where industry plays a role in improving the standard of living and make future developments faster and more efficient. According to Nielsen [27] there are ways to measure the living standards in every country using the criteria of the quality and level of education, sustainability, and income level, for instance. Further, Nielsen [27] draws a line between developing and developed countries based on various economic factors and other living standards. The political and ethical values are major obstacles that developing countries need to overcome [22]. Lipton [22] further points out that the greatest enemy of development is the politics of a nation. The corrupt politics maintain a distinct divide between rich and poor. Therefore, the poverty level in ta country will remain the same forever, unless a more viable solution is sought to address social problems such as poverty. According to a United Nations (2014)'s article, a country's standard of living has been measured by its per capita gross national income. Accordingly, the countries are further categorized under four levels of income: high-income, upper middle income, lower middle income and lower income.

## 9 Sustainability in Sri Lanka

Sri Lanka is geographically located in the Indian Ocean and is part of South Asia. It was formally known as Ceylon. According to an Economics [8] article, the country's population was 22.6 million in 2014, a significant increase of 108% over 50 years. Further, population growth was 0.93% from 2012 to 2015. Sri Lanka is a developing nation according to United Nations' criteria. The country's main industry is agriculture [6]. Further, the literature discusses the importance of introducing new technology to farmers so that primary industry production is easier and more efficient. By maintaining a well-established communication link with farmers using ICT, this could help to eradicate wastage and other issues resulting from the wastage. Wastage has severe negative impacts on society in terms of cost and time. The positive side of the effort will be having adequate production to meet demand and then they will be able to reduce excessive usage of fertilizers and water for crops. All those outputs are currently contributing to the maintaining of the ecological balance. According to Sandell [29], the Sri Lankan farmers had been farming sustainably by using greener fertilizers since the 1950s. Hence, their awareness of the importance of sustainability began a long time ago.

The construction industry has begun to perform well in the country due to rapid implementation of sustainability measures, although wastage and waste disposal is still a common problem. The research carried out by Sandell [29] has identified the core issue for the wastage is that the negligence and attitude of different groups of workers are disrupting the sustainability process. The researcher has concluded that lack of awareness and training is the main impediment to achieving sustainability goals as planned. This generation's thinking and attitude is different compared to sustainable thinking in the 1950s. The country needs to conduct more training and raise awareness of sustainability and why it is important for us. This can be easily achieved by adding this to school and university curricula and making it a compulsory subject.

Theoretically, the sustainability measures and importance of achieving sustainable targets has been confined to documents and reports. Lack of controls of the sustainability and awareness in some industries is preventing the achievement of sustainability goals and targets. This clearly indicates that Non-Governmental Organizations (NGO) involvement in protecting many natural resources in the country [17]. Involvement of NGOs indicates the level of severity of the subject, on many occasions the authorities were trying to cause damage nature to gain economic benefits such as deforestation and logging in tropical rainforest. Due to political influence the general public cannot withstand, they have to seek the help of NGOs to thwart government efforts. Hence, NGOs have taken full responsibility and they have taken the lead in the country, of raising awareness of sustainability [18]. This has been due to poor political agendas and lack of knowledge around the world; hence, politicians have caused severe damage to the country's economy. Since the researcher is Sri Lankan, he is well-placed to closely observe the level of understanding in sustainability of those politicians who are controlling the country. The majority of politicians do not come from well-educated diverse backgrounds, and when they make decisions, their morals and attitude do have a greater impact on the society and the whole nation. This can be clearly demonstrated with the recent ICT implementation proposals that are being pushed aside due to political reasons and the influence of politics has been proven by the research carried out by Kapurubandara and Lawson [19]. Kapurubandara and Lawson [19] further demonstrate how has the political barriers have hindered the adoption of new technology, which could support the country's economy and have an effect on it to a certain extent. Kapurubandara and Lawson [19] maintain that political and legal barriers are more powerful than social barrier as statistics show in Table 3.

According to the figures shown in Table 3 above, an unstable political situation does have a direct influence on changing and tightening up policies and regulations, however, technology has come into play to support sustainability. As a direct effect of political decisions, the infrastructure will be less efficient and distribution of information and knowledge to the general public will be limited if politicians have not been educated about sustainability.

The country's new president elected in 2005, managed to bring sustainability awareness to the public by taking part in global efforts to achieve sustainability, and raising awareness. According to Sri Lankan drafted sustainability report indicates the achievements in past years under the new government agenda. Achievements range from awareness to a new education front as shown in Fig. 3.

In a nutshell, the sustainability level in Sri Lanka has progressed to a new level through various channels in society. New developments undertaken in the country

**Table 3** External barriers in Sri Lanka [19]

| | Mean | n | % |
|---|---|---|---|
| Cultural barriers | | | |
| – Lack of popularity for online marketing and sales | 3.19 | 19 | 51.4 |
| Infrastructure barriers | | | |
| – Inadequate speed and quality of infrastructure | 3.23 | 19 | 54.3 |
| – Low Internet penetration in the country | 3.24 | 20 | 54.1 |
| Political barriers | | | |
| – Unstable economic climate in the country | 3.16 | 18 | 50.7 |
| – Changing regulations with each government change | 3.21 | 20 | 52.6 |
| Social barriers | | | |
| – No one-stop shop facility for services | 3.37 | 21 | 60 |
| – Lack of information on e-commerce | 3.32 | 23 | 62.2 |
| – Senior management in another sector lack ICT knowledge | 3.00 | 15 | 44.1 |
| Legal and regulatory | | | |
| – Little support and policies for SMEs from government and industry associations | 3.64 | 27 | 75 |
| – Inadequate legal framework for businesses using e-commence | 3.47 | 21 | 62.8 |
| – No simple procedures and guidelines | 3.38 | 24 | 64.9 |
| – Lack of suitable software standards | 3.21 | 20 | 52.6 |

N = number of organizations

**Fig. 3** Haritha (Green) Lanka programme [9]

**Box 2.1 Novel and Innovative Initiatives towards sustainable development**

- Established *"The National Council For Sustainable Development"* (NCSD)
- Developed *Sustainable Human Development Index* (SHDI) in 2008
- Launched "*Haritha (Green) Lanka*" in 2009
- Developed "*National Environmental Action Plan*" under the Haritha Lanka Programme
- Developed "*Sri Lanka Strategy For Sustainable Development"* in 2009
- Developed "*National Green Accounting Mechanism*" in 2011
- Established "*National Cleaner Production Centre*" (*NCPC*) & *Sri Lanka Carbon Fund* (SLCF)
- Established "*Green Job Awards Programme*" (2009) and "*National Green Reporting System*" (2011)

must comply with the latest sustainability regulations and procedures. But the level of awareness still needs to be addressed because the attitude to sustainability determines the state of the natural environment. This is due to the social backgrounds and beliefs of certain groups of people. Therefore, those social elements also need to be addressed effectively. The recent influx of foreign investment in the country is a good example of the success of sustainability.

## 10 Advantages and Disadvantages of Survey Outcome

Advantages of the survey outcome would be the key to eliminating barriers, by identifying issues. By analyzing survey results, the reader and stakeholders can obtain an insight into the current progress of sustainability in Sri Lanka in brief by examining the current state of sustainability awareness in the country. Further, since the research survey was conducted online, it was easy to reach a larger number of potential participants. The survey was carried out to determine the level of IT literacy that people have and their level of awareness of sustainable best practices. All the collected data is statistically collected under factors by means of the IBM SPSS (Version 22) software. Also, the open-ended questions allowed participants to voice their own opinions freely. All the quantified results are easy to statistically analyze and compare with each other in order to identify any trends or patterns. The outcome of the survey will be based on the individual responses of survey participants. There is no way of measuring whether or not the answers are genuine and truly meant. Further, users are restricted in giving their opinions on many elements. Therefore, some option selections may not be applicable to users even though they might make a choice. Moreover, the level of understanding of the questionnaire's language may also lead to misunderstanding of the questionnaire. Table 4 below illustrates the main advantages and disadvantages of conducting an online survey.

**Table 4** Advantages and disadvantages – prepared by Gihan Dhanushka

| Advantages |
|---|
| • High representativeness<br>• Low cost<br>• Convenient data gathering<br>• Good statistical significant<br>• Little or no observer subjectivity<br>• Precise results |
| **Disadvantages** |
| • Inflexible design<br>• Not ideal for controversial issues<br>• Possible inappropriateness of questions |

## 11 Research Method and Question

Many research methods have been used to understand and analyze philosophies. Research methods can vary based on the research subject. The way in which the research is conducted in order to obtain further knowledge of the subject matter entirely depends on the level of quality and accuracy of information that the researcher wants to obtain or receive. According to Scheurich [30], the research can be conducted in two ways, qualitative and quantitative. The method of the research can differ according to the subject matter and the nature of the research; it can be a mixed-methods research approach as well [5], which comprises both qualitative and quantitative methods. Later in this dissertation, the qualitative and quantitative research methods will be discussed in more detail.

In quantitative research, researchers are often use established theories to explain answers to assigned research questions [4]. Quantitative research basically assesses the situation using the collected data, in a statistical or mathematical form to support the theory or the philosophy. Further, Creswell [4] explains that the process of measurement is the core of the quantitative research because it provides a broader knowledge base for the subject matter. Quantitative researches are more interested in elements such as sampling, issues of design and measurement because the deductive approach is more concerned with careful planning prior to commencing data collection and analysis of the collected data [26]. Quantitative research uses numerical data and statistical values to analyze the subject matter to determine or to find the true nature of the matter. Further, Neuman [26] maintains that quantitative and qualitative researches are somewhat complementary. Carefully collected data, after being filtered, is used to understand the patterns of social life or any philosophy. The deviation comes from the nature of the data as in all numerical values and statistical data in percentage are counted as quantitative. On the other hand, soft data, such as expressions, words, symbols and photos fall under the category of qualitative research. Literature further confirms that quantitative researchers basically rely on the positivist approach used in the social sciences [26].

To deliver well justified outcome of the case given, sustainability awareness in Sri Lankan organization, have chosen Qualitative approach and Deductive approach to analyze and evaluate the current situation in the country in regards to sustainability awareness. As mentioned in above paragraphs, the Quantitative approach has been used to quantify facts in terms of gaining knowledge in particular segments to cater much broader insight precisely of research questionnaire. The data collection method should be planned prior to conducting information gathering process. This is the deductive approach. Since the deductive approach plan of its execution method in boarder sense, this research has been designed by targeting an appropriate audience to address the selected research questions and then the survey had been designed to target four segments (Age, Gender, Literacy and Awareness). Thereafter, gathered data sets will be evaluated by segments, to gain statistical and percentage overviews of selected segments, to quantify elements and to draw a conclusion regarding the research objective.

The primary research question has been developed based on the current situation in Sri Lanka and considering the fact that the country has an increasing number of industries in various sectors. To have an understanding of sustainability awareness, the designed survey questionnaire contributes considerable insight about their knowledge of sustainability. The question is to discuss upon gathering needed information, to consider and deliver fair research based on results received. The question proposed is as follow: *What are the opportunities and risk of sustainability awareness in the Sri Lanka organizations?*

The primary objective is the end result which will be achieved by addressing the primary research question. The primary research question concerns the opportunities and risks associated with achieving sustainability. The objective is to acquire a thorough insight into the issue of sustainability and the impediments to its implementation in Sri Lanka. According to the literature, there is a brighter future for achieving sustainability measures in Sri Lanka.

## 12 Results

The research was carried out to identify the level of sustainability awareness in Sri Lankan organizations. Therefore, to identify the current level of awareness, the research was further narrowed down to focus on the opportunities and risks associated with awareness of sustainability. The survey questionnaire was designed and developed accordingly to target the elements of opportunities and risks in organizations. The method chosen for the research was the online survey. Further, it examined the statistical process used to extract valid information from the data. Of the 312 potential participants, the responses of 179 were used for analysis as each of these respondents completed the entire questionnaire Therefore, greater accuracy was assured. Table 5 further illustrates the number of participants according to gender.

The ages of participants clearly demonstrate the present generation those who are actively involved in economic activities. Further, based on participants' ages, it is highly important to ascertain the type of economic activities that they are engaged in and their understanding and use of technology and their knowledge of the ethics associated with technology. Table 6 shows the participants' age ranges.

It clearly indicates that over 50% of the country's current active workforce with new technological literacy is aged between 25 and 30; those in the 21–24 age-group are the future leaders of the country. Therefore because of their age and

**Table 5** Workforce participation rate – prepared by Gihan Dhanushka

| Answer | Response | % |
|---|---|---|
| Male | 182 | 58% |
| Female | 130 | 42% |
| Total | 312 | 100% |

**Table 6** Age of participants – prepared by Gihan Dhanushka

| Answer | Response | % |
|---|---|---|
| 17 years and under | 2 | 1% |
| 18–20 | 6 | 2% |
| 21–24 | 98 | 31% |
| 25–30 | 166 | 52% |
| 31–35 | 22 | 7% |
| 36–40 | 9 | 3% |
| 41–45 | 4 | 1% |
| 46–50 | 4 | 1% |
| 51–55 | 3 | 1% |
| 56–60 | 3 | 1% |
| 61–65 | 0 | 0% |
| Over 65 | 0 | 0% |
| Total | 317 | 100% |

**Table 7** Main field of study – prepared by Gihan Dhanushka

| Answer | Response | % |
|---|---|---|
| Accounting | 88 | 30% |
| Business Law | 2 | 1% |
| Economics and Finance | 20 | 7% |
| Information Systems | 10 | 3% |
| Information Technology | 18 | 6% |
| Computer Science | 7 | 2% |
| Management | 70 | 24% |
| Marketing | 19 | 7% |
| Health Sciences | 3 | 1% |
| Humanities | 3 | 1% |
| Science and Engineering | 26 | 9% |
| Art and Design | 3 | 1% |
| Others – please specify | 22 | 8% |
| Total | 291 | 100% |

their knowledge of technology, it is vital that they have an understanding and awareness of sustainability.

Each participant's main field of education clearly indicates their career profession and also demonstrates the highest level education reached. This can be seen in Tables 7 and 8.

Table 7 clearly indicates that most of the participants' areas of study are in the fields of accounting and Management, both of which are directly involved in the economic activities of industries. Table 8 below shows the highest level of

**Table 8** Education level – prepared by Gihan Dhanushka

| Answer | Response | % |
|---|---|---|
| Primary Education | 5 | 2% |
| Higher Secondary/Pre-University | 8 | 3% |
| Professional Certificate | 43 | 15% |
| Diploma | 17 | 6% |
| Advanced/Higher/Graduate Diploma | 32 | 11% |
| Bachelor's Degree | 127 | 44% |
| Post Graduate Diploma | 13 | 5% |
| Master's Degree | 34 | 12% |
| PhD | 2 | 1% |
| Others – please specify | 5 | 2% |
| Total | 286 | 100% |

**Table 9** Number of hours spent on computer – prepared by Gihan Dhanushka

| Answer | Response | % |
|---|---|---|
| Less than an hour | 18 | 6% |
| Up to 5 h | 87 | 30% |
| 5–10 h | 124 | 43% |
| 10–20 h | 56 | 19% |
| Over 20 h | 4 | 1% |
| Total | 289 | 100% |

education reached by participants, this as well as the organizational sector in which they are involved are highly relevant to the findings.

Table 8 above clearly indicates that most participants hold a Bachelor Degree, followed by Professional Certification and Master Degree as their highest level of education. This is the most promising result for the sustainability awareness because the higher level of education and their current job status in the industry can have a direct influence on the organization's actions, as they are keys to the success of sustainability. Since the majority of employee's literacy level is ranging from Master's degree to Bachelor's Degree, after analyzing the level of awareness, it is highly important point to introduce or increase sustainability studies in those educational streams. Corcoran and Wals [3] also discuss the importance of sustainability education at all levels in order to achieve a more sustainable generation in the future through education.

Furthermore, the survey collected data about every individual's level of usage of information technology such as computers, access time, and accessing media and demographic of Information Technological elements. The following table shows the behaviour of participants in the organizational activities and discipline in usage of technology. Tables 9, 10, 11 and 12 show the number of hours spent on computers and how many hours are spent on internet activities.

**Table 10** Number of hours spent on internet activities – prepared by Gihan Dhanushka

| Answer | Response | % |
|---|---|---|
| Less than an hour | 52 | 18% |
| Up to 5 h | 132 | 46% |
| 5–10 h | 62 | 22% |
| 10–20 h | 37 | 13% |
| Over 20 h | 1 | 0% |
| Total | 284 | 100% |

**Table 11** Hours spent on Internet – prepared by Gihan Dhanushka

| Answer | Response | % |
|---|---|---|
| Less than an hour | 181 | 64% |
| Up to 5 h | 84 | 30% |
| 5–10 h | 9 | 3% |
| 10–20 h | 9 | 3% |
| Over 20 h | 1 | 0% |
| Total | 284 | 100% |

**Table 12** Hours spent on Internet or email – prepared by Gihan Dhanushka

| Answer | Response | % |
|---|---|---|
| Less than an hour | 186 | 66% |
| Up to 5 h | 62 | 22% |
| 5–10 h | 23 | 8% |
| 10–20 h | 11 | 4% |
| Over 20 h | 1 | 0% |
| Total | 283 | 100% |

The results presented in Tables 9 and 10 shows that the average time spent on computers is between 5 and 10 h per day. According to Tables 11 and 12 shows that the majority of users use the internet only for 1 hour for social networking and also 1 hour for checking emails. The percentage is 64–66% respectively. Further, Table 13 shows that the majority of participants use the internet for purposes such as studying, shopping, working and email. The results were quite compromising because sustainability education has been incorporated into higher education but when it comes to practical implementation, the situation is alarming. But the use of the internet and the time spent online is also high. With increased awareness of sustainability, this could be halved in the future see Table 14.

It is very encouraging that the majority of employees believe that sustainability can be achieved by raising awareness. This can clearly be seen in Table 15. This is a good indication that the workforce has identified the underlying issue of current sustainability measures and environmental impact of economic activities.

**Table 13** Purpose of internet usage – prepared by Gihan Dhanushka

| Answer | Response | % |
|---|---|---|
| Email | 255 | 89% |
| Playing games | 48 | 17% |
| Studying | 221 | 78% |
| Working | 190 | 67% |
| Shopping online | 125 | 44% |
| Chatting | 138 | 48% |
| Researching hobbies | 88 | 31% |
| Banking online | 115 | 40% |
| Buying goods or services | 90 | 32% |
| Buying stocks or investing online | 22 | 8% |
| Researching travel information or making reservations | 77 | 27% |
| Others – please specify | 9 | 3% |

**Table 14** Sustainability education learnt – prepared by Gihan Dhanushka

| Answer | Response | % |
|---|---|---|
| School | 103 | 38% |
| Higher education | 155 | 57% |
| Internet | 110 | 40% |
| Books | 46 | 17% |
| Magazine | 44 | 16% |
| News media | 53 | 19% |
| Conferences | 18 | 7% |
| Others – please Specify | 6 | 2% |

**Table 15** Ways of making sustainability alive – prepared by Gihan Dhanushka

| Answer | Response | % |
|---|---|---|
| Training | 127 | 47% |
| Education | 160 | 59% |
| Awareness | 188 | 69% |
| Workshop | 84 | 31% |
| Internet | 141 | 52% |
| T.V. | 82 | 30% |
| Social Networking | 178 | 65% |

The above manipulated samples from survey results help to understand the current level of sustainability awareness in the country's organizations, the level of technological literacy and level of education. Statistics shown in the mentioned tables above, demonstrate the following key elements respectively. The linkage is highly essential in order to put in place solutions for current shortfalls because education and best practices are already in place. Therefore, having awareness through proper channels is required. Those elements are:

- Educational level
- Position held in corporate
- Literacy of technological elements
- Ethical viewpoint of technological elements
- Awareness of sustainability and
- Awareness of importance of sustainability.

The following will demonstrate all findings in a much broader sense by analyzing statistically, to obtain a general understanding of the current standing of the sustainability concept within the organizational sector and its associated opportunities and risks. Further, all findings are compared using selected statistical measures like Alpha, Kaiser-Meyer-Olkin (KMO) and Bartlett's tests. KMO measures the level of performance by comparing each observed element's correlation coefficient to the degree of partially associated coefficients [12]. According to Zaiontz [38], the outcome of the KMO has been categorized as indicated in Table 16 below.

According to Table 16 above, the KMO results presented in Table 17 fall under the category of marvelous and excellent for opportunities and risks respectively at 0.937 and 0.939. Therefore, the survey results show a positive outcome for opportunities and risks at an excellent rate in organizations in Sri Lanka. The statistics of Cronbach's Alpha gives promising results too. The Alpha is basically a common measure of internal reliability. This measure is used mostly to measure reliability of, for example, Likert questionnaires [21]. According to research carried out and current survey on opportunity and risk, Alpha values are excellent, being 0.956 and 0.955 respectively. The Bartlett's test results are also very promising and show a positive outcome as shown in Table 17 below.

**Table 16** KMO statistic interpretation [38] – prepared by Gihan Dhanushka

| KMO | Interpretation |
|---|---|
| 0.9 and above | Marvelous |
| 0.8–0.9 | Meritorious |
| 0.7–0.8 | Middling |
| 0.6–0.7 | Mediocre |
| 0.5–0.6 | Miserable |
| Under 0.5 | Unacceptable |

**Table 17** Survey analyze statistics – prepared by Gihan Dhanushka

| | Cronbach's Alpha | KMO sampling accuracy | Bartlett's test of sphericity |
|---|---|---|---|
| Opportunities | 0.956 | 0.937 | $X^2 = 3018.277$; $df = 276$ P < 0.000 |
| Risks | 0.955 | 0.939 | $X^2 = 2206.548$; $df = 78$ P < 0.000 |

## 12.1 Opportunity Analysis

The opportunity element will demonstrate the degree of opportunity in sustainability awareness in the organizations. Therefore, using the IBM SPSS version 22, all data has been computed based on various statistical terms. As shown in Table 18, the Communalities values are acceptable and satisfactory. Any values under 0.50 have been excluded because according to Hakstian et al. [13], if the factor content falls 50%, it will be the half of the original variable. Therefore, those items are removed from the analysis and the remaining communalities are those shown in the following table. Table 18 shows the combination of two variables and the results are very significant for determining the opportunity. Table 18, the reduction of pollution has a higher value, which is 0.814. This is a very promising factor because it indicates that the general public is aware of the pollution issue.

**Table 18** Opportunity communality – prepared by Gihan Dhanushka

| | Initial | Extraction |
|---|---|---|
| Reduce energy and water usage | 1.000 | 0.686 |
| Reduce paper usage | 1.000 | 0.662 |
| Reduce consumption of raw materials | 1.000 | 0.705 |
| Increase efficiency | 1.000 | 0.627 |
| Increase cost-effectiveness | 1.000 | 0.413 |
| Enhance reputation | 1.000 | 0.512 |
| Improve corporate social responsibility | 1.000 | 0.698 |
| Reduce risk management | 1.000 | 0.354 |
| Differentiate businesses | 1.000 | 0.599 |
| Increase green strategy | 1.000 | 0.734 |
| Improve community investments | 1.000 | 0.569 |
| Improve human rights | 1.000 | 0.598 |
| Improve social responsibility investing | 1.000 | 0.622 |
| Increase triple bottom line – People, Planet and Profit | 1.000 | 0.577 |
| Increase productivity | 1.000 | 0.558 |
| Attract quality employees | 1.000 | 0.675 |
| Satisfy customer needs | 1.000 | 0.684 |
| Meet stakeholder expectations | 1.000 | 0.706 |
| Attract new opportunities | 1.000 | 0.725 |
| Create new jobs | 1.000 | 0.576 |
| Reduce carbon footprint | 1.000 | 0.743 |
| Reduce emissions | 1.000 | 0.807 |
| Reduce pollution | 1.000 | 0.814 |
| Reduce health hazards | 1.000 | 0.691 |

**Table 19** Total variance – prepared by Gihan Dhanushka

| Total variance explained | | | | | |
|---|---|---|---|---|---|
| Component | Initial eigenvalues | | | Extraction sums of squared loadings | |
| | Total | % of variance | Cumulative % | Total | % of variance |
| 1 | 12.239 | 50.996 | 50.996 | 12.239 | 50.996 |
| 2 | 1.702 | 7.093 | 58.089 | 1.702 | 7.093 |
| 3 | 1.393 | 5.806 | 63.894 | 1.393 | 5.806 |

**Table 20** Rotational sum of squared loading – prepared by Gihan Dhanushka

| Component | Extraction sums of squared loadings | Rotation sums of squared loadings | | |
|---|---|---|---|---|
| | Cumulative % | Total | % of variance | Cumulative % |
| 1 | 50.996 | 6.334 | 26.392 | 26.392 |
| 2 | 58.089 | 6.030 | 25.123 | 51.516 |
| 3 | 63.894 | 2.971 | 12.378 | 63.894 |

To gain further knowledge of the variances regarding the opportunity factor, the variances have been generated as shown in Table 19. Table 19 shows the variation division, by 24 components. Those components are elements which were mentioned in the survey regarding opportunity.

According to Table 19, the above variances are generated based on the main three factors which the researcher has chosen to deliver the outcome for opportunities. According to Table 19, one can clearly see the accumulation of the factors like percentage variance of the second factor is being added to the cumulative of the first factor. It clearly shows the total variance of 63.894% which is a highly acceptable outcome.

Table 20 above further demonstrates the proportion of variances according to the selected three factors. All three factors have a total value of to 63.894%, which is closer to 70% and therefore this indicates the level of achievability and the awareness being prompted and identified by the participants in general.

### 12.1.1 Rotated Component Matrix

The rotated component matrix table is the finalized version of the aggregated statistical effort. Table 21 below shows the selected 24 opportunity elements and the values for each component under three factors. In order to implement and also to take future initiatives to achieve and optimize sustainability awareness, only the higher values for every component have been selected. Those values are highlighted and the selected values are given new labels. Those selected components are considered for future optimizations and current implementations. Nevertheless, those are the indicators of level

**Table 21** Rotated component matrix – prepared by Gihan Dhanushka

| Rotated component matrix | | | |
|---|---|---|---|
| | Component | | |
| | 1 | 2 | 3 |
| *Meet stakeholder expectations* | 0.803 | 0.118 | 0.218 |
| *Satisfy customer needs* | 0.776 | | 0.268 |
| *Attract new opportunities* | 0.762 | 0.356 | 0.132 |
| *Attract quality employees* | 0.718 | 0.362 | 0.166 |
| Improve human rights | 0.677 | 0.366 | |
| Create new jobs | 0.657 | 0.379 | |
| Increase productivity | 0.655 | 0.294 | 0.207 |
| Differentiate businesses | 0.607 | 0.322 | 0.356 |
| Improve social responsibility investing | 0.583 | 0.504 | 0.165 |
| Improve community investments | 0.546 | 0.495 | 0.163 |
| Increase cost-effectiveness | 0.417 | 0.394 | 0.291 |
| *Reduce pollution* | 0.194 | 0.858 | 0.199 |
| *Reduce emissions* | 0.206 | 0.833 | 0.265 |
| *Reduce carbon footprint* | 0.212 | 0.802 | 0.234 |
| *Reduce health hazards* | 0.327 | 0.751 | 0.140 |
| *Increase green strategy* | 0.388 | 0.728 | 0.232 |
| Improve corporate social responsibility | 0.482 | 0.666 | 0.148 |
| Increase efficiency | 0.507 | 0.550 | 0.259 |
| Increase triple bottom line – People, Planet and Profit | 0.470 | 0.484 | 0.348 |
| Enhance reputation | 0.381 | 0.465 | 0.387 |
| *Reduce consumption of raw materials* | 0.152 | 0.107 | 0.819 |
| *Reduce energy and water usage* | 0.137 | 0.355 | 0.736 |
| *Reduce paper usage* | 0.191 | 0.347 | 0.710 |
| Reduce risk management | 0.417 | | 0.422 |

**Table 22** Higher efficiency and higher standard – prepared by Gihan Dhanushka

| New label | Variable name | Factor loading |
|---|---|---|
| Higher efficiency and higher standard | Meet stakeholder expectations | 0.803 |
| | Satisfy customer needs | 0.776 |
| | Attract new opportunities | 0.762 |
| | Attract quality employees | 0.718 |

of awareness of sustainability in the organizations. Those are the current sustainability awareness level existing in Sri Lankan organizations.

The following Tables 22, 23, and 24 are the selected factors under new labels, after being identified by the higher values from Table 21 above.

**Table 23** Green environment – prepared by Gihan Dhanushka

| New label | Variable name | Factor loading |
|---|---|---|
| Green Environment | Reduce pollution | 0.858 |
| | Reduce emissions | 0.833 |
| | Reduce carbon footprint | 0.802 |
| | Reduce health hazards | 0.751 |
| | Increase green strategy | 0.728 |

**Table 24** Improve sustainability standards – prepared by Gihan Dhanushka

| New label | Variable name | Factor loading |
|---|---|---|
| Improve sustainability standards | Reduce consumption of raw materials | 0.819 |
| | Reduce energy and water usage | 0.736 |
| | Reduce paper usage | 0.710 |

## *12.2 Risk Analysis*

The risk analysis will demonstrate the degree of risk factors that have been identified and also the level of variance compared to other risk elements. To statistically compare the collected data, the survey answers for 14 risks are outlined. The following tables will show the degrees of variance and, based on the variance values, what are the key values which have been selected as risk factors. Table 25 below shows the level of risk identification rate. The highlighted value in Standard Deviation, which has a value of 1.247, clearly indicates the level of increase in fraudulent acts in the country. This high value could be due to several reasons such as poor politics or lack of authority. This result and the participation level clearly indicate the extent of awareness of risk factors.

The following communality Table 26 displays the communal statistics based on the abovementioned 14 risk questions. Based on responses received, they have been further statistically processed to examine the variation between correlated values. As discussed above, values less than 50% can be eliminated because the other variable will have more than 50% of the weight. Further, as highlighted in the following table, the increase in scandals by perceived environmental irresponsibility has been identified with the highest value of 0.882. This clearly indicates that the low level of political involvement and lack of authoritative powers are harming the natural environment.

As discussed above in the chapter under the opportunity, the same applies to this as well because the only difference is having 14 components as they are targeting the answers for risk. The mechanism of the calculation is the same but the values are different. According to Table 27, the total cumulative percentage is 76.177% for two factors. This is a very good indication of risk awareness, as it is more than 50%. The variance between first factor and the second factor is only 10.251%.

**Table 25** Descriptive statistics – prepared by Gihan Dhanushka

| Descriptive statistics | | | |
|---|---|---|---|
| | Mean | Std. Deviation | Analysis N |
| Increase fraud | 2.68 | 1.247 | 170 |
| Increase litigation and compliance breaches | 2.79 | 1.212 | 170 |
| Increase governance failure | 2.64 | 1.200 | 170 |
| Increase transaction failure due to environmental liabilities | 2.70 | 1.166 | 170 |
| Increase security and systems failures caused by environmental problems | 2.68 | 1.144 | 170 |
| Increase scandals by perceived environmental irresponsibility | 2.68 | 1.102 | 170 |
| Increase marketing failure by perceived environmental irresponsibility | 2.67 | 1.155 | 170 |
| Increase supply chain crises due to suppliers' environmental problems | 2.71 | 1.122 | 170 |
| Inflate costs | 2.76 | 1.173 | 170 |
| Increase interest rates | 2.57 | 1.059 | 170 |
| Increase insurance crises due to environmental disasters | 2.74 | 1.063 | 170 |
| Increase number of new regulations including environmental | 3.29 | 1.234 | 170 |
| Increase competition for and cost of raw materials | 2.95 | 1.056 | 170 |

**Table 26** Risk communality – prepared by Gihan Dhanushka

| | Initial | Extraction |
|---|---|---|
| Increase fraud | 1.000 | 0.757 |
| Increase litigation and compliance breaches | 1.000 | 0.705 |
| Increase governance failure | 1.000 | 0.823 |
| Increase transaction failure due to environmental liabilities | 1.000 | 0.827 |
| Increase security and systems failures caused by environmental problems | 1.000 | 0.874 |
| Increase scandals by perceived environmental irresponsibility | 1.000 | 0.882 |
| Increase marketing failure by perceived environmental irresponsibility | 1.000 | 0.830 |
| Increase supply chain crises due to suppliers' environmental problems | 1.000 | 0.783 |
| Inflate costs | 1.000 | 0.763 |
| Increase interest rates | 1.000 | 0.733 |
| Increase insurance crises due to environmental disasters | 1.000 | 0.693 |
| Increase number of new regulations including environmental | 1.000 | 0.556 |
| Increase competition for and cost of raw materials | 1.000 | 0.677 |

**Table 27** Total variance explain – prepared by Gihan Dhanushka

| Total variance explained | | | | | |
|---|---|---|---|---|---|
| Component | Initial eigenvalues | | | Extraction sums of squared loadings | |
| | Total | % of variance | Cumulative % | Total | % of variance |
| 1 | 8.570 | 65.926 | 65.926 | 8.570 | 65.926 |
| 2 | 1.333 | 10.251 | 76.177 | 1.333 | 10.251 |

**Table 28** Rotation sum of square loading – prepared by Gihan Dhanushka

| Component | Extraction sums of squared loadings | Rotation sums of squared loadings | | |
|---|---|---|---|---|
| | Cumulative % | Total | % of variance | Cumulative % |
| 1 | 65.926 | 6.025 | 46.346 | 46.346 |
| 2 | 76.177 | 3.878 | 29.831 | 76.177 |

**Table 29** Risk rotated component matrix – prepared by Gihan Dhanushka

| Rotated component matrix | | |
|---|---|---|
| | Component | |
| | 1 | 2 |
| *Increase scandals by perceived environmental irresponsibility* | 0.879 | 0.332 |
| *Increase security and systems failures caused by environmental problems* | 0.873 | 0.334 |
| *Increase governance failure* | 0.865 | 0.273 |
| *Increase marketing failure by perceived environmental irresponsibility* | 0.846 | 0.337 |
| *Increase transaction failure due to environmental liabilities* | 0.837 | 0.356 |
| *Increase fraud* | 0.826 | 0.273 |
| *Increase supply chain crises due to suppliers' environmental problems* | 0.768 | 0.439 |
| *Increase litigation and compliance breaches* | 0.734 | 0.407 |
| *Increase interest rates* | 0.249 | 0.819 |
| *Inflate costs* | 0.343 | 0.803 |
| *Increase competition for and cost of raw materials* | 0.275 | 0.776 |
| *Increase insurance crises due to environmental disasters* | 0.424 | 0.716 |
| Increase number of new regulations including environmental | 0.282 | 0.691 |

The results are displayed in Table 28 which clearly shows the total values for both factors are 65.926% and 76.177% respectively. The total variance for both factors is 46.346% and 76.177% respectively

The rotated component matrix further analyses 14 risks results collected from the survey and they are further statistically reduced to two factors. The outcome of the analysis was the selection of only the higher values from each factor table. Thereafter, the selected new values are given new labels to sum up the risk factor and then to use as a benchmark for future reference and current implementations. Table 29 below displays the results.

**Table 30** Poor economic management – prepared by Gihan Dhanushka

| New label | Variable name | Factor loading |
|---|---|---|
| Poor economic management | Increase interest rates | 0.819 |
| | Inflate costs | 0.803 |
| | Increase competition for and cost of raw materials | 0.776 |
| | Increase insurance crises due to environmental disasters | 0.716 |

**Table 31** Corruption and negligence – prepared by Gihan Dhanushka

| New label | Variable name | Factor loading |
|---|---|---|
| Corruption and Negligence | Increase scandals by perceived environmental irresponsibility | 0.879 |
| | Increase security and systems failures caused by environmental problems | 0.873 |
| | Increase governance failure | 0.865 |
| | Increase marketing failure by perceived environmental irresponsibility | 0.846 |
| | Increase transaction failure due to environmental liabilities | 0.837 |
| | Increase fraud | 0.826 |
| | Increase supply chain crises due to suppliers' environmental problems | 0.768 |
| | Increase litigation and compliance breaches | 0.734 |

According to Table 29 above, the highlighted values in those two factors are the highest vales which are selected to develop new values. The identified value elements are indicating highly essential aspects of risk awareness and the lagging points of the sustainability in Sri Lankan organizations. Tables 30 and 31 below show the recreated variables under new labels.

In summary, this section demonstrates the factual findings statistically by using various tools and techniques to display the degree of every elements compatibility. The survey findings were further refined using Cronbach's Alpha, KMO and Bartlett's test to deliver an accurate overview of the level of awareness of sustainability opportunities and risk factors. Then the data was subjected to further processing to identify, the degree of variance in every single element in the opportunities and risks categories. The opportunity and risks are then further identified under new labels based on higher values. Those values can be used as a benchmark of the sustainability awareness in Sri Lankan Organizations sector.

## 13 Discussion and New Significance

It is very important to have an understanding and awareness of sustainability world-wide due to current changes in the world environment and the rapid growth of population. The environmental changes have been identified in various ways and then those changes have been addressed. Global warming is the main reason for the world to discuss the climate change aspects due to rapid movement and growth of industrial elements in the economic sector. Therefore, extensive research has been carried out in order to identify the root cause of the current environmental issues. This topic of excessive use of natural resources first emerged during the seventeenth century. But, over time, the need for awareness has become more vital because of the ever increasing pollution factor and threat of the extinction of species. The literature review clearly demonstrated that the extinction of entire civilizations has been caused by natural disasters, due to poorly unplanned human activities. Then, in the nineteenth century there emerged a great need to protect nature and bio diversity from our activities in order to prevent extinction. Therefore, the first world summit of the United Nations took place in Rio de Janeiro in 1992 [31]. The eco-friendly summits have attempted to address ways to maintain ecological balance, reduce poverty in the world, and eliminate unsustainable economic practices. Therefore various policies and procedures have been established during those summits to address the key issues. The drafted frameworks have been constantly re-evaluated and long-term objectives have been formulated.

## 14 Limitations

There were less obvious limitations that the researcher identified during the research process and when conducting the survey. The identified limitations are: limited to only Colombo city, circulation was within the top level of the organization and lack of interest due to lack of awareness.

- ***Limited to the city of Colombo***
  The survey questionnaire was distributed mainly via emails and the Face-book social network. The majority of participants were residing in the city of Colombo, the nation's capital. The survey was passed from one peer to another within the same organization and its sub-divisions. Due to technical inefficiency (browser issues and internet bandwidth) and lack of understanding of the subject it was a great challenge to reach out-skirts of Colombo. Therefore to obtain the acceptable level of survey results which is about 312 and then to have 178 fully completed results from a total of 312 participants, required considerable effort. To achieve this, the researcher had to send out a considerable number of emails and social media requests.

- ***Circulation within top layer of the hierarchy***
  The majority of participants were generally at the top level in their organization. They ranged from executives to bottom level managers. This was due to the governance factor in the organization, but also the time factor caused this limitation because when it comes down to the hierarchy, the top level managers are more curious about sustainability measures and share their knowledge with their equals within the company. When the survey reaches the lower management level, the time factor hinders the further circulation of the survey.
- ***Lack of awareness and negligence***
  The negligence was apparent in the number of mails which had to be sent to achieve the target of 312 responses from a considerable number of emails. This indicates that the only people who are interested in this issue are those who have previously undertaken studies on sustainability or those who are willing to make a change in the world. This has been clearly reflected in the survey results, and the survey responses indicate that the majority of respondents were made aware of sustainability during their tertiary studies and they are currently working as top level managers as indicated by the demographic information pertaining to job and level of education.

## 15 Conclusion

This chapter provides a new perspective on the identified opportunities and risk factors with new labels for the readers to understand the final outcome of the research. Further, this chapter discusses the future opportunities and future possible benefits due to countries' political changes because government authorities are required to establish and regulate any measures and implementations. Therefore, it is very important to have an understanding of a country's political environment and its sustainability. This has been clearly reflected throughout the research. Due to new policies and procedures in the country, the public's awareness of sustainability has risen, and new policies have been adopted by organizations. Further, to address the limitations of this research as outlined above, future research could target a much broader population and include more sectors, using the current findings as a benchmark.

## References

1. Baumgartner RJ, Ebner D (2010) Corporate sustainability strategies: sustainability profiles and maturity levels. Sustain Dev 18(2):76–89
2. Bromley DW (2007) Environmental regulations and the problem of sustainability: moving beyond "market failure". Ecolog Econ 63(4):676–683
3. Corcoran PB, Wals AE (2004) Higher education and the challenge of sustainability. Springer, the Netherlands

4. Creswell JW (2013) Research design: qualitative, quantitative, and mixed methods approaches. Sage Publications Inc., USA
5. Creswell JW, Clark VLP (2007) Designing and conducting mixed methods research. Sage Publications Inc., USA
6. De Silva C, Weatherhead E, Knox JW, Rodriguez-Diaz J (2007) Predicting the impacts of climate change – a case study of paddy irrigation water requirements in Sri Lanka. Agric Water Manag 93(1):19–29
7. Deutz P, Himley M, Smith M, Hargroves KC, Desha C (2012) Cents and sustainability: a panel on sustainable growth, politics and scholarship. Geogr J 178(2):188–192
8. Economics T (2015) Sri Lanka Population 1960-2015. Trading Economics. http://www.tradingeconomics.com/sri-lanka/population. Accessed 1 June 2017
9. Environment Mo (2012) Sri Lanka's Middle Path to Sustainable Development through 'Mahinda Chintana – Vision for the Future' (trans: Development S). Sustainable Development. Ministry of Environment Division, Sri lanka
10. Farmer JG (1994) Environmental change and the chemical record in Loch Lomond sediments. Hydrobiologia 290(1):39–49. doi:10.1007/BF00008951
11. Glasson J, Therivel R, Chadwick A (2013) Introduction to environmental impact assessment. Routledge, Abingdon, Oxon OX14 4SB, UK
12. Hagedoorn J, Cloodt M (2003) Measuring innovative performance: is there an advantage in using multiple indicators? Res Policy 32(8):1365–1379
13. Hakstian AR, Rogers WT, Cattell RB (1982) The behavior of number-of-factors rules with simulated data. Multivar Behav Res 17(2):193–219
14. Hawken P, Lovins AB, Lovins LH (2013) Natural capitalism: the next industrial revolution. Taylor & Francis Ltd, UK
15. Heinberg R (2010) What is sustainability? *The Post Carbon Reader*. pp 11–19
16. Huesemann MH (2004) The failure of eco-efficiency to guarantee sustainability: future challenges for industrial ecology. Environ Prog 23(4):264–270
17. Ishwaran N, Erdelen W (1990) Conserving Sinharaja: An experiment in sustainable development in Sri Lanka. Ambio 19(5):237–244
18. Jayawardane A (2006) Disaster mitigation initiatives in Sri Lanka. Faculty of Engineering, University of Moratuwa, Sri Lanka. http://iiirr.ucalgary.ca/files/iiirr/Kochifullpaper-final.pdf. Accessed 1 June 2017
19. Kapurubandara M, Lawson R (2006) Barriers to Adopting ICT and e-commerce with SMEs in developing countries: an exploratory study in Sri Lanka. Proceedings of the 2006 Collector Conference on Electronic Commerce, Adelaide, 2006
20. Kuhlman T, Farrington J (2010) What is sustainability? Sustainability 2(11):3436–3448
21. Leard (2013) Cronbach's Alpha (α) using SPSS Statistics. Laerd. https://statistics.laerd.com/spss-tutorials/cronbachs-alpha-using-spss-statistics.php. Accessed 24 Nov 2015
22. Lipton M (1977) Why poor people stay poor: urban bias in world development. Scand J Econ 80(3). doi:10.2307/3439587
23. McDonald KA, Brown JH (1992) Using montane mammals to model extinctions due to global change. Conserv Biology 6:409–415
24. McLamb E (2011) The Ecological Impact of the Industrial Revolution. Ecology Communications Group. http://www.ecology.com/2011/09/18/ecological-impact-industrial-revolution/. Accessed 24 Nov 2015
25. Morgan R (1980) Soil erosion and conservation in Britain. Prog Phys Geogr 4(1):24–47
26. Neuman WL (1991) Qualitative and quantitative approaches. Social research methods, 4th edn. Pearson Education Limited, UK
27. Nielsen L (2011) Classifications of countries based on their level of development: how it is done and how it could be done. International Monetary Fund Working Paper. WP/11/31. IMF, Washington, DC
28. Philipsen DP (2010) Sustainability. Green business: an A-to-Z guide. Sage, Thousand Oaks. doi:10.4135/9781412973793

29. Sandell K (1988) Ecostrategies in theory and practice: Farmer's perspectives on water, nutrients and sustainability in low-resource agriculture in the dry zone of Sri Lanka. Linköping University, Arid regions, Sweden
30. Scheurich JJ (1997) Research method in the postmodern Falmer Press Inc., Philadelphia
31. Sherman K, Duda A (1999) An ecosystem approach to global assessment and management of coastal waters. Mar Ecol Progr Ser 190:271–287
32. Simon DR, Eitzen DS (2002) Elite deviance. Allyn and Bacon, Boston
33. Slaper TF, Hall TJ (2011) The triple bottom line: what is it and how does it work? IBR 86 (1):4–8
34. UN (2005) General Assembly. United nations. Available via UNAIDS. http://data.unaids.org/Topics/UniversalAccess/worldsummitoutcome_resolution_24oct2005_en.pdf. Accessed 24 Nov 2015
35. Vitousek PM, Mooney HA, Lubchenco J, Melillo JM (1997) Human domination of Earth's ecosystems. Science 277(5325):494–499
36. Watson RT, Noble IR, Bolin B, Ravindranath N, Verardo DJ, Dokken DJ (2000) Land use, land-use change and forestry: a special report of the Intergovernmental Panel on Climate Change. Cambridge University Press, UK
37. Willard B (2010) 3 sustainability models. Sustainability advantage. Available via Sustainability Advantage. http://sustainabilityadvantage.com/2010/07/20/3-sustainability-models/. Accessed 24 Nov 2015
38. Zaiontz C (2013) Validity of correlation matrix and sample size. Real statistic using Excel. http://www.realstatistics.com/multivariate-statistics/factor-analysis/validity-of-correlation-matrix-and-sample-size/. Accessed 1 June 2017

# The Advantages and Risks of Sustainability Awareness in the Indian Higher Education Sector

**Nidhin Francis**

**Abstract** In response to the global climatic changes, loss of biodiversity and the immense human encroachment upon nature, sustainability development is being promoted worldwide. Sustainability development through education is being fostered by international organizations worldwide in efforts to make the younger generation more aware of sustainability issues. In this chapter, we explore the development of sustainability through education in the socio-cultural context of India. Through the online survey, research explores the advantages and risks associated with sustainability awareness. Findings of the research were new factors generated using the statistical analysis tool SPSS (Statistical Package for Social Science). The new factors generated for the advantages were 4R, increased effectiveness and efficiency; meeting the responsibilities of the society, and the stakeholder expectations. The new factors generated for the risks were market failure crisis, increased fraud and failure in governance, more regulations and market scarcity of materials. In a further discussion, the research compares the new factors with the findings from the literature review. The thesis also suggests the future research areas and outlines the limitations of this project.

## 1 Introduction

In this twenty-first century, the world is changing as a result of complex and interlinked developments which present a number of lifestyle challenges and problems. The challenges always arise from the changing values and practices which create an unsustainable society. We need to have a common goal for educating and empowering ourselves and our society for such changes. Therefore the quality of education and awareness should be such that it builds the knowledge, values, skills and competency to practise a sustainable lifestyle and benefit society [6]. Education for sustainability embraces a vison for the newer society, not only

N. Francis (✉)
School of Information Systems, Curtin University, Perth, Western Australia, Australia
e-mail: NidhinFranz@gmail.com



T. Issa et al. (eds.), *Sustainability, Green IT and Education Strategies in the Twenty-first Century*, Green Energy and Technology, DOI 10.1007/978-3-319-57070-9_9

ecologically – it also provides social and economic sustainability. Educational institutions are formative places where the younger generation acquires knowledge, values and skills, and also are taught to become contributing members of the society. Other factors such as family, peers, laws and media to name a few, also play a role in socialisation and education. Since every person has the right to an education, educational institutions are the primary areas where our goals and visions are being conveyed. Education for sustainability is intended to effect a total change in vision and a mission of personal and social change through the modern approach, which is teaching and learning the integrated goals of conservation, social justice, appropriate development and democracy. The educational programs aim to educate and empower all citizens and, through them, to bring about a positive change in society; that is, to achieve the transition from the current unsustainable lifestyle to a greener and more sustainable lifestyle. The advantages of having an awareness of sustainability education are quite evident, along with the high risk factors associated with its implementation. The research was conducted in India as it has a high potential for development amongst the third world countries; moreover, it has one of the strongest economies in the world. According to the Balasubramanian [2] and Fortress research team, India has the world's third largest population after the U.S. and China. Given these facts, the research has been conducted in India. This research aims to discuss the advantages and risks of educating society about sustainability in India. This chapter discusses the research topic and its background. Later in this section, an outline of all subsequent chapters will be given.

## 2 Sustainability

There are many definitions of sustainability which are influenced by people's values and cultures. The most popular among them is the UN definition, commonly known as the *Brundland definition*: Sustainable development is development that meets the needs of the present without compromising the ability of future generations to meet their own needs [3]. "Sustainability is also defined as the use of the vital functions (possible uses) of our biophysical surroundings in such a way that they remain indefinitely available" [11]. Most of the first world countries have adopted this definition to shape the objectives of local conditions when developing an approach to education that integrates the notion of sustainability. In general, there are non-sustainability recipes available to bring about the changes, but it always depends on a number of factors which affect this change in every community. Sustainability is a continuous learning process in which stakeholders actively participate in directing the new changes. Many of the current practices are devastating. Questioning our own actions and reflecting on the current decisions by challenging our own thinking in unconventional ways can help to achieve sustainability. The most important practice to accomplish the goal of sustainability is to adopt new behaviours in order to avoid future socio-economic problems, rather than being reactive and not leaving anything for the future. This is the point where the importance of education comes into the context of sustainability. The United Nations has declared this as the decade for education on sustainability development. The founding value of this education for sustainability development is the notion of respect: respect for us and for future generations, and respect for our planet and what it has provided us [16]. Education has been recognized as fundamentally important as a means of addressing the critical global challenges which we encounter. The development of education for sustainability has complex and far reaching links with social, economic, cultural and environmental concerns. The scenarios for the different parts of the world are different as the factors discussed take various forms and shapes. Most of the first world countries have already prioritised their action plans for introducing a sustainable lifestyle and they are moving towards achieving it. Education for sustainability is important to reorient the thinking of people so that they become more innovative and find more solutions in their daily lives.

This is not the case in third world countries. As former colonies, most secured their independence in the twentieth century, and are in the process of achieving their development goals. It is a fact that given the current situations in these countries, it is difficult for development and sustainability to go hand in hand. If there is a lack of appropriate planning and sound investments, both of these can cause mutual damage. According to Gray [9] report, the ecological indicator which denotes the impact of human actions on the environment with relation to the natural resources available to serve the ecosystem services, shows that humanity has already used one third of the world's natural resources. Factors such as rapid growth in population, diversity of cultural values and ideas, regional differences, political views and many more factors are reason enough to unite the world under a common goal. Therefore, education is an important tool which can be used to resolve these problems to a certain extent by showing individuals the importance of interdependency between humans and every single element of ecology. Traditional India has been a sustainable society. A large part of the Indian population still maintains the lifestyle which revolves around the basic Three-R principles of sustainability (Reduce, Recycle, and Reuse). Even though it is a personal choice, it is also driven by strong economic considerations. The government of India had already integrated the principles of sustainability in the policies and strategies of their development programs. India's development strategies work within a 5-year-plan framework and when it completed the Eleventh 5-year plan, it had a strong focus on the education sector. In order to promote the value of sustainable development in education, the government has directed that environmental education be integrated into all curricula. Moreover, India is the only country whose Supreme Court has directed that all educational boards make environmental studies compulsory in all courses. Even though the Indian higher education sector is doing well in terms of environmental studies and making the younger generation aware of sustainability issues, there is still room for improvement. To improve the system, it is vital to know the strengths and weaknesses. A better understanding of the advantages and risks of sustainability education can help to formulate creative ideas and this will certainly help India to move closer to the ultimate goal. Finally, this research focuses on examining the advantages and risks associated with sustainability awareness in India's higher education sector.

## 3 Research Methods and Question

Understanding the importance of sustainability in the present and future world, and the impact that the higher education sector can make on the awareness of sustainability, is vital in order for the universities and other education providers to realign the courses being offered. The changes in the course should consider the issues which are being faced by the prevailing unsustainable practices. In order to produce changes in the structure of education courses, there should be a good understanding of the risks and advantages of the awareness in the higher education sector. Knowing the risks associated with changing the system is as equally important as recognising the advantages. Also, there should be an analysis of the cultural, ethical, political and geographical factors in order to ascertain the extent of the changes that should be made in the existing system. The term "sustainability" is simple, but the concept of sustainability has different levels of applications and different levels of understanding. Training the next generation's leaders according to their particular domain of study is a challenge. The precise understanding of the area of education, effects of the change and how to incorporate this into the existing system of education is most important.

Hence, the question that drives this research is: *What are the advantages and risks associated with sustainability awareness in the Indian higher education sector?* This research question is aiming to understand both the advantages and risks of the sustainability awareness for the stakeholders such as students, educators, government and non-government education providers, universities etc. related to the higher education sector. Examining and analysing this question from different perspectives can clarify the different ways in which awareness programs need to be arranged according to the cultural differences among students and staff. The different disciplines of studies need different strategies and should analyse the different perspectives of sustainability. The advantages for different disciplines are different and this is the same for the risks. And so, this question basically discusses the positives and negatives of implementing a change which will include the study of sustainability in tertiary institutions. In their article, Cotton et al. [5] doubt the extent to which the notion of sustainability can be translated into practical changes to the higher education curriculum. The possibilities of organizational changes for including sustainability in higher education courses was discussed from the perspective of what national and EU governments can do to accelerate the changes needed in order to include sustainability in tertiary education, and how it can be made safe for courageous students to take educational paths different from traditional tracks, even if technical options exist to do so [1]. These statements establish the context of this research which intends to determine how well

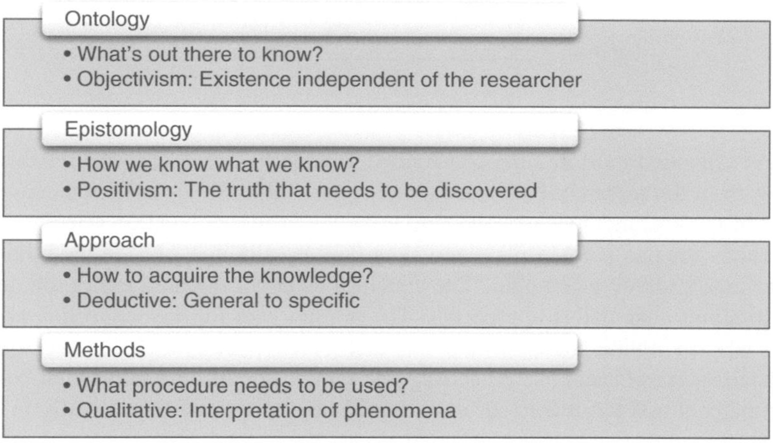


**Fig. 1** The research philosophy, methodology and methods which have been used in this research

the concept of sustainability is understood by current students; it also raises the question of how to reduce the risk and increase the advantages of introducing awareness of sustainability into the higher education sector.

This research adopts objectivism as the ontology as the knowledge that will be acquired through the research is independent from the researcher. Regarding epistemology, the research adopts positivism as the outcome(s) will follow from the analysis of the survey results; until the analysis has been conducted, results and conclusions cannot be established. The survey approach used in this research is deductive rather than inductive; therefore, the survey has been designed in order to move from the general to the specific. The quantitative method of data collection will be used (see Fig. 1).

This research method is largely based on numerical data. Mostly, with this method the statistical explanations which have been generated from the data collected substantiate the theory, hypothesis or the research question formulated by the researcher. The quantitative research method analyzes the problem using the numerical data or the statistics that have been generated by the collected data. This method is often used to draw conclusions from a large group of samples. It is a highly effective and efficient method of collecting opinions or information about behavior. As the data collected is measurable, it is easy to draw conclusions and also the method is more structured than the qualitative approach. Data collection by means of the quantitative research method can be done in many ways, the most important of which is the survey. Surveys can be done through a variety of means including online, telephone, paper and many more. Other important methods are interviews, observation, vote/poll [18, 15] The primary data collection tool for this research was the online survey. Online surveys are one of the most commonly used techniques for collecting data in research fields as they are easy to distribute and respond to [21]. Mostly, the research being done is in related to current issues so

**Table 1** Likert Scale (prepared by author)

| Strongly disagree | Disagree | Neutral | Agree | Strongly agree |
|---|---|---|---|---|
| 1 | 2 | 3 | 4 | 5 |

the data collection must be time-constrained in order to derive the most reliable conclusion from the research. For many reasons, the online survey has been considered as one of the best methods of collecting opinions from a population sample. One of the reasons for this is that online surveys offers possibilities outside the other self-administrated survey techniques. The conclusions about research topics are derived from different quantitative approaches. The primary tool for data collection as mentioned was the online survey which was distributed through different social media and email such as Facebook, WhatsApp and Gmail. As the communication between the participant and the researcher easy, it will be easier to keep constantly in touch with the participants. Once the survey is done, the researcher is able to download it in different formats and analyses it using various software. According to Issa [12], it is easier to analyze the results when the data is downloaded to an Excel spreadsheet which makes it easier to handle the collected data.

The survey is an effective tool for collecting the opinions of the sample group; however, the design should be simple and understandable. The survey which has been used for this particular research consisted of three sections: the first section was intended to collect the demographic data; the second section consisted of questions about the advantages and risks associated with sustainability awareness in the Indian higher education sector. In the first section, respondents were given several options from which they chose an answer; the second section had a Likert scale index that ranged from strongly disagree to strongly agree. The Likert scale was used in the second section to obtain opinions from the sample group [13]. The third section of the survey gave respondents the opportunity to elaborate on answers or to bring up additional relevant questions and issues (Table 1).

The unit of analysis is the entities which are being analyzed as part of the research. It can vary from individuals, groups, books, and geographical data to name a few. As the primary question of this research concerns the advantages and risks associated with sustainability awareness in the Indian higher education sector, the survey targeted a sample group selected from India who are enrolled in various tertiary courses. The literature review was guided by the primary and secondary questions in order to discover the results and examine the conclusions of previous authors who have conducted research in this area. This study is intended to discover the opinions of the sample group regarding the advantages and risks associated with sustainability awareness in the Indian higher education sector. It also analyzes the same to determine whether the gender of the respondents is a significant factor in determining the types of responses.

The target population was selected from Kerala state in India which has a high literacy rate and there is gender equality in the education sector. The survey targeted 100 selected adults from different age groups, who are related to the education sector in one way or another. The ages range from 17 to 65. The target population selected

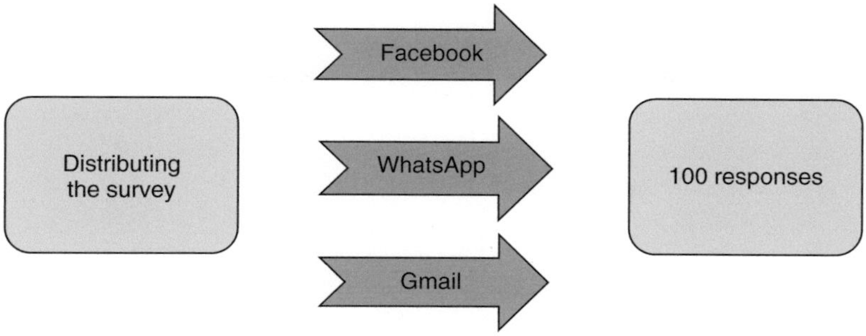


**Fig. 2** Distribution of survey (prepared by author)

was equal for both genders, to make it easier to ascertain the similarities and differences of their ideas about sustainability awareness. Of the 104 valid survey results, 51% were from males and 49% from females.

As described in the above sections, quantitative data collection can be done in many different ways. In this research, data collection was conducted via a web/online survey. As the sample group targeted resided in India, the online survey was the best means of collecting the data. The survey was launched using a specific data collection tool developed by Qualtrics. After completion of the design, the survey was distributed among the potential sample group through Facebook, Gmail and WhatsApp. Once the target had been achieved, the survey was closed and the data collected was exported as an Excel file for further analysis (see Fig. 2).

The data collected through Qualtrics was analyzed to generate new factors which pertained to the advantages and the risks associated with the sustainability awareness in the Indian higher education sector. SPSS was tool used for the analysis as it is a well-known statistical tool used in social science. With the help of SPSS version 22, the factor analysis was conducted in order to discover new factors which have been generated other than those factors which emerged from the literature review. In order to ensure the reliability and validity of the survey results, the survey questions were carefully formulated for clarity and ease of understanding. This helps to minimize or overcome any chance of confusion or ambiguity due to misinterpretation of the survey by the participants. Any incomplete surveys were eliminated prior to the analysis in order to achieve a more reliable outcome and conclusion.

## 4 Results

The survey was distributed among current and future students and people who are in one way or another related to the higher education sector. The group for the study was randomly selected from a state in India which has the highest rate of literacy. The group selected has sound technological knowledge, ages ranged from

**Table 2** Statistics of the sample group and the total responses (prepared by author)

| Statistic | Value |
|---|---|
| Min value | 1 |
| Max value | 2 |
| Mean | 1.49 |
| Variance | 0.25 |
| Standard deviation | 0.50 |
| Total responses | 222 |

18 to 65 years. The statistical package generally used for social science (SPSS) was used for the analysis (see Table 2).

Of the 222 respondents, 51% (113) responses were from males and 49% (109) were from females. According to the online survey statistics, most of the respondents are aged between 21 and 24 which constitute 50% of all the respondents. The responses from the 25–30 years age group accounted for 19% which is slightly higher than 18% from respondents aged between 18 and 20. The remaining respondents were aged between 31 and 65 years, accounting for 12% (see Table 3). The online survey respondents who had completed the bachelor's degree as their highest level of education comprised 56%, which is slightly less than double the second largest group of master's degree holders at 34%. The professional certificate holder's accounted for 10%, followed by the graduate diplomas and post-graduate diplomas at 6% and 4% respectively. Individuals with higher secondary education as their highest educational level made up 3%; 1% had a PhD (See Table 3).

Table 4 shows the major areas of study for the participants of the online survey. This table shows the survey was collected from diverse areas, even though the participants are mostly from the engineering and computer science domains at 24% and 34% respectively. Information technology participant's account for 12% and all other fields mentioned in the surveys such as accounting, business law, economics and finance, management, marketing, health science, humanities are less than 5%.

It was found that about 41% of the respondents spend between 1–5 h on average on computer each day, whereas individuals using computer between 5–10 h and less than an hour are 34% and 16% respectively. Even though most of the respondents fall within the above groups, 8% use it for 10–20 h each day. From Table 5, it is evident that 51% of the respondents spend up to 5 h on the Internet. Respondents who spend less than an hour and between 5–10 h account for 22% and 21% respectively. The group spending 10–20 hours on the Internet each day is 4%, whereas 1% of respondents spend more than 20 h a day on the Internet, which is the smallest group. The time spent on email and social networking each day seems to be comparable for all the groups. Most respondents spend less than an hour on emails and social networking. Moreover, 85% of respondents spent time on email, 69% spent time on social networking. Of those spending 5–10 h, it is evident that 26% of respondents spent this time on social networking while only

**Table 3** Online survey statistics – India (prepared by the author)

| Number and percentage of online survey | |
|---|---|
| Questionnaires distributed | 222 |
| Valid survey results | 110 |
| Valid response rate | 49.5% |
| *Gender* | |
| Male respondents | 113 (51%) |
| Female respondents | 109 (49%) |
| *Age* | |
| 17 years and under | 1 (0%) |
| 18–20 | 41 (18%) |
| 21–24 | 112 (50%) |
| 25–30 | 43 (19%) |
| 31–35 | 10 (4%) |
| 36–40 | 5 (2%) |
| 41–45 | 0 (0%) |
| 46–50 | 2 (1%) |
| 51–55 | 4 (2%) |
| 56–60 | 1 (0%) |
| 61–65 | 2 (1%) |
| Over 65 | 5 (2%) |
| *Qualifications* | |
| Primary Education | 0 (0%) |
| Higher Secondary/Pre-University | 6 (3%) |
| Professional Certificate | 21 (10%) |
| Diploma | 4 (2%) |
| Advanced/Higher/Graduate Diploma | 13 (6%) |
| Bachelor's Degree | 113 (56%) |
| Post Graduate Diploma | 8 (4%) |
| Master's Degree | 34 (17%) |
| PhD | 3 (1%) |
| Others – please specify | 1 (0%) |

12% spent the time on the email. Three percent spent time on email for 5–10 h, whereas 4% spent time on social networking (see Table 5)

Table 6 shows that respondents use the Internet mainly for studying and chatting which is 74% and 68% respectively which are the highest percentages among the group. Fifty-nine percent use the Internet for chatting, 53% use it for online banking and 50% use the Internet for working and for shopping online. In total, 44% of respondents do research on travel, 41% buy goods and services, 37% research their hobbies, and 31% play games. Furthermore, 18% use the Internet for investment and share trading activities.

**Table 4** Participants' main area of study (prepared by the authors)

| Answer | Response | % |
|---|---|---|
| Accounting | 11 | 5 |
| Business law | 4 | 2 |
| Economics and finance | 7 | 3 |
| Information system | 3 | 1 |
| Information technology | 24 | 12 |
| Computer science | 70 | 34 |
| Management | 9 | 4 |
| Marketing | 6 | 3 |
| Health science | 4 | 2 |
| Humanities | 2 | 1 |
| Science and engineering | 49 | 24 |
| Arts and design | 3 | 1 |
| Others | 11 | 5 |

**Table 5** Technology use by Indian users (prepared by the authors)

| Answer | Hours spent on the computer per day | Hours spent on the Internet per day | Hours spent on email per day | Hours spent on social networking per day |
|---|---|---|---|---|
| | Response % | Response % | Response % | Response % |
| Less than an hour | 36 (16%) | 40 (22%) | 154 (85%) | 125 (69%) |
| Up to 5 h | 84 (41%) | 92 (51%) | 21 (12%) | 47 (26%) |
| 5–10 h | 69 (34%) | 38 (21%) | 6 (3%) | 8 (4%) |
| 10–20 h | 17 (8%) | 8 (4%) | 0 (0%) | 0 (0%) |
| Over 20 h | 3 (1%) | 2 (1%) | 0 (0%) | 1 (1%) |

**Table 6** Internet usage by Indian users (prepared by the authors)

| Answer | Response | % |
|---|---|---|
| Email | 124 | 68 |
| Playing games | 57 | 31 |
| Studying | 135 | 74 |
| Working | 91 | 50 |
| Shopping online | 91 | 50 |
| Chatting | 126 | 59 |
| Research hobbies | 68 | 37 |
| Banking online | 96 | 53 |
| Buying goods and services | 74 | 41 |
| Buying stocks and investing online | 33 | 18 |
| Researching travel information or making reservation | 80 | 44 |
| Others | 2 | 1 |

It was found that the 74% of the sample group depend on smart phone, whereas laptops are in second position as 65% of respondents use them. Desktop computers are used by 42% of the respondents, while only 27% use personal computers. Workstations and tablets are less than 20%, while netbooks and PDAs are less than 10% (see Table 7).

The concept of sustainability and green information technology for 56% of the respondents was primarily introduced through the Internet, whereas other media and higher education have made 42% respondents aware of the issues of sustainability and green IT. Schools first introduced this concept to 32% of respondent, 27% of them learned about the issue independently from books and magazines. Conferences have made 13% of the sample population aware of the sustainability issue (see Table 8).

The sample group mostly uses Google devices which is 41%, whereas the number of respondents using Dell devices is 23%. Nineteen percent of survey respondents are using other devices from Samsung, Sony, Microsoft, Nokia etc. Those using Apple devices and IBM devices are 13% and 4% respectively (see Table 9).

It is evident from the statistic that 46% of the sample size does not tend to read the sustainability report of the vendors before buying a device. Furthermore,

**Table 7** Devices usage by Indian users (prepared by the author)

| Answer | Response | % |
|---|---|---|
| PC | 48 | 27 |
| Desktop | 74 | 42 |
| Laptop | 115 | 65 |
| Netbook | 13 | 7 |
| PDA | 15 | 8 |
| Workstation | 22 | 12 |
| Tablet | 32 | 18 |
| Smartphone | 141 | 74 |
| Others | 1 | 1 |

**Table 8** First introduced to the concepts of sustainability and green information technology by Indian users (prepared by the author)

| Answer | Response | % |
|---|---|---|
| School | 55 | 32 |
| Higher education | 72 | 42 |
| Internet | 97 | 56 |
| Books | 47 | 27 |
| Magazine | 51 | 29 |
| News media | 72 | 42 |
| Conference | 23 | 13 |
| Others | 1 | 1 |

**Table 9** Indian user's devices (prepared by the author)

| Answer | Response | % |
|---|---|---|
| Apple | 22 | 13 |
| Google | 73 | 41 |
| Dell | 40 | 23 |
| IBM | 7 | 4 |
| Others | 34 | 19 |

**Table 10** Indian users who read the sustainability report of the company before buying a device (prepared by the author)

| Answer | Response | % |
|---|---|---|
| Yes | 44 | 22 |
| No | 80 | 46 |
| Maybe | 48 | 28 |
| Not at all | 2 | 1 |

**Table 11** Frequency of Indian users changing their devices (prepared by the author)

| Answer | Response | % |
|---|---|---|
| Every 6 months | 1 | 1 |
| Every 12 months | 8 | 5 |
| Every 18 months | 16 | 9 |
| Every 24 months | 25 | 15 |
| Every 30 months | 30 | 18 |
| Every 36 months | 44 | 26 |
| Every 42 months | 23 | 14 |
| Others | 22 | 13 |

28% state that they may or may not understand the sustainability report (see Table 10).

It was found that ta quarter of the total sample prefer to change their devices every 36 months, whereas 18% change them within 30 months. Fourteen percent change devices every 42 months, 15% change every 24 months. Thirteen percent of respondents prefer to change their device once that have had the maximum use from it or when it no longer meets their needs (see Table 11).

The major reason for respondents changing their devices is to keep up to date with the technology and functionalities. Fifty-six percent of respondents mentioned that speed was the deciding factor when buying new devices. The size of the device matters to 28% of participants. Four percent of respondents retain their devices until they are damaged or no longer serviceable (see Table 12).

**Table 12** Indian user's responses to "Why do you change your device?" (prepared by the author)

| Answer | Response | % |
|---|---|---|
| Size | 48 | 28 |
| Speed | 95 | 56 |
| Functionality | 121 | 71 |
| Keeping up with technology | 114 | 67 |
| Other – please specify | 7 | 4 |

**Table 13** Indian users' responses to "changing device frequently will cause damage to our planet" (prepared by the author)

| Answer | Response | % |
|---|---|---|
| Yes | 83 | 48 |
| No | 23 | 13 |
| Maybe | 64 | 37 |
| Not at all | 4 | 2 |

**Table 14** Indian users' responses to: "Can we change the mindset of designers and users regarding sustainability?" (prepared by the author)

| Answer | Response | % |
|---|---|---|
| Training | 86 | 50 |
| Education | 92 | 53 |
| Awareness | 145 | 84 |
| Workshop | 75 | 43 |
| Internet | 88 | 51 |
| TV | 55 | 32 |
| Social network | 114 | 66 |
| Others | 1 | 1 |

The survey attempted to discover the attitude of Indians towards sustainability issues caused by the frequent changing of devices. Forty-eight percent of respondents believed that the frequent changing of devices can cause serious damage to the planet, which shows their awareness of sustainability; on the other hand, only 13% believe that it does not affect the planet. "Maybe" was the response of 37%, while 2% believe the frequent change of devices will affect the planet (see Table 13).

The survey examined the Indian users' recommendation for the easiest and most effective way to change the mindset of users in terms of sustainability. The survey concluded that awareness and social networks can bring about a change in mind-set; this was supported by 84% and 66% respectively. Fifty percent of respondents believe that the Internet, education and training can bring change (see Table 14).

**Table 15** Cronbach's Alpha test for reliability – the risk and advantages factors (prepared by author)

| | No of items | Cronbach's alpha | KMO sampling adequacy | Bartlett's test of sphericity |
|---|---|---|---|---|
| Advantages | 24 | 0.950 | 0.872 | $\chi^2$ = 1568.364; df = 276; sig = 0.000 |
| Risks | 13 | 0.945 | 0.901 | $\chi^2$ = 998.886; df = 78; sig = 0.000 |

A total of 104 survey results were used for the factor analysis. The analysis was conducted separately for the advantages and risks which contained 24 and 13 questions respectively. To further examine the survey responses, the axial factoring method was used to extract new factors.

To measure the sampling adequacy, Cronbach's Alpha, Kaiser-Meyer-Olkin and Bartlett's test of sphericity were carried out by the researcher. Table 15 shows the statistical results for the Alpha, KMO and Bartlett's test. The Cronbach's Alpha for 24 variables for advantages is 0.950 and for 13 variables for risk it is 0.945, indicating an excellent internal consistency of the items [8]. A Kaiser-Meyer-Olkin (KMO) measure of sampling adequacy of 0.872 for advantages and 0.901 for risks indicates that an excellent sample size was obtained for the analysis [10]; hence, KMO results are appropriate and acceptable for this study. The Bartlett's test of sphericity is highly significant, $\chi^2$ = 1568.364; df = 276; sig = 0.000 for advantages and $\chi^2$ = 998.886; df = 78; sig = 0.000 for risks , indicating that the items of the scale are sufficiently correlated for factors to be found [24]. Therefore, results shown in Table 15 indicate the validity of the survey conducted.

Factor analysis was used by the researcher to reduce the variables for the advantages and associated with sustainability awareness in the Indian higher education sector. Factor analysis performs the group factoring and suggests new factors by means of the data reduction technique.

There are different methods for the extraction of data. According to MacCallum et al. [14], if the data is relatively normally distributed, maximum likelihood is the best method.

However, if the data violates the multivariate normality rule, it is preferred to do the principle axis factors [7]. In general, both these methods give the best results. In this data extraction, the researcher has used the principle axis factor. The rotation method for data extraction which has been used here is the orthogonal instead of the oblique, specifically the Varimax rotation. In this method, the factors are uncorrelated. In SPSS, the rotated matrix for factors is interpreted after the orthogonal rotation [4]. Communalities are the variance associated with each factor which together add up 1. Initial communalities represent the total accounted for by all the factors or components. Extraction communalities are estimated to be the variance of each factor. Communalities with small values are not strong factors (see Table 16).

When variables or factors do not account for a significant part for the factor analysis, this results in small communalities. Table 16 shows the communalities

**Table 16** Communalities for the advantages (generated from SPSS)

| Communalities | | |
|---|---|---|
| | Initial | Extraction |
| Reduce energy and water usage | 1.000 | 0.674 |
| Reduce paper usage | 1.000 | 0.578 |
| Reduce consumption of raw materials | 1.000 | 0.763 |
| Increase efficiency | 1.000 | 0.653 |
| Increase cost-effectiveness | 1.000 | 0.631 |
| Enhance reputation | 1.000 | 0.508 |
| Improve corporate social responsibility | 1.000 | 0.589 |
| Reduce risk management | 1.000 | 0.450 |
| Differentiate businesses | 1.000 | 0.521 |
| Increase green strategy | 1.000 | 0.654 |
| Improve community investments | 1.000 | 0.531 |
| Improve human rights | 1.000 | 0.527 |
| Improve social responsibility investing | 1.000 | 0.595 |
| Increase triple bottom line – People, Planet and Profit | 1.000 | 0.543 |
| Increase productivity | 1.000 | 0.462 |
| Attract quality employees | 1.000 | 0.586 |
| Satisfy customer needs | 1.000 | 0.370 |
| Meet stakeholder expectations | 1.000 | 0.596 |
| Attract new advantages | 1.000 | 0.582 |
| Create new jobs | 1.000 | 0.527 |
| Reduce carbon footprint | 1.000 | 0.771 |
| Reduce emissions | 1.000 | 0.747 |
| Reduce pollution | 1.000 | 0.731 |
| Reduce health hazards | 1.000 | 0.598 |

from the factor analysis test. In factor analysis, stronger data means uniformly high communalities [23, 17]. If the communalities' values are greater than 0.8 then they are considered "high" according to Vellicer and Jackson [22] but this rarely occurs in the real data. The low to moderate values are from 0.4 to 0.7. If the item value is less than 0.4, it may not be related to the other factors, which means another factor is to be considered. This will help the researcher whether or not to consider similar factors for future research [4]. In the Table 16 the factor with low significance and which is not to be considered in future research has been highlighted.

The principal component method was used for the correlation matrix. The Eigen value shows the number of factors which account for the correlation. For the advantages for the awareness of sustainability, Table 17 show the total variance with a total of 59.11% variation. The initial Eigen values which are more than 1 are being considered. Table 17 shows the factors which have met this measure [19]. The three factors explain 59.11% of the total variance (see Table 17).

**Table 17** The total variance explained for the advantages (generated from SPSS)

| Total Variance Explained | | | | | |
|---|---|---|---|---|---|
| Component | Initial Eigenvalues | | | Extraction Sums of Squared Loadings | |
| | Total | % of Variance | Cumulative % | Total | % of Variance |
| 1 | 11.292 | 47.049 | 47.049 | 11.292 | 47.049 |
| 2 | 1.492 | 6.218 | 53.267 | 1.492 | 6.218 |
| 3 | 1.403 | 5.844 | 59.111 | 1.403 | 5.844 |
| 4 | 1.074 | 4.476 | 63.588 | | |
| 5 | 0.998 | 4.160 | 67.748 | | |
| 6 | 0.855 | 3.565 | 71.313 | | |
| 7 | 0.788 | 3.282 | 74.595 | | |
| 8 | 0.742 | 3.092 | 77.687 | | |
| 9 | 0.612 | 2.552 | 80.238 | | |
| 10 | 0.597 | 2.489 | 82.728 | | |
| 11 | 0.578 | 2.406 | 85.134 | | |
| 12 | 0.509 | 2.120 | 87.254 | | |
| 13 | 0.472 | 1.966 | 89.220 | | |
| 14 | 0.394 | 1.641 | 90.862 | | |
| 15 | 0.380 | 1.583 | 92.444 | | |
| 16 | 0.334 | 1.392 | 93.836 | | |
| 17 | 0.299 | 1.245 | 95.081 | | |
| 18 | 0.254 | 1.057 | 96.138 | | |
| 19 | 0.247 | 1.030 | 97.168 | | |
| 20 | 0.181 | 0.754 | 97.923 | | |
| 21 | 0.162 | 0.676 | 98.599 | | |
| 22 | 0.143 | 0.597 | 99.196 | | |
| 23 | 0.112 | 0.468 | 99.664 | | |
| 24 | 0.081 | 0.336 | 100.000 | | |

The rotated component matrix has been formed with the Varimax rotation method which is an orthogonal process. The regression method was used to obtain the coefficient. According to Costello [4], the cleanest fact structure should be considered. According to Stevens' rule of thumb, when the sample size is above 100, the factors with factor loadings below 0.5 can be eliminated [20]. Table 18 shows the factors with factor loadings for each of the new factors generated.

From the rotated component matrix above, the factors that belong to each of the components have been grouped and labelled below (see Table 19).

Table 20 show the communalities according to the risk factors or variables that have been discussed in the survey. It shows the initial and extracted values. The communalities with values lower than 0.4 are not considered as their components will not be related. As shown in the table, all the 13 variables have a communality value over 0.6 which is a very good value [4]. Increase in fraud has the highest

**Table 18** Factor loading for each component (generated from SPSS and edited by author)

| Rotated component matrix[a] | | | | |
|---|---|---|---|---|
| | Component | | | |
| | 1 | 2 | 3 | 4 |
| Reduce carbon footprint | 0.803 | | | |
| Reduce emissions | 0.787 | | | |
| Reduce pollution | 0.764 | | | |
| Reduce health hazards | 0.702 | | | |
| Increase green strategy | | | | |
| Improve community investments | | 0.673 | | |
| Reduce consumption of raw materials | | 0.665 | | |
| Improve social responsibility investing | | 0.660 | | |
| Reduce risk management | | | | |
| Increase triple bottom line – People, Planet and Profit | | | | |
| Enhance reputation | | | | |
| Differentiate businesses | | | | |
| Increase productivity | | | | |
| Increase efficiency | | | 0.753 | |
| Increase cost-effectiveness | | | 0.726 | |
| Reduce paper usage | | | 0.605 | |
| Reduce energy and water usage | | | | |
| Satisfy customer needs | | | | |
| Attract quality employees | | | | |
| Improve corporate social responsibility | | | | |
| Meet stakeholder expectations | | | | 0.771 |
| Attract new advantages | | | | 0.689 |
| Create new jobs | | | | 0.637 |
| Improve human rights | | | | |

Extraction Method: Principal Component Analysis.
Rotation Method: Varimax with Kaiser Normalization.
[a]Rotation converged in 7 iterations.

value of extraction value of 0.852 and increased governance failure has an extraction value of 0.681 which is the lowest among the risks.

The total variance for the risks is shown in Table 21. The principle component method was used to generate the correlation matrix. The Eigen value shows the number of factors which account for the correlation. For the risks associated with the awareness of sustainability, the table gives the total variance with a total of 74.46% variation. Table 21 shows the factors which have met this measure [19]. The three factors explain 74.46% of the total variance.

The rotated component matrix was been formed with the Varimax rotation method which is an orthogonal process. The regression method was used to obtain the coefficient. According to Costello [4], the cleanest fact structure should

**Table 19** The components generated with new labels (prepared by author)

| Label | Variable name | Factor loading |
|---|---|---|
| 4Rs | Reduce carbon footprint | 0.803 |
| | Reduce emissions | 0.787 |
| | Reduce pollution | 0.764 |
| | Reduce health hazards | 0.702 |
| Increased effectiveness and efficiency | Increase efficiency | 0.753 |
| | Increase cost-effectiveness | 0.726 |
| | Reduce paper usage | 0.605 |
| Meet responsibilities to society | Improve community investments | 0.766 |
| | Reduce consumption of raw materials | 0.703 |
| | Improve social responsibility investing | 0.654 |
| Stakeholders' expectations | Meet stakeholder expectations | 0.771 |
| | Attract new advantages | 0.689 |
| | Create new jobs | 0.637 |

**Table 20** The communalities for risks (generated by SPSS)

| Communalities | | |
|---|---|---|
| | Initial | Extraction |
| Increase fraud | 1.000 | 0.852 |
| Increase litigation and compliance breaches | 1.000 | 0.673 |
| Increase governance failure | 1.000 | 0.681 |
| Increase transaction failure due to environmental liabilities | 1.000 | 0.766 |
| Increase security and systems failures caused by environmental problems | 1.000 | 0.705 |
| Increase scandals by perceived environmental irresponsibility | 1.000 | 0.723 |
| Increase marketing failure by perceived environmental irresponsibility | 1.000 | 0.804 |
| Increase supply chain crises due to suppliers' environmental problems | 1.000 | 0.788 |
| Inflate costs | 1.000 | 0.753 |
| Increase interest rates | 1.000 | 0.709 |
| Increase insurance crises due to environmental disasters | 1.000 | 0.792 |
| Increase number of new regulations including environmental | 1.000 | 0.729 |
| Increase competition for and cost of raw materials | 1.000 | 0.704 |

be considered. According to Stevens' rule of thumb, when the sample size is above 100, the factors with factor loadings below 0.5 can be eliminated [20]. Table 22 shows the factors with a factor loading for each of the new factors generated. The results were quite interesting when three factors were generated.

From the rotated component matrix above, the factors that belong to each of the components have been grouped and labeled below (see Table 23).

**Table 21** The total variance explanation (generated by SPSS)

| Total variance explained | | | | | |
|---|---|---|---|---|---|
| Component | Initial eigenvalues | | | Extraction sums of squared loadings | |
| | Total | % of Variance | Cumulative % | Total | % of Variance |
| 1 | 7.935 | 61.038 | 61.038 | 7.935 | 61.038 |
| 2 | 0.972 | 7.480 | 68.517 | 0.972 | 7.480 |
| 3 | 0.773 | 5.943 | 74.461 | 0.773 | 5.943 |
| 4 | 0.632 | 4.862 | 79.323 | | |
| 5 | 0.551 | 4.241 | 83.564 | | |
| 6 | 0.485 | 3.734 | 87.299 | | |
| 7 | 0.414 | 3.182 | 90.481 | | |
| 8 | 0.298 | 2.294 | 92.775 | | |
| 9 | 0.240 | 1.842 | 94.617 | | |
| 10 | 0.228 | 1.755 | 96.372 | | |
| 11 | 0.208 | 1.598 | 97.971 | | |
| 12 | 0.155 | 1.194 | 99.165 | | |
| 13 | 0.109 | 0.835 | 100.000 | | |

**Table 22** The factors generated for risk (generated by SPSS and edited by author)

| Rotated component matrix[a] | | | |
|---|---|---|---|
| | Component | | |
| | 1 | 2 | 3 |
| Increase marketing failure by perceived environmental irresponsibility | 0.861 | | |
| Increase supply chain crises due to suppliers' environmental problems | 0.766 | | |
| Increase transaction failure due to environmental liabilities | 0.689 | | |
| Increase scandals by perceived environmental irresponsibility | | | |
| Increase security and systems failures caused by environmental problems | | | |
| Increase fraud (including environmental actions and environmental compliance failures) | | 0.867 | |
| Increase governance failure (including environmental actions and environmental compliance failures) | | 0.754 | |
| Increase litigation and compliance breaches (including environmental actions and environmental compliance failures) | | 0.633 | |
| Increase competition for and cost of raw materials | | | 0.800 |
| Increase number of new regulations including environmental | | | 0.789 |
| Increase interest rates | | | 0.622 |
| Increase insurance crises due to environmental disasters | | | |
| Inflate costs | | | |

Extraction Method: Principal Component Analysis.
Rotation Method: Varimax with Kaiser Normalization.
[a]Rotation converged in 6 iterations.

**Table 23** The components generated with new labels (prepared by author)

| Label | Variable name | Factor loading |
|---|---|---|
| Market failures and crisis | Increase marketing failure by perceived environmental irresponsibility | 0.861 |
| | Increase supply chain crises due to suppliers' environmental problems | 0.766 |
| | Increase transaction failure due to environmental liabilities | 0.689 |
| Increased frauds and failures in governance | Increase fraud (including environmental actions and environmental compliance failures) | 0.867 |
| | Increase governance failure (including environmental actions and environmental compliance failures) | 0.754 |
| | Increase litigation and compliance breaches (including environmental actions and environmental compliance failures) | 0.633 |
| More regulations and scarcity in market | Increase competition for and cost of raw materials | 0.800 |
| | Increase number of new regulations including environmental | 0.789 |
| | Increase interest rates | 0.622 |

## 5 Discussion and New significance

This chapter examined the sustainability awareness in the Indian Higher Education Sector and from the online survey a new factors were generated from the Indian perspective. Figures 3 and 4 illustrates the findings of the research in reference to the advantages and risks of sustainability awareness and the sub-variables. A total of four factors have been extracted from the analysis of the survey. These are the advantages of sustainability awareness in the Indian higher education sector namely:

- **4Rs:** From the analysis of the data collected, the variable generated by the factor analysis was 4R. It has four sub-variables for which "reduce the carbon footprint" had a 0.8 factor loading and the remaining three "reduce emission", "reduce pollution", "reduce health hazards" each had 0.7 as a factor loading. All these sub-variants indicate that sustainability awareness is highly correlated with reduced pollution and health hazards.
- **Increased effectiveness and efficiency:** Increased effectiveness and efficiency was the factor generated from the three sub-variables. "Increased efficiency", "increased cost effectiveness" each had a 0.7 factor loading and "reduced paper usage" had a factor loading of 0.6. This indicates that one advantage is better resource management.
- **Meet the responsibilities to society:** The sub-variables of this label are "increased community investment" with a factor loading of 0.7 and "increased social responsibility investment" which had a factor loading of 0.6. The third variable was

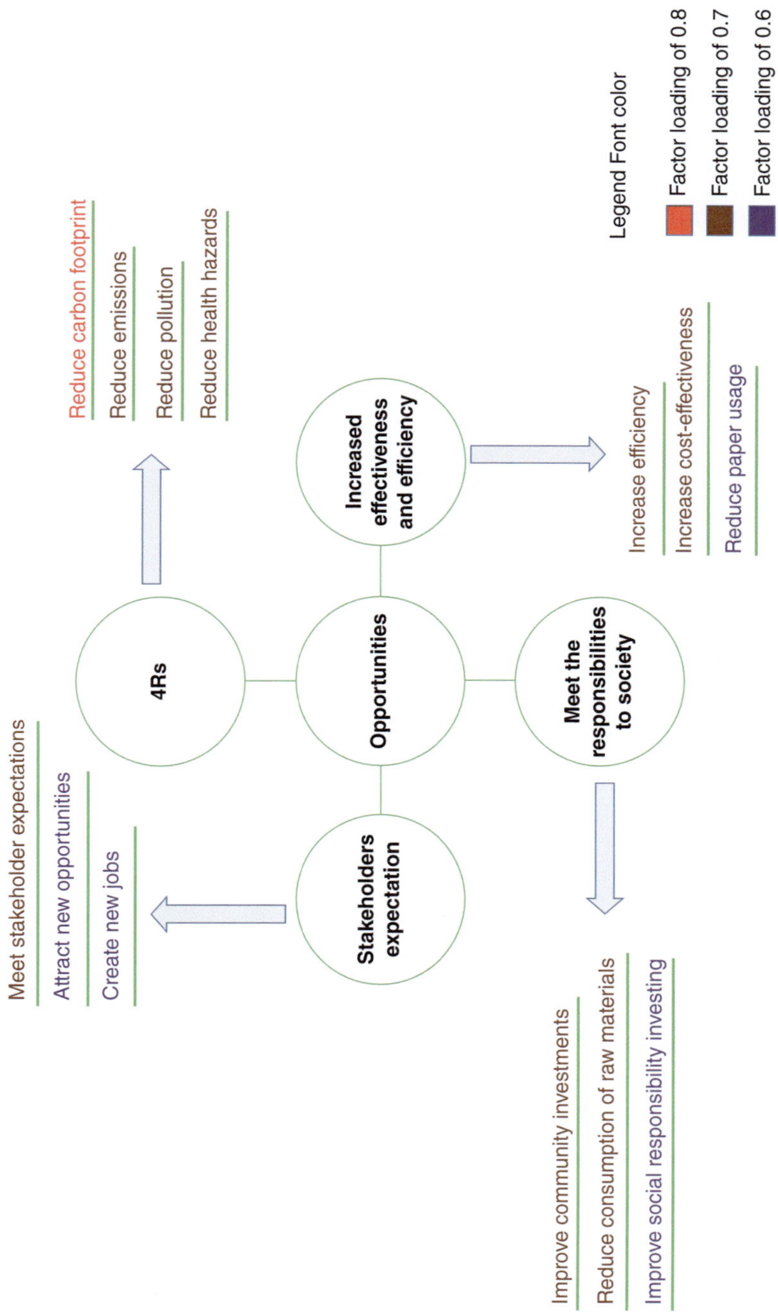


**Fig. 3** Advantages of sustainability awareness in higher education (prepared by author)

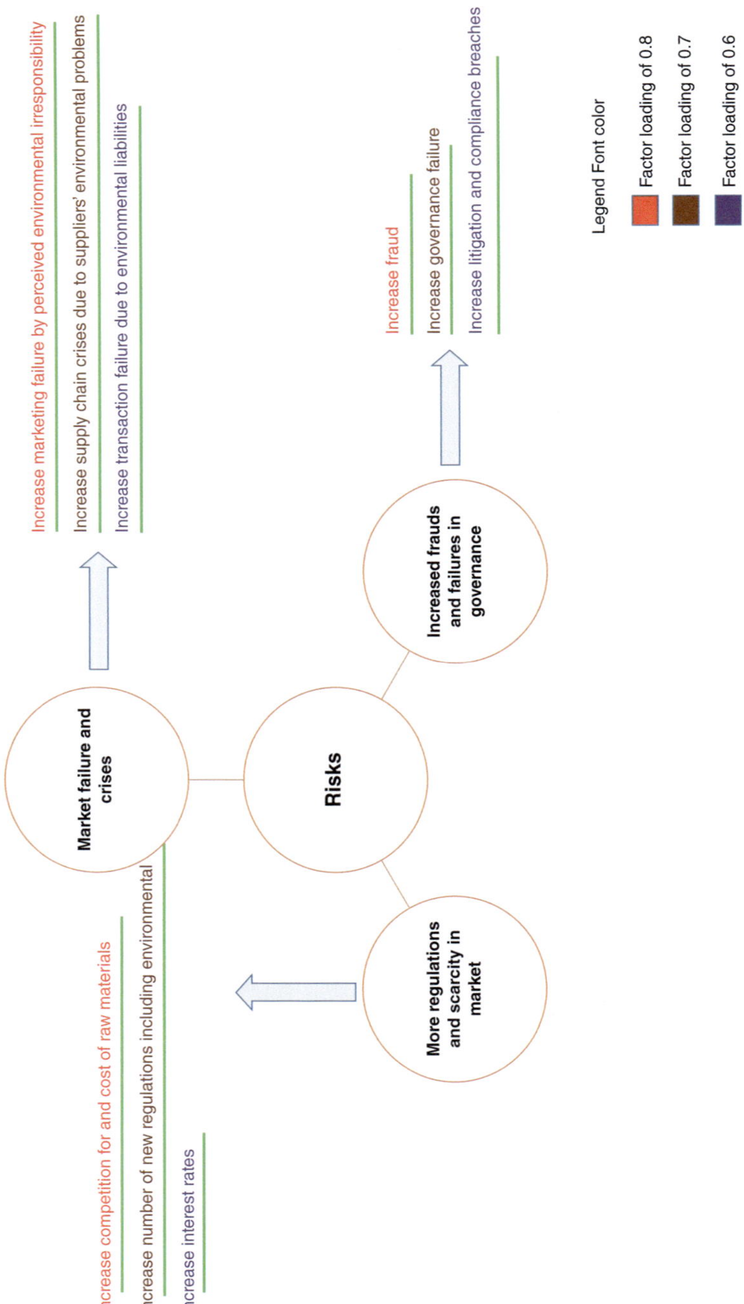


**Fig. 4** Risks of sustainability awareness in higher education (prepared by author)

"reducing the use of raw material" which had h a factor loading value of 0.7. All these sub-variables point to the advantages of social responsibilities through sustainability awareness.
- **Stakeholders' expectation:** a stakeholder expectation briefly denotes the expectations of the different stakeholders involved in the higher education systems. The sub-variables include "meet the stakeholder expectation" with a factor loading value of 0.7. "Attracting new opportunities" and "creating new jobs" are the other two sub-variables with a loading value of 0.6. All of these factors indicate the advantages for the different stakeholders, especially the institution.

Furthermore, the risks of sustainability awareness in the Indian higher education sector namely:

- **Market failure and crises:** The lack of environmental responsibility is considered the major risk or threat from the analysis. The factor "the increased market failure by perceived environmental irresponsibility" had the highest factor loading with 0.8. The crisis in the supply chain due to suppliers' environmental problems and increased transactional failures due to environmental liabilities, are the other major sub-variables under this factor.
- **Increased frauds and failures in governance:** Environmental actions and the environmental compliance failures due to the fraud activities and the poor governance emerged as potential risks in sustainability awareness. Increase in the breach of laws was also indicated in the results. The "increased fraud" had a factor loading of 0.8; the other two factors "increased governance failure" and "increased litigation and compliance breach" had factor loadings of 0.7 and 0.6 respectively.
- **More regulations and scarcity in market:** An increase in the number of regulations and the scarcity of raw materials in the market emerged as two of the factors generated. Increased competition and cost of raw materials had a factor loading of 0.8 and the increased number of environmental regulations was also considered as a potential risk. The factor loading for the increase in the number of environmental regulations was 0.7. Also, the increased interest rate was another sub-variable with a 0.6 factor loading.

## 6 Limitations

Several limitations emerged in the course of this research related to the time and scope of the research topic. Moreover, the lack of cooperation of the participants and the interviewer's inability to clarify the survey questions for respondents were also some of the main limitations. The inability reach the sample population caused a delay in obtaining reliable data. The limitations are discussed in the following section.

**Time:** As this research was conducted over the course of a semester, there was limited time to collect the data, and therefore the number of respondents was restricted to 100. Therefore, the depth of the research is limited by the small amount of the data collected, which was one of the major limitations.

**Scale of research:** as mentioned, the research was limited to 100 participants due to limited time and other constraints. The reliability of the data from a low number of the respondents is another limitation. This opens up an opportunity to broaden the survey in the future in order to obtain a greater range of responses, thereby increasing the validity of the results.

**Lack of cooperation of the participants:** Although the online survey in many fields can achieve higher response rate than the traditional survey techniques, the lack of cooperation of the participants due to information overload makes it difficult to obtain the best results from participants. When potential participants are constantly receiving spam messages, there is a high probability that the survey will be ignored. So the need to constantly communicate with participants to obtain their opinions was also a limitation.

**Lack of interviewer:** as the survey topic was new to participants, confusions regarding the survey questions could adversely affect the survey result. If a survey is not understood, then the survey results will be unreliable. The term "sustainability" is not familiar to many people due to the diversities in language and it was hard to accurately convey the meaning of the word.

**Inability to reach the sample populations:** as the research is about the higher education sector in India, conducting the survey, from Australia, of Indian participants was a tedious procedure. The lack of accessibility to the sample group was another limitation and resulted in a poor response from the sample group. Since most of the participants have English as their second or third language, they experienced problems understanding the survey.

## 7 Conclusion

In India, efforts to promote sustainability have been made in different areas, from industries to the smallest community programs. Various sustainability projects have been undertaken to educate the workforce in the industrial sector, mainly in information and technology sector, about introducing sustainability changes in their operations. In tertiary institutions, environmental studies have been introduced in all courses and sustainability-embedded universities provide a real-life example for the younger generation of how to become sustainable, and give them an insight into how sustainability works across different diverse sectors of the society. In order to incorporate sustainability education in tertiary institutions, and indeed at all levels of education, implementation methods and tools, political and economic considerations, and the foundation and purpose of education must be reconsidered.

In order to create a sustainable generation, individuals and organizations need to have the knowledge, skills, values and capacity to live a sustainable life. The acquisition of knowledge is not just about the theory; it is basically a change in lifestyle. The understanding of the advantages and risks for each actor in implementing the sustainability awareness in this sector is very important to make the better tomorrow. Perfect comprehensive understanding of the strengths and weaknesses, advantages and threats of sustainability awareness is important for the actors involved the higher education sector in order to bring about the desired change to a more sustainable lifestyle. The stakeholders including students, educators, teachers, government, shareholders and the other organizations which are related to this sector need to understand what is needed by analysing the advantages and risks.

Even though India is doing exceptionally well in terms sustainability awareness compared with other developing countries, various factors have impeded this process. The diverse nature of Indian society with respect to the economy, religion, language, politics, education, states and geographic regions, has made it difficult to implement sustainability awareness and practical measures. The quality of the awareness program and the lack of design and planning also present hurdles. On the other hand the well-regulated legal structure and the fast growing economy with the potential high quality human resource give added strength to the Indian higher education sector. The well-educated and trained professionals in this area with a strong cultural infrastructure enrich the education sector.

However, the diverse nature of the regional imbalance makes it hard to improve sustainability awareness. In terms of higher education institutions, the lack of infrastructure, strong and complex regulatory frameworks makes the situation worse. Lack of green field investments in the education sector, and poor implementation of the policies and guidelines also impede the implementation of measures to improve sustainability awareness.

The advantages which sustainability awareness creates are wide-ranging and will be different according to each actor's perspective. As the advantages for all the actors are interlinked, everyone will obtain advantages in one way or other. Potential economic stability is one of the major advantages which in turn will help to eradicate poverty and provide a better quality of life for everyone throughout the country. IT also strengths the democracy, human rights and brings peace to the nation. Effective and efficient resource management is another advantage. All the above mentioned advantages are interlinked and the actors obtain the advantages in different ways depending on their role and influence in the education sector.

The risks or threats to the actors compared to the advantages they obtain from sustainability awareness are negligible. The initial implementation stages can bring confusion to the stakeholders and this can cause the threat of not having appropriate investors. The time lag due to the diverse nature of Indian society is a potential threat to sustainability awareness in the education sector. The decline in the quality of education as a result of the sudden change of the curriculum may adversely affect the students and other stakeholders.

This research focused on Indian society which is diverse in many different ways, was aiming to understand the different advantages and risks in sustainability awareness in the higher education sector. The aim of the research was to understand the underlying advantages and risks for the stakeholders, especially the education providers. The advantages and risks which have been discovered through this research will help to increase an awareness of the newly evolved risks and advantages when introducing sustainability awareness in the higher education sector. This research will help the universities, government, educators and other actors to bring about changes in the higher education sector which is currently undergoing a transformation in order to become more sustainable. It also helps the beneficiaries of the education system to have a better understanding of the newer sustainable lifestyle.

## References

1. Ashford AN (2004) Major challenges to engineering education for sustainable development: What has to change to make it creative, effective, and acceptable to the established disciplines? Int J Sustain High Educ 5(3):239–250
2. Balasubramanian A (2012) A snapshot of the Indian education sector. *The Hindu*, 21 Nov 2012
3. Brundtland GH (1987) Report of the World Commission on environment and development: our common future. UN
4. Costello AB (2009) Getting the most from your analysis. Pan 12(2):131–146
5. Cotton D, Warren M, Maiboroda O, Bailey I (2007) Sustainable development, higher education and pedagogy: a study of lecturers' beliefs and attitudes. Environ Educ Res 13(5):579–597
6. Education for sustainability (2009). The Australian Research Institute in Education for Sustainability, Australia http://aries.mq.edu.au/publications/aries/efs_brochure/pdf/efs_brochure.pdf. Accessed 1 Nov 2015
7. Fabrigar LR, Wegener DT, MacCallum RC, Strahan EJ (1999) Evaluating the use of exploratory factor analysis in psychological research. Psychol Methods 4(3):272
8. Gliem RR, Gliem JA (2003) Calculating, interpreting, and reporting Cronbach's alpha reliability coefficient for Likert-type scales. In: 2003 Midwest Research to Practice Conference in Adult, Continuing, and Community Education, Columbus, Ohio: Ohio State University
9. Gray S (2011) State of the world 2010: transforming cultures from consumerism to sustainability. Soc Environ Account J 31(2):180–181
10. Hill BD (2011) The sequential Kaiser-Meyer-Olkin procedure as an alternative for determining the number of factors in common-factor analysis: a Monte Carlo simulation. Oklahoma State University, USA
11. Hueting R, Reijnders L (1998) Sustainability is an objective concept. Ecol Econ 27(2):139–148
12. Issa T (2013) Online survey: best practice. In: Pedro I, Miguel, Baptista N (eds) Information systems research and exploring social artifacts: approaches and methodologies. IGI Global, Hershey, PA, USA, pp 1–19. doi:10.4018/978-1-4666-2491-7.ch001
13. Likert R (1932) A technique for the measurement of attitudes. Arch Psychol 22(140):55
14. MacCallum RC, Widaman KF, Zhang S, Hong S (1999) Sample size in factor analysis. Psychol Methods 4(1):84
15. Mertens DM (1998) Research methods in education and psychology: Integrating diversity with quantitative & qualitative approaches. Sage Publications, California
16. Mula I, Tilbury D (2009) A United Nations decade of education for sustainable development (2005–14) What difference will it make? J Educ Sustain Dev 3(1):87–97
17. Mulaik SA (1990) Blurring the distinctions between component analysis and common factor analysis. Multivar Behav Res 25(1):53–59
18. Neuman WL (2005) Social research methods: quantitative and qualitative approaches, vol 13. Allyn and Bacon, Boston
19. Pallant J, Manual SS (2007) A step by step guide to data analysis using SPSS for windows version 15. Open University Press, Milton Keynes, UK

20. Stevens JP (2012) Applied multivariate statistics for the social sciences. Routledge, New york
21. Van Selm M, Jankowski NW (2006) Conducting online surveys. Qual Quant 40(3):435–456
22. Vellicer WF, Jackson DN (1990) Component analysis versus common factor analysis: some further observations. Multivar Behav Res 25(1):97–114
23. Widaman KF (1993) Common factor analysis versus principal component analysis: differential bias in representing model parameters? Multivar Behav Res 28(3):263–311
24. Williams B, Brown T, Onsman A (2012) Exploratory factor analysis: A five-step guide for novices. Australas J Paramed 8(3):1

# Sustainability in Organizations: Bhutan's Perspective

**Sonam Delma**

**Abstract** Sustainability and sustainable development are two buzzwords gaining significance globally because of glaring challenges that are threatening the sustainability of the earth in the form of climate change, shortage of energy resources, food security, global warming, etc. Unlike conventional sustainability that considers only the environmental perspective, the modern view of sustainability is three-dimensional comprising of social values, stewardship of the natural environment, and economic benefits. This study examines the factors influencing the behavior of organizations in Bhutan towards sustainability explicated in terms of opportunities and risks in order to understand Bhutan's perspective on sustainability or sustainable development. A deductive approach and quantitative online survey method were used in this study. One hundred and seventeen eligible respondents from various public organizations in Bhutan responded to the online survey questionnaires. Email was used as the main source of dissemination of the survey. Factor analysis was performed to discover underlying unobserved factors that are impacting on the observed variables. The results indicated that Bhutan places relatively high importance on the protection of the environment and reduction of health hazards, and also believes that sustainable practices in organizations can satisfy stakeholders' expectations and improve business performance. On the other hand, the study also found that Bhutan apparently feels that integrating sustainability practices in organizations' business models might increase fraud and environmental responsibility, which could escalate cost and investment, increase the number of regulations. In a nutshell, it can be concluded that Bhutan is aware of sustainability but its perception regarding sustainability is inclined more towards the environmental elements.

S. Delma (✉)
School of Information Systems, Curtin University, Perth, Western Australia, Australia
e-mail: sdelma55@yahoo.com

 
T. Issa et al. (eds.), *Sustainability, Green IT and Education Strategies in the Twenty-first Century*, Green Energy and Technology, DOI 10.1007/978-3-319-57070-9_10

# 1 Introduction

The term 'sustainable development' was first coined and introduced in 1987 by Brundtland in the 'Our Common Future' report that was published by the World Commission on Environment and Development [11]. Since then, the words 'sustainability' and 'sustainable development' have become pervasive because of the undeniable fact that our planet Earth, which provides the basic resources for our survival, is slowly losing its ability to produce these resources [27]. It has also been indicated that unsustainability of Earth is mainly caused by unlimited and unsustainable human activities carried out to fulfill our greed, not our needs. Although sustainability and sustainable development are two terms with relatively different meanings, they are used interchangeably in most of the literature [2]. Thus, in this paper they will also be used in a similar way. Brundtland describes sustainable development as utilizing resources prudently for our needs and also preserving them for the future generations. Traditional sustainability is presumably applied in relation to activities carried out to protect the environment and exercise stewardship of the environment. However, the modern concept of sustainability includes other dimensions such as societal responsibility and economic growth [27, 32]. In a similar way, Gobble also asserts that sustainability is more than about protecting the environment; it is also about considering the 'triple bottom line' suggested by Elkington (1997): 'People [47]', 'Planet (environment)', and 'Profit (Economic)' [Elkington 1997, cited in 17]. The triple bottom line (TBL) was introduced into organizations to promote corporate social responsibility (CSR) so that the business model concentrating only on profit is shifted to a broader paradigm that integrates society and environment as well. Although the world has been witnessing huge economic development recently, it is also confronted with glaring issues such as climate change, global warming, poverty, pollution, greenhouse gases (GHGs), water shortages, energy resources depletion, etc. [54] caused by an unsustainable lifestyle and development activities. It is also known that the Earth is losing its ability to produce abundant resources thereby creating disproportion between demands of human society and what the Earth is able to provide.

Developed countries have already started undertaking programs to support sustainability. However, developing countries and underdeveloped countries are still lagging behind in achieving the sustainable development goals. Therefore, this research is undertaken to comprehend Bhutan's attitude towards sustainability and sustainable development. Since organizations play important roles in sustainability, the study is focused on organizations in Bhutan so as to understand the perspective of these organizations in relation to opportunities and risks of adopting sustainable practices in their business operations. It is vital for organizations to understand the effects of their activities on our natural environment and adopt innovative solutions using new technologies to enable sustainable approaches.

For instance, investors or financial providers are now concerned about sustainability and they tend to show interest only in those companies that have integrated sustainability in their business strategies, and this could be one of the reasons why sustainability is gaining visibility amongst organizations irrespective of their size or the nature of their business. By incorporating sustainability in their business strategies, companies can become environmentally and socially responsible, and also perform better in terms of increasing profit, subsequently attracting potential investors and customers.

The United Nations is taking a lead role in shaping the future of humanity particularly in terms of sustainability. For instance, the UN conference on Sustainable Development (Rio +20 or Earth Summit 2012) in 2012 agreed to create a 'universal sustainable development goals (SDGs)' that would help countries to take a holistic approach to set targets and monitor progress in all aspects of sustainable development [49]. Further, UN led the negotiation on post-2015 sustainable development agenda, 'which aims at creating a universal agenda for all countries, both rich and poor, and includes commitments on all three dimensions of sustainable development – the social, the economic and environmental sustainability' [49]. Another initiative is the COP21 (21st Conference on Parties), a Climate Change Summit held in Le Bourget, France from 30 November to 11 December 2015 [53] to discuss measures to combat climate change.

The main purpose of the research is to examine and understand the factors, expressed in terms of opportunities and challenges, influencing the Bhutanese organizations' attitudes towards sustainability.

## 2 Sustainability

In the era of IT, a plethora of information is generated by individuals as well as by companies/organizations and the amount of information will continue to increase because of more usage of ICT devices. The information created or generated is stored in the data centers of IT companies and these centers need a large amount of power energy for their operations. A research study indicated that data centers account for approximately 1.3% of the world's electricity consumption [57]. Data centers not only consume more energy, but also generate greenhouse gases that have more impact on the environment. Organizations predominantly depend on IT, which in turn requires data centers to support their business operations and applications, and organizations in Bhutan are no exception. Although private organizations are a major contributor to economic development, they tend to consume a large portion of the available resources (energy, fuel, water, natural resources, etc.) to operate their business, which in turn affects the sustainability of the planet.

Bhutan, being a developing country, needs to carry out several developmental activities for its economic growth and also to alleviate poverty in the country. In the process of carrying out such activities, Bhutan has largely ignored issues pertaining to sustainability. The development initiatives are carried out in line with the philosophy of 'Gross National Happiness', which is middle-path to development covering social well-being, preservation of environment, and socio-economic growth [8]. Although Bhutan, as a country, understands sustainable development and has taken initiatives to incorporate the concept in its policies, a majority of the people is still not aware of sustainability and even most of the organizations do not understand the significance of becoming sustainable. Hence, the main reason for this research is to provide organizations, particularly those in Bhutan, with a better understanding of sustainability and the need to embed sustainability in their strategies, highlighting the opportunities and also presenting the challenges or risks when adopting sustainability. The significance, therefore, has two parts: the practical implication of the research, and the theoretical significance.

According to Robinson et al., implementation of sustainable practices in industries can reduce operating costs and an increase profits [Robinson et al. 2006, quoted in 50]. Furthermore, an empirical study conducted in the UK construction industry also revealed that sustainability performance reporting and sustainable strategies do help the contractors to establish a particular image and gain a competitive advantage [Adetunji et al. 2003, quoted in 50]. In addition, a research conducted by Tan et al. [50] finds that international construction companies with high performance in terms of sustainability are more likely to be more successful in the international construction market. Hence, most companies have been producing sustainability reports, abiding by the guidelines issued by the Global Reporting Initiative as such reports are beneficial to the stakeholders as well as internal management for long-term strategies [25]. However, the integrated report is now becoming a trend for companies because it includes the 'Triple Bottom Line' achievements rather than reporting only one aspect of sustainability. For example, companies such as Sony, Nova Nordisk, and Hyundai have started using integrated reporting [25].

The global problems encompassing climate change, pollution, social inequalities, poverty, and growing population are impacting on humanity and these problems are interrelated. In order to resolve these problems, it is the responsibility of every individual and private/public organization to embrace a common approach that contains sustainability concepts covering all three aspects (social, economic, and environment). This joint effort will meet the current demands while preserving resources for future generations.

It is also important to have an understanding of the opportunities of sustainability for organizations so that they will make strategic decisions to incorporate sustainability in their business strategies. Thus, this research with its findings will attempt to help practitioners and organizations that are in the process or thinking of adopting sustainable solutions to have better insights into the opportunities provided by using sustainable approaches. Since most of the public organizations in Bhutan are government agencies, the research will provide recommendations that would help the policy makers to develop policies that would promote sustainable practices in the organizations and encourage them to work towards achieving sustainable development goals [43].

As there is little literature on the advantages of sustainability to public organizations and the challenges encountered by these organizations in accepting sustainability, the research address the existing gap in the literature. In addition, the findings from the research would help researchers or academies or university students, who are conducting research on sustainability or are interested in studying sustainability in public organizations in the future, providing a better understanding of the advantages that public organizations can obtain after adopting sustainable strategies and the kinds of challenges that organizations would face when adopting sustainability concepts. Moreover, the findings will highlight and present Bhutan's perspective on sustainability.

Before discussing sustainability, it is essential to know what the term actually means. Sustainability is interchangeably used with 'sustainable development' or 'environmentally sound development' or 'ecological sustainability' [33, 37]. The literal meaning of sustainability is anything that lasts over an implicit period of time. Diverse definitions of sustainability and sustainable development are found in the literature. These definitions are defined by different entities. However, the most generally accepted is the definition given by the UN. Some of the definitions are cited below: According to the Brundtland Commission, sustainable development is defined as 'the development that meets the needs of the present without compromising the ability of future generations to meet their own needs' [10, 11, 14, 27, 30, 54]. In other words, the available resources must be utilized consciously to fulfil the present generation's needs while conserving them for future generations.

The Department of Economics and Social Development defines sustainable development as 'the bridge between environmental, economic and social goals; between north and south; between governments, civil society and business; between science and policy and between policy and action.' [DESA 2013, quoted in 54]. As defined by the International Union for Conservation of Nature and Natural Resources in 1980, sustainability is 'the integration of conservation and development to ensure that modifications to the planet do indeed secure the survival and well-being of all people' [37]. It is also defined as 'the long-term maintenance of systems according to environmental, economic and social considerations.' [Crane and Matten 2007, quoted in 10].

In 1713, the word 'Nachhaltigkeit (the German term for sustainability)' was known to be used for the first time in Europe by German forester and scientist Hans Carl von Carlowitz in the book called 'Sylvicultura Oeconomica' [23, 30]. Nevertheless, it gained recognition only in 1987 after the issuance of the 'Our Common Future', which is also known as the Brundtland report, defining the term 'sustainable development' [23, 30, 37]. In the early 1970s, a book titled 'The Limits to Growth' was published by a research group discussing the limits of exponential growth of anything – be it population, pollution, food, energy, etc. [36, 54]. The main argument of the book concerned the limit of the growth that would turn into catastrophe if we do not take appropriate actions towards proper management of the resources. It also discussed the global challenges that are impacting on human society. According to the United Nations, some of the global challenges faced in this twenty-first century are 'climate change, clean water, population growth, rich-poor gap, health issues, peace and conflict, energy' [54]. As these challenges are interlinked, addressing one would effect a systematic improvement in others.

In the late 1980s, a new discourse began; the argument shifted from 'limits to growth' to 'sustainable development', which was discussed in two reports viz. 'the World Conservation Strategy (1980)' and 'Our Common Future (1987)' [36]. These two reports were also recognized in *Agenda 21* of the 1992 Rio Earth Summit [36]. The 'Our Common Future' report was published in 1987 by the WCED, a subsidiary of the United Nations [30, 35]. Economists and environmentalists held contradicting views. Economists, on one hand, argued that there is a need for more development to eliminate poverty in the under-developed countries; the environmental activists, on the other hand, advocated that there be limits to development activities. Thus, in order to link these two contrasting opinions, Gro Harlem Brundtland, who led WCED, made an urgent call for a new approach that would address both limits and growth [36, 46] and thereby created the concept of sustainable development [37]. However, some authors proclaim that the origin of the concept came from the renewable natural resources such as forest and fisheries [30, 33], where it means not reaping more than what the forest produces [30] and protecting resources for future generations.

Even though some were skeptical that the concept would become only a catchword, it gained prominence as a developmental policy or philosophy for international aid sponsors, development partners and government agencies [35], learning subject for academics, and watchword for environmental advocates [33]. The Brundtland report also emphasizes how development activities can ensure the welfare of society without endangering the environment for future generations. In other words, it covers the economic [4] and the environmental [30] considerations.

Since its inception, sustainability has evolved over a period of time and, today, sustainability does not only mean environmental challenges but also incorporates other dimensions like social responsibility and economic growth. Considering these three dimensions in sustainability is also known as 'triple bottom line' perspective of sustainability or presented as 3Ps – People [47], Planet (environment) and Profit (economic). The concept of Triple Bottom Line was created by Elkington who aimed to promote corporate social responsibility in business organizations [17]. As per Elkington [1998, cited in 26], TBL includes three elements: 'economic properity, environmental quality and – the element which business had preferred to overlook – social justice'. The economic aspect ('Profit') deals with the financial growth and its effect on the society and the environment [12] or gross domestic product [30] whereas the social component ('people') relates to human rights, equity, health, poverty, and employment; environment is about protecting the natural world [12]. The triple bottom line concept is mainly used to implement sustainability in organizations by considering all three aspects when measuring performances [2].

Bhutan is a small landlocked country with a small population of 757,042 [5] and sandwiched between two giant countries, China in the north and India in the south, east, south-west. In 1907, the First King of Bhutan was crowned with the unanimous support from the people and from then the country was ruled by the Wangchuck dynasty until 2008, when the Fourth King initiated a democratic system. Bhutan is now a Democratic Constitutional Monarchy with an elected parliament and government and the King is the Head of State. Bhutan is the least urbanized country in the world and it started its first development plan activities in 1961 [13, 52] by building basic infrastructures such as roads, telecommunications, and transports and since then Bhutan has achieved substantial economic growth in education, health and other sectors. The development plan activities are guided by a development philosophy of Gross National Happiness (GNH) [52], which was coined by His Majesty the Fourth King of Bhutan in the early '70s. This concept captures the essence of sustainable development stating that true economic development must not be carried out at the cost of environment and social wellbeing of the people. It is considered as a 'middle path' to development balancing all aspects that include 'society', 'economy', 'culture', and 'environment' [52]. It also stresses the importance of gross national happiness rather than gross national product, which is very much consistent with the thoughts of a great Greek philosopher Aristotle recognized happiness as the ultimate goal of life [14]. With this philosophy, Bhutan places environment and its people at the center of development and considers measuring the happiness of the people along with GDP [52]. Bhutan included sustainability aspects in the GNH concept by incorporating sustainable and equitable socio-economic development, conservation and sustainable utilization and management of the environment as some of the dimensions or pillars of GNH [13].

Bhutan's development plan is for every 5 years and every plan has a theme in order to drive its focus towards achieving the objectives set out in the plan. Accordingly, the theme of the tenth 5-year plan for 2008–2013 was to reduce poverty [6] and for the eleventh 5-year plan, it was to promote green socio-economic development [13], which incorporates the social aspects of sustainability into development activities. Poverty elimination and improving the livelihood of people or communities are the priority goals in South Asian countries [52]. Bhutan made enormous progress in reducing the poverty line from 25% in 2007 to 12.7% in 2012 [7]. Further, Bhutan also integrates the UN Millennium Development Goals into its national development planning process [6]. Additionally, Bhutan has established a Sustainable Development Secretariat Office, an overarching authority to promote sustainable development in the country and to implement a sustainable development agreement within the framework of Agenda 21 of the Rio Declaration on Environment and Development (SDS). Although Bhutan's commitment to the natural environment is clearly guided by GNH, it is further supported by the Constitution mandating to maintain 60% of total land as forest coverage at all times [8]. As a result of this mandate, currently 70% of its land is still covered by forest.

Bhutan is also taking active roles in the international arena to support sustainable development. As an initiative, Bhutan proposed 'an inclusive, equitable and balanced approach to economic development that promotes environmental sustainability, well-being, and happiness as overarching goals' during the 65th Session of the United Nations General Assembly in July 2011 [9]. This proposal was passed as a resolution, which was adopted by 193 member countries [9]. Bhutan also made a commitment to combat climate change by remaining 'carbon neutral' at the UNFCCC 15th Session of Conference of Parties in Copenhagen in 2009, which means the country will ensure that greenhouse gas emissions levels do not 'exceed the sequestration capacity of its forests' [9, 45]. As such, the country was recognized as one of the few countries in the world with negative greenhouse gas emission. Further, Bhutan is a party to several multilateral agreements, and in this context it has made progress in strengthening policies and frameworks to achieve sustainable development goals. The country has integrated the concepts of green economy in its economic development policy [9].

Taking advantage of being mountainous and having an abundance of water from its rivers, Bhutan has committed to using this renewable resource for electricity production (hydropower) and minimizing consumption of fossil fuels. However, Bhutan also uses fuelwood as a principal energy source in both urban and rural areas accounting for 70% of total energy usage [45] and some households use fossil fuels such as kerosene and liquid petroleum gas (LPG) [42]. However, despite the abundant usage of fuelwood, Siebert and Belsky [45] assert that the forest coverage increased from 72% to 81% between 1995 and 2010. Bhutan is also vulnerable to the impacts of climate change because most of its economy depends on hydropower industries, which depend on renewable water resources. Notwithstanding its fine achievements and evident move towards sustainability, Bhutan does lack implementation of sustainability strategies due to resource constraints, human capacity, and legal frameworks.

The world has shown incredible achievements in alleviating poverty, increasing life expectancy, reducing infant mortality, increasing the number of children at school, increasing the adult literacy rate, and producing global food [11]. However, on the other hand, society is also facing global challenges including increased rate of poverty in some areas, shortage of safe drinking water, the widening gap between rich and poor, doubts about a secure food supply due to increasing population, climate change and global warming due to greenhouse gases, shortage of natural energy sources, productive land turning into desserts, human health issues, toxic waste, etc. [11, 54]. Virakul [54] argues that there is disparity between the demands made by the population and the amount of resources supplied by the planet. He also states that the current generation has consumed most of the resources and, if action is not taken, then future generations will be adversely affected. This is supported by a survey carried out by Accenture in 2013 highlighting that by 2050 the world will need a huge amount of resources, equivalent to 2.3 planets, to meet the needs of population of more than 9 billion and by 2030 it will suffer 40% shortage of water and also 11% energy reduction per annum until 2050 [22]. Several other research papers also mention the unsustainable consumption of natural resources and the need to change our habits [54].

Although access to affordable clean energy is necessary for improving the livelihood of society, studies have shown that 1.4 billion of the world population still lacks access to electricity [28, 42] thereby compelling them to use alternative energy sources. This may result in using methods that produce carbon emissions, which in turn cause global warming and climate change [42]. The scientific community has also agreed that human activities are causing global warming rapidly leading to climate change which poses a significant threat to the sustainability of the earth [24, 48]. Hence, the Kyoto protocol was created in 1997 based on the assessment carried out by the Intergovernmental Panel on Climate Change (IPCC) establishing a commitment to reduce greenhouse gas emissions [24]. Further, IPPC has stated that the increased energy consumption is increasing carbon emissions [34]. In regard to energy consumption and carbon emission, China was the highest consumer of energy as well the largest contributor to carbon emission in the world during the year 2010 [Amco 2011, cited in 28, 34].

In addition, the World Business Council for Sustainable Development reports that there is a wide gap between the rich and poor, especially in the developing countries as 20% of the population in the world holds 75% of the wealth and only 2% of the world's wealth is owned by 20% of the poorest population [14]. Fonseca [14] asserts that climate change is one of the greatest challenges facing humanity and developing countries will suffer more from the consequences of climate change. Besides, it is estimated that developing countries will be contributing 91% of global greenhouse gas (GHG) emissions by 2030 because of their increased consumption of coal and global energy [44]. This may be because developing countries are giving priority to economic development rather than GHG abatement. Further, India and South Africa are dependent on coal for energy generation and they account for more than two-third of GHG emissions. Similarly, in Brazil the GHG emissions are mostly from 'agriculture, land-use and forest management practices' [44]. The global energy consumption is forecasted to increase from 12,467 million tons of oil equivalent (Mtoe) in 2008 to 14,765 Mtoe by 2020 leading to increase in energy related carbon emissions and the major growth will be in the developing countries [28] The International Energy Agency [40] has predicted that the introduction of carbon capture and storage (CCS) technology in developing countries will reduce one-fifth of global GHG emissions by 2050 and various CCS projects and initiatives have already been implemented in Algeria, China, Malaysia, Indonesia, and in several Middle Eastern countries. Nevertheless, there are many challenges that need to be addressed before implementation of CCS initiatives such as supporting infrastructure and technologies, and also amendments of laws to support CCS projects [44].

Furthermore, it is understood that organizations are become more responsive to sustainability concerns because of the impact of their business operations on the environment, society and economy [1]. At the same time, these organizations, as the main players, have the potential to combat global issues and make a difference. Thus, the leading companies are designing sustainable strategies to address the growing global problems [54]. In other words, the global challenges are compelling organizations to become sustainable. Nevertheless, a survey showed that the business leaders from 103 countries generally believe that business is not striving hard to solve global sustainability challenges [22]. Even though most of the companies or consumers consider sustainable practices as reducing only environmental damages, they can also add value to society by initiating social responsibility.

The whole world is responding to the issues that are threatening sustainability. Various initiatives are being taken at both local and global levels around the world.

For instance, the ABC NEWS (9 November 2015) reported that 'Australia's property and banking industries' are making moves to increase efficiency and minimize carbon emissions by switching to LED lighting in commercial buildings. This indicates that there is a strong support from business communities to address climate change and make the society carbon free. The United Nations Conference on Climate Change, COP21, held in Le Bourget, France from 30 November to 11 December 2015 [53] is an indication of unification of all nations, be it developed or developing countries, to achieve one common goal i.e. to plan appropriate measures in addressing climate change. Similarly, China also made a commitment during the UN Copenhagen Conference on Climate Change in 2009 to reduce its carbon footprint below 40–45% by 2020 [34].

Everything has a limit in this world and the same goes for our environment and the planet Earth. We do not realize that our unlimited human greed is exceeding our needs and pushing the limits of what our Earth can offer. Human actions are directly or indirectly causing climate change, land degradation, global warming, poverty, famine, water shortages, pollution, etc. Consequently, the earth will become unsustainable if we do not change and continue doing what we have been doing till now. Therefore, it is high time that we act together, not as individuals but as global partners, to become more sustainable so as to reduce impacts at societal level, improve quality of life, waste less, pollute less, and save resources for our future generations. In doing so, the earth, in return, will provide the materials and renewable resources that are required for our survival.

Conventional business organizations were focused only on making profits. Mackey (2005) argues that the modern organization is made up of many stakeholders, and companies with a focus on profit generation by considering 'customer satisfaction as a means to an end' do not inspire employees or satisfy stakeholders and could cause damage to the environment as well as to the communities in which they are operating [Mackey 2005, cited in 37]. However, with the introduction of sustainable development in 1987, organizations gradually have come to understand the importance of sustainability for their future survival and they are taking various initiatives to integrate sustainability concepts in their business operation and become green business. For instance, a survey conducted by KPMG in 2008 has implied that sustainability is a vital factor for companies to become innovative [10]. Since many stakeholders are involved in comtemporary companies, it is also important for these companies to focus on social and environmental concerns rather than becoming profit-centric. This approach can leverage the companies to improve their reputation, creating brand and profit at the same time [37]. Further, Hamel suggests that the holistic perspective to make profits through sustainable principles would promote innovation, thus gaining 'long-term competitive sustainable advantage' [Hamel 2000, cited in 37]

The extant literature widely discusses the importance of sustainable development and sustainability in higher education but few researches examine the factors that are driving or preventing organizations from integrating sustainability in their businesses. Yet, in regards to advantages, some authors put forward that adoption of sustainable practices can lead to cost reduction, increase in revenue, improved reputation, and greater access to markets [Robinson et al. 2006, cited in 50]. In addition, Fonseca [14] also mentions that differentiating products and services through innovation is driving Portuguese organizations to implement sustainability. Similarly, desire to reduce cost and management risk, improve quality, reuse water, satisfy customers, use of renewable energy source, provide employment, satisfy investor's requirement can also be some factors encouraging organizations to move towards sustainability [2]. Sustainable practices in organizations also attract quality employees [Drews 2010, cited in 37] and enhance employee's emotional attachment [Allen and Meyer 1996; Rego et al. 2010, cited in 37] thereby reducing 'employee turnover' and absenteeism [Meyer et al. 2002, cited in 37], and also showing 'customer-oriented behaviors' [Chang and Lin 2008, cited in 37].

Further, Adetunji et al. also assert that a sustainability strategy with efficient stakeholder reporting helps to build a company's creditability or reputation [Adetunji et al. 2003, cited in 50]. For instance, sustainability reporting, which is also known as triple bottom line (TBL) reporting is used as a mean to report to the stakeholders showing how the organization is performing in terms of sustainability [26]. Sustainability reporting is a response made by companies, being a major contributor to greenhouse gases, to demonstrate their corporate responsibility towards the environment [48]. Sustainability reporting includes a climate change reporting element [48]. In order to promote sustainability reporting, an autonomous organization, the Global Reporting Initiative [10] was established in 1997, which is responsible for the design of a reporting framework for organizations [26]. In recent times, stakeholders have put pressure on companies to produce sustainability reporting [3, 26]. Because of this, most companies are making moves to integrate sustainable practices in their strategies to demonstrate corporate social responsibility (CSR). In contrast, a study indicates that companies adopt CSR with an intention to please investors or to promote their image to their stakeholders [3]. CSR is becoming a part of every company and it is an attempt to build brand value and profit through enhancing social responsibility of the company towards its stakeholders including customers, suppliers, employees, and investors or towards to the communities in which it is operating [32, 37]. Nevertheless, CSR considers the social dimension but ignores the environmental responsibility [37]. The Dow Jones Sustainability Indexes (DJSI), which was created in 1999 by the Dow Jones Index, take a holistic approach to measure financial performance and drive companies to adopt sustainable business strategies [39].

Moreover, sustainable practices also help in maintaining competitive position [2] resulting in increased productivity, new opportunities, and ultimately improving the company's economic growth [50]. In addition, investors are investing in those companies that are socially and environmentally sustainable [39], thereby encouraging companies to adopt sustainable practices and begin reporting their sustainability performance. With the intention of creating social values by improving the quality of life of people in developing countries, organizations are providing financial assistance or initiating programs in these countries [29].

It was noted from the existing literature that there are many benefits that organizations can harness if they take a holistic approach to becoming sustainable. The benefits are identified in terms of opportunities and grouped into four broad categories namely financial, brand and reputation, human resources and shareholders, and environmental as shown in Fig. 1.

On the other hand, there are also risks or challenges associated with sustainable endeavors as sustainability is a complex and complicated concept that comprises numerous factors. The existing literature rarely discusses the risks encountered by organizations when integrating sustainable solutions. Nevertheless, some researchers claim that because of the strict enforcement of environmental regulations, companies are burdened with the additional costs of implementing environmental protection practices and thus retarding the economic growth of the company [50]. Although companies are taking initiatives to become green businesses, they need more capital. McCarthy stated that $3.4 billion was predicted to be the investment in green technologies in companies from 2006 to 2009 [McCarthy 2006, cited in 39]. Likewise, Tan et al. [50] state that the benefits of sustainable practices decrease after a peak point because of the increased cost of investment. Ahi and Searcy [2] also mention that the increased cost due to penalties for non-compliance with legal requirements can be a deterring factor for sustainability.

In order to reduce carbon emissions, there has been an increase in regulations, such as those established early in 2012 by the European Union Emission Trading Scheme (EU ETS), which includes 27 EU members in addition to Norway, Iceland and Liechtenstein [24]. Organizations may have to incur costs to comply with these regulations and this might discourage organizations from embracing sustainability. In addition, stakeholders expect companies to abide by laws and behave in an ethical manner [32]. Similarly, Ahi and Searcy [2] assert that work related crisis or accidents, customer dissatisfaction, regulations and increased competition challenges imposed by the competitors discourage companies from becoming sustainable. Other challenges include the high cost of renewable energy, cost involved in disposing of hazardous waste, environmental liabilities, and supply chain crises due to breach of environmental legal requirements.

Similar to opportunities, the risk factors are also classified into failure and cost categories as presented in the Fig. 2.

## 3 Research Questions and Methods

The main purpose of the research is to examine the factors influencing public organizations in Bhutan to adopt sustainability, and these factors are expressed in terms of opportunities and challenges. In order to fulfill the objectives, the research is guided by a primary research question. The primary research question is directly based on the purpose of the research and it is broader, in the sense that it looks at the opportunities and challenges faced by public organizations particularly in Bhutan. Although there are many literatures related to sustainability in

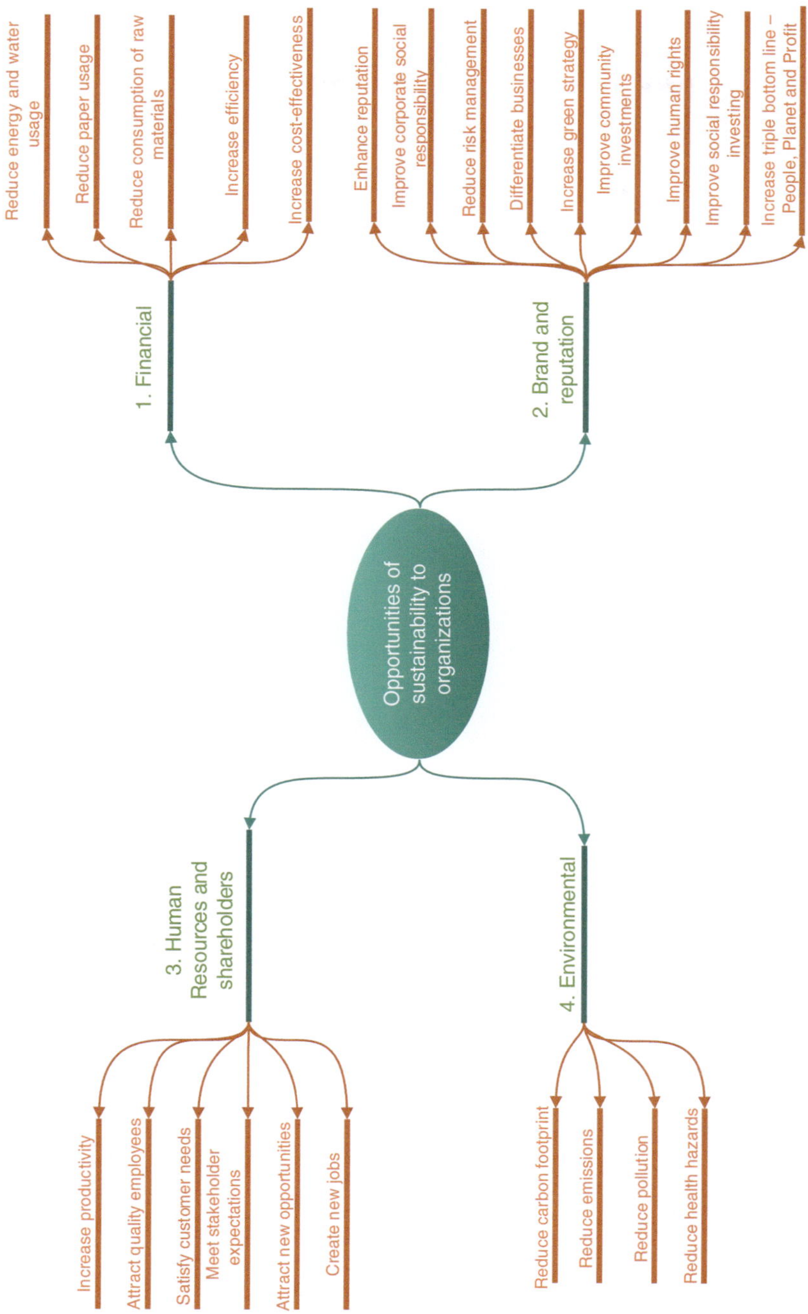


**Fig. 1** Opportunity factors

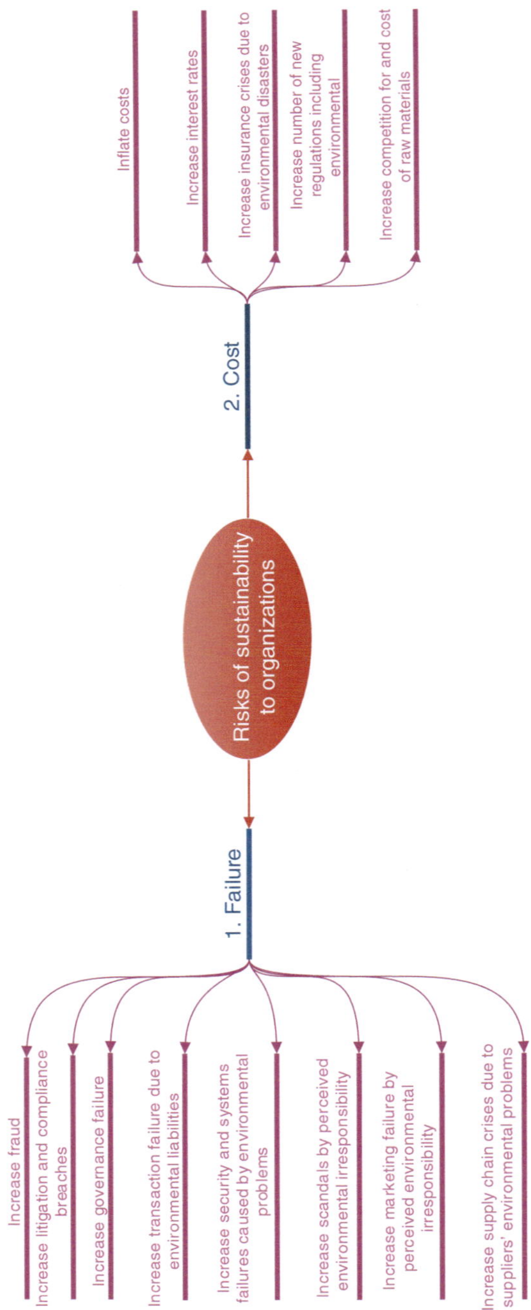


**Fig. 2** Factors related to risks of sustainability

business organizations, no research has been carried out specifically in Bhutan in relation to sustainability in public organizations. Thus, in order to fill the gap in the literature, the primary research question is: What are the factors that are influencing the attitude of Bhutanese public organizations towards sustainability in terms of opportunities and challenges? What is Bhutan's perspective on sustainability?

From the extent literature, it is known that those organizations that take sustainability as an opportunity and incorporate into their strategies can become more innovative and are able to stand tough competition imposed by competitors in the market [19]. They also mention that organizations should consider all facets of sustainability (social, economic, and environmental) in order to have long-term success. For instance, Savtiz and Weber [2006, quoted in 15] mention the companies under the Business Council for Sustainable Development were performing better in terms of their stock exchanges.

Even though top managements and academics agree that sustainability contributes to the success of both companies and communities in which they are operating, the drive towards becoming a sustainable organization depends on the leaders [15]. Despite the opportunities offered by sustainability, there are also some challenges faced by organizations when initiating sustainable endeavors. Often, the complexity of the organizational culture poses an immense challenge for the leaders who are making an effort to create awareness among the employees of the importance of sustainability for the organization's success and the wellness of our planet [15].

The chosen research method for the current study was the *quantitative* method using the online survey technique to gather data. The study was based on the *positivist* paradigm and adopted a *deductive* approach whereby the study intends to analyze data collected from public organizations in Bhutan to find out whether the data substantiate research question or objective in regard to factors, in terms of opportunities and challenges, influencing organizations' intentions to become sustainable. The positivist paradigm was chosen because the author believes that there are factors in the world that could affect the attitudes of organizations regarding their adoption of sustainability. The quantitative method was adopted for this study because the author intents to gather numerical data that fairly represent the population to a certain extent. The qualitative method was not applied because it is expensive and difficult to collect from a larger sample group and also it is not appropriate for numerical data. The online survey questionnaire technique was selected for collecting data from a selected sample of respondents. The online survey involved self-completed questionnaires, which are completed by the respondents without the presence of the researcher. This technique seems to be cost effective and faster than any other techniques [21]. It is cost effective in terms of overhead cost incurred in printing questionnaires, postages, and data-entry. Online surveys are flexible and highly flexible because they can be accessed from anywhere at any time regardless of geographical location. Such surveys guarantee anonymity [31], which might encourage respondents to share their true feelings or opinions without fear of being judged. In addition, online surveys help to provide answers that are less influenced by the survey administrator [20] as opposed to offline surveys.

However, on the other hand, the drawback of online surveys is that they are limited to only specific groups in the target population – those who have access to computers and the Internet. Furthermore, although, online surveys can reach a wider group of the target population, there is no assurance that all recipients will take part and finish the survey questionnaires, which may lead to a low rate of valid responses [31]. Also, if the participant needs clarification on some questions, there is no way of asking for further explanation. Consequently, this might result in wrong interpretation or misunderstanding of the questions. Moreover, additional information cannot be provided to support responses given by respondents. Online surveys may discourage users from participating if the presentation and design (user interface) of the questionnaires are not aesthetically pleasing. Additionally, users may discard the email containing the link to the survey due to the fear of a computer virus [20] and phishing emails. The chosen research method for the current study was the *quantitative* method using the online survey technique to gather data. The study was based on the *positivist* paradigm and adopted a *deductive* approach whereby the study intends to analyze data collected from public organizations in Bhutan to find out whether the data substantiate research question or objective in regard to factors, in terms of opportunities and challenges, influencing organizations' intentions to become sustainable. The positivist paradigm was chosen because the author believes that there are factors in the world that could affect the attitudes of organizations regarding their adoption of sustainability. The quantitative method was adopted for this study because the author intents to gather numerical data that fairly represent the population to a certain extent. The qualitative method was not applied because it is expensive and difficult to collect from a larger sample group and also it is not appropriate for numerical data. Finally, The online survey outcomes were analyzed by using SPSS software version 24.

# 4 Results

The target population is the employees working in public organizations in Bhutan. Email was used as the main means of distributing and disseminating the online survey questionnaires. Apart from emails, Facebook and WeChat, an instant messaging app, were also used for disseminating the survey. On 22nd July 2015, the survey was circulated to 85 employees from various organizations in Bhutan and it was further forwarded to another 130 employees on 16th August 2015, making a total of 215 participants. Out of 215, only 117 respondents completed the whole survey resulting in a response rate of 54.42% (See Table 1). Table 1 also shows that more than half (59.72%) of the respondents were males.

Table 2 illustrates the number of respondents according to their age groups and qualifications. As shown in the table, the respondents' ages range from 21 to 60 years and the majority of respondents were aged between 25 and 35 years, accounting for a total of 109 respondents (25–30 years = 58 and 31–35 years = 51) and representing an aggregated percentage of 75% (25–30 years = 40% and

**Table 1** Statistics of respondents from Bhutan (prepared by the author)

| Number of surveys and response rate | |
|---|---|
| Questionnaires distributed | 215 |
| Questionnaires returned | 144 |
| Valid responses | 117 |
| Response rate | 54.42% |
| *Gender* | |
| Male | 86 (59.72%) |
| Female | 58 (40.28%) |

**Table 2** Number of Bhutanese respondents according to age and qualification (prepared by the author)

| Answer | Number of respondents | % |
|---|---|---|
| *Age* | | |
| 17 years and under | 0 | 0 |
| 18–20 | 0 | 0 |
| 21–24 | 2 | 1 |
| 25–30 | 58 | 40 |
| 31–35 | 51 | 35 |
| 36–40 | 19 | 13 |
| 41–45 | 11 | 8 |
| 46–50 | 1 | 1 |
| 51–55 | 2 | 1 |
| 56–60 | 2 | 1 |
| 61–65 | 0 | 0 |
| Over 65 | 0 | 0 |
| *Qualification* | | |
| Primary Education | 1 | 1 |
| Higher Secondary/Pre-University | 1 | 1 |
| Professional Certificate | 2 | 1 |
| Diploma | 9 | 6 |
| Advanced/Higher/Graduate Diploma | 1 | 1 |
| Bachelor's Degree | 52 | 36 |
| Post Graduate Diploma | 11 | 8 |
| Master's Degree | 66 | 46 |
| PhD | 0 | 0 |
| Others | 2 | 1 |

31–35 years = 35%). Further, it demonstrates various educational backgrounds of the respondents and it can be seen that most of the respondents, representing 45% (see Table 2), have a master's degree qualification followed by bachelor's degree for 36% of the total respondents.

**Table 3** The usage of computer, Internet, SNS, email by Bhutanese respondents (prepared by the author)

| Answer | Hours spent on the computer per day | | Hours spent on the Internet per day | | Hours spent on the social networking sites per day | | Hours spent on the Internet for emails per day | |
|---|---|---|---|---|---|---|---|---|
| | No. of responses | % | No. of responses | % | No. of responses | % | No. of responses | % |
| Less than an hour | 1 | 1 | 12 | 8 | 88 | 61 | 84 | 58 |
| Up to 5 h | 27 | 18 | 76 | 52 | 51 | 35 | 52 | 36 |
| 5–10 h | 100 | 68 | 45 | 31 | 5 | 3 | 8 | 6 |
| 10–20 h | 16 | 11 | 12 | 8 | 0 | 0 | 0 | 0 |
| Over 20 h | 2 | 1 | 0 | 0 | 0 | 0 | 0 | 0 |

**Table 4** Demonstrating the purpose of using Internet in their workplace amongst Bhutanese users (prepared by the author)

| Answer | No. of responses | % |
|---|---|---|
| Email | 137 | 94 |
| Playing games | 9 | 6 |
| Studying | 109 | 75 |
| Working | 129 | 89 |
| Shopping online | 33 | 23 |
| Chatting | 79 | 54 |
| Researching hobbies | 50 | 34 |
| Banking online | 58 | 40 |
| Buying goods or services | 19 | 13 |
| Buying stocks or investing online | 1 | 1 |
| Researching travel information or making reservations | 34 | 23 |
| Others | 9 | 6 |

Table 3 illustrates the number of hours spent by the Bhutanese respondents on computer, Internet, social networking sites, and emails in a day. It can be seen that 68% of the respondents spend 5–10 h on the computer in a day while 52% spend up to 5 h on the Internet. In addition, respondents also spend less than an hour on social networks representing 61%, whereas 58% spend less than an hour on checking emails per day.

Additionally, the survey has confirmed that a majority (94%) of Bhutanese respondents use the Internet for accessing emails. Another 89% and 75% use the Internet for work and studying respectively (see Table 4).

Furthermore, the survey result also confirms that laptops and smartphones are the two most popular electronic devices used by the Bhutanese users to access Internet, indicating 85% for laptops and 73% for smartphones (see Table 5).

**Table 5** Electronic devices used to access Internet in Bhutan

| Answer | No. of responses | % |
|---|---|---|
| PC | 43 | 30 |
| Desktop | 62 | 44 |
| Laptop | 120 | 85 |
| Netbook | 1 | 1 |
| PDAs | 2 | 1 |
| Workstation | 3 | 2 |
| Tablet | 28 | 20 |
| Smartphone | 104 | 73 |
| Others | 1 | 1 |

**Table 6** Various means through which the concepts of sustainability and green IT are first introduced to the Bhutanese users

| Answer | No. of responses | % |
|---|---|---|
| School | 20 | 14 |
| Higher education | 45 | 32 |
| Internet | 86 | 61 |
| Books | 22 | 15 |
| Magazine | 18 | 13 |
| News media | 55 | 39 |
| Conferences | 35 | 25 |
| Others | 12 | 8 |

**Table 7** Demonstrating the companies associated with devices used by the Bhutanese respondents

| Answer | No. of responses | % |
|---|---|---|
| Apple | 49 | 35 |
| Google | 30 | 21 |
| Dell | 49 | 35 |
| IBM | 0 | 0 |
| Others | 14 | 10 |

Similarly, Table 6 illustrates how the Bhutanese respondents were first introduced to the concepts of sustainability and green IT. It is evident from the table that these concepts were first introduced to the respondents via Internet, news media, and higher education indicating 61%, 39%, and 32% respectively (see Table 6). On the other hand, 8% of the respondents had never heard of sustainability or green IT concepts.

Furthermore, the survey also indicated both Apple and Dell are popular brands among the Bhutanese users in terms of devices each accounting for 35% (see Table 7). However, the survey result indicates that IBM is an unpopular brand amongst Bhutanese users.

**Table 8** Respondents who read sustainability reports of a company

| Answer | No. of responses | % |
|---|---|---|
| Yes | 26 | 18 |
| No | 69 | 49 |
| Maybe | 35 | 25 |
| Not at all | 11 | 8 |

**Table 9** Showing frequency of changing devices by the Bhutanese respondents

| Answer | No. of responses | % |
|---|---|---|
| Every 6 months | 1 | 1 |
| Every 12 months | 5 | 4 |
| Every 18 months | 12 | 9 |
| Every 24 months | 23 | 16 |
| Every 30 months | 5 | 4 |
| Every 36 months | 17 | 12 |
| Every 42 months | 27 | 19 |
| Others | 51 | 36 |

**Table 10** Reasons for changing devices in Bhutan

| Answer | No. of responses | % |
|---|---|---|
| Size | 15 | 11 |
| Speed | 35 | 25 |
| Functionality | 102 | 74 |
| Keeping up with technology | 70 | 51 |
| Others | 11 | 8 |

Table 8 indicates whether the Bhutanese respondents read the sustainability reports produced by the companies. It can be seen that 49% (see Table 8) do not care about sustainability or are not interested in reading the sustainability reports. However, only 18% (see Table 8) say they read the report before buying any electronic devices, which could indicate that more awareness should be created on sustainability.

Further, the survey result showed that 19% of the Bhutanese respondents change their devices every 42 months, 26% every 24 months, and 12% every 36 months (see Table 9).

As indicated in the Table 10, a majority of the respondents, representing 74%, has provided functionalities as the main reason for changing their devices. Another 51% (see Table 10) denoted that they change their devices to keep up with technology.

Similarly, Table 11 demonstrates whether the respondents are aware of the damages caused to our planet due to frequent change of devices. A majority (78%) of the respondents agree that their habit of changing devices is causing damage to our planet. On the other hand, 19% of the respondents are not sure of the effect of their actions on our planet earth.

**Table 11** Respondents' understanding of the damages to our planet

| Answer | No. of responses | % |
|---|---|---|
| Yes | 108 | 78 |
| No | 3 | 2 |
| Maybe | 27 | 19 |
| Not at all | 1 | 1 |

**Table 12** Various methods of changing the mindsets of designers and users

| Answer | No. of responses | % |
|---|---|---|
| Training | 57 | 40 |
| Education | 84 | 60 |
| Awareness | 122 | 87 |
| Workshop | 51 | 36 |
| Internet | 83 | 59 |
| T.V. | 71 | 50 |
| Social networking | 92 | 65 |
| Others | 4 | 3 |

Table 12 illustrates various methods of changing the mindsets of designers and users. From the result of the survey, it can be confirmed that creating awareness (87%) would play a significant role in changing the mindset of designers and users. This can be achieved through education (60%) using tools such as social networking sites (65%) and Internet (59%).

Descriptive analysis was performed on all variables measuring opportunities of adopting sustainability in the organizations from Bhutan in order to present descriptive statistics including mean and standard deviation.

It can be observed from Table 13 that statistical means of all variables are populated between the scale of 3 and 4, which are 'neutral' and 'agree' as per the Likert's 5-point scale used in the survey. The highest mean was 4.32 with a standard deviation of 0.927. Likewise, the lowest mean was 3.45 with a standard deviation of 1.007. There were 10 variables with mean greater than 4. This clearly shows that most of the respondents either strongly agree or agree that some of the variables could influence the organizations in Bhutan to become sustainable. In addition, it also demonstrates that some of the respondents neither agree nor disagree with some of the variables.

Likewise, Table 14 also presents the descriptive statistics for the variables measuring the risks of adopting sustainability in the organizations in Bhutan. According to the table, the means range from 2 to 3 and the highest mean was 3.48 with a standard deviation of 1.072 from the mean and also the lowest was 2.57 with a standard deviation of 1.077. Although it can be seen statistically that respondents neither disagree nor agree with a majority of the variables identified as deterring factors for sustainability adoption, they also tend to disagree with some items.

**Table 13** Descriptive statistics for opportunity factors (prepared by the author)

| Descriptive statistics | | | | |
|---|---|---|---|---|
| Items | N | Mean | Std. deviation | Variance |
| Reduce energy and water usage | 116 | 4.16 | 0.956 | 0.915 |
| Reduce paper usage | 117 | 4.23 | 0.986 | 0.972 |
| Reduce consumption of raw materials | 117 | 4.06 | 0.985 | 0.971 |
| Increase efficiency | 115 | 4.03 | 1.047 | 1.096 |
| Increase cost-effectiveness | 117 | 3.97 | 1.046 | 1.094 |
| Enhance reputation | 116 | 3.90 | 0.917 | 0.841 |
| Improve corporate social responsibility | 116 | 4.15 | 0.916 | 0.839 |
| Reduce risk management | 116 | 3.67 | 0.912 | 0.831 |
| Differentiate businesses | 116 | 3.53 | 0.889 | 0.790 |
| Increase green strategy | 117 | 4.26 | 0.948 | 0.899 |
| Improve community investments | 116 | 3.71 | 0.951 | 0.905 |
| Improve human rights | 116 | 3.49 | 1.083 | 1.174 |
| Improve social responsibility investing | 116 | 3.91 | 0.992 | 0.984 |
| Increase triple bottom line – People, Planet and Profit | 117 | 3.79 | 0.979 | 0.958 |
| Increase productivity | 116 | 3.53 | 1.059 | 1.121 |
| Attract quality employees | 115 | 3.64 | 1.010 | 1.021 |
| Satisfy customer needs | 116 | 3.50 | 1.000 | 1.000 |
| Meet stakeholder expectations | 116 | 3.45 | 1.007 | 1.015 |
| Attract new opportunities | 116 | 3.73 | 0.898 | 0.806 |
| Create new jobs | 116 | 3.70 | 1.023 | 1.047 |
| Reduce carbon footprint | 116 | 4.28 | 0.958 | 0.918 |
| Reduce emissions | 117 | 4.32 | 0.927 | 0.859 |
| Reduce pollution | 115 | 4.28 | 0.987 | 0.975 |
| Reduce health hazards | 115 | 4.10 | 1.095 | 1.200 |
| Valid N (listwise) | 100 | | | |

The Cronbach's Alpha and KMO and Bartlett's Test were conducted on a total of 117 valid responses to ascertain the reliability of the data collected and adequacy of the sample dataset.

It is important to determine Cronbach's Alpha coefficient when the Likert Scale is used as a measurement instrument in a research [16]. Alpha is expressed in terms of coefficient values between 0 and 1 and it gives 'a measure of internal consistency of a test or a scale' [51]. The most acceptable coefficient values of Alpha ranges from 0.70 to 0.90 according to Tavakol and Dennick [51]. The nearer the coefficient value is to 1, the higher is the internal consistency within the items. In other words, high Alpha values imply high reliability of test data and less measurement error. However, a high Alpha value does not determine whether

**Table 14** Descriptive statistics for risk factors of sustainability

| Descriptive statistics | | | | |
|---|---|---|---|---|
| Items | N | Mean | Std. deviation | Variance |
| Increase fraud (including environmental actions and environmental compliance failures) | 116 | 2.66 | 1.237 | 1.529 |
| Increase litigation and compliance breaches (including environmental actions and environmental compliance failures) | 117 | 2.77 | 1.255 | 1.576 |
| Increase governance failure (including environmental actions and environmental compliance failures) | 117 | 2.58 | 1.176 | 1.383 |
| Increase transaction failure due to environmental liabilities | 116 | 2.65 | 1.144 | 1.309 |
| Increase security and systems failures caused by environmental problems | 116 | 2.71 | 1.103 | 1.218 |
| Increase scandals by perceived environmental irresponsibility | 117 | 2.70 | 1.124 | 1.263 |
| Increase marketing failure by perceived environmental irresponsibility | 117 | 2.57 | 1.077 | 1.161 |
| Increase supply chain crises due to suppliers' environmental problems | 117 | 2.76 | 1.080 | 1.166 |
| Inflate costs | 117 | 3.02 | 1.067 | 1.138 |
| Increase interest rates | 115 | 2.91 | 1.048 | 1.098 |
| Increase insurance crises due to environmental disasters | 117 | 2.87 | 1.071 | 1.147 |
| Increase number of new regulations including environmental | 117 | 3.48 | 1.072 | 1.148 |
| Increase competition for and cost of raw materials | 117 | 3.27 | 1.096 | 1.200 |
| Valid N (listwise) | 112 | | | |

**Table 15** Alpha test results for opportunity and risk aspects of sustainability (prepared by the author)

| Group | Cronbach's Alpha |
|---|---|
| Opportunities | 0.956 |
| Risks | 0.951 |

the scale is one-dimensional, so factor analysis must be applied to understand the dimensionality of the scale [16].

Table 15 shows the Cronbach's Alpha test performed on 37 items measuring both opportunities and risks in relation to sustainability in Bhutanese organizations. The test results demonstrated very high Alpha coefficient values of **0.956** (excellent) for opportunities and **0.951** (excellent) for risks (see Table 15) indicating high reliability [16]. From these two test results, it can be deduced that the attributes identified to measure opportunities and risks that influence the behavior of the organizations in adopting sustainability are reliable and consistent.

**Table 16** The KMO and Bartlett's test for both opportunities and risks aspects of sustainability (prepared by the author)

| Group | KMO measure of sampling adequacy | Bartlett's test of sphericity | | |
|---|---|---|---|---|
| | | Approx. Chi-Square | Df | Sig. (p value) |
| Opportunities | 0.903 | 2140.516 | 276 | 0.000 |
| Risks | 0.912 | 1363.543 | 78 | 0.000 |

Before applying factor analysis, the dataset was subjected to measures such as the Kaiser-Meyer Olkin (KMO) measure of sampling adequacy and Bartlett's test of sphericity. KMO is a coefficient value measured between 0 and 1 and it 'compares the size of the observed correlation coefficients with the magnitude of the partial correlation coefficients' [18]. A high KMO value (0.7 to 0.9) indicates that the sample data is highly adequate and a value less than 0.5 indicates that the data is not appropriate for factor analysis. Likewise, the Bartlett's test for sphericity is also used for testing the significance of correlations among the variables and it is considered significant if the value is less than 0.05 [Hair et al. 1998, cited in 18].

The results of the KMO and Bartlett's test carried out on 37 items measuring both opportunities and risks of sustainability in Bhutanese organizations are shown in Table 16. The test results demonstrated high KMO coefficient values of **0.903** for opportunities and **0.912** for risks (see Table 16) signifying excellent sample size for carrying out analysis (Hill 2012). Further, it indicated a high Bartlett's significant level (p = **0.000** < **0.05**) for both opportunities and risks (see Table 16) clearly showing that adequate correlation is found within the sample dataset in order to continue with factor analysis [18]. Therefore, it is statistically confirmed that factor analysis can be performed on this dataset.

Factor analysis is a multivariate analysis carried out to examine highly intercorrelated observed variables that could be measuring similar constructs and grouping them under newly identified underlying factors [38]. In this way, it focuses on important key factors by reducing dimensionality or the number of variables that could be insignificant [56]. The underlying factors are normally unobserved and not measured but they nevertheless influence respondents' answers to observed variables [38]. Factor analysis was performed mainly to study a set of observed variables measuring opportunities and risks related to sustainability in organizations in Bhutan and to examine whether these observed variables share a common construct. Further, highly correlated variables within a subset of these variables were grouped together under unobserved labeled factors. By doing so, the number of variables was also reduced. Factor analysis was performed separately for items measuring opportunities and risks of sustainability in the Bhutanese organizations. The analysis resulted in various outputs, which are explained below.

Communalities are extracted using the principal component analysis technique which is used to reduce data or variables from a sample dataset [56]. The communality is also described as 'the variance in the observed variables which are accounted

**Table 17** Results of extraction of communalities for opportunities of sustainability (prepared by the author)

| Communalities (opportunities) | |
|---|---|
| Items | Extraction |
| Reduce energy and water usage | 0.703 |
| Sustainability – reduce paper usage | 0.599 |
| Reduce consumption of raw materials | 0.732 |
| Increase efficiency | 0.783 |
| Increase cost-effectiveness | 0.658 |
| Enhance reputation | 0.573 |
| Improve corporate social responsibility | 0.715 |
| Reduce risk management | 0.607 |
| Differentiate businesses | 0.480 |
| Increase green strategy | 0.749 |
| Improve community investments | 0.750 |
| Improve human rights | 0.727 |
| Improve social responsibility investing | 0.793 |
| Increase triple bottom line – People, Planet and Profit | 0.710 |
| Increase productivity | 0.659 |
| Attract quality employees | 0.812 |
| Satisfy customer needs | 0.808 |
| Meet stakeholder expectations | 0.793 |
| Attract new opportunities | 0.729 |
| Create new jobs | 0.654 |
| Reduce carbon footprint | 0.853 |
| Reduce emissions | 0.903 |
| Reduce pollution | 0.901 |
| Reduce health hazards | 0.753 |

Extraction method: Principal Component Analysis.

for by a common factor or common variance' [41, 56]. The communality is measured between 0 and 1.0, with higher values demonstrating that the variance in a particular item is explained more by the extracted factors [41]. Table 17 shows the result of communalities for items measuring opportunity aspects of sustainability in Bhutanese organizations. Variables with high communality values are extracted. However, in this case, none of the variables was excluded as every variable has variance greater than **0.5**. From the table, it can be seen that there were five variables having high communality values of **0.9** and **0.8** (see Table 17). Thus, it can be construed that Bhutanese organizations may be adopting sustainability with an aim to decrease pollution by reducing emissions and their carbon footprint.

In a similar way, Table 18 displays the results of communalities for risks posed to Bhutanese organizations when adopting sustainability practices. In this case,

**Table 18** Communalities extracted for risks of sustainability (prepared by the author)

| Communalities (risks) | |
|---|---|
| Items | Extraction |
| Increase fraud (including environmental actions and environmental compliance failures) | 0.772 |
| Increase litigation and compliance breaches (including environmental actions and environmental compliance failures) | 0.773 |
| Increase governance failure (including environmental actions and environmental compliance failures) | 0.835 |
| Increase transaction failure due to environmental liabilities | 0.802 |
| Increase security and systems failures caused by environmental problems | 0.792 |
| Increase scandals by perceived environmental irresponsibility | 0.842 |
| Increase marketing failure by perceived environmental irresponsibility | 0.823 |
| Increase supply chain crises due to suppliers' environmental problems | 0.770 |
| Inflate costs | 0.798 |
| Increase interest rates | 0.789 |
| Increase insurance crises due to environmental disasters | 0.777 |
| Increase number of new regulations including environmental | 0.766 |
| Increase competition for and cost of raw materials | 0.862 |

Extraction method: Principal Component Analysis.

none of the variables was excluded as every variable has high variance indicating that variances for all variable items can be explained by the extracted factors.

Table 18 illustrates five variables with a high communality value of **0.8**. This implies that Bhutanese organizations are not incorporating sustainable strategies in their business models because of perceived risks or failures associated with environmental responsibility and increased competition and also the cost of raw materials.

Table 19 presents all significant factors for opportunities of sustainability along with eigenvalues. Further, the percentage of variance attributable to each factor is also computed in this procedure. It can be seen from Table 19 that the total variance is 72.679% of variation satisfying the criterion of 60% or more for total variance. Table 19 also displays 24 components and the variance is divided among these components. According to *Kaiser's criterion* [Kaiser 1960, cited in 56], only those factors with eigenvalues greater than 1 will be extracted or retained. An eigenvalue represents the 'amount of variance in all of the items that can be explained by a given principal component or factor' [41]. Similarly, Jolliffe recommends retaining factors with values greater than 0.70 [Jolliffe 1986, cited in 56]. Nevertheless, in this case, four components including the fourth factor with eigenvalue less than 1 are displayed (see Table 19), which will be extracted for those variables because the system was forced to extract four factors.

Likewise, Table 20 illustrates all extracted factors in relation to the risks aspect of sustainability adoption by Bhutanese organizations along with eigenvalues and

**Table 19** Total variances with extractable factors for opportunities of sustainability

| Total variance explained | | | | | | | | | |
|---|---|---|---|---|---|---|---|---|---|
| Component | Initial eigenvalues | | | Extraction sums of squared loadings | | | Rotation sums of squared loadings | | |
| | Total | % of variance | Cumulative % | Total | % of variance | Cumulative % | Total | % of variance | Cumulative % |
| 1 | 12.037 | 50.153 | 50.153 | 12.037 | 50.153 | 50.153 | 4.687 | 19.529 | 19.529 |
| 2 | 2.658 | 11.075 | 61.228 | 2.658 | 11.075 | 61.228 | 4.424 | 18.434 | 37.963 |
| 3 | 1.784 | 7.434 | 68.662 | 1.784 | 7.434 | 68.662 | 4.380 | 18.248 | 56.211 |
| 4 | 0.964 | 4.016 | 72.679 | .964 | 4.016 | 72.679 | 3.952 | 16.468 | 72.679 |

Extraction method: Principal Component Analysis.

**Table 20** Total variance for the risks aspects of sustainability

| Total variance explained | | | | | | | | | |
|---|---|---|---|---|---|---|---|---|---|
| Component | Initial eigenvalues | | | Extraction sums of squared loadings | | | Rotation sums of squared loadings | | |
| | Total | % of Variance | Cumulative % | Total | % of Variance | Cumulative % | Total | % of Variance | Cumulative % |
| 1 | 8.300 | 63.845 | 63.845 | 8.300 | 63.845 | 63.845 | 4.381 | 33.702 | 33.702 |
| 2 | 1.337 | 10.286 | 74.131 | 1.337 | 10.286 | 74.131 | 3.573 | 27.481 | 61.183 |
| 3 | 0.764 | 5.880 | 80.011 | 0.764 | 5.880 | 80.011 | 2.448 | 18.828 | 80.011 |

Extraction Method: Principal Component Analysis.

**Table 21** Rotation matrix of opportunities of sustainability (prepared by the author)

| | Component | | | |
|---|---|---|---|---|
| | 1 | 2 | 3 | 4 |
| **Reduce energy and water usage** | | | | **0.658** |
| **Reduce paper usage** | | | | **0.636** |
| **Reduce consumption of raw materials** | | | | **0.725** |
| **Increase efficiency** | | | | **0.671** |
| **Increase cost-effectiveness** | | | | **0.748** |
| Enhance reputation | | | | |
| Improve corporate social responsibility | | | | |
| Reduce risk management | | | | |
| Differentiate businesses | | | | |
| Increase green strategy | | | | |
| **Improve community investments** | | | **0.773** | |
| **Improve human rights** | | | **0.717** | |
| **Improve social responsibility investing** | | | **0.801** | |
| Increase triple bottom line – People, Planet and Profit | | | | |
| Increase productivity | | | | |
| **Attract quality employees** | | **0.777** | | |
| **Satisfy customer needs** | | **0.845** | | |
| **Meet stakeholder expectations** | | **0.837** | | |
| Attract new opportunities | | | | |
| Create new jobs | | | | |
| **Reduce carbon footprint** | **0.829** | | | |
| **Reduce emissions** | **0.856** | | | |
| **Reduce pollution** | **0.871** | | | |
| **Reduce health hazards** | **0.826** | | | |

also the percentage of variance of each factor prior to rotation. The first column of the table shows 13 factors with all eigenvalues irrespective of being less than 1 or greater than 1. In regards to risks of sustainability, three factors including a factor with eigenvalue less than 1 were retained after extraction. As seen from the table, the first factor accounts for 63.84% of the variance, the second factor for 10.28%, and the third factor for 5.88%.

The rotation method is used to make the interpretation of analysis easier and clearer [56]. Table 21 shows the rotation matrix for opportunities of sustainability in Bhutanese organizations and it displays four factors along with factor loadings for all variables. Variables might have high loadings on key factors and low loadings on less important factors. According to Yong and Pearce [56], it is necessary to have at least three variables for each factor.

**Table 22** Labels for opportunity factors

| Factor 1 | | |
|---|---|---|
| Variables | **Factor loadings** | **Label Name** |
| Reduce carbon footprint | 0.829 | **4Rs** |
| Reduce emissions | 0.856 | |
| Reduce pollution | 0.871 | |
| Reduce health hazards | 0.826 | |
| Factor 2 | | |
| Satisfy customer needs | 0.845 | **Stakeholders' satisfaction** |
| Meet stakeholders expectations | 0.837 | |
| Attract quality employees | 0.777 | |
| Factor 3 | | |
| Improve social responsibility investing | 0.801 | **Investment & corporate social responsibility** |
| Improve community investments | 0.773 | |
| Improve human rights | 0.717 | |
| Improve corporate social responsibility | 0.625 | |
| Factor 4 | | |
| Increase cost-effectiveness | 0.748 | **Efficiency and cost reduction** |
| Reduce consumption of raw materials | 0.725 | |
| Increase efficiency | 0.671 | |
| Reduce energy and water usage | 0.658 | |
| Reduce paper usage | 0.636 | |

As depicted in the table, factor 1 is loaded with four variables having high factor loadings, factor 2 with three variables, factor 3 with three, and factor 4 with five variables. After the factors have been extracted by factor analysis, the factors can be labelled or placed in a meaningful category. Thus, variables with high factor loadings were combined and labeled as shown in Table 22.

Similarly, Table 23 displays the rotation matrix for risks of sustainability in Bhutanese organizations and it shows three risk factors along with factor loadings from all variables. It is essential to have at least three variables for every factor extracted. Thus, as seen from the table below, factor 1 is loaded with five variables having high factor loadings, factor 2 with three variables, and factor 3 also has three variables.

From the rotation matrix, variables corresponding to a factor with high factor loadings were grouped together and given labels as shown in Table 24.

**Table 23** Rotated matrix for risks of sustainability (prepared by the author)

| | Component | | |
|---|---|---|---|
| | 1 | 2 | 3 |
| **Increase fraud (including environmental actions and environmental compliance failures)** | **0.860** | | |
| **Increase litigation and compliance breaches (including environmental actions and environmental compliance failures)** | **0.807** | | |
| **Increase governance failure (including environmental actions and environmental compliance failures)** | **0.797** | | |
| Increase transaction failure due to environmental liabilities | | | |
| Increase security and systems failures caused by environmental problems | | | |
| **Increase scandals by perceived environmental irresponsibility** | **0.721** | | |
| **Increase marketing failure by perceived environmental irresponsibility** | **0.743** | | |
| **Increase supply chain crises due to suppliers' environmental problems** | | **0.609** | |
| **Inflate costs** | | **0.812** | |
| **Increase interest rates** | | **0.726** | |
| **Increase insurance crises due to environmental disasters** | | | **0.665** |
| **Increase number of new regulations including environmental** | | | **0.767** |
| **Increase competition for and cost of raw materials** | | | **0.891** |

**Table 24** Labels for risk factors

| Factor 1 | | |
|---|---|---|
| Variables | **Factor loadings** | **Label name** |
| Increase fraud | 0.860 | **Environmental compliance & responsibility** |
| Increase litigation and compliance breaches | 0.807 | |
| Increase governance failure | 0.777 | |
| Increase scandals by perceived environmental irresponsibility | 0.721 | |
| Increase marketing failure by perceived environmental irresponsibility | 0.743 | |
| Factor 2 | | |
| Inflate costs | 0.812 | **Cost inflation & supply chain crisis** |
| Increase interest rates | 0.726 | |
| Increase supply chain crises due to suppliers' environmental problems | 0.609 | |
| Factor 3 | | |
| Increase competition for and cost of raw materials | 0.891 | **Competition, regulations and insurance** |
| Increase number of new regulations including environmental | 0.767 | |
| Increase insurance crises due to environmental disasters | 0.665 | |

## 5 Discussion and New Significance

Based on the research findings (see Figs. 3 and 4) and also comparing the findings with those from the literature review, it is evident that new insights and information on sustainability have been generated from Bhutan's perspective. In relation to the opportunities provided by the integration of sustainability in Bhutanese organizations, the research reveals that Bhutan places relatively high importance on the protection of the environment and the reduction of health hazards as these factors showed high factor loadings (see Fig. 3). This finding aligns with Bhutan's commitment enshrined in its *Constitution* to maintain 60% of its land as forest coverage at all times [8]. In addition, Bhutan has made a declaration to become a carbon neutral country in an attempt to combat the pressing issue of climate change and global warming mostly caused by greenhouse gases. This declaration was made during the UNFCCC held in Copenhagen in 2009 [9]. The finding is also in line with the development philosophy or ideology of 'Gross National Happiness', which takes a holistic approach to development by promoting environmental sustainability and social well-being in addition to socio-economic growth [9]. However, the finding disagrees with the statement made by Walsh [55], that developing countries 'may conflict with environmental sustainability goals that seek to restrict resources use relative to its availability.' Bhutan is striving hard to achieve sustainable development goals by every possible means, although the country is challenged by a constant growing problem of economic crisis, widening gap between rich and poor, and increasing poverty. Further, the research indicated that Bhutan also wishes to meet stakeholders' expectations, promote corporate social responsibility, improve efficiency and reduce costs (see Fig. 3) as these are some of the advantages of becoming a sustainable society. In simple terms, it can be deduced that Bhutan believes that sustainable practices in organizations could satisfy stakeholders and improve business performance by enhancing efficiency and effectiveness. The findings also illustrate that Bhutan agrees with other research findings that integration of sustainability in organizations can minimize operational costs and enhance social welfare [50].

On the other hand, in regards to the risk factors of sustainability, the research findings have shown that Bhutan apparently believes that integrating sustainability practices in organizations' strategies might increase fraud and environmental responsibility, escalate cost, increase competition, and increase the number of regulations (see Fig. 4). These factors were supported with high factor loadings. This may be because Bhutan, as mentioned earlier, gives priority to the environment, which thereby might result in the creation of stringent environmental rules and regulations. Due to this, organizations might incur additional costs as a result of environmental irresponsibility and non-compliance with environment regulations. There might be an increase in fraud and corruption cases due to manipulation of documents by organizations to show compliances with environmental regulations or to meet environmental obligations. Furthermore, Bhutan also feels

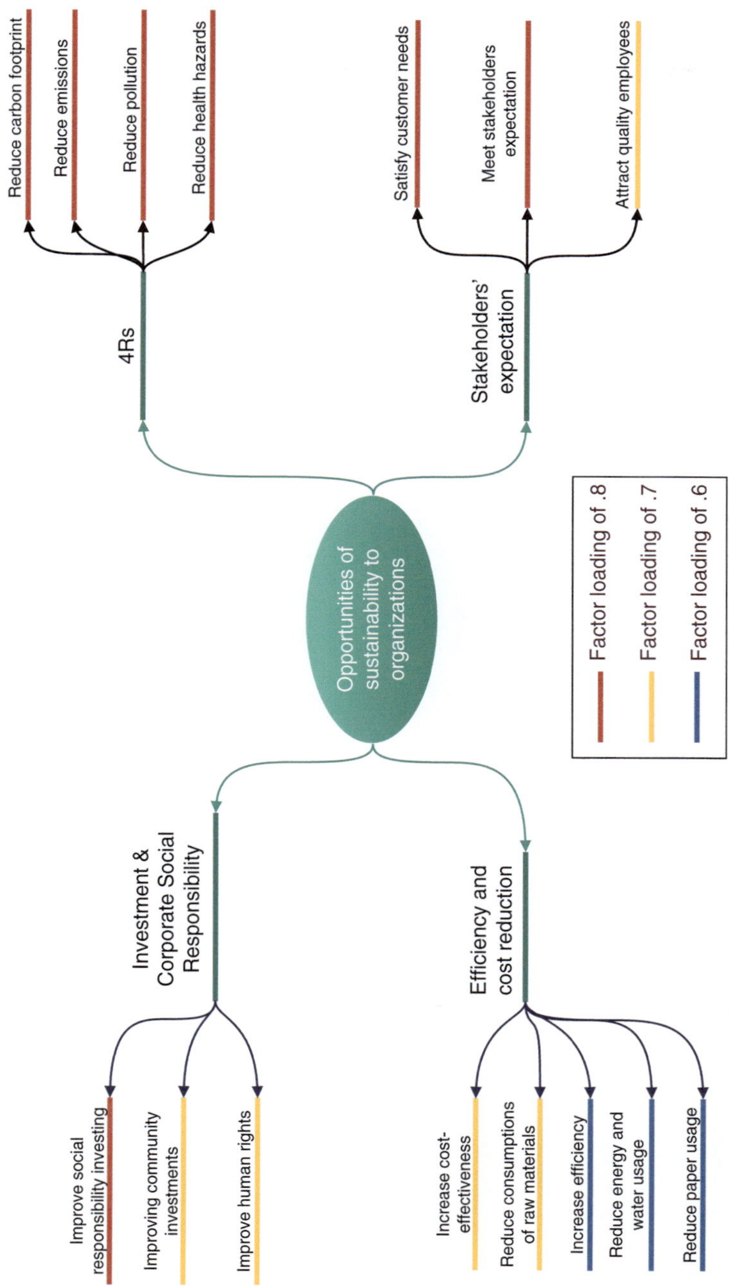


**Fig. 3** Significant opportunities of sustainability in relation to Bhutanese organizations

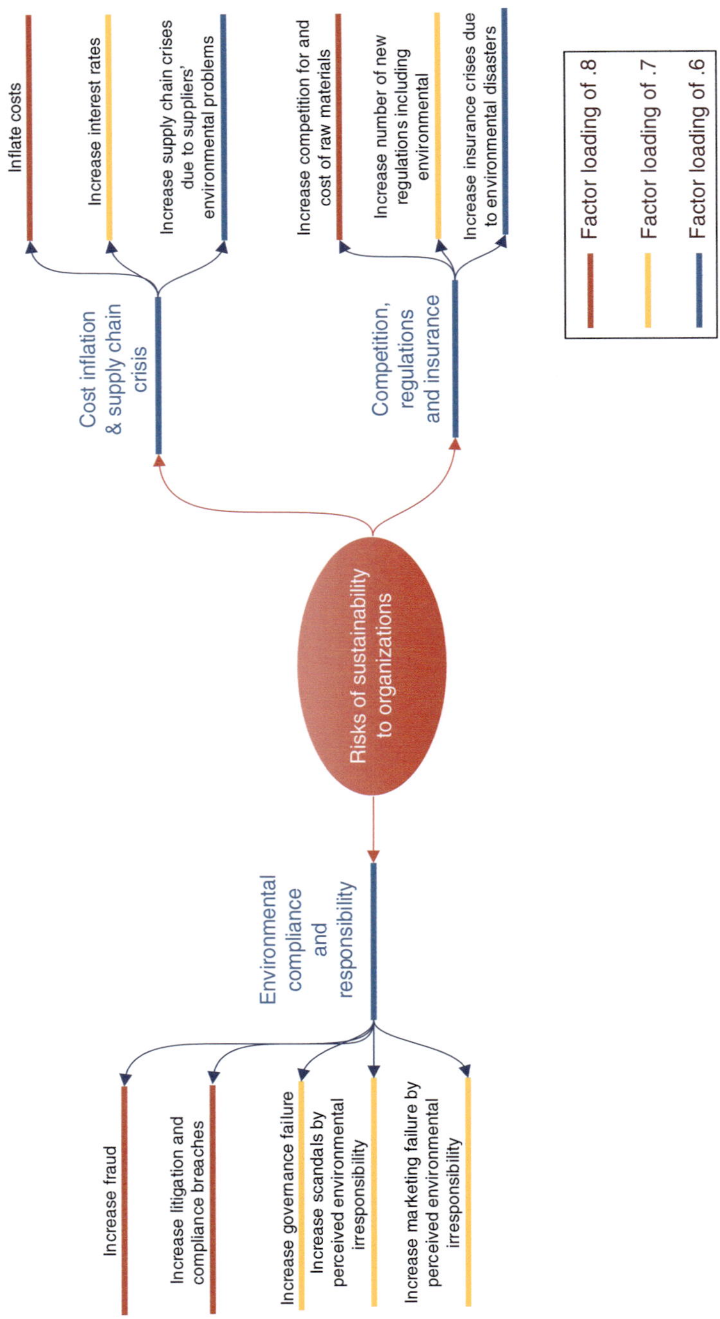


**Fig. 4** Significant risks of sustainability in relation to Bhutanese organizations

that increased competition amongst the organizations and increased investment in sustainability initiatives could be some of the risks that could discourage organizations from adopting sustainability practices. It is clear from the findings that Bhutan is aware of sustainability but the understanding is inclined more towards the environment dimension. In other words, Bhutan is more focused on resolving issues related to environment and more programs are undertaken to protect the environment rather than balancing the three dimensions of sustainability. Thus, the findings answer the research question 'What are the factors that are influencing the attitudes of Bhutanese public organizations towards sustainability in terms of opportunities and challenges? – What is Bhutan's perspective on sustainability.'

## 6 Limitations

Time constraints were the major limitation to this study, which inevitably have affected the quality of the thesis. Because of limited time, the research targeted only a small sample size consisting of 117 respondents. Thus, making generalizations to a larger representation (population) of Bhutan based on this sample size could be another drawback of the research. Also, the research focus was narrow because it considered only public organizations and did not include private organizations, political bodies, non-governmental organizations, civil society organizations, and higher education. Although the survey was distributed to 215 employees working in Bhutanese public organizations, only 117 completed the survey. This might have been due to the fact that the research was carried out offshore where the researcher was not physically present at the time of data collection. The response rate would have been higher if the researcher had gone to the field to physically collect the data since face-to-face interaction does make a difference. Lack of information related to sustainability in Bhutan was one of the significant limitations faced during the course of the research. There was very little information available as no prior research studies have been carried out in Bhutan, which specifically focus on sustainability and sustainable development. The research was designed to collect quantitative data, although there were few options where the respondents could provide comments and remarks. However, during the analysis, the qualitative data (comments) was not taken into consideration because there were only a few comments made by the respondents, which were insufficient for carrying out any analysis. The outcome of the research would have given more insights and dimensions if the research were carried using the mixed-methods approach (both quantitative and qualitative). Although 144 participants made an attempt to respond to the survey, only 117 participants completed the whole survey. Some of the participants did not respond to one or more survey questions.

## 7 Conclusion

The main purpose of the research is to examine and understand the factors, expressed in terms of opportunities and risks, influencing the Bhutanese organizations' attitudes towards sustainability. It is essential for organizations to become sustainable in this modern and competitive era by adopting strategies that include maximization of profit along with a focus on corporate social responsibility and reduction of damages to the environment from their business operations. Understanding sustainability is not only vital to organizations but also significant to every individual as there is no other planet to supply our basic necessary resources. Additionally, sustainability is not a new concept but it is an emerging trend, which has changed over a period of time since its inception. The concept of sustainability has changed over time to take into account social and economic elements rather than considering environmental responsibility only. The world has achieved incredible progress in terms of economic development but this has caused damage to the environment. Sustainability goes beyond protecting the environment; it takes into consideration the 'triple bottom line' that includes the 3Ps – Planet, People, and Profit. Human activities are directly or indirectly causing climate change, land degradation, global warming, energy shortages, etc. These problems are impacting on the sustainability of the Earth. Thus, it is very important to understand the global issues and create innovate solutions or initiate activities that support sustainability. This research study selected the quantitative research method, which tends to be more positivist and deductive in nature. In addition, an online survey questionnaire was conducted to collect data from public organizations in Bhutan. The online survey questionnaire seemed to be the best option for the gathering of quantitative data because of cost and time constraints. The survey was designed using Qualtrics software and was distributed via email and Facebook and WeChat, an instant messaging app. The research outcomes identified four opportunities and three risks of sustainability in relation to Bhutanese organizations. Further research will be carried out in the future to include more organizations perception toward sustainability to strength the research objectives and goals.

## References

1. Adams CA, Muir S, Hoque Z (2014) Measurement of sustainability performance in the public sector. Sustain Acc Manage Policy J 5(1):46–67. doi:10.1108/SAMPJ-04-2012-0018
2. Ahi P, Searcy C (2015) Assessing sustainability in the supply chain: a triple bottom line approach. Appl Math Model 39(10–11):2882–2896. doi:http://dx.doi.org/10.1016/j.apm.2014.10.055
3. Amran A, Haniffa R (2011) Evidence in development of sustainability reporting: a case of a developing country. Bus Strat Environ 20(3):141–156
4. Asian Development Bank (2013) Vietnam environment and climate change assessment
5. National Statistical Bureau Bhutan (2015) Bhutan at a glance

6. Planning Commission of Bhutan (2007) Bhutan Millennium Development Goals Need Assessment and Costing Report 2006–2015
7. National Statistical Bureau of Bhutan and World Bank (2014) Bhutan Poverty Assessment 2014
8. Bhutan, in pursuit of sustainable development (2012) United Nation. http://www.undp.org/content/undp/en/home/presscenter/articles/2012/09/12/bhutan-s-gross-national-happiness.html. Accessed 29 Nov 2015
9. National Environment Commission of Bhutan (2012) Bhutan: in pursuit of sustainable development. National Report for the United Nation Conference on Sustainable Development 2012
10. Bonn I, Fisher J (2011) Sustainability: the missing ingredient in strategy. J Bus Strat 32(1): 5–14. doi:10.1108/02756661111100274
11. Brundtland GH (1987) Report of the world commission on environment and development: our common future. UN, Oslo
12. Chou DC, Chou AY (2012) Awareness of Green IT and its value model. Comput Stand Interfaces 34(5):447–451. doi:10.1016/j.csi.2012.03.001
13. Gross National Happiness Commission (2013) Eleventh Five Year Plan Document. Royal Government of Bhutan 1(0)
14. Fonseca LM (2015) Strategic drivers for implementing sustainability programs in Portuguese organizations-let's listen to Aristotle: from triple to quadruple bottom line. Sustain J Rec 8(3): 136–142
15. Galpin T, Whitttington JL, Bell G (2015) Is your sustainability strategy sustainable? Creating a culture of sustainability. Corp Gov 15(1):1–17. doi:10.1108/CG-01-2013-0004
16. Gliem JA, Gliem RR (2003) Calculating, interpreting, and reporting Cronbach's alpha reliability coefficient for Likert-type scales. In: Midwest Research-to-Practice Conference in Adult, Continuing, and Community Education
17. Gobble MM (2012) Innovation and sustainability. Res Technol Manag 55(5):64–66
18. Grace D, O'Cass A (2004) Examining service experiences and post-consumption evaluations. J Serv Mark 18(6):450–461. doi:10.1108/08876040410557230
19. Günerergin M, Penbek Ş, Zaptçıoğlu D (2012) Exploring the problems and advantages of Turkish SMEs for sustainability. Procedia Soc Behav Sci 58:244–251. doi:http://dx.doi.org/10.1016/j.sbspro.2012.09.998
20. Gunter B, Nicholas D, Huntington P, Williams P (2002) Online versus offline research: implications for evaluating digital media. Aslib Proc 54(4):229–239. doi:10.1108/00012530210443339
21. Hair JFJ, Babin B, Money AH, Samouel P (2003) Essentials of business research methods. Wiley, New York
22. Hayward R, Lee J, Keeble J, McNamara R, Hall C, Cruse S, Gupta P, Robinson E (2013) The UN global compact-accenture CEO study on sustainability 2013. UN Global Compact Rep 5(3):1–60
23. Heinberg R (2010) What is sustainability? The Post Carbon Reader11–19
24. Hoen KM, Tan T, Fransoo JC, van Houtum GJ (2014) Effect of carbon emission regulations on transport mode selection under stochastic demand. Flex Serv Manuf J 26(1–2):170–195. doi:http://dx.doi.org/10.1007/s10696-012-9151-6
25. James ML (2015) The benefits of sustainability and integrated reporting: an investigation of accounting majors' perceptions. J Leg Ethical Regul Issues 18(1):1–20
26. Joseph C (2013) Understanding sustainable development concept in Malaysia. Soc Responsib J 9(3):441–453. doi:http://dx.doi.org/10.1108/SRJ-03-2012-0024
27. Kandachar P (2014) Chapter 7 – materials and social sustainability. In: Rognoli EKP (ed) Materials experience. Butterworth-Heinemann, Boston, pp 91–103. doi:http://dx.doi.org/10.1016/B978-0-08-099359-1.00007-2
28. Kaygusuz K (2012) Energy for sustainable development: a case of developing countries. Renew Sustain Energy Rev 16(2):1116–1126. doi:http://dx.doi.org/10.1016/j.rser.2011.11.013
29. Knife K, Haughton A, Dixon E (2014) Measuring sustainability and effectiveness of social value creation by social sector actors/social enterprises, within developing countries. Acad Entrepreneurship J 20(1):1–22

30. Kuhlman T, Farrington J (2010) What is sustainability? Sustainability 2(11):3436–3448
31. Kumar M (2007) Mixed methodology research design in educational technology. Alberta J Educ Res 53(1):34–44
32. Law KM (2010) Factors affecting sustainability development: high-tech manufacturing firms in Taiwan. Asia Pac Manag Rev 15(4):619–633
33. Lele SM (1991) Sustainable development: a critical review. World Dev 19(6):607–621
34. Lu W, Chen C, Su M, Chen B, Cai Y, Xing T (2013) Urban energy consumption and related carbon emission estimation: a study at the sector scale. Front Earth Sci 7(4):480–486. doi:http://dx.doi.org/10.1007/s11707-013-0363-1
35. Mebratu D (1998) Sustainability and sustainable development: historical and conceptual review. Environ Impact Assess Rev 18(6):493–520
36. Mitcham C (1995) The concept of sustainable development: its origins and ambivalence. Technol Soc 17(3):311–326
37. Nastanski M, Baglione SL (2014) Sustainability: benefits of social responsibility to brand value & profit. J Int Manag Stud 9(2):164–173
38. O'Rourke N, Psych R, Hatcher L (2013) A step-by-step approach to using SAS for factor analysis and structural equation modeling, 2 edn. SAS Institute, Cary
39. Page RA, Collins KA (2010) The green MBA: a competing values matrix. Competit Rev 20(1): 62–71. doi:http://dx.doi.org/10.1108/10595421011019984
40. Petraşcu D, Tieanu A (2014) The role of internal audit in fraud prevention and detection. Procedia Econ Finance 16:489–497. doi:http://dx.doi.org/10.1016/S2212-5671(14)00829-6
41. Pett MA, Lackey NR, Sullivan JJ (2003) Interpreting factors and generating factor scores. In: Pett MA, Lackey NR, Sullivan JJ (eds) Making sense of factor analysis. SAGE, Thousand Oaks, pp 17–226. doi:http://dx.doi.org/10.4135/9781412984898.n7
42. Rahut DB, Behera B, Ali A (2016) Household energy choice and consumption intensity: empirical evidence from Bhutan. Renew Sustain Energy Rev 53:993–1009. doi:http://dx.doi.org/10.1016/j.rser.2015.09.019
43. Ravi F (2012) Sustainable globalization and implications for strategic corporate and national sustainability. Corp Gov 12(4):579–589. doi:http://dx.doi.org/10.1108/14720701211267883
44. Román M (2011) Carbon capture and storage in developing countries: a comparison of Brazil, South Africa and India. Global Environ Change 21(2):391–401. doi:http://dx.doi.org/10.1016/j.gloenvcha.2011.01.018
45. Siebert SF, Belsky JM (2015) Managed fuelwood harvesting for energy, income and conservation: an opportunity for Bhutan. Biomass Bioenergy 74:220–223. doi:http://dx.doi.org/10.1016/j.biombioe.2015.01.013
46. Sneddon C, Howarth RB, Norgaard RB (2006) Sustainable development in a post-Brundtland world. Ecol Econ 57(2):253–268
47. Social Watch (2012) Unsustainability: causes and consequences. http://www.socialwatch.org/node/14008. Accessed 30 May 2016
48. Solomon JF, Solomon A, Norton SD, Joseph NL (2011) Private climate change reporting: an emerging discourse of risk and opportunity? Acc Audit Account J 24(8):1119–1148. doi:http://dx.doi.org/10.1108/09513571111184788
49. Sustainable Development (2015) Helping stakeholders shape new gloab goals for humanity's future. http://www.sustainabledevelopment2015.org/index.php/intergovernmental-processes/sustainable-development-goals. Accessed 29 Nov 2015
50. Tan Y, Ochoa JJ, Langston C, Shen L (2015) An empirical study on the relationship between sustainability performance and business competitiveness of international construction contractors. J Clean Prod 93:273–278. doi:http://dx.doi.org/10.1016/j.jclepro.2015.01.034
51. Tavakol M, Dennick R (2011) Making sense of Cronbach's alpha. Int J Med Educ 2:53–55. doi:10.5116/ijme.4dfb.8dfd
52. Uddin SN, Taplin R, Yu X (2007) Energy, environment and development in Bhutan. Renew Sustain Energy Rev 11(9):2083–2103. doi:http://dx.doi.org/10.1016/j.rser.2006.03.008

53. United Nations Conference on Climate Change (2015) http://www.cop21.gouv.fr/en/. Accessed 29 Nov 2015
54. Virakul B (2015) Global challenges, sustainable development, and their implications for organizational performance. Eur Bus Rev 27(4):430–446. doi:10.1108/EBR-02-2014-0018
55. Walsh PR (2011) Creating a "values" chain for sustainable development in developing nations: where maslow meets porter. Environ Dev Sustain 13(4):789–805. doi:http://dx.doi.org/dbgw.lis.curtin.edu.au/10.1007/s10668-011-9291-y
56. Yong AG, Pearce S (2013) A beginner's guide to factor analysis: focusing on exploratory factor analysis. Tutor Quant Methods Psychol 9(2):79–94
57. Zachary Woodruff J, Brenner P, Buccellato APC, Go DB (2014) Environmentally opportunistic computing: a distributed waste heat reutilization approach to energy-efficient buildings and data centers. Energy Build 69:41–50. doi:http://dx.doi.org/10.1016/j.enbuild.2013.09.036

# Sustainability Awareness in Singapore's Higher Education Sector

**Nimitra Revi Chandran**

**Abstract** Sustainable development is not merely a passing trend but a movement in transforming our world into a better place. This study explores and examines sustainability awareness in the Singaporean higher education sector. As deduced from the literature review, sustainability plays an important role in improving the environmental, economic and social performance of individuals and businesses. This study examines the advantages and risks in relation to sustainability awareness in Singapore's tertiary education institutions. It is evident that strides are being made in sustainable development on a global scale. Singapore was the first South East Asian country to recognise the potential importance of sustainability. However, Singapore's higher education sector has not fully integrated sustainability into its education programs. This research will act as a tool of information and a foundation for further research studies on sustainability awareness and the role of higher education. This research was conducted using an online survey questionnaire which was distributed to the general public in Singapore. The data collected was analysed in accordance with the research objectives and questions. The newly derived factors were formed based on the data collected in relation to the advantages and risks associated with sustainability in Singapore. A few key recommendations were provided such as collaboration with all stakeholders, environmental issues need to be a top priority for the education system, and constant monitoring on environmental efforts. This research aims to provide information about the role of higher education in cultivating sustainability awareness for the future young leaders in Singapore.

## 1 Introduction

Self-driving cars, global internet access and robots will most likely be prominent by mid-century. At the same time, it is anticipated that the human population will increase to 9 billion, consumption of the world's energy will triple alongside

N.R. Chandran (✉)
School of Information Systems, Curtin University, Perth, Western Australia, Australia
e-mail: nimitra.revichandran@postgrad.curtin.edu.au



T. Issa et al. (eds.), *Sustainability, Green IT and Education Strategies in the Twenty-first Century*, Green Energy and Technology, DOI 10.1007/978-3-319-57070-9_11

global carbon emissions and pollution, food and water scarcity would have doubled, rain forests would face annihilation, and climate change will cause frequent natural disasters [12]. One of the most critical developments for humankind today is sustainable development. There has been an increasing effort to encourage academic institutions to include sustainability in their curricula. Universities around the world have taken an initiative in restructuring their vision, courses and research programmes [5]. The tertiary education sector has become aware of environmental impacts not only in regards to limited and scarce resources but also in the way they equip their graduates to deal with sustainability issues [5]. In today's society, the higher education sector has the responsibility to create awareness on sustainability and change the mind-sets of future leaders [5].

Singapore and sustainable development are interlinked, as its growth and economic development are heavily dependent on the implementation of sustainability. Singapore gained its independence on 9th of August 1965 and as an island city-state with limited and scarce resources, sustainability plays an important role in providing a clean environment and better quality of life [2]. However, with the increasing level of population, economy growth, and limited resources, pressure to take action has increased in a renewed emphasis on sustainable development.

This dissertation will critically explore the higher education sector in Singapore in relation to the risks and advantages associated with the adoption of sustainability. Higher education is viewed as a catalyst in creating awareness of sustainability around the world. This research will also explore the dimension of gender difference in regards to views on sustainability's advantages and disadvantages. The questions highlighted in this research will indicate the level of awareness of sustainability in Singapore's tertiary students.

## 2 Background of the Research

Sustainability is a global phenomenon that is changing every aspect of human life, as with time, our needs and wants in this material world increasingly change as we try to accommodate the growing human population. The basic elements of human life, which are water, food production, health, shelter and environment, are being threatened by climate change, and political and social issues. The current globalised world that we live in is interconnected and volatile, which makes the situation precarious. The United Nations created the World Commission on Environment and Development (WCED) to combat the global environmental issues; and to meet the present generation needs without compromising the future generation needs.

Organizations are now increasingly adopting sustainability as an organizational imperative [13]. The term "sustainability" has become a corporate buzzword that is another term for corporate social responsibility [8].

Sustainability has three components: social (people), environmental (planet), and economic (profit) responsibilities. UN has given a meaning for sustainable development, however it failed to provide guidance as to how businesses can incorporate sustainability into their daily operations and effectively balance organizational responsibilities and the expectations of multiple stakeholders. Sustainability is a current goal for businesses, non-profit organizations and governments; however, there is a lack of appropriate measurements to identify the degree to which an organization is being sustainable.

Hence, John Elkington created a framework to measure sustainability in corporate America during the mid-1990s [10]. This framework is known as the Triple Bottom Line (TBL). TBL can be defined as an accounting framework which integrates three essential dimensions of performance which are social, environmental and financial [10]. TBL plays a vital role in supporting sustainability objectives by understanding the investment results in accordance with the correlated dimensions of profit, people and the planet.

## *2.1 Sustainability in Education*

In many countries today, there has been a strong upward trend in making education increasingly relevant to social, economic and environmental changes. is important to acknowledge that quality education for sustainable development will strengthen the individual's responsibility as a global citizen to protect this planet we call home [9]. The role of education is crucial in nurturing many potential leaders of the world for the current and future generations. In 2002, the United Nations executed the Decade of Education of Sustainable Development and appointed UNESCO as the leading implementing agency [9]. The main objective is to address and integrate the principles and practices of sustainable development and the role of education and learning. Besides that, since 1990 the International Association of Universities (IAU) has been urging universities to promote sustainable development [9]. Subsequently, the Kyoto Declaration became a means of addressing all important issues in relation to the role of universities in promoting sustainable development [9].

However, this effort was not sufficient in implementing sustainable development. In 2014, UNESCO developed a programme for sustainable development called global action programme also known as GAP for education on sustainable development [9]. The GAP mainly highlighted the criticality of adopting education for sustainable development and transforming academic institutions by instilling sustainable development values and principles in their daily activities [9]. An International Conference by Nagoya Universities was held to address the need for higher education to adopt a whole-institution approach which includes transformative leadership and the assessment of the institution's sustainability [9].

In five decades, Singapore is the only South-East Asian country to have grown rapidly from a developing nation to a developed country [4]. According to the chairman of the World Urban Campaign Steering Committee at UN – Habitat,

"Singapore is at the leading edge of sustainability" [14]. Singapore is the fourth biggest financial hub and one of the leading commercial centres in addition to being one of the top five busiest ports [4]. The human population in Singapore has exceeded over 5 million, making it one of the most densely populated cities [4]. One of the main reasons that Singapore has made sustainability its main priority is that, as an island city, its only way to survive is through sustainable living [14]. Siemen's Green City Index for Asia has recognized Singapore as being the best in creating a sustainable environment [14]. Although economic growth is its main priority, Singapore has always understood that its growth might have an impact on the environment and society [14]. Singapore has one of the most advanced technologies when it comes to green city planning. Although, Singapore might be one of the emerging and leading economies in Asia, yet it is still one of the most vulnerable cities in regards to the alarming effects of climate change. Singapore envisioned a vibrant economy, providing essential sustainable quality living for all by taking accountability for the environment, and efficient usage of land and sea space.

Before climate change was even a major issue, Singapore prioritised sustainability in the early '1960s. At that time, Singapore faced major issues such as high unemployment rates, lack of infrastructure, high level of poverty and poor sanitation [2]. Singapore's first Prime Minister, Mr Lee Kuan Yew, wholeheartedly believed that "in an urban jungle of concrete like Singapore, it will destroy human spirit and we need the greenery of nature to lift our spirit" [3]. He planted the first tree in 1963 and Singapore Garden City was developed. Moreover, in 1968 the Keep Singapore Clean Campaign was launched. Singapore made a decision in 1971 to relocate the residential areas away from the polluting industries [2]. Singapore paved its road to sustainability in 1992 by developing the first Green Plan which was initiated by the Ministry of Environment soon after participating in the UN Conference on Environment and Development [2]. This Green Plan was then reinforced in 2012 with a new 10-year plan which had proper integration in planning and precise attention to detail [2].

Furthermore, the Ministry of Environment and Water Resources implemented the Sustainable Singapore Blueprint in 2009 which defined the key strategies and targets in regards to sustainable development until 2030 [2]. Singapore's National Climate Change Strategy was published in 2012 which provided a guideline for reducing carbon emissions, prioritizing energy efficiency, protecting the coastline, managing risk involving floods, reducing usage of water, and developing relationships with international partners in accordance with global sustainable development.

In addition to that, Singapore has been participating actively in multilateral negotiations involving the UN Framework Convention on Climate Change (UNFCC), Kyoto Protocol and others [2]. Moreover, Singapore has implemented the Singapore Cooperation Programme which involves over 7000 participants from more 170 developing countries who have been trained in various areas of sustainability such as urban development, energy efficiency, water management and reduction in pollution.

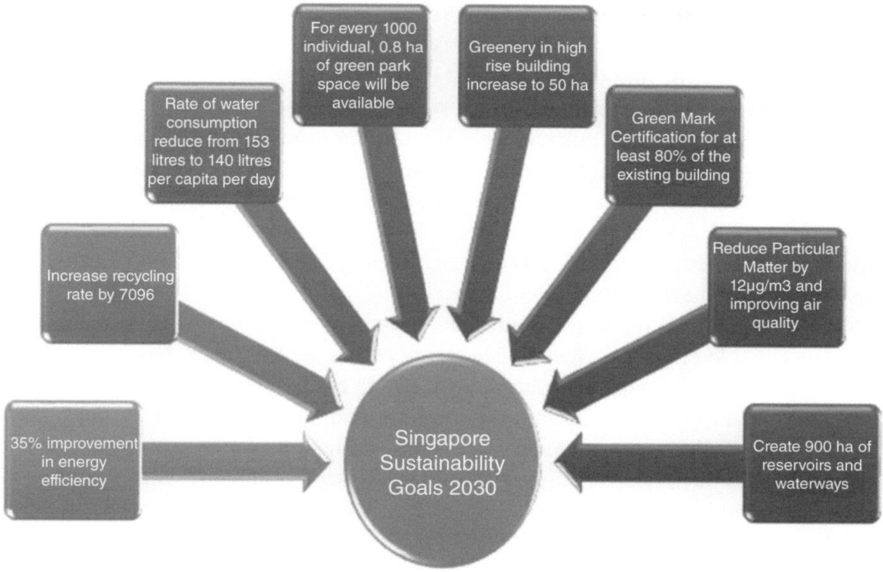


**Fig. 1** Singapore Sustainability Goals 2030 (Prepared by the Author 2016)

In 2015, the Sustainable Singapore Blueprint was updated with new key strategies [2]. Figure 1 illustrates the goals and targets that need to be achieved in accordance to the UN Sustainable Development Goals [2].

Singapore is a small and developed nation-state in Southeast Asia with a diverse multilingual and multi-religious population. Education has facilitated a meritocratic ideology of pragmatism and is a vital key in Singapore's rapid development from third world developing nation to a first world cosmopolitan city-state [15]. The Post-Secondary Education for Sustainability in Asia was held at Nanyang Technological University (NTU) in Singapore [1].

This conference was held to address concerns, opportunities and solutions in relation to education sustainability practitioner and supporters in Asia. In 2014, a conference on Sustainability in Education: Pedagogical Themes and Practices in Asian Countries was held in Singapore [1].

The main purpose of this workshop was to investigate the practice and pedagogical themes in relation to sustainability education in Asian countries [1]. Besides that, this conference addressed the various ways that sustainability is studied in different subjects such as science, social science, economics and engineering.

In Singapore, environmental and sustainability education is infused into mainstream disciplines [15]. In this way, students have various opportunities to explore fossil formation in science or explore land-use change in geography [15]. However, although this infusion process allows students to be exposed to environmental education, it is difficult for students to develop a connection between theoretical application and their experience in the real world. Environmental education has not been very successful in Singapore as curriculum is directed towards performance outcomes [15]. Environmental education is not a stand-alone subject; hence, its teachers do not really devote much time to it [15].

Singapore is a conservative country, which makes administrators and teachers conform to the national curriculum [15]. Environmental education is mainly focused on creating environmental awareness, knowledge and attitudes [15]. Informal education which includes workshops and guided tours provided by government and non-profit organizations have become a big part of Singapore's environmental education [15]. However, it is not enough to facilitate and foster positive environmental behaviour.

Our Home, Our Environment and Our Future are Singapore's vision in becoming a sustainable city-state. Sustainable development for the future has been set out in the Sustainable Singapore Blueprint. Singapore continuous its efforts to improve the environment and create an environmentally responsible community [11]. Eco-smart town, car-lite, green economy, a zero-waste nation and an active sustainable community are the objectives highlighted in the Blueprint [11].

Research and development in Singapore is currently working on strategies to improve ways to meet the targets by 2030 [11]. Singapore aims to be a Zero Waste Nation by 2030. Singapore has realised that investing in technology, infrastructure and policies can only sustain to a certain level [11]. The most critical part of progressing towards sustainability is society's view and approach towards sustainability by making changes in lifestyle and habits [11]. Hence, the public sectors play a vital role in enhancing and demonstrating the advantages of a sustainable environment.

Singapore's continuing growth in economy and population will have to meet the increasing demand for resources, which makes it even more crucial to embrace sustainable development. It will require a creative way of managing resources and developing innovative resolutions [11]. In short, to sustain new areas of economic growth and create a better home for future generations, the Singapore government aims to collaborate and cooperate with all stakeholders.

## 3 Research Method and Question

The main research question of this research is: What are the advantages and risk of sustainability awareness in adopting sustainability in the tertiary education sector? In this era of digital information, sustainability awareness plays a critical role in creating a sustainable future for this generation and the next. This research will be able to provide a comprehensive view of Singapore's approach to sustainability. It also questions whether Singaporean's are aware of the criticality of adopting the fundamental principles of sustainability. Based on the findings of this research, academic institutions and business will be able to increase Singaporeans' level of awareness of sustainability and green IT.

The research method is the method used to collect data, which involves a specific instrument. There are three types of research methods: quantitative, qualitative and mixed-methods research. This research is focused on addressing the awareness on sustainability in Singaporeans the higher education sector. The data is collected by adopting a quantitative research method for the purpose of this research. Quantitative research is a research strategy that views social reality as an external and objective reality. Primarily quantitative research method focuses mainly on deductivism. Deductivism involves the testing of theories [6]. This research implements an objectivist approach, as the sample population chosen is the general Singaporean public. The tool used to gather data for this research is a web-based survey which was designed to collect views in regards to the level of awareness of sustainability in Singapore. Analytical methods and tools are used to analyse the data gathered via web-survey.

# 4 Results

The data was distributed via social media, mainly Facebook (100 message via Facebook), WhatsApp (30 message via WhatsApp) and LinkedIn (22 messages via LinkedIn).

This section collected demographic information from respondents. Table 1 indicates the overall online survey rate of responses in Singapore. The total number of responses collected through the survey was 152 and out of the 152, only 94 were considered valid. Hence, the rate of response was 61.8%. Table 2 shows the demographic details of the respondents in terms of gender, age and qualifications. The response rate from the female respondents was 51.6% and 48.3% from the male respondents. The majority of the respondents were aged between 21 and 24 years. Moreover, most of the respondents had a bachelor's degree, which was 51.61% of the overall participants.

Table 3 indicates that most of the respondents spend 5–10 h on the computer per day. Furthermore, 39% of the respondents spend 5–10 h on the internet and 34% of the respondents spend up to 5 h on a social network per day. Respondents spend less than an hour on emails which equates to 42.5% of the overall participants.

In addition, the online survey confirms that 82.58% of the respondents use the internet to check on their email, 65.8% use it for academic purposes and 64.5% use it for work tasks (see Table 4).

Table 5 shows that the majority of the respondents use smartphones to access the internet which equates to 85.81% of the respondents. Moreover, 75.5% of the respondents use laptops and 44.5% use tablets to access the internet.

Table 6 shows that the majority of the respondents were first introduced to sustainability through higher education which accounts for 54.8% of the respondents.

**Table 1** Online survey rate of response

| | |
|---|---|
| Questionnaires distributed | 152 |
| Questionnaires Returned | 94 |
| Rate of Response | 61.8% |

**Table 2** Demographic details of the respondents

| Factors | % | Number |
|---|---|---|
| *Gender* | | |
| Male respondents | 48.34 | 73 |
| Female respondents | 51.66 | 78 |
| *Age* | | |
| 17 years and under | 0.64 | 1 |
| 18–20 | 4.49 | 7 |
| 21–24 | 33.97 | 53 |
| 25–30 | 33.33 | 52 |
| 31–35 | 11.54 | 18 |
| 36–40 | 4.49 | 7 |
| 41–45 | 5.13 | 8 |
| 46–50 | 2.56 | 4 |
| 51–55 | 1.28 | 2 |
| 56–60 | 2.56 | 4 |
| 61–65 | 0.00 | 0 |
| Over 65 | 0.00 | 0 |
| *Qualifications* | | |
| Primary Education | 0.00 | 0 |
| Higher Secondary/Pre-University | 3.87 | 6 |
| Professional Certificate | 2.58 | 4 |
| Diploma | 14.84 | 23 |
| Advanced/Higher/Graduate Diploma | 9.68 | 15 |
| Bachelor's Degree | 51.61 | 80 |
| Post Graduate Diploma | 6.45 | 10 |
| Master's Degree | 8.39 | 13 |
| PhD | 2.58 | 4 |
| Others | 0.00 | 0 |

**Table 3** Usage of technology by Singaporean users

| Time spent | Hours spent on the computer per day | Hours spent on the internet per day | Hours spent on social network per day | Hours spent on email per day |
|---|---|---|---|---|
| | Rate of response | Rate of response | Rate of response | Rate of response |
| Less than an hour | 14 (9.03%) | 5 (3.23%) | 42 (27.10%) | 66 (42.58%) |
| Up to 5 h | 40 (25.81%) | 52 (33.55%) | 54 (34.84%) | 28 (18.06%) |
| 5–10 h | 68 (43.87%) | 61 (39.35%) | 38 (24.52%) | 46 (29.68%) |
| 10–20 h | 30 (19.35%) | 34 (21.94%) | 21 (13.55%) | 15 (9.68%) |
| Over 20 h | 3 (1.94%) | 3 (1.94%) | 0 (0.00%) | 0 (0.00%) |

**Table 4** Internet usage by Singaporean users

| Types of activities | Rate of responses | Number |
|---|---|---|
| Email | 82.58% | 128 |
| Playing games | 38.06% | 59 |
| Studying | 65.81% | 102 |
| Working | 64.52% | 100 |
| Shopping online | 54.84% | 85 |
| Chatting | 58.06% | 90 |
| Researching hobbies | 42.58% | 66 |
| Banking online | 53.55% | 83 |
| Buying goods or services | 41.94% | 65 |
| Buying stocks or investing online | 25.16% | 39 |
| Researching travel information or making reservations | 50.97% | 79 |
| Others – please specify | 6.45% | 10 |

**Table 5** Devices used by Singaporean users

| Type of device | Rate of responses | Number |
|---|---|---|
| PC | 37.42% | 58 |
| Desktop | 32.26% | 50 |
| Laptop | 75.48% | 117 |
| Netbook | 7.74% | 12 |
| PDAs | 5.16% | 8 |
| Workstation | 6.45% | 10 |
| Tablet | 44.52% | 69 |
| Smartphone | 85.81% | 133 |
| Others | 0.00% | 0 |

**Table 6** First introduced to the concepts of sustainability and green information technology

| Type of medium | Rate of responses | Number |
|---|---|---|
| School | 43.87% | 68 |
| Higher education | 54.84% | 85 |
| Internet | 43.87% | 68 |
| Books | 25.16% | 39 |
| Magazine | 19.35% | 30 |
| News media | 22.58% | 35 |
| Conferences | 7.10% | 11 |
| Others – please specify | 2.58% | 4 |

**Table 7** Brands associated with devices that are widely used by Singaporeans

| Brand | Number of respondents | Rate of responses (%) |
|---|---|---|
| Apple | 79 | 50.97% |
| Google | 53 | 34.19% |
| Dell | 14 | 9.03% |
| IBM | 0 | 0.00% |
| Others – please specify | 9 | 5.81% |

**Table 8** Singaporean users reading sustainability report of the company before purchasing a device

| Choice | Rate of responses | Number |
|---|---|---|
| Yes | 31.61% | 49 |
| No | 37.42% | 58 |
| Maybe | 24.52% | 38 |
| Not at all | 6.45% | 10 |

**Table 9** Singaporean users switching or changing their devices

| Months (time) | Rate of responses | Number |
|---|---|---|
| Every 6 months | 0.64% | 1 |
| Every 12 months | 9.62% | 15 |
| Every 18 months | 8.33% | 13 |
| Every 24 months | 45.51% | 71 |
| Every 30 months | 14.74% | 23 |
| Every 36 months | 9.62% | 15 |
| Every 42 months | 9.62% | 15 |
| Other | 1.92% | 3 |

It can be concluded that the most popular companies are Apple, which represents 50.9% of the overall respondents, and Google with 34.19% of the respondents based on Table 7. Table 8 indicates the extent to which Singaporeans read a company's sustainability report before purchasing a device.

Based on Table 8 it shows that 37.42% do not read the report before purchasing the device. However, 31.61% read the report, 24.52% might read the report and 6.45% do not even attempt to read the report. Singaporean users change their device every 18–42 months; with percentiles ranging from 9% to 45% respectively (see Table 9).

The survey explored the reasons why Singaporean uses choose to change their device. Table 10 shows that the majority of respondents opted to change their devices because of functionality which equated to 71.43%, speed with 52.60%, keeping up with technology with 41.56%, and size with 40.91%.

The online survey explored the extent to which Singaporeans are aware that changing devices frequently will cause damage to our planet. It can be concluded that 62.34% of the respondents acknowledge that changing devices frequently will cause damage to our planet, 24.37% of the respondents said "maybe", 12.34% said "no" and 1.30% chose "not at all" (see Table 11).

**Table 10** Reasons for Singaporean users to change their devices

| Reason | Rate of responses | Number |
|---|---|---|
| Size | 40.91% | 63 |
| Speed | 52.60% | 81 |
| Functionality | 71.43% | 110 |
| Keeping with technology | 41.56% | 64 |
| Others – please specify | 3.90% | 6 |

**Table 11** Singaporean view on whether changing devices frequently will cause damage to our planet

| Choice | Rate of responses | Number |
|---|---|---|
| Yes | 62.34% | 96 |
| No | 12.34% | 19 |
| Maybe | 24.03% | 37 |
| Not at all | 1.30% | 2 |

**Table 12** Singaporeans' view on whether we can change the mindset of designers and users regarding sustainability via different mediums

| Type of media | Rate of responses | Number |
|---|---|---|
| Training | 57.14% | 88 |
| Education | 73.38% | 113 |
| Awareness | 79.22% | 122 |
| Workshop | 42.86% | 66 |
| Internet | 62.34% | 96 |
| T.V. | 46.10% | 71 |
| Social networking | 56.49% | 87 |
| Others – please specify | 3.25% | 5 |

Moreover, the online survey investigated how we could change the mind-set of designers and users regarding sustainability. The majority of the respondents chose awareness, education, and the internet as main means of changing the opinions about sustainability (see Table 12).

A total of 94 valid responses were analysed using Factor Analysis. This analysis is used to investigate the awareness of sustainability in Singapore's higher education sector. The level of awareness was measured by examining the Singaporeans' view of the risks and advantages related to sustainability.

As previously mentioned, this research uses Likert-type scales to elicit and analyse the view of individuals in regards to sustainability advantages and risks. Hence, it is vital to calculate and report Cronbach's Alpha coefficient to ensure the internal consistency reliability of any scales.

Table 13 highlights the reliability statistics for advantages of sustainability in Singapore. It can be seen that the Cronbach's Alpha is 0.954 and number of items is 24. The value of 0.954 indicates that the items in the scales have a good level of internal consistency. The number of items indicates the number of each individual opportunity listed on a Likert-type of scale. Looking at the results presented in the

**Table 13** Cronbach's Alpha on advantages for sustainability

| Reliability statistics | |
|---|---|
| Cronbach's Alpha | N of items |
| 0.954 | 24 |

**Table 14** Cronbach's Alpha on risk for sustainability

| Reliability statistics | |
|---|---|
| Cronbach's Alpha | N of items |
| 0.961 | 13 |

**Table 15** KMO and Bartlett Test for advantages on sustainability

| KMO and Bartlett's Test | | |
|---|---|---|
| Kaiser-Meyer-Olkin Measure of Sampling Adequacy | | 0.884 |
| Bartlett's Test of Sphericity | Approx. Chi-Square | 1936.612 |
| | df | 276 |
| | Sig. | 0.000 |

table above, 24 items were calculated to determine the consistency in regards to the advantages of sustainability in Singapore. The results show that the views regarding advantages of sustainability in Singapore are very consistent.

Table 14 shows the reliability statistics relating to risks associated with sustainability in Singapore. Based on the table above, it can be seen that the Cronbach's Alpha is 0.961 and the number of items is 13. Hence, for each of the 13 items listed under risk of sustainability in Singapore, the results are consistent. Overall, based on Tables 1 and 2, it can be stated that survey results in regards to reliability statistics show that the risk and advantages of sustainability in Singapore are excellent.

A large number of relationships can usually be determined more simply through exploratory factor analysis (EPA) and principal component analysis (PCA) [7]. Confirmatory factor analysis is a test which shows how variables are linked to underlying constructs, and requires additional software.

KMO is the test used for assumptions [16]. From Table 15, it can be seen that the KMO measure of sampling adequacy is 0.884. This value indicates the measure of adequacy and 0.8 indicates that the adequacy is meritorious. Hence, the evidence recommends that the sample size be large enough to yield reliable factors. Moreover, the Bartlett's test of sphericity is used to determine whether the correlation matrix is different from the identity matrix where all variables are zero [16]. It can be seen that the significant value is 0.000 which is less than 0.5.

Hence, this means that the variables are related to one another and can therefore be used for factor analysis in the research. In conclusion, KMO and Bartlett's Test for sustainability advantages in Singapore indicate that the data in this research is adequate for factor analysis. Based on Table 16, these initial commonalities signify the link between the variables and all other variables. These commonalities table shows the initial commonalities before rotation. Moreover, the table shows that the initial commonalities for all the factors are (<0.30) which is good.

**Table 16** Communalities – advantages

| Commonalities | |
|---|---|
| | Initial |
| Reduce energy and water usage | 0.597 |
| Reduce paper usage | 0.758 |
| Reduce consumption of raw materials | 0.831 |
| Increase efficiency | 0.827 |
| Increase cost-effectiveness | 0.604 |
| Enhance reputation | 0.725 |
| Improve corporate social responsibility | 0.731 |
| Reduce risk management | 0.680 |
| Differentiate businesses | 0.836 |
| Increase green strategy | 0.824 |
| Improve community investments | 0.830 |
| Improve human rights | 0.804 |
| Improve social responsibility investing | 0.854 |
| Increase triple bottom line – People, Planet and Profit | 0.711 |
| Increase productivity | 0.812 |
| Attract quality employees | 0.805 |
| Satisfy customer needs | 0.754 |
| Meet stakeholder expectations | 0.817 |
| Attract new advantages | 0.752 |
| Create new jobs | 0.769 |
| Reduce carbon footprint | 0.590 |
| Reduce emissions | 0.818 |
| Reduce pollution | 0.834 |
| Reduce health hazards | 0.578 |

Extraction method: Principal Axis Factoring.

The Total Variance Explained in the Table 17 illustrates how the variance is divided among all the possible 24 factors. Initial eigenvalues indicate a measure of the explained variance.

For this research, four factors were specified for the rotation sums of squared loadings. Hence, four factors emerged from the rotation sum of squared loadings. This factor analysis test used an orthogonal rotation (varimax). This means that each factor's information is independent of another factors' information. The total variance of the four factors shows that the cumulative percentage of sustainability's advantages in Singapore is 66.49%. This value indicates that it is acceptable to provide an explanation of the variance in the extracted factors.

The rotated factor matrix is a type of matrix used to interpret the results for analysis. The main reason that factors are rotated is that they are easier to interpret. The Table 18 demonstrates how the factors are grouped into three different

**Table 17** Total variance explained – advantages

| Total variance explained | | | | | | |
|---|---|---|---|---|---|---|
| Factor | Initial eigenvalues | | | Rotation sums of squared loadings | | |
| | Total | % of variance | Cumulative % | Total | % of variance | Cumulative % |
| 1 | 11.856 | 49.399 | 49.399 | 4.792 | 19.965 | 19.965 |
| 2 | 2.183 | 9.096 | 58.495 | 3.774 | 15.725 | 35.690 |
| 3 | 1.913 | 7.970 | 66.465 | 3.699 | 15.412 | 51.102 |
| 4 | 1.245 | 5.188 | 71.653 | 3.694 | 15.394 | 66.496 |

Extraction method: Principal Axis Factoring.

**Table 18** Rotated factor matrix advantages

| Rotated factor matrix[a] | | | | |
|---|---|---|---|---|
| | Factor | | | |
| | 1 | 2 | 3 | 4 |
| Attract quality employees | 0.809 | | | |
| Increase productivity | 0.745 | | | |
| Satisfy customer needs | 0.706 | | | |
| Meet stakeholder expectations | 0.650 | 0.423 | | |
| Attract new advantages | 0.618 | 0.499 | | |
| Create new jobs | 0.564 | 0.486 | | |
| Increase triple bottom line – People, Planet and Profit | 0.514 | | 0.433 | |
| Reduce emissions | | 0.854 | | |
| Reduce pollution | | 0.813 | | |
| Reduce carbon footprint | | 0.677 | | |
| Reduce health hazards | | 0.506 | | |
| Differentiate businesses | | | 0.732 | |
| Improve community investments | 0.447 | | 0.732 | |
| Improve human rights | 0.483 | | 0.680 | |
| Increase green strategy | | 0.430 | 0.651 | |
| Improve social responsibility investing | 0.541 | | 0.590 | |
| Enhance reputation | | | 0.515 | 0.447 |
| Reduce consumption of raw materials | | | | 0.826 |
| Increase efficiency | 0.464 | | | 0.707 |
| Reduce energy and water usage | | | | 0.652 |
| Reduce paper usage | | 0.421 | | 0.642 |
| Increase cost-effectiveness | | | | 0.545 |
| Improve corporate social responsibility | | 0.419 | 0.421 | 0.529 |
| Reduce risk management | 0.432 | | | 0.475 |

Extraction Method: Principal Axis Factoring.
Rotation Method: Varimax with Kaiser Normalization.
[a]Rotation converged in 10 iterations.

**Table 19** New advantages for sustainability for Singapore

| New factors | Variables |
|---|---|
| Improve business efficiency | • Attract quality employees |
| | • Increase productivity |
| | • Satisfy customer needs |
| | • Meet stakeholder expectations |
| | • Attract new advantages |
| Increase eco-efficiency or (3Rs) | • Reduce emissions |
| | • Reduce pollution |
| | • Reduce carbon footprint |
| Meeting social or business needs | • Differentiate businesses |
| | • Improve community investments |
| | • Improve human rights |
| | • Increase green strategy |
| Improve usage of resources | • Reduce consumption of raw materials |
| | • Increase efficiency |
| | • Reduce energy and water usage |
| | • Reduce paper usage |

**Table 20** KMO and Bartlett Test for disadvantages on sustainability

| KMO and Bartlett's Test | | |
|---|---|---|
| Kaiser-Meyer-Olkin Measure of Sampling Adequacy | | 0.934 |
| Bartlett's Test of Sphericity | Approx. Chi-Square | 1136.625 |
| | Df | 78 |
| | Sig. | 0.000 |

categories according to the highest loading on each item. As each item in the table has a loading value of at least 0.6, any loadings less than 0.3 are excluded from the output. This research established loading ranges from 0.6 to 0.8 which are highly preferable. The contents of these items are analysed to ensure that the items with high loading values are group together.

The Table 19 presents the newly developed factors together with their variables. The table shows that the four main factors do support the advantages of sustainability in Singapore.

Table 20 shows that the sampling adequacy for the disadvantages of sustainability is 0.934 which is greater than 0.70. Hence, the sample is adequate for this research. The Bartlett's Test significance is less than 0.5, which signifies that the variables are associated and therefore are suitable for factor analysis.

**Table 21** Communalities – disadvantages

| Commonalities | |
|---|---|
| | Initial |
| Increase fraud (including environmental actions and environmental compliance failures) | 0.602 |
| Increase litigation and compliance breaches (including environmental actions and environmental compliance failures) | 0.647 |
| Increase governance failure (including environmental actions and environmental compliance failures) | 0.796 |
| Increase transaction failure due to environmental liabilities | 0.787 |
| Increase security and systems failures caused by environmental problems | 0.760 |
| Increase scandals by perceived environmental irresponsibility | 0.697 |
| Increase marketing failure by perceived environmental irresponsibility | 0.780 |
| Increase supply chain crises due to suppliers' environmental problems | 0.731 |
| Inflate costs | 0.635 |
| Increase interest rates | 0.732 |
| Increase insurance crises due to environmental disasters | 0.831 |
| Increase number of new regulations including environmental | 0.763 |
| Increase competition for and cost of raw materials | 0.745 |

Extraction method: Principal Axis Factoring.

**Table 22** Total variance explained – disadvantages

| Total variance explained | | | | | | |
|---|---|---|---|---|---|---|
| Factor | Initial eigenvalues | | | Rotation sums of squared loadings | | |
| | Total | % of variance | Cumulative % | Total | % of variance | Cumulative % |
| 1 | 8.844 | 68.034 | 68.034 | 3.428 | 26.367 | 26.367 |
| 2 | 0.833 | 6.406 | 74.440 | 3.302 | 25.400 | 51.768 |
| 3 | 0.694 | 5.342 | 79.782 | 2.894 | 22.260 | 74.027 |

Extraction method: Principal Axis Factoring.

Table 21 shows the initial commonalities before rotation. Table 21 shows that all the factors had initial commonalities above 0.30.

Table 22 illustrates the total variance explained is divided among 13 factors. The cumulative percentages of the total variance explained for the four factors of

**Table 23** Rotated factor matrix disadvantages

| Rotated factor matrix[a] | | | |
|---|---|---|---|
| | Factor | | |
| | 1 | 2 | 3 |
| Increase insurance crises due to environmental disasters | 0.721 | 0.477 | 0.324 |
| Increase interest rates | 0.720 | | 0.420 |
| Increase number of new regulations including environmental | 0.712 | 0.455 | |
| Increase competition for and cost of raw materials | 0.698 | 0.433 | |
| Inflate costs | 0.579 | | 0.549 |
| Increase fraud (including environmental actions and environmental compliance failures) | 0.462 | 0.399 | 0.423 |
| Increase governance failure (including environmental actions and environmental compliance failures) | 0.341 | 0.793 | |
| Increase transaction failure due to environmental liabilities | 0.385 | 0.675 | 0.444 |
| Increase litigation and compliance breaches (including environmental actions and environmental compliance failures) | 0.344 | 0.648 | 0.314 |
| Increase security and systems failures caused by environmental problems | | 0.629 | 0.523 |
| Increase marketing failure by perceived environmental irresponsibility | | 0.466 | 0.733 |
| Increase scandals by perceived environmental irresponsibility | 0.405 | 0.363 | 0.650 |
| Increase supply chain crises due to suppliers' environmental problems | 0.363 | 0.441 | 0.627 |

Extraction Method: Principal Axis Factoring.
Rotation Method: Varimax with Kaiser Normalization.
[a]Rotation converged in 11 iterations.

risk are 74.2% which is adequate enough to provide an explanation of the variance of the extracted factors.

The Rotated Factor Matrix concerns the extent to which a simple structure has been achieved for the data. Table 23 shows that the analysis has sorted the 13 factors into three overlapping groups of items as shown in the highlighted section. Every item has a loading weight; however, loadings that are less than 0.30 are excluded from the output. This research set a high loading range from 0.6 to 0.8. These items were then group and the content of these items were analysed to ensure that factors that had a high loading fit together.

Table 24 illustrates the new factors that emerged alongside the grouped variables. Based on the results derived from the online survey, these variables contribute to risk in sustainability based on the level of awareness in Singapore.

**Table 24** New disadvantages for sustainability for Singapore

| New factors | Overlapping variables |
|---|---|
| Increase cost in economic conditions | • Increase insurance crises due to environmental disasters |
| | • Increase interest rates |
| | • Increase number of new regulations including environmental |
| | • Increase competition for and cost of raw materials |
| Lack of government support | • Increase governance failure (including environmental actions and environmental compliance failures) |
| | • Increase transaction failure due to environmental liabilities |
| | • Increase litigation and compliance breaches (including environmental actions and environmental compliance failures) |
| | • Increase security and systems failures caused by environmental problems |
| Poor conduct of business operations | • Increase marketing failure by perceived environmental irresponsibility |
| | • Increase scandals by perceived environmental irresponsibility |
| | • Increase supply chain crises due to suppliers' environmental problems |

# 5 Discussion and New Significance

Research studies involving sustainable awareness in higher education have been conducted in a general context but have omitted to consider the advantages and risks associated with sustainability. This research study focuses mainly on sustainability awareness in terms of its risks and advantages from the perspective of Singaporeans in Singapore's higher education.

This research contributes to the literature on sustainable education and awareness. Various studies have been conducted to examine sustainable business practices. However, there has been limited research on sustainability awareness in the higher education sector. Singapore has always supported sustainability well before the United Nations coined the term "sustainable development". However, since Singapore's transformation into a developed, urban city-state, education on sustainability has become much more important to Singaporeans. The diagram below illustrates the newly derived factors in relation to sustainability based on data obtained from Singapore's higher education sector. In conclusion, the main aim of this research is to address sustainability awareness in relation to Singapore's higher education sector. This research aims to examine the advantages and risks of sustainability adoption in Singapore's higher education sector Subsequently, based on the research findings, new factors were derived which can be seen in Figs. 2 and 3.

As Singapore aims to be the first environmentally-friendly city-state at the cutting edge of sustainable development, this research will provide Singaporeans

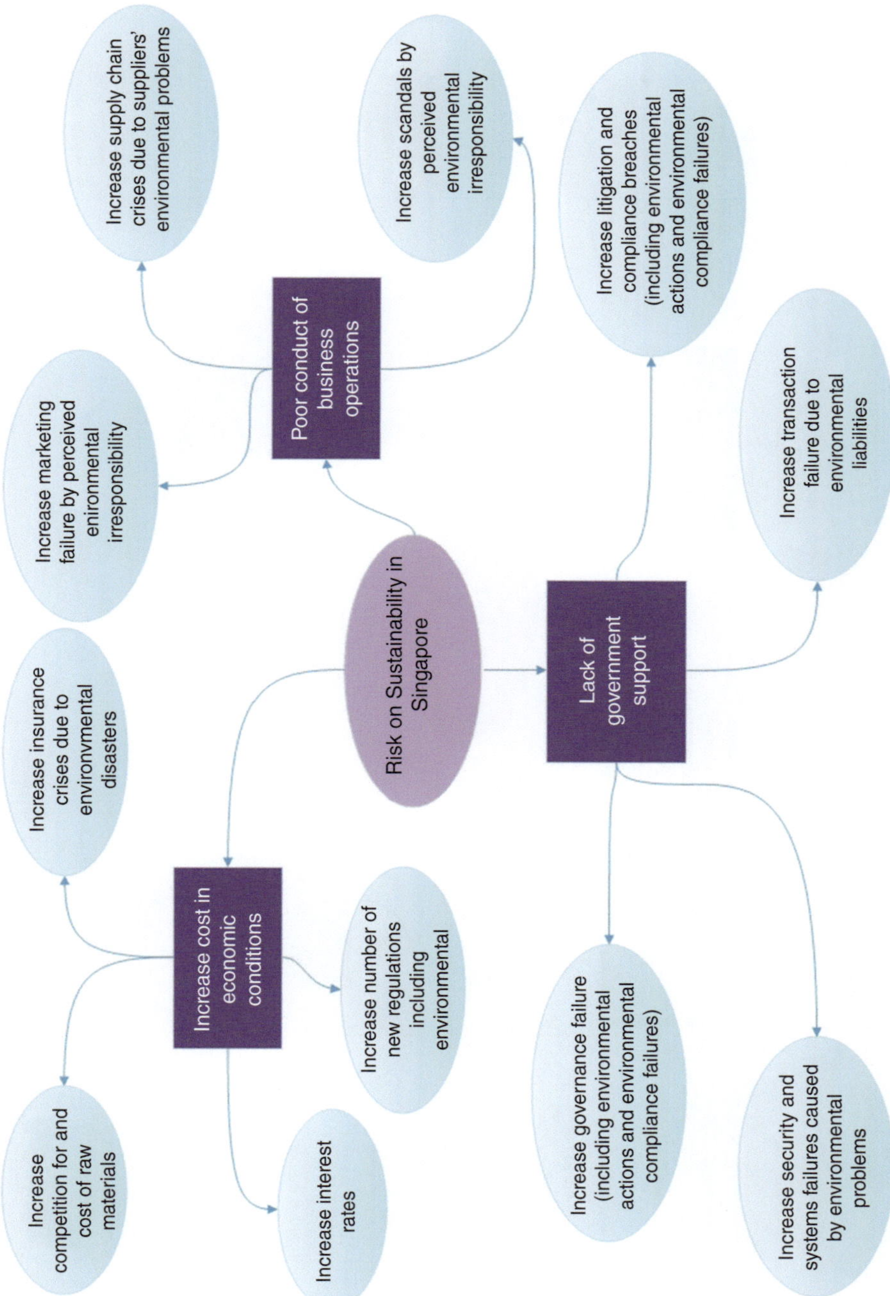


**Fig. 2** Risk on sustainability in Singapore

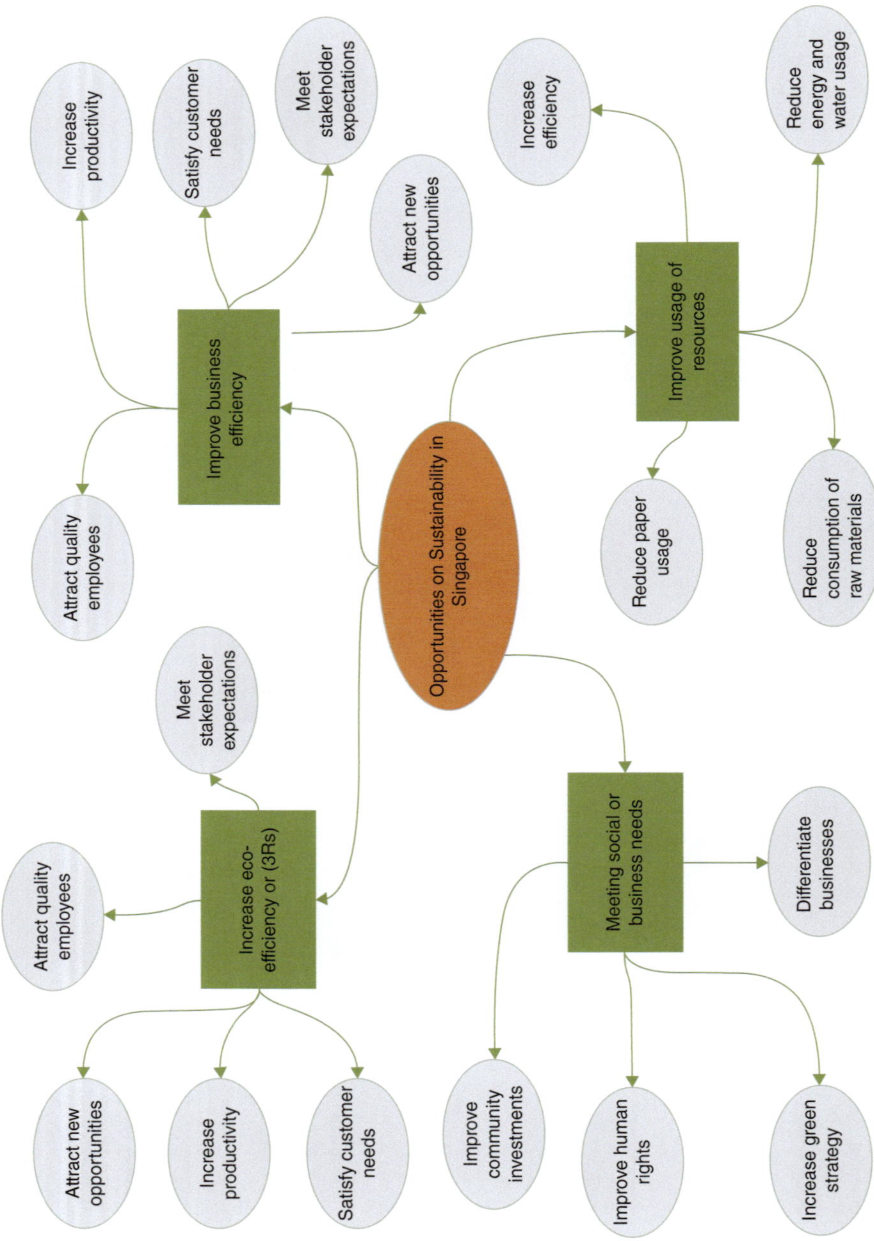


**Fig. 3** Advantages on sustainability in Singapore

with a better understanding of how their citizens view sustainability. Higher education plays a critical role in nurturing future leaders. Hence, this research gives Singaporean tertiary students guidelines for leading the nation and ways to improve their education system. The Singaporean government has been supportive in transforming Singapore into a leading green nation in Southeast Asia.

It can be seen Singaporeans are aware of the importance of sustainability and its consequences. However, the fundamental question that has to be highlighted is: "Is it enough"? Is this effort substantial enough to create sustainable awareness? Singapore's higher education sector has taken into account environmental education; however, it has failed to foster a thorough awareness of sustainability or change the behaviour of Singaporean tertiary students although Singapore is aspiring to be a Zero Waste Nation by 2030.

Although this research will provide a better understanding of sustainability awareness in Singapore, it has several limitations which can be addressed in future research. The small sample size played a role in limiting the scope of the research. Most of the survey responses were from the general public in Singapore, but many could not be used because they were incomplete. The main objective of this research is to ascertain the extent to which there is an awareness of risk and advantages with the adoption of sustainability in Singapore. However, due to the limited time given for this research, only 152 responses were received.

Besides that, the online survey had provided comment boxes. Ideally, this comment section was intended to give the researcher a better understanding of each individual's opinion about sustainability. Social media such as Facebook, Twitter, and LinkedIn played a critical role in collecting data. Hence, the sample was limited to respondents who actively engaged in social media.

Time constraint was a major issue for this research. Further research is required to acquire a wider source of information in relation to Singaporean awareness of sustainability in Singapore's higher education sector. The research emphasises the importance of sustainability awareness in higher education. There should be more research dedicated to business practices and the education sector. In future, there should be research conducted in primary school sector and the ways in which sustainability can be incorporated into the curriculum.

This research provides recommendations that will assist individuals, business, educators and communities to integrate sustainability into the higher education system on a much larger scale. Hence, these recommendations will assist in tackling the risks associated with the adoption of sustainability, and increase the advantages of sustainability in Singapore.

Critical to the transformation of Singapore into a sustainable nation is the communication between and collaboration of all stakeholders. Environmental issues are not given high priority in businesses and the education system. All members of society need to cooperate and work together in bring about a change in the way we look at these issues. School management at the highest level, teachers, students and parents need to cooperate and consciously integrate environmentalism into their daily lives. Besides that, the constant review and monitoring of environmental efforts will help direct Singapore in the right direction. Singapore is a small island; hence, sustainability plays an important role in the lives of its people. Every single environmental effort needs to be recorded, and the data gathered from these efforts need to be analysed so that Singapore can achieve its vision.

## 6 Conclusion

The most important issues of our time are environmental issues. Every year, scientists worldwide have been urging us to be aware of environmental issues and the consequences of global warming. Humankind has embraced materialism and we have failed to realise, let alone meet, our social and environmental obligations. We have been warned about climate change and the importance of the sustainability for centuries. The only difference is that now, we are knowingly acting irresponsible and it is on a much larger scale. However, there is hope and we can redirect our steps in order to protect this planet.

Awareness and education play a crucial role in sustainable development for the current and future generation. We need to transform the way we live our lives in terms of our behaviours and what we consume. Sustainable development requires a collective and conscious effort from every single individual around the globe.

Singapore aims to transform its cities into sustainable and liveable places. Food, water, energy and shelter are the basic ingredients necessary for humans to live with dignity. New technology and design, mobility efficiency, a new management approach and investing in sustainable education are basic elements of creating sustainable leadership and cities that ensure a sustainable use of the planet's limited resources. The younger generation need to be aware of their actions not only for themselves, but also for their children. The only planet we know is Earth and that is our common home. Hence, we need to protect and conserve it for the future.

## References

1. Anonymous (2014) Sustainability in education: pedagogical themes and practices in asian countries 2014. Nanyang Technological University: School of Humanities and Social Science. http://www.hss.ntu.edu.sg/Research/Clusters/EnvironmentandSustainability/efsasia/Pages/2014-Conference.aspx. Accessed 1 June 2017
2. Anonymous (2015) Sustainable Singapore blueprint 2015. Ministry of the Environment and Water Resources Ministry of National Development. http://www.mewr.gov.sg/ssb/files/ssb2015.pdf. Accessed 1 June 2017
3. Daipi H (2015) Tribute to Mr Lee Kuan Yew. Heart Talk. Sembawang Town Council. http://sbtc.org.sg/uploads/files/Hearttalk_June%202015.pdf. Accessed 1 June 2017
4. Frangoul A (2013) Sustainable Singapore: a model to be replicated? http://www.cnbc.com/2013/12/17/sustainable-singapore-a-model-to-be-replicated.html. Accessed 1 June 2017
5. Hales D (2008) Sustainability and higher education. N Engl J High Educ 23(2):23–24
6. Harvey L (2012) Social research glossary. Quality Research International. http://www.qualityresearchinternational.com/socialresearch/csr.htm. Accessed 1 June 2017

7. Jolliffe I (2002) Principal component analysis Encyclopedia of Statistics in Behavioral Science: John Wiley & Sons, Ltd. doi:10.1002/0470013192.bsa501
8. Little D (2014) Defining sustainability in meaningful ways for educators. J Sustain Educ 7:1–16
9. Mohamedbhai G (2015) What role for higher education in sustainable development? http://www.universityworldnews.com/article.php?story=20150108194231213. Accessed 1 June 2017
10. Slaper TF, Hall, T (2011) The triple bottom line: what is it and how does it work? http://www.ibrc.indiana.edu/ibr/2011/spring/article2.html. Accessed 1 June 2017
11. Soon YT (2010) Sustainable development: challenges and opportunities. https://www.cscollege.gov.sg/Knowledge/Ethos/Issue%207%20Jan%202010/Pages/Sustainable-Development-Challenges-and-Opportunities.aspx. Accessed 2 June 2017
12. Sterbenz CaEB (2015) 15 ways the world will be terrifying in 2050. https://www.businessinsider.com.au/waysthe-world-will-be-terrifying-in-2050-2015-3?r=US&IR=T#the-number-of-people-living-in-cities-will-likely-triple-1. Accessed 2 June 2017
13. Stern N (2006) Stern review on the economics of climate change. Cambridge. http://unionsforenergydemocracy.org/wp-content/uploads/2015/08/sternreview_report_complete.pdf. Accessed 2 June 2017
14. Webb F (2012) Sustainable cities: innovative urban planning in Singapore. https://www.theguardian.com/sustainable-business/sustainable-cities-innovative-urban-planning-singapore. Accessed 2 June 2017
15. Wee B (2008) Moving toward sustainability? the face of environmental education in Singapore. Green Teach 83:35–38
16. Zaiontz C (2016) Validity of correlation matrix and sample size. Real statistics using excel. http://www.real-statistics.com/multivariate-statistics/factor-analysis/validity-of-correlation-matrix-and-sample-size/. Accessed 1 June 2017

# Examining the Opportunities and Risks Associated with Sustainability Awareness in Higher Education in Pakistan

**Muhammad Umar Javed and Umera Imtinan**

**Abstract** This study aims to examine the risks and opportunities associated with adopting sustainability in higher education, focusing on Pakistan. The issues related to climate change and global warming have surpassed shocking levels to the extent that it is being discussed in major forums on a frequent basis. Most recently a summit organised in Paris, United Nations conference on climate change (COP21), aims to address these issues. More than 150 Heads of States and Governments are due to participate in this conference, making it one of the largest diplomatic conferences ever organised, aside from the United Nations General Assembly sessions. Literature review shows us that many steps have been taken in order to address these issues, however, as we are reminded by Albert Einstein "…it is important to avoid doing the same thing over and over again and expecting different results". Pakistan, being a developing country, is at a cross-road as a wide range of issues bear down upon it. Maintaining sustainability and stabilising the sustainability agenda is not one of the primary focus for its government. As it is known that poverty and sustainability go hand-in-hand, Pakistan must first adhere to the agenda of the Rio Convention of 1992 for eradication of poverty in order to increase the production and sustainability so that the needs of the majority of people can be met. Pakistan currently stands at number 8 on the list of 10 countries that will bear the worst implications of climate change. It is estimated that current pledges to cut carbon emissions will cost developing countries 50% more in the near future as compared to previous years. Higher education around the world is of prime focus as the forces of change present various challenges. These challenges are multifaceted, interconnected and are reinforced by social, economic, cultural and environmental developments. This setting puts higher education in a leadership role which calls for development of future leaders who are fully equipped to handle these challenges effectively.

M.U. Javed (✉)
School of Information Systems, Curtin University, Perth, Western Australia, Australia
e-mail: umor@hotmail.com

U. Imtinan
Department of Computer Science, COMSATS Institute of IT, Lahore, Pakistan
e-mail: Umera_imtinan@yahoo.com



T. Issa et al. (eds.), *Sustainability, Green IT and Education Strategies in the Twenty-first Century*, Green Energy and Technology, DOI 10.1007/978-3-319-57070-9_12

# 1 Introduction

Global warming and climate change has been a topic of discussion and study for quite some time now. Communities worldwide are becoming more aware of the potential issues related to sustainability and are showing strong interest in protecting the environment. The mitigation of global warming is increasingly gaining popularity and is becoming one of the most globally debated topics involving various industries, governments and educational institutions.

The best way to counter global warming, as perceived by experts, is to reduce the consumption of non-renewable energy sources and fossil fuels and cutting down on greenhouse gas emissions. As the world's population increases at an exponential rate, it is crucial to understand that the conventional means of generating energy must be revised, if natural and environmental hazards are to be prevented. This study argues that sustainability implies learning challenges to higher education both in terms of the paradigm and the provisions available. The underlying principle regarding sustainability is to integrate sustainable thinking into the curriculum of the universities, who help prepare the future leaders. This integration process may require certain changes to be made in the context of sustainability which not only requires sustainability to be embedded in the course structures in the universities, but also in the policy-making decisions of government bodies.

Sustainability does not simply require an addition to the existing structures and curriculum, but mandates a change of fundamental epistemology in culture, educational thinking, and practice. The key issue related to sustainability is that of responsibility which denotes that how far the educational institutions and governments as a whole are able to respond to the crisis and grasp the available opportunities. These responsibility measures will be tackled differently according to a nation's economic status; a developed country would be more prone to pushing these issues and providing various mitigation measures, whereas a developing country would be more focused on its economic rather than environmental health.

This study aims to "examine sustainability awareness the opportunities and risks associated with sustainability awareness in the Pakistani higher education sector?" requires a substantial review of available literature globally and specifically the research that applies to Pakistan.

# 2 Sustainability Awareness

The term sustainability was first introduced officially in 1987, in a report published by the World Commission on Environment and Development (Brundtland Commission). The report defined sustainable development as: development which meets the needs of current generations without compromising the ability of future generations to meet their own needs [38]. The term has its origins in Germany and was first coined by Hans Carl von Carlowitz in his book Sylvicultura Oeconomica published in 1713, where he was working out ways to guarantee a permanent supply of timber without cutting more trees than could be replaced [24].

Because of the array of ideas and varying terminologies put forward by various authors in regards to sustainability, it is ideal to keep a broad perspective. Sustainability should not be viewed as a single concept; rather, it should be thought of as an approach that emphasises the need to integrate environmental, social and economic issues in a long-term perspective. These three imperatives are commonly referred to as the triple bottom line which was further defined by Robinson [39]:

- The ecological/environmental imperative is to stay within the biophysical capacity of the planet;
- The economic imperative is to provide an adequate material standard of living for all; and
- The social imperative is to provide systems, processes and values which people want to live by.

Sustainability is both a strategy and a mind-set. The literature points out a profound need for transformation of thinking, behaviour, practices, competencies and leadership in the management of people. By integrating sustainability in the higher education curricula, the society will produce more enlightened leaders and the development of the general community will be more focused on saving the resources for future generations.

When talking about sustainability, carbon footprints, greenhouse gases and ecosystems usually come to mind. However, these deal with only the environmental part of the definition. Other factors that contribute to sustainability are economic and social factors. Combined, these three pillars of sustainability are often termed people, planet and profit. In short, they constitute the triple bottom line (3BL).

The concept of triple bottom line was coined by Elkington [14], and originates from the domain of management sciences. Depending on how 3BL is perceived, it carries a different meaning depending on the perspective. The goals of a business may be different from those of a public institution in that businesses are driven by a desire for profit. The concept of profit in a government organisation relates to the generation of revenue although the government is a not-for-profit entity. However, the measurement of profit would be expressed in terms of the GDP the nation accumulates [29].

The advantages to be derived from this approach has revolutionised the way businesses, not-for-profit organisations and government bodies measure sustainability and their relative performance on projects and policies. Companies have proven to do better even in depressions, when their strategy revolves around the sustainability perspective. Chasing short-term goals by forgoing long-term objectives has negative ramifications both for businesses and the planet [48]. Resource management and conservation is the essence of 3BL, where conservation and equity in the social arena are maximised in liaison with reduced overall cost [22]. In the contemporary market, the primary focus of an organisation is to ensure transference in dynamics from merely the creation of wealth through competitive economic performance to the inclusion of social and environmental parameters coupled with monetary benefit in order to achieve high levels of sustainable output and performance. As discussed earlier, long-term sustainability objective leads to a competitive edge.

As the proverb goes, knowledge is power. What people learn when they are young will be invaluable when they grow old. For future generations to thrive in this world, education must lead the way – by teaching and by example. The twenty-first century will be characterised by a massive and rapid change, a change which poses a grave danger as well as an opportunity to live in a world which is more stable, peaceful and sustainable. Institutions of higher education will be the vessels in which both individual and societal responses to these challenges are shaped. Today's education institutions have demonstrated the inability of long-term planning to address these challenges. This is because they are led by inconsistent and special interests. The performance of these institutions in terms of human and eco-systems' well-being has not been convincing [21].

In order for higher education sectors to become enablers and prepare students adequately for the future, all stakeholders must work as a team. All the stakeholders involved must be educated themselves and take strong initiatives which the students can follow. Furthermore, sustainability does not come cheap. One of the major barriers to sustainability is finances. This argument was substantiated by the study conducted in a Canadian university by Wright and Wilton [49], where students and faculty members were asked a series of questions relating to sustainability. The results showed that although many respondents had implemented the sustainability initiative, it was not their top priority mainly due to financial constraints.

As pointed out by Hales [21], higher education must move beyond the responsibility to prepare the students to live in this world as it will be; instead, they should be educated and prepared to shape the world in which they will live. He further goes on to say that the task at hand is difficult but necessary; education is the only way to enlighten, enable and empower the choices our future generations make.

After poverty, foreign debt has largely become a prevalent issue in the majority of developing countries, and Pakistan is no exception. It has been observed that the poorest countries are heavily in debt which means that most of the revenue earned from exports goes to paying off those debts. As a result of this debt servicing, there are cuts in expenditure on healthcare, education and general social welfare [8]. As outlined above, Pakistan faces the same fundamental dilemmas as any other developing nation. Economic and political stability is a prerequisite for any sustainability model to take effect. Although sustainability is embedded in the curriculum of a handful of private schools in Pakistan, the majority of government institutions ignore the issue [8].

Sustainable development requires institutional transformation which still remains a considerable challenge for educational institutions. To meet this challenge, the Copernicus Charta was formed by the European Rectors' Conference in 1994. To date, more than 300 European universities have signed the charter and accepted its obligations [3]. However, the majority of universities in developing nations are yet to take a step forward.

A typical journey for an educational institution towards sustainability starts from drawing attention to the fact that integrating sustainability into higher education requires theoretical reflection as well as practical approaches to show that sustainability can be articulated, demonstrated and put into action. However, issues may arise during the integration process which needs to be addressed both on a micro-level – diversity of disciplinary cultures, and the macro-level – the higher education system as a social construct [41].

Higher education systems have a moral responsibility to increase the knowledge, awareness, skills, and values needed to create a just and sustainable future [9]. This is because they prepare most of the professionals who will play a key role in the adaptation of our society to a sustainable model of living. In agreement with Cortese (cited in Lukman and Glavič [33]) state that we can have a sustainable future only if universities actively promote sustainability. Moreover, there has been an increasing interest shown in the incorporation of sustainability by a variety of institutions around the world. To help these institutions reach their goals, various tools are available which assist with the process. Cortese [9] identified four independent dimensions namely: Education, Research, Operations and Community Outreach. This was later modified by Lozano and Young [32], who added a fifth and crucial element, Assessment and Reporting.

Based on the above model, Urquiza Gómez et al. [47] conducted a research that focused on three main elements: institutional commitment, example setting, and advancing sustainability. When applied, this model indicated that the first two elements were most favourable with a score of 36% and 38% respectively, whereas the latter scored only 25%. This demonstrates how important it is for institutions to instil and encourage commitment to the sustainability agenda in all their relevant stakeholders. While this study sees it as a shortcoming, sustainability experts view it as an opportunity to build upon.

Institutions and corporations are resorting to sustainable practices mainly due to ever-increasing pressure. However, this might not be the only case; as reported by Kiron et al. [27], most managers state that their company's sustainability activities have increased company profits. This, however, can be achieved only by changing or altering the business model and securing the commitment from all the stakeholders. In the competitive dimensions, financial and investment analysts have started to include sustainability as a vital evaluation criterion for companies.

Companies struggle when trying to integrate sustainability in their long-term strategies. This may be due to certain obstacles such as capital budgeting, financial goals not aligning with sustainability objectives, and uncertainty about implementation. However, companies that adopt these practices benefit by expanding their target markets globally, in countries where a company's sustainability track record is accounted for [45].

Environmental sustainability is of core importance to companies, as it not only helps the general environment but also improves the business processes. Moving towards environmental sustainability does not present any risk to business operations. As can be seen from the literature, environmental sustainability is becoming a heavily regulated issue, which indicates that it is desirable that companies start incorporating sustainable practices as soon as possible, before they lose their competitive edge [50].

Universities play a key role in the development of the society. With the introduction of sustainability reporting, universities can now choose voluntarily to disclose their steps and actions taken towards sustainability. This encourages the students to follow in the same direction [12]. By reporting the key matrices like water and energy consumption, recycling, and carbon emissions, the institutions can benefit by increasing their triple bottom line. Reporting also allows for regular monitoring which in turn saves cost due to energy conservation and propels the move towards a green strategy.

With the development of more sustainable practices, such as the conservation of resources, operations become more efficient and streamlined, resulting in lowered costs. Sustainable practices allow firms to reduce their overhead costs due to reduced wastage and usage of raw materials, thereby achieving comprehensive economic returns [23]. Companies enjoy increased cost-effectiveness because of the prolonged life cycle of the product, re-usability, and recycling of raw materials.

Companies that rate better on the sustainability index will attract quality employees due to an increased awareness. People want to work with companies that are proactive with corporate social and environmental programmes. These programmes also ensure employee loyalty and encourage an ethical corporate culture. The same principle applies to the consumers as well. Consumers today are more likely to purchase products from companies who are mindful of their impacts on the environment and society. The adoption of sustainable practices can also enhance the corporate reputation of the firms, allowing them to secure business contracts and new market segments.

One of the major failures of sustainable development in general is that it does not differentiate between the needs of the developed and those of the developing nations. Much of the development has already taken place in the developed nations that have already consumed an extravagant share of the world's resources, whereas the developing world has not seen the same rate of progress. In order for sustainable development to take effect, this pattern must be reversed, which implies that the sustainability in developing nations be measured depending on their consumption of natural resources [6]. Following the same argument, they further went on to add that given the trend, it is more acceptable to deduce that lesser developed countries may be more sustainable than their more developed counterparts, whereas the world as a whole is not sustainable.

Continuing in the same vein, education for a sustainable society requires the development of techniques which can be used to impart knowledge to pre-college students. However, these students are constrained by the resources available to their educational system. There is a risk of failure if the institution providing the education is limited by knowledge and ideas itself, which is largely dependent on the geographical location of the higher education institution [2]. These educating institutions possess the ability to confront global sustainability by means of collaboration, sharing best practices and helping the under-privileged institutions in the developing world [28].

The Brundtland report produced a holistic definition of sustainability. However, differences in teaching methodologies have given rise to varying individual perspectives. Where the embedding of sustainability in coursework is viewed as a positive indication, it also suggests how the coursework can have a significant impact on the students' attitude and thinking regarding sustainability.

An understanding of how greener practices reduce costs through decreased consumption of energy and less wastage, promotes the cost factor as the driver of sustainable development. However, the training and education costs associated with implementing a sustainability agenda are perceived as an obstacle to the process.

The extent to which cost-benefit analysis can adequately account for environmental concern varies in accordance with the weight given to the most significant environmental effects. The main bearers of these associated costs are the future generations and these costs are already being incurred in the form of rapidly depleting natural resources, not sanctioning long-term investments like forestry, and shifting the distant costs of things such as nuclear waste disposal to future generations [5]. There are, however, measures in place to ensure sustainable development. These measures form part of the government policies which envision a sustainable future based on an increasing contribution of renewables. Policies such as carbon tax and subsidies for the use of renewable energy discourage users from opting for non-renewables forms of energy, thus saving the environment and resources for future generations [31]. Apart from generous subsidies, sustainability is gaining popularity in the financial arena as well. Socially responsible investments are encouraged and the companies' portfolios are judged according to their strengths or weaknesses in terms of the three dimensions of the triple bottom line [25].

Sustainability is expensive. It costs the public their money, in the form of taxes, and their jobs to some extent. When governments subsidise the renewable energy, it drains revenue per individual/business. Taking an example from Spain, when the government started to pay an exorbitant amount of money to everyone who installed solar panels, this in turn affected Spain's economy in the long-run and prompted heavy cuts to the subsidies [35]. This goes to show that sustainability is not sustainable until or unless appropriate strategies and techniques are implemented in accordance with the operating environment.

Sustainability and the green economy present great challenges and opportunities – and sustainable leadership is what will turn the tide. Sustainable business models need to be created that enable companies, people, and the planet to flourish and thrive now and in the future. This can be achieved by inspiring, motivating, and developing leaders and managers to succeed in a triple-bottom-line world. Given the holistic nature of the definition for sustainability, individual perspectives must not be restricted to a confined framework. A study carried out by Fisher and McAdams [16] took into account the same dichotomy and made three recommendations: Sustainability programs in universities should expose the students to a wide variety of scholarly approaches to sustainability.

Sustainability should be presented to students in a way in which disciplinary focus is situated within a larger framework; and the approach towards sustainability should come from different viewpoints and angles, including social, cultural, political and artistic.

## 3 Research Question and the Chosen Research Method

The research question was formulated after broaden extensive review of the existing literature. It focuses on and targets what this study is intended to achieve. This study aims to address the following research question: "What are the opportunities and risks associated with sustainability awareness in the Pakistani higher education sector?" Sustainability awareness still remains a key issue in Pakistan's higher education sector. The perspectives captured by means of this research will provide vital insights into the extent to which there is awareness and understanding of sustainability in Pakistan. The answer to this question will help the Pakistani community at large to identify the issues and improve sustainability through practical implementation.

This study takes the objectivist approach and the respondents in the sample size are social actors for the research. The data and information for this study will be collected through an online survey without the researcher physically participating or being a member of the institutions under study. The research does not intend to reveal social phenomenon from the perceptions and subsequent actions of social actors. In this study, there is no social interaction within the research environment and the researcher does not physically take part in analysing the details of the situation to understand and interpret events and the reality working behind them [40]. The research paradigm for this study follows a deductive approach where a theory or hypotheses are developed before designing the research strategy. The inductive approach relies on the empirical verification of general conclusions derivable from a finite number of observations, i.e. if an event occurs for a specific number of times, it shall continue to occur ceteris-peribus [46].

The deductive approach states that there are established universal laws which are mainly based on hypotheses which require continuous testing against the laws themselves [36]. In other words, the deductive approach is commonly used to answer questions like "what", "who", "when", "how much", and "how many".

In this research, a deductive approach is employed whereby a general set of propositions relating to a given phenomenon is reduced to a specific set of testable hypotheses or a single testable hypothesis. A survey research strategy is used in order to examine the risks and advantages associated with sustainability in higher education in Pakistan. This deductive approach draws conclusions from a set of variable factors.

Researchers are often interested in the opinions of a group of people relating to a particular issue or topic. A number of questions are asked in the survey, all relating to that particular issue in order to find answers. Information is collected from a sample of the population to describe some aspects or characteristics such as abilities, opinions, beliefs, and knowledge. The main way in which information is collected is by asking questions; the answers to these questions constitute the data for the study [18]. In the survey approach, participants are asked a series of questions which could be face-to-face, telephonic or web-based questionnaires.

This study uses the quantitative research method because data is collected online. This research paradigm examines the relationship between the stated theory and the proposed research and at the same time it is a discovery-oriented approach intended to obtain the participants' opinions about sustainability in higher education in Pakistan.

According to Saunders [40], the survey strategy allows the collection of data which can be analysed quantitatively using descriptive and inferential statistics, revealing possible reasons for relationships between variables.

There are several benefits of using a survey in a research study. Collection of data by means of a survey is efficient because it can be administered to many individuals in person, via the Internet, over the phone, or through direct mail. In addition, data is usually easy to analyse and can be used to describe the population of interest. Methods of survey research such as the Internet are inexpensive and help to keep the budget reasonable. It gives control over the research process and the sample can generate findings representative of the whole population at less cost. Therefore, it is feasible to obtain a large sample that makes the results statistically significant even when analysing multiple variables [1].

The main drawback of survey research is that the data obtained from participants is a self-report and may not be completely accurate. For example, in the case of this study, the respondents might provide a response relating to sustainability that they think the researcher wants rather than their true feelings [1]. Another concern is the low response rate. Surveys distributed via mail or e-mail can easily be discarded, thus reducing the response rate [17].

Technological advances have made administering the surveys on the Internet quite common and have provided an ease and comfort for both the respondent and the researcher. Increasingly, researchers and students are turning to e-mail or web-based software and services to collect survey data from their target population. Services such as Survey Monkey and Qualtrics allow users to design their own surveys free of cost. Additional services like survey administration and analysis can be purchased at a minimal cost [11]. This platform offers multiple advantages which include greater convenience, lower costs, faster turnaround times, multimedia interface, and mobile administration and reduced data entry [1].

A respondent can be given a hyperlink to a website containing the survey and can return it by email as an attachment or just simply submit it online through the abovementioned services [10]. The respondents can take more time to respond at their convenience and still maintain anonymity. Additionally, it can reach large numbers and a wide variety of potential respondents. Moreover, it has the ability to directly download the data for immediate analysis in SPSS programs and can present the results in the form of charts, tables and graphs [42].

Disadvantages include lower response rate, erroneous data entry and bias. Another major limitation of the web-based surveys is to determine whether the sample members have Internet access which may be due to low coverage [13]. There is a great possibility of technical failure and hacking into the web-based survey. Moreover, the data collected may lack credibility, validity and reliability [19].

The survey designed for this study contains a welcome page covering the purpose of the survey, the number of questions available, time required to complete the survey, deadline and an undertaking assuring the participant that data privacy and confidentiality will be maintained [44].

The survey consists of three sections: Demography, Risks, and Opportunities associated with sustainability in the higher education sector in Pakistan. The survey commences by collecting demographic information about the respondent, and then asks questions regarding their Internet usage. Finally, there are scaled response questions with regards to sustainability which are measured on a five-point Likert scale as shown below: 5 = Strongly Disagree; 4 = Disagree; 3 = Neutral; 2 = Agree; 1 = Strongly Agree. The survey contains both scaled responses and open-ended questions. Open-ended questions allow for more individualised responses, whereas close-ended questions restrict the respondent to choosing from the offered alternatives. But at the same time, open-ended questions are sometimes hard to interpret, score and sometimes are not appropriate for some respondents [18].

Qualtrics, a web-based survey tool was utilised for designing, distributing and analysing the responses. As mentioned earlier, it is automatic software that can be used to analyse the data at a minimal cost. The respondents found that the software provided a convenient means of completing the survey.

The first section of the survey is intended to gather the respondents' demographic information which includes gender, age, job title, qualification. The second section focused on Internet usage and how that time is being spent, such as time spent on social networking, working studying etc. This section also provided for the modes of using the Internet, be it on a personal computer or handheld devices. The last section of the survey asks users' opinions about sustainability.

Statistically, the unit of analysis is defined as the "who" or "what" for which information is analysed and conclusions are drawn. According to Sedgwick [43], the common unit of analysis ranges from individuals, groups, artefacts to geographical units social interactions.

The unit of analysis for this study is ungrouped because the research intends to capture individual opinions and not organisations. As per the research question, the unit of analysis for this research aims to elicit the opinions of users regarding the risks and opportunities associated with sustainability in higher education.

The data analysis follows a five-point Likert scale model. This scale allows the users to express their agreement/disagreement with the posed questions. The Likert scale is a non-comparative scaling technique and is one-dimensional in nature [26]. Five categories are appointed for each item which range from "Strongly Agree" at one end, to "Strongly Disagree" at the other, whereas "Neither Agree nor Disagree" are centred [37]. This allows the survey to obtain the data with regards to respondents' degree of consent to the stated question.

The population refers to the group of users from whom the survey is seeking to obtain the information. It is a near-impossible task to study all members of a population, so a sample represents the population which makes it easier, rather than dealing with the entire population. According to McGaghie and Crandall [34], a sample is defined as the systematic selection of targeted people for the research purpose and is an important factor in the survey process.

The target population for this study comprises students from different universities in Pakistan and several professionals who have recently finished their degrees, regardless of their disciplines of study.

The proposed study was carried out using the primary data obtained by using an online survey tool, Qualtrics, which was provided by Curtin University of Technology, as it uses an intuitive interface. The survey was distributed using various social networking tools such as Facebook. The hyperlink for the survey was also distributed via email and WhatsApp to enable respondents to complete the survey on their hand-held devices.

A follow-up reminder was sent using the same mediums of communication. The respondents were informed about the importance of the survey and any problems encountered while completing the survey were promptly addressed through the corresponding channels.

The main purpose of data analysis is to obtain findings from the gathered data. Once all the information has been collected, the next step is to examine it. This examination of the data helps the researcher to answer the question that was the catalyst for the investigation.

The data collected for this study was recorded in a Word document in order to keep track of the changes. The collection method was by means of web-based surveys and the analysis was conducted using the Statistical Package for Social Sciences (SPSS) version 22, which is a software developed by IBM. This tool is used for factor analysis as a means of understanding individuals' opinions on sustainability issues. Since factor analysis requires a large sample size, regular follow-ups were undertaken in order to obtain a reasonable response of at least 100 users.

Reliability deals with the quality and consistency of measurements, whereas validity concerns the appropriateness of the methodology employed and whether the concept being investigated is actually being tested or measured [30]. A survey questionnaire must be reliable in order to be valid but being reliable is not sufficient for the interpretation as participants' understandings differ from one another [40]. Therefore, reliability needs to be consistent with the findings at different times, situations, and conditions. Given all the conditions, the online survey tool, Qualtrics was chosen with much consideration as it provides a clear layout and ensures the respondents are informed when answering the questions, increasing the reliability of data.

## 4 Results

The data collection method is discussed prior to presenting the results of the survey. Survey web links generated by Qualtrics were distributed through various student forums on Facebook, text messages through WhatsApp & Viber, and the survey links were also distributed through emails to the participants.

There were queries from the respondents regarding the survey questionnaire such as whether it is mandatory to provide basic demographic information and to differentiate opportunities from risks. All the respondents' queries were dealt with and responded to clearly via the participants' preferred mode of communication. The minimum requirement for carrying out the factor analysis was satisfied with a valid sample size of 109 out of 202 respondents. Ninety-three questionnaires were screened out due to missing values.

The demographic part of the survey asked users for personal information regarding gender, age, job title, educational qualifications, and the number of hours spent daily on the Internet and computer, time spent on social networking emails and other purposes, devices used for accessing the Internet, and finally how the users were introduced to the concept of sustainability.

Figure 1 indicates that approximately 48% of the respondents were male and 52% female. There is no evidence of gender bias as the survey was distributed to everyone uniformly.

Figure 2 shows the age of participants. From the results of the survey, majority of the respondents are in the 25–30 age group, comprising nearly 35% of the overall population. This is followed by the age group of 21–24 accounting for 26% of the participants. The minority of 2% are in the 46–50 age group.

Figure 3 indicates that the majority of respondents belong to the Management sector, consisting of approximately 35% of the population. Following were the student with 25% and then participants related to the teaching profession with around 20%. Respondents involved in technical and administration jobs constituted nearly 9% and 7% respectively.

Figure 4 indicates that the majority of the participants' field of education was Management, which comprises of 18% of the total population. Second was Humanities with 12%, followed by Marketing at 11%, Accounting and Science & Engineering both at 8%, Economics and Finance 6%, Information Systems, Information Technology, and Computer Science at 5%, 3%, and 4% respectively.

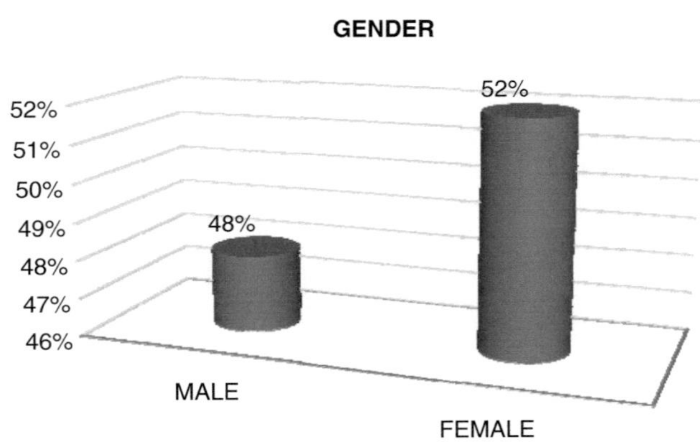


**Fig. 1** Participant gender (prepared by author)

Others accounted for 10% which were mainly constituted of Literature and Languages.

Figure 5 indicates that the majority of respondents have a Bachelor's Degree at 31% of the total population, followed by Master's Degree at 30%; Higher Secondary/Pre-University and Post Graduate Diploma both at 7% followed by Diploma and PhD both at 6%.

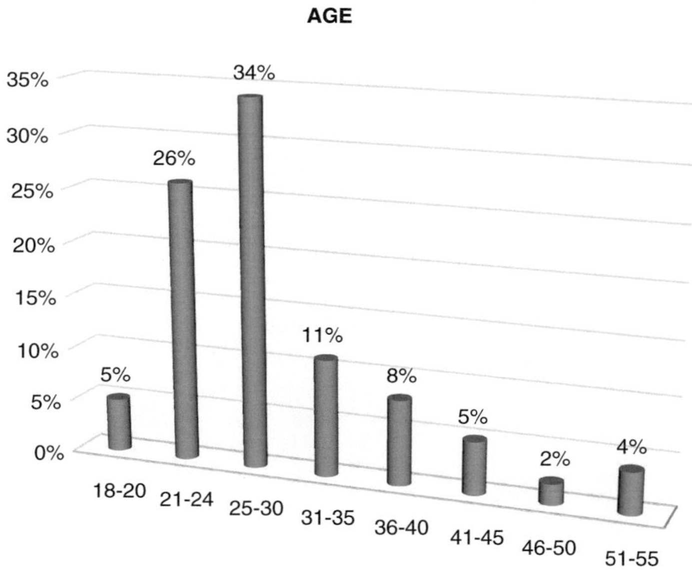


**Fig. 2** Participant Age (prepared by author)

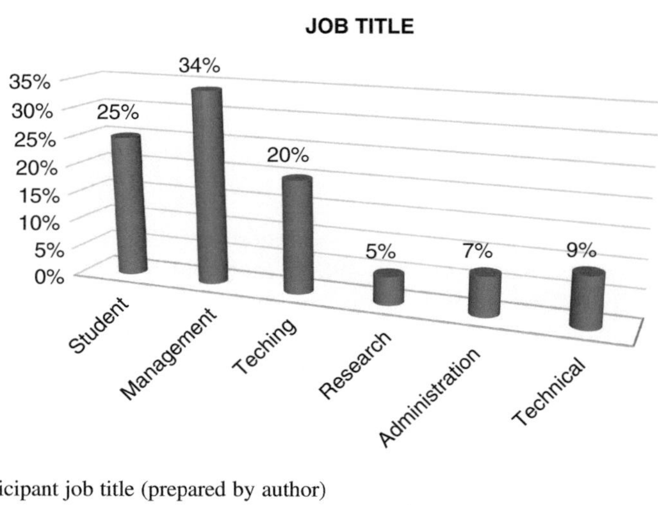


**Fig. 3** Participant job title (prepared by author)

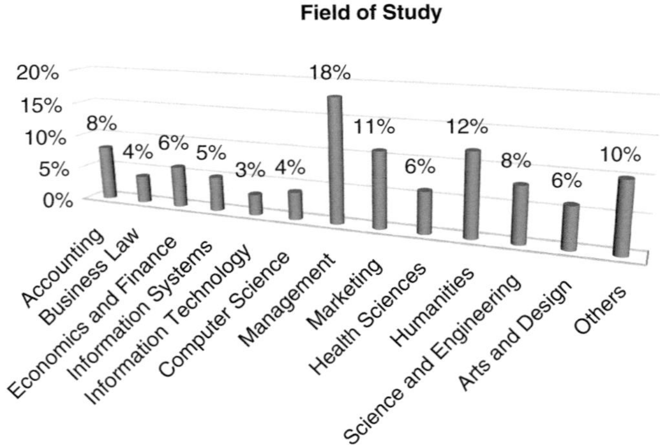


**Fig. 4** Participant field of study (prepared by author)

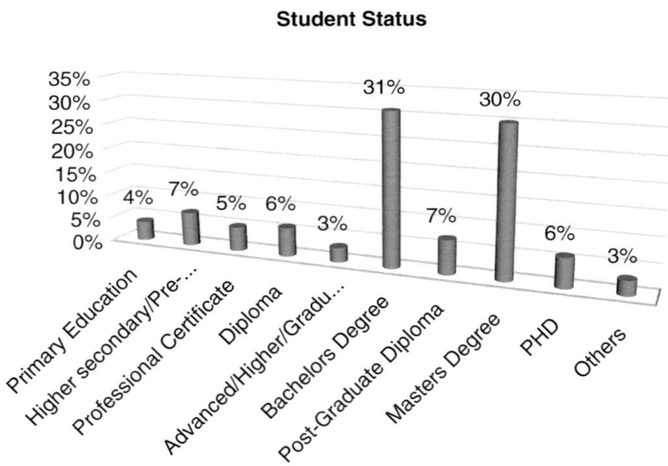


**Fig. 5** Participant student status (prepared by author)

The survey indicates (see Fig. 6) that the majority of respondents spend up to 5 h on the computer each day making it 40% of the population, followed by 5–10 h at 29%. The "Less Than an Hour" category was at 17% which was followed by 14% of the respondents who use the computer for 10–20 h a day. Respondents who use the computer for more than 20 h represent only 1% of the population.

Figure 7 indicates that 39% of the respondents spend up to 5 h on the Internet per day, followed by 27% of respondents who spend 5–10 h. Respondents who spend 10–20 h on the Internet comprised of 17% and less than an hour 14%. Respondents who spend over 20 h are merely at 2%.

Fig. 6 Number of hours spent on the computer (prepared by author)

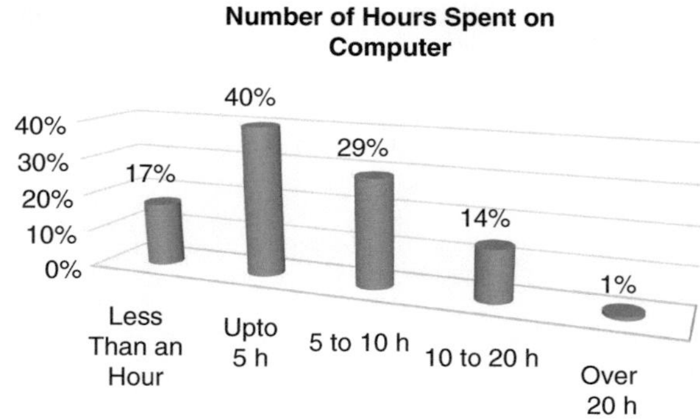


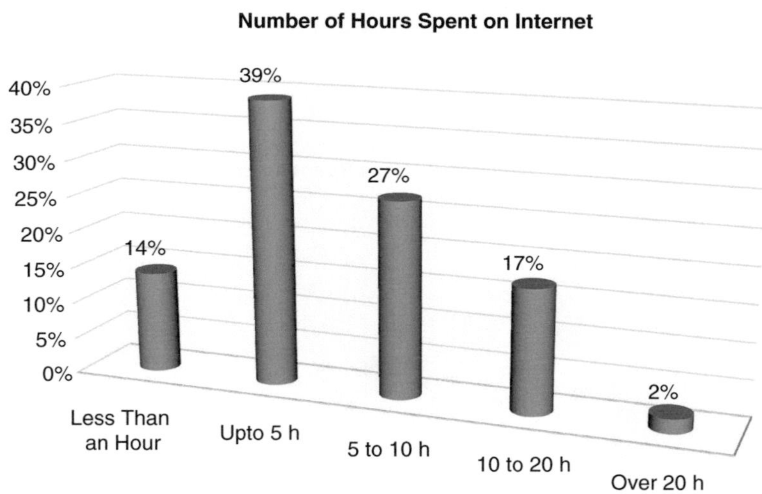


Fig. 7 Number of hours spent on the Internet (prepared by author)

Figure 8 shows that the majority of respondents spend up to 5 h on social networking per day, followed by respondents who spend less than an hour at 34%. The 5–10 category accounted for 19% followed by 10–20 h at 10%.

Figure 9 indicates that the majority of respondents spend less than an hour on email per day, followed by up to 5 h at 24%, 5–10 h at 8%, 10–20 at 5%, and none of the respondents spend over 20 h on email.

The survey produced the following results for the respondents' usage of Internet for various purposes: email at 59%, playing games at 20%, studying at 44%, working at 61%, shopping online at 31%, chatting at 44%, researching hobbies at 36%, banking online at 23%, buying goods or services at 24%, buying

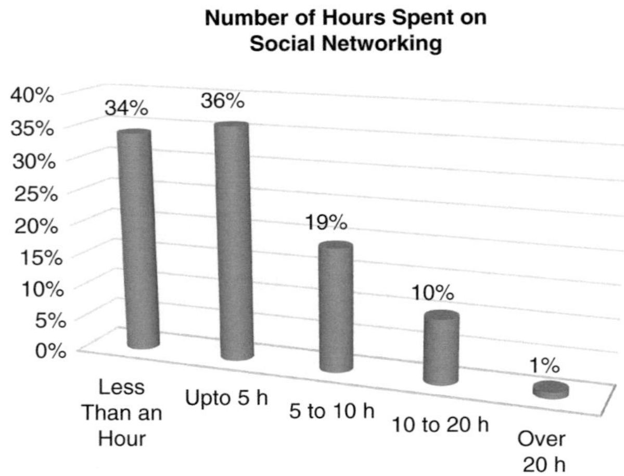


**Fig. 8** Number of hours spent on social networking (prepared by author)

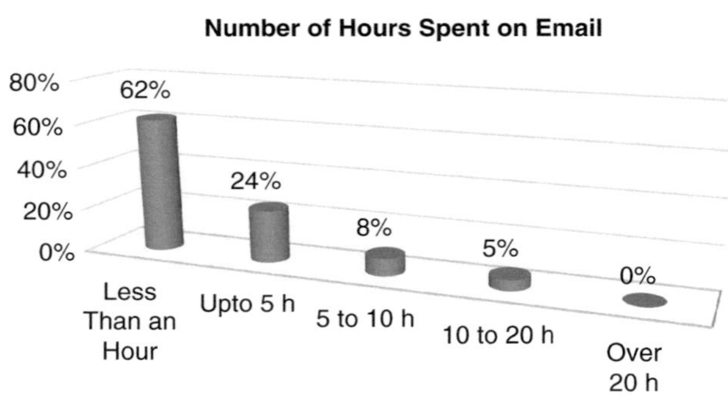


**Fig. 9** Number of hours spent on email (prepared by author)

stocks or investing online at 7%, researching travel information or making reservations at 24%, and other at 4% which included watching movies and T.V shows and news (see Fig. 10).

The results in Fig. 11, show the types of devices used by respondents to access the Internet, these are: PC at 20%, desktop at 15%, laptop at 70%, netbook at 5%, PDAs at 1%, workstation at 4%, tablet at 21%, smartphones at 71%.

Figure 12 indicates how the respondents were first introduced to the concept of sustainability and green information technology. The majority (43%) of the respondents indicated that they obtained the information initially through the Internet. The other sources and their respective percentages are as follows: School at 36%, higher education at 30%, books at 9%, magazines at 12%, news media at

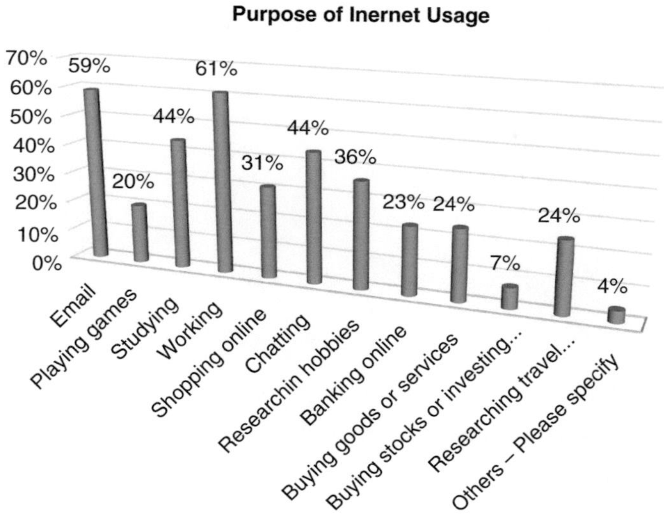


**Fig. 10** Purpose of Internet usage (prepared by author)

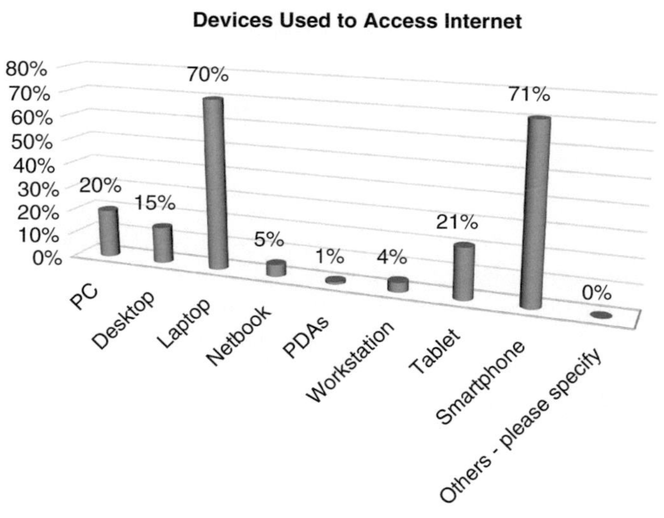


**Fig. 11** Devices used to access the Internet (prepared by author)

20%, conferences at 8%, and others at 4% which include the exposure to various advertisements.

Figure 13 indicates the company that is associated with each respondent's device. The majority of respondents prefer Apple devices, comprising of 42% of the population, followed by Google at 35%, Dell at 11%, IBM at 3%, and Others at 10% which includes HP, Samsung, Sony, LG, Lenovo, Asus, and Haier.

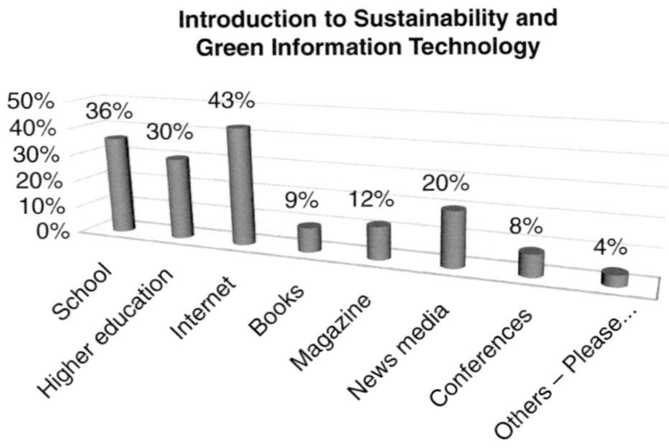


**Fig. 12** Introductions to sustainability and green information technology (prepared by author)

**Fig. 13** Company associated with device (prepared by author)

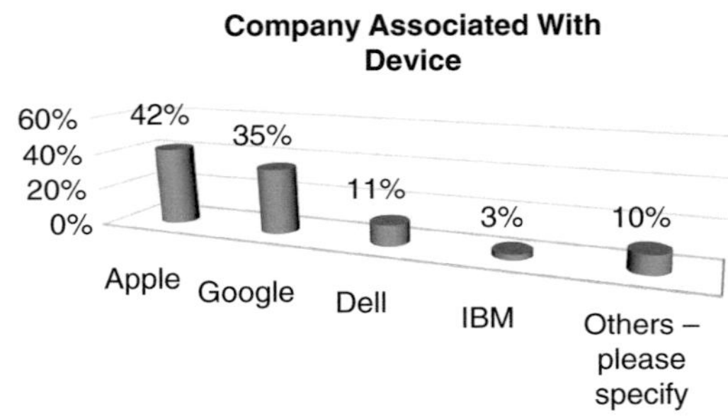


Respondents were asked if they read the company's sustainability report before buying the device (see Fig. 14). The majority of respondents comprising 50% of the sample population answered No. This was followed by respondents who were not sure and answered with a Maybe at 27%. Respondents who read the sustainability report answered "Yes" which accounts for 20% of the population and respondents who answered "Not at all" comprise only 5% of the sample.

The respondents were asked how often they change their device (see Fig. 15), to which 22% replied with every 12 months, followed by 19% for 24 months and 17% for 18 months. Respondents who change their device on a more frequent basis, that is, every 6 months were only 9% of the total population. This statistic was the same for people who change their device every 24 months.

The majority of respondents which comprises of 55% of the sample population change their device for functionality reasons; however 53% said the reason was to keep up with technology, which is justified as new updates are announced every year by big companies. The speed of the device was the reason given by 29% whereas 11% said it was because of the size. Other responses accounted for the

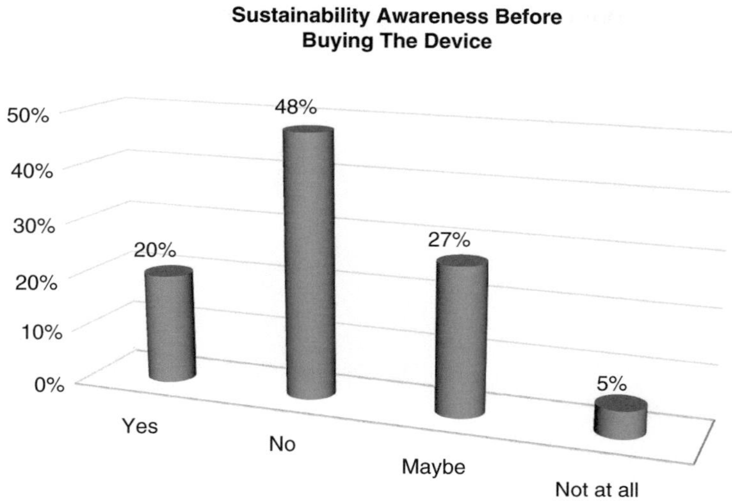


**Fig. 14** Sustainability awareness before buying the device (prepared by author)

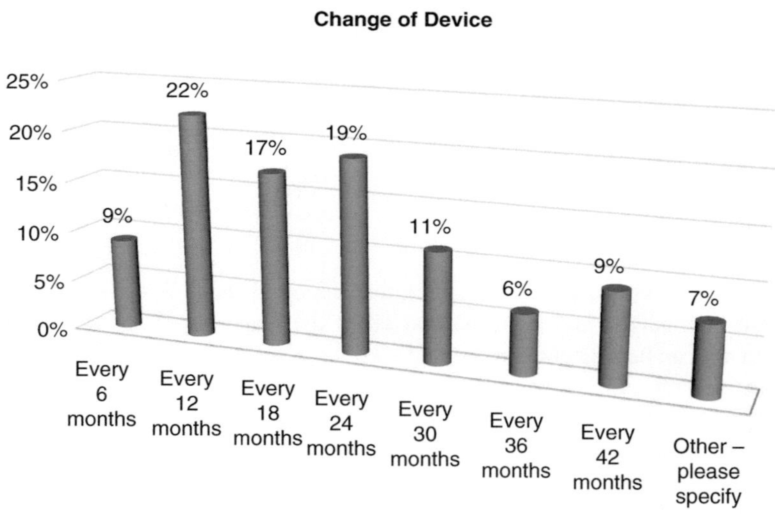


**Fig. 15** Change of device (prepared by author)

remaining 5%, whose reasons included changing trends, budget, and changing simply because of to the attractiveness of a different device (see Fig. 16).

Respondents were asked if they think changing their devices would cause damage to the planet, to which the Yes and No responses were 28% and 29% respectively. This shows the thought process of the respondents which was almost equally polarised. People who responded with Maybe were at 39%, and only 4% of respondents said Not at All (see Fig. 17).

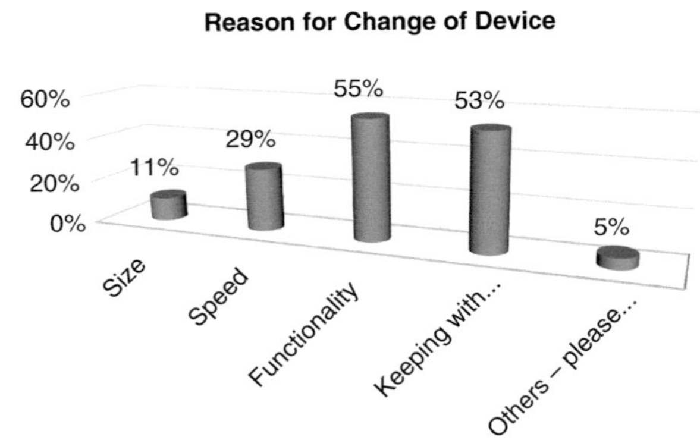


**Fig. 16** Reason for change of device (prepared by author)

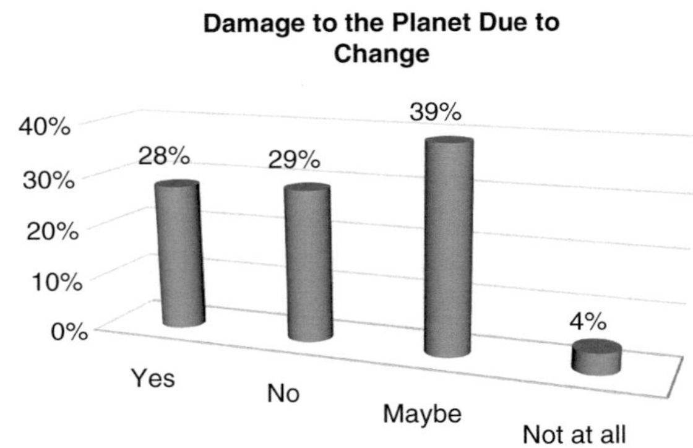


**Fig. 17** Damage to the planet due to change of device (prepared by author)

Respondents were asked if they think that the mind-set of designers and consumers can be changed and how. Almost all of the respondents agreed, with 60% saying that it can be changed through more awareness. Fifty-four percent emphasized social networking, 48% – education, 44% – Internet, 35% – training, 30% – T.V, and 25% said workshops. Four percent of the respondents believed there were other ways to achieve change, which included enforcement of laws and government policies (see Fig. 18).

To analyse the data, factor analysis was used. Factor analysis is a procedure used to determine the extent of shared variance that exists among a set of variables. It is essentially a process whereby the number of variables is reduced by determining which variables can be clustered together, and factors are the groupings of variables that are measuring some common entity or construct [15].

Factor analysis is used to develop, refine, and evaluate the tests, scales, and measures. It is a common method used to interpret self-reporting questionnaires and reduce large number of variables to a smaller set of variables or factors [20]. In order to carry out factor analysis, the variable should be quantified at interval level and the required sample size for a reliable factor analysis is at least five times

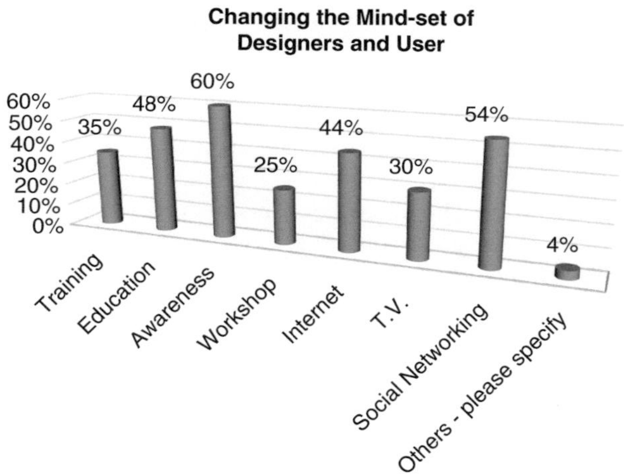


**Fig. 18** Changing the mind-set of designer and user (prepared by author)

per variable or at least 100 for the sample size. There are five basic requirements and conditions that must be met in order to carry out factor analysis. These are explained below. The same processes were carried out for analysing the data gathered for this study.

**Step 1: Sample Size:** In order to carry out factor analysis, the sample size should be greater than 100 and the larger the sample, the more accurate the results will be [15]. The Kaiser-Meyer-Olkin (KMO) is used to assess the suitability of the respondent data for factor analysis. The KMO index ranges from 0 to 1 and 0.5 is considered acceptable to carry out factor analysis. Moreover, with Bartlett's Test of Sphericity, the significant value should be lower than ($p < 0.05$) which signifies that there is a relationship between variables [15].

According to Armstrong [4], a KMO close to 1.0 indicates that a factor analysis is useful from the gathered data and that the data can be grouped into a smaller set of underlying factors. Marvellous 0.90s; Meritorious 0.80s; Middling 0.70s; Mediocre 0.60s; Miserable 0.50s; Unacceptable Below 0.50

**Step 2: Choosing a Factor Extraction Method:** There are a number of ways to extract factors; however, the most commonly used is Principal Component Analysis (PCA). It is also the default method of extraction used in statistical packages such as SPSS. However, if the data is distributed normally, then maximum likelihood is the best choice, whereas if the normality test is violated, then principal axis factoring (PAF) has to be considered as the preferred method for extraction [15].

**Step 3: Criteria for Determining Factor Extraction:** The main purpose of data extraction is to reduce a large number of items to a smaller set of factors. This is followed by the decision about the number of factors to be retained for rotation. Both under-extraction and over-extraction will have adverse effects on the results. The default number of factors retained in almost all the statistical software packages have Eigen values greater than 1.0 [15].

**Step 4: Selection of Rotation Method:** Rotation is used to clarify and simplify data. The most common rotation method used is Varimax rotation in factor analysis, producing factors that are not uncorrelated and making it possible to identify a variable with a factor [15].

**Step 5: Interpretation:** The researcher is responsible for deciding the variables to be attributed to a factor and names that factor accordingly. In order to form a new factor, which groups the majority of responses, it must have more than two variables in order to provide a meaningful interpretation [15].

A Principal Component Analysis (PCA) of 37 Likert scale questions together with Varimax rotation approach are factor extraction methods used for both opportunities and risks of adopting sustainability in higher education in Pakistan, and were applied to analyse the data from 109 participants.

In order to determine whether or not the variables are correlated, reliability testing needs to be conducted. Cronbach's Alpha test is the most popular test used to check the internal consistency and to measure the scale for reliability. An Alpha value above 0.70 is considered as reliable and an Alpha value above 0.60 is considered probably reliable; however, an Alpha value below 0.59 is considered not reliable [7]. Table 1 shows the reliability value for both opportunities and risks. The Alpha values for both are greater than 0.70, which means that the data is reliable.

An examination of the Kaiser-Meyer-Olkin (KMO) measure of sampling adequacy suggested that the sample was factorable for both the opportunities and risks. The KMO value for opportunities is 0.815 which means that variance in the data can be explained "Meritoriously" by the factors. However, the KMO value for risks is 0.925, which signifies a "Marvellous" data set. In addition to this, the Sig. value of Bartlett's test of Sphericity is less than 0.05 for both the opportunities and risks, which indicates that the data is suitable for factor analysis. Moreover, it also signifies that there is sufficient correlation between variables. An anti-image correlation matrix was used to check the correlation of the data matrix for appropriateness of factor analysis. Several factors were extracted and retained after conducting the tests which represent a substantial portion of the variation. Twenty-four variables of Opportunities were extracted into three factors; which means that the Component Matrix has retained three factors for variables for the Opportunities. From the percentage of variance, the cumulative percentage of the total variance for the three factors for Opportunities is 54% (Table 2), which is quite adequate to explain the variances of the extracted factors.

**Table 1** Alpha, KMO, and Bartlett's test (prepared by author)

| Items | Alpha | KMO | Bartlett's test |
|---|---|---|---|
| Opportunities | 0.907 | 0. 815 | 1265.815 |
| Risks | 0. 958 | 0.925 | 1138.023 |

As can be seen from Table 3, the cross-loading variables were deleted one by one and then a check was done to determine whether or not this improved the factor's cross-loading. The following three factors were generated: factor 1 comprised five variables, factor 2 had four variables, and factor 3 had two variables. The highest loading variable is *Reduce Pollution*, which shows that the adoption of sustainability in higher education in Pakistan will help in the reduction of pollution.

Factor 1 comprises the following variables and it has been named *Stakeholders' Needs and Investments* (Table 4).

**Table 2** Total variance explained for opportunities (prepared by author)

| Total variance explained | | | | | | |
|---|---|---|---|---|---|---|
| Component | Extraction sums of squared loadings | | | Rotation sums of squared loadings | | |
| | Total | % of variance | Cumulative % | Total | % of variance | Cumulative % |
| 1 | 7.788 | 32.449 | 32.449 | 5.002 | 20.843 | 20.843 |
| 2 | 3.366 | 14.027 | 46.476 | 4.547 | 18.945 | 39.788 |
| 3 | 1.813 | 7.555 | 54.031 | 3.418 | 14.243 | 54.031 |

Extraction method: Principal Component Analysis.

**Table 3** Factors for opportunities (prepared by author)

| Rotated component matrix[a] - OPPORTUNITIES | | | |
|---|---|---|---|
| | Component | | |
| | 1 | 2 | 3 |
| Increase efficiency | | | 0.710 |
| Increase cost-effectiveness | | | 0.686 |
| Improve community investments | 0.788 | | |
| Improve human rights | 0.759 | | |
| Improve social responsibility investing | 0.711 | | |
| Attract quality employees | 0.737 | | |
| Satisfy customer needs | 0.734 | | |
| Reduce carbon footprint | | 0.750 | |
| Reduce emissions | | 0.786 | |
| Reduce pollution | | 0.815 | |
| Reduce health hazards | | 0.768 | |

[a]Rotation converged in 3 iterations.

**Table 4** Factor 1 – opportunities (prepared by author)

| Variables | Factor name |
|---|---|
| Improve community investments | Stakeholders' Needs and Investments |
| Improve human rights | |
| Improve social responsibility investing | |
| Attract quality employees | |
| Satisfy customer needs | |

**Table 5** Factor 2 – opportunities (prepared by author)

| Variables | Factor name |
|---|---|
| Reduce carbon footprint | 4Rs |
| Reduce emissions | |
| Reduce pollution | |
| Reduce health hazards | |

**Table 6** Factor 3 – opportunities (prepared by author)

| Variables | Factor name |
|---|---|
| Increase efficiency | Productivity |
| Increase cost-effectiveness | |

**Table 7** Total variance explained for risks (prepared by author)

| Total variance explained | | | | | | |
|---|---|---|---|---|---|---|
| Component | Extraction sums of squared loadings | | | Rotation sums of squared loadings | | |
| | Total | % of variance | Cumulative % | Total | % of variance | Cumulative % |
| 1 | 7.788 | 32.449 | 32.449 | 5.688 | 23.701 | 23.701 |
| 2 | 3.366 | 14.027 | 46.476 | 5.466 | 22.775 | 46.476 |

Extraction method: Principal Component Analysis.

Factor 2 consists of four variables and it has been named *4Rs*, where R stands for reduction and these variables reflect a reduction in environmental emissions (Table 5).

Factor 3 comprises of two variables and has been named *Productivity* (Table 6).

On the other hand, Table 7 examine the sustainability risks, the cumulative percentage of the total variance explained for the two factors of the Risk is 46.4%, which is not sufficient enough to explain the variances of the extracted factors. Therefore, it is possible that the respondents are not really concerned with the risks associated with sustainability awareness in higher education. However, it can be deduced that they are more concerned with the opportunities rather than the risks.

The same checks were applied for Risks, whereas cross-loading factors were deleted in order to improve the factor' cross-loading. The following two factors were generated: factor 1 comprised four variables, and factor 2 comprised four variables. The highest loading factors were *Increase Fraud* and *Increase Litigation*, followed by *Inflate Costs* (Table 8).

Factor 1 comprises 4 variables and has been named *Compliance, Fraud & Governance* (Table 9).

Factor 2 consists of four variables named the *4Is*, which stands for Inflation, Interest, Insurance, and Increased regulations (Table 10).

**Table 8** Factors for risks (prepared by author)

| Rotated component matrix[a] - RISKS | Component | |
|---|---|---|
| | 1 | 2 |
| Increase fraud (including environmental actions and environmental compliance failures) | 0.833 | |
| Increase litigation and compliance breaches (including environmental actions and environmental compliance failures) | 0.838 | |
| Increase governance failure (including environmental actions and environmental compliance failures) | 0.742 | |
| Increase transaction failure due to environmental liabilities | 0.703 | |
| Inflate costs | | 0.826 |
| Increase interest rates | | 0.792 |
| Increase insurance crises due to environmental disasters | | 0.723 |
| Increase number of new regulations including environmental | | 0.768 |

[a]Rotation converged in 3 iterations.

**Table 9** Factor 1 – risks (prepared by author)

| Variables | Factor name |
|---|---|
| Increase fraud (including environmental actions and environmental compliance failures) | Compliance, Fraud & Governance |
| Increase litigation and compliance breaches (including environmental actions and environmental compliance failures) | |
| Increase governance failure (including environmental actions and environmental compliance failures) | |
| Increase transaction failure due to environmental liabilities | |

**Table 10** Factor 2 – risks (prepared by author)

| Variables | Factor name |
|---|---|
| Inflate costs | 4Is |
| Increase interest rates | |
| Increase insurance crises due to environmental disasters | |
| Increase number of new regulations including environmental | |

## 5 Discussion and New Significance

Sustainability has gained widespread recognition globally in the contemporary world. Awareness of sustainability and the adoption of its practices present both challenges and prospects for change and betterment. This can be seen from the recent conference on sustainability in Paris, where 151 world leaders joined hands

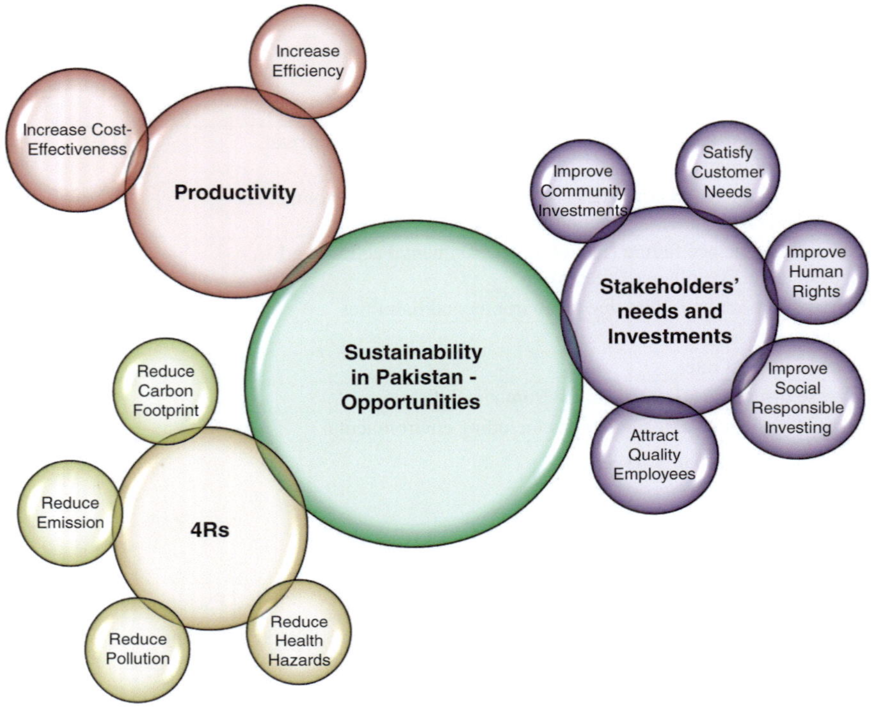


**Fig. 19** Sustainability opportunities in Pakistan (prepared by author)

and showed commitment to pursue this phenomenon jointly. This move emphasizes the extent of the problems that nations are facing currently and more so the problems that the future generations may encounter. The problem is just as big for the developed nations as it is for the third world countries which, after the carbon tax is applied, will face the biggest set-back in terms of cost.

The research question for this study was based pertaining to these issues, where the study looks to find some answers. The primary research question is: *What are the opportunities and risks associated with sustainability awareness in the Pakistani higher education sector?*

Different research methodologies were outlined and the chosen research method was vindicated. The deductive research approach and a quantitative research method were identified for this study since the relationships between variables were to be examined. The survey data was measured numerically and was analysed using statistical SPSS software.

The target population for this study was mainly the students of Pakistan. The sample size for this study was 109 participants who completed the surveys. The participants came from different backgrounds and demographics. The survey research approach was applied for this study as it is convenient and comfortable for both the respondent and the researcher. The survey was distributed electronically

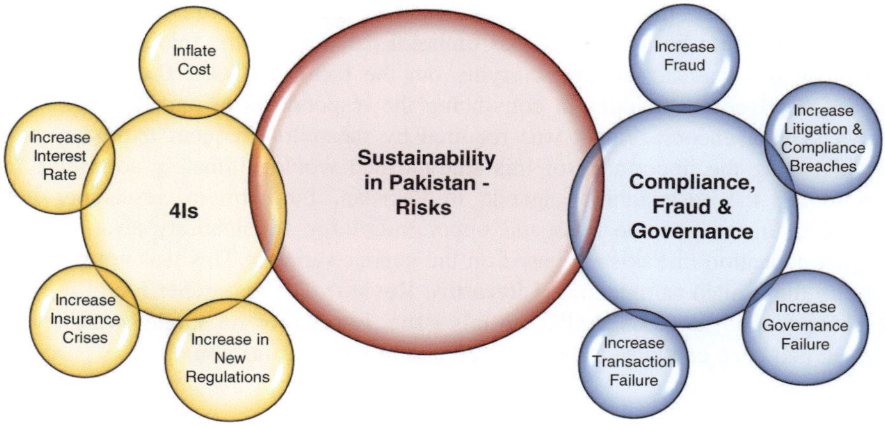


**Fig. 20** Sustainability risks in Pakistan (prepared by author)

using the web-link generated by the survey software, Qualtrics. All the results from the surveys were analysed using IBM SPSS version 22 and were presented in the form of tables and graphical charts.

The reasons provided by the participants for their willingness to adopt sustainability in higher education in Pakistan are: increase in efficiency and cost effectiveness; improvements in community investments, human rights, socially responsible investments; and reductions in carbon footprint, emissions, pollution, and health hazards. On the other hand, participants state that adoption may result in: increased fraud, litigation and compliance breaches, governance failures, and transaction failures due to environmental liabilities. They also believed that it would increase the interest rates, inflate costs, and increase insurance crises due to environmental disasters. This information is represented in the Figs. 19 and 20.

## 6 Research Limitations

Several limitations were identified during the course of this study, and they are explained in this section. **Time:** Due to the time constraint, this research could be conducted over a period of only 3 months. The data collection and analysis took longer than was initially planned which prevented the researcher from covering the research topic more comprehensively. Furthermore, it was not possible to distribute the survey to a wider population and involve a larger demography. **Sampling:** The researcher was able to obtain the sample only from selected universities in Pakistan which did not enable to cover a larger geographical area to distribute the survey more evenly. **Incomplete Surveys:** Out of the 202 responses generated from the survey, only 109 were deemed to be valid as many respondents did not complete several sections of the survey. This could be either intentional, or simply because the respondents did not feel that the section was relevant to them, or it could be due to careless omissions. However, the minimum requirement of 100 was satisfied for carrying out the factor analysis. **Data Size:** The researcher faced difficulties in convincing the respondents to undertake the surveys. Ample encouragement was required by means of frequent reminders and emphasizing the importance of this study which would ultimately contribute to improving the sustainability agenda in Pakistan. Furthermore, researcher had intended to examine the risks and opportunities for sustainability awareness in higher education in Pakistan, based on the gender variable. This was not possible due to the limited sample size. **Literature Review:** The researcher found it challenging to find previous studies on this issue. Limited or no studies were found which had been conducted in this regard and in a similar fashion.

## 7 Future Research

There is a big gap in sustainability awareness in the Pakistani higher education system. This research was able to obtain only a limited number of responses from the pupils; however, it is equally important to secure the participation of teachers as well, since they are the ones who impart the knowledge. Further research can include teachers and educators as part of the survey in order to assess their willingness to adopt sustainability practices.

Furthermore, the research was limited to people with Internet access and only accounted for the urban areas. Obtaining the perspectives of the rural inhabitants will be ideal as they constitute the majority of the population of the country.

## 8 Conclusion

This study aims to examine the sustainability awareness in higher education in Pakistan. It was found that out of the 109 respondents from Pakistan 48% were male and 52% female. The age group with the highest number of participants was 25–30 years. Of the 109 participants, the majority were Bachelor's and Master's Degree holders and were working in the field of management. Around 40% of the respondents spend up to 5 h a day on a computer, and of these, the majority use the Internet mainly for social networking and working. When the respondents were asked about the use of email, 62% of them replied that they spend less than an hour a day on email correspondence. Laptops and smartphones were the evident choice of devices to access the Internet, which goes hand-in hand-with the next question that asked them how they were first introduced to the concept of sustainability, to which 43% said Internet. However, 36% of the respondents were first introduced to the concept of sustainability in school. This further signifies the importance of the research question and identifies the need to implement sustainability in the school curriculum, as stated in the literature review. However, the majority of the respondents (48%) do not read the company's sustainability report before buying a device. This shows that there is an urgent need for sustainability awareness in Pakistan to change the consumer behaviour, thereby making the consumer more conscious of making the "right" decision when purchasing devices.

The second section of the results highlights that the reliability testing in SPSS is used to determine whether there is a strong correlation between the variables. Cronbach's Alpha test was used to check the internal consistency and to ensure the validity of the data. The reliability values for both opportunities and risks were greater than 0.70, which indicates that all the variables are reliable.

The factor analysis results of the opportunities for sustainability awareness in higher education in Pakistan generated three new factors with highly correlated variables. The first new factor generated was *Stakeholders' Needs and Investments*, this factor comprised of five new variables, namely: *Improve Community Investments, Improve Human Rights, Social Responsible Investing, Attract Quality Employees*, and *Satisfy Customer Needs*. This means that the respondents believe that an increase in sustainability awareness will cater to the ever-increasing stakeholders' needs and create more opportunities for investments. The second factor generated was *4Rs*, which is made up of four variables, namely; *Reduce Carbon Footprint, Reduce Emissions, Reduce Pollution*, and *Reduce Health Hazards*. This indicates that the sample population believes that the introduction and awareness of sustainability in higher education will lead to reduction in all the four variables. This could be due to the fact that future generations will be better equipped and trained to manage effectively. The third factor generated was *Productivity*, which further comprises two variables: *Increase Efficiency*, and *Increase Cost-Effectiveness*. This factor indicates the respondents' belief that productivity is directly proportional to sustainability awareness.

The factor analysis result for risks generated two new factors with correlated variables. The first factor generated was *Compliance, Fraud & Governance*, which comprises four variables, namely: *Increase Fraud, Increase Litigation and Compliance Breaches, Increase Governance Failure*, and *Increase Transaction Failure*. This indicates that the respondents feel that with the increase in sustainability awareness, more avenues will open for people to become involved in fraud and corruption, which will lead to failures in governance and breaches of compliance with laws and regulations. The second factor generated was *4Is*, which comprises four variables, namely: *Inflate Costs, Increase Interest Rates, Increase Insurance Crises*, and *Increase in New Regulations*. This shows that the respondents fear that with the increase in sustainability awareness, there would be price inflation of products, followed by an increase in interest rates due to an increase in borrowing by the developing countries. There is also the risk of an increase in insurance crises due to either fraud or breaches which means that the government might increase and impose new regulations.

Finally, further research will be carried out in the future with more varied groups of students in Pakistan to strengthen the research findings and aims.

## References

1. Aaron L (2012) Survey research. Radiol Technol 84(2):190–192
2. Abraham M (2005) Sustainability education. Environ Prog 24(4):343–344. doi:10.1002/ep.10116
3. Adomssent M, Godemann J, Michelsen G (2007) "Sustainable university": holistic approach to sustainability in higher education institutions. Int J Sustain High Educ 8(4)
4. Armstrong D (2000) Factor analysis. Nurs Stand 14(48):25
5. Barbier EB, Markandya A, Pearce DW (1990) Environmental sustainability and cost-benefit analysis. Environ Plan A 22(9):1259–1266
6. Bebbington J, Gray R (2001) An account of sustainability: failure, success and a reconceptualization. Critical Perspectives Account 12(5):557–588
7. Bonett DG, Wright TA (2015) Cronbach's alpha reliability: interval estimation, hypothesis testing, and sample size planning. J Organizational Behav 36(1):3–15. doi:10.1002/job.1960
8. Chandia KE, Javid AY (2013) An analysis of debt sustainability in the economy of Pakistan. Procedia Econ Finance 5:133–142. doi:10.1016/S2212-5671(13)00019-1
9. Cortese AD (2003) The critical role of higher education in creating a sustainable future. Plan High Educ 31(3):15–22
10. Couper MP, Kapteyn A, Schonlau M, Winter J (2007) Noncoverage and nonresponse in an Internet survey. Soc Sci Res 36(1):131–148. doi:10.1016/j.ssresearch.2005.10.002
11. Couper MP, Traugott MW, Lamias MJ (2001) Web survey design and administration. Public Opin Q 65(2):230–253
12. del Mar Alonso-Almeida M, Marimon F, Casani F, Rodriguez-Pomeda J (2015) Diffusion of sustainability reporting in universities: current situation and future perspectives. J Clean Prod 106:144–154
13. Dillman DA, Phelps G, Tortora R, Swift K, Kohrell J, Berck J, Messer BL (2009) Response rate and measurement differences in mixed-mode surveys using mail, telephone, interactive voice response (IVR) and the Internet. Soc Sci Res 38(1):1–18. doi:10.1016/j.ssresearch.2008.03.007
14. Elkington J (1994) Towards the sustainable corporation: win-win-win business strategies for sustainable development. Calif Manag Rev 36(2):90–100
15. Ertel S (2011) Exploratory factor analysis revealing complex structure. Personality Individ Differ 50(2):196–200. doi:10.1016/j.paid.2010.09.026
16. Fisher PB, McAdams E (2015) Gaps in sustainability education: the impact of higher education coursework on perceptions of sustainability. Int J Sustain High Educ 16(4):407–423. doi:10.1108/IJSHE-08-2013-0106
17. Fowler FJ (2014) The problem with survey research. Contemp Soc 43(5):660–662. doi:10.1177/0094306114545742f
18. Freankel JR, Wallen NE, Hyun HH (2012) How to design and evaluate research in education, 8th edn. McGraw-Hill, New York
19. Gordon J, McNew R (2008) Developing the online survey. Nurs Clin N Am 43(4):605
20. Hair JF, Black WC, Babin BJ, Anderson RE, Tatham RL (2006) Multivariate data analysis, 6th edn. Pearson Education, New Jersey
21. Hales D (2008) Sustainability and higher education. N Engl J High Educ Fall: 23–24
22. Halpern BS, Klein CJ, Brown CJ, Beger M, Grantham HS, Mangubhai S, Ruckelshaus M, Tulloch VJ, Watts M, White C, Possingham HP (2013) Achieving the triple bottom line in the face of inherent trade-offs among social equity, economic return, and conservation. Proc Natl Acad Sci U S A 110(15):6229. doi:10.1073/pnas.1217689110
23. Hartnet B (2012) Sustainability. *Superfunds Magazine* (373):14
24. Heinberg R, Daniel L (eds) (2010) The post carbon reader: managing the 21st century's sustainability crises. Watershed Media, Healdsburg
25. Herzel S, Nicolosi M, Stărică C (2012) The cost of sustainability in optimal portfolio decisions. Eu J Finance 18(3–4):333–349. doi:10.1080/1351847X.2011.587521

26. Kim K (2011) Likert scale. Korean J Fam Med 32(1):1. doi:10.4082/kjfm.2011.32.1.1
27. Kiron D, Kruschwitz N, Reeves M, Goh E (2013) The benefits of sustainability-driven innovation. MIT Sloan Manag Rev 54(2):69
28. Koehn P, Uitto J (2014) Evaluating sustainability education: lessons from international development experience. Int J High Educ Res 67(5):621–635. doi:10.1007/s10734-013-9669-x
29. Kuhlman T, Farrington J (2010) What is sustainability? vol 2. MDPI AG, Basel, doi:10.3390/su2113436
30. Lassalle G, Bourdaud P, Saint-Béat B, Rochette S, Niquil N (2014) A toolbox to evaluate data reliability for whole-ecosystem models: application on the Bay of Biscay continental shelf food-web model. Ecol Model 285:13–21. doi:10.1016/j.ecolmodel.2014.04.002
31. Linklater N (2014) The cost of sustainability. Sutton. http://utilityweek.co.uk/news/the-cost-ofsustainability/990562#.WTFIQcYlHIU. Accessed 2 June 2017
32. Lozano R, Young W (2013) Assessing sustainability in university curricula: exploring the influence of student numbers and course credits. J Clean Prod 49:134–141. doi:10.1016/j.jclepro.2012.07.032
33. Lukman R, Glavič P (2007) What are the key elements of a sustainable university? Clean Technol Environ Policy 9(2):103–114. doi:10.1007/s10098-006-0070-7
34. McGaghie W, Crandall S (2001) Population and sample. Acad Med 76(9):934–935
35. Michaels PJ (2015) Sustainability is too expensive. Bloomberg Business. https://malk.com/sustainability-is-too-expensivethe-costs-of-renewable-energy-and-production-processes-and-green-building-materials-exceed-their-value-tothe-environment-pro-or-con/. Accessed 2 June 2017
36. Monsen KA, Westra BL, Yu F, Ramadoss VK, Kerr MJ (2009) Data management for intervention effectiveness research: comparing deductive and inductive approaches. Res Nurs Health 32(6):647–656. doi:10.1002/nur.20354
37. Moser CA, Kalton G (1989) Survey methods in social investigation. Gower Publishing, Hampshire
38. Mullins GR (2002) Sustainability. Ann N Y Acad Sci 969(1):1–3. doi:10.1111/j.1749-6632.2002.tb04342.x
39. Robinson J (2004) Squaring the circle? Some thoughts on the idea of sustainable development. Ecol Econ 48(4):369–384
40. Saunders M, Philip L, Adrian T (2007) Research methods for business students, 4th edn. Pearson Education Limited, Harlow
41. Schneidewind U, Augenstein K (2012) Analyzing a transition to a sustainability-oriented science system in Germany. Environ Innovation Soc Transitions 3:16–28. doi:10.1016/j.eist.2012.04.004
42. Schonlau M (2002) Conducting research surveys via e-mail and the web. In: M Schonlau, RD. Fricker, Jr, MN Elliott (eds) Rand, Santa Monica
43. Sedgwick P (2014) Unit of observation versus unit of analysis. Br Med J 348:g3840. doi:10.1136/bmj.g3840
44. Smyth JD, Dillman DA, Christian LM, O'Neill AC (2010) Using the Internet to survey small towns and communities: limitations and possibilities in the early 21st century. Am Behav Sci 53(9):1423–1448. doi:10.1177/0002764210361695
45. Soyka PA (2012) Creating a sustainable organization: approaches for enhancing corporate value through sustainability. FT Press. http://ptgmedia.pearsoncmg.com/images/9780132874403/samplepages/0132874407.pdf. Accessed 2 June 2017
46. Spector P, Rogelberg S, Ryan A, Schmitt N, Zedeck S (2014) Moving the pendulum back to the middle: reflections on and introduction to the inductive research special issue of Journal of Business and Psychology. J Bus Psychol 29(4):499–502. doi:10.1007/s10869-014-9372-7
47. Urquiza Gómez F, Sáez-Navarrete C, Rencoret Lioi S, Ishanoglu Marzuca V (2015) Adaptable model for assessing sustainability in higher education. J Clean Prod 107:475–485. doi:10.1016/j.jclepro.2014.07.047

48. Wirtenberg J (2012) Triple-Bottom-Line. Leadersh Excell 29(6):8
49. Wright TSA, Wilton H (2012) Facilities management directors' conceptualizations of sustainability in higher education. J Clean Prod 31:118–125. doi:10.1016/j.jclepro.2012.02.030
50. Zhou J, McCreanor PT, Montalto F, Erdal ZK (2011) Sustainability. Water Environ Res 83(10):1414–1438

# Sustainability Awareness in Saudi Arabia

**Abdulaziz Albahlal, Sulaiman Alqahtani and AL-Hassan Al-Muqri**

**Abstract** Sustainability has become essential concerns throughout the world due the high damage to the environment. Education is one of the aware tool that have the ability to create initiatives to lead future generations to a new path in which sustainability in general can be attainable. This chapter is to investigate how education can play a major role in the transition from the current situation for developing countries toward sustainability, especially in the areas of energy generation and consumption. Also, this chapter examines common problems that have already been identified around the globe and attempts to make a contribution to the debate on the awareness of sustainability and its needs, especially in terms of the education of Saudi's students and the general Saudi's society based on age. The conducting survey questionnaires were designed to investigate Saudi students' attitudes towards sustainability. This chapter highlights the great importance of educating the current and future generations in the Kingdom of Saudi Arabia in order to avoid economic and environmental catastrophe. Additionally, this chapter has identified the knowledge gap pertaining to sustainability and sustainable development in the Kingdom of Saudi Arabia. The survey has 1821 participants were all Saudi nationals, mainly students with tertiary degree, students who are currently studying in universities in Saudi Arabia or abroad, and Saudi's who participates socially. The results show that responded has a basic knowledge of sustainability, and sustainability has various opportunities such as create new jobs and improve corporate social responsibility. However, participants indicated the main risks are increase governance failure and increase litigation and compliance breaches (including environmental actions and environmental compliance failures).

A. Albahlal (✉) · S. Alqahtani · A.-H. Al-Muqri
Curtin University, Perth, Western Australia, Australia
e-mail: a.albahlal@postgrad.curtin.edu.au

S. Alqahtani
e-mail: s.alqahtani8@postgrad.curtin.edu.au

A.-H. Al-Muqri
e-mail: alhassan.muqri@postgrad.curtin.edu.au



T. Issa et al. (eds.), *Sustainability, Green IT and Education Strategies in the Twenty-first Century*, Green Energy and Technology, DOI 10.1007/978-3-319-57070-9_13

## 1 Introduction

In April of 2015 deputy crown of Saudi Arabia prince Mohammed bin Salman announces a new Saudi's vision called "Saudi Vision 2030". The main target of the new vision is removing the Saudi's dependence on oil. The Saudi Vision 2030 target a renewable energy by producing 9.5 gigawatt (GT) in 2023 [2]. The main aim for this chapter is investigates the opportunities and risks of sustainability awareness in Saudi Arabia education sector. Also, based on gender, investigates the risks and opportunities of sustainability awareness in Saudi Arabia from all areas. Thus, the chapter aims to investigate how education and social area can play a major role in the transition from the current situation toward sustainability, especially in regards to energy consumption, oil & gas, etc. Therefore, this study targets a sample population selected from Saudi Arabian students, academics and social. A survey was distributed among Saudi students and public social to elicit their thoughts and understandings.

One important aim of this literature review is to examine Saudi Arabian students' understandings about sustainability and the level of public awareness. It also focuses on the current situation regarding the sustainability level in the higher education sector and public social in Saudi Arabia and gives several examples from the developed nations. Finally, it shows the advantages and risks associated with the adoption of sustainability practices. The chapter lay-out, firstly gives brief sustainability background, secondly methodology used in this chapter, thirdly analysis the data, lastly Discussion of Findings and new significance.

## 2 Sustainability Background

The word "Nachhaltigkeit" which is the German word for sustainability was first used in 1713, and referred only to preserving natural resources like native flora and fauna [11]. Furthermore, the concept of sustainability first appeared in modern times in 1960 and 1972 as an ecological issue rose by Goldsmith, who considered sustainability in the context of ecosystems which are vital to support human existence. Moreover, the 1972 United Nations Conference on Human Environment (UNCHE) suggested that the population of the world needs to be controlled and its capacity not exceeded, as it would impact on and lead to the inevitable depletion of Earth's natural resources [9].

Sustainability is a critical term in developing countries, most of the business investments in developing countries in unsustainability resources. Developing countries get past developed countries in total climate footprint emission, usually in business [7]. Saudi Arabia considered as one of developing countries that has a high environmental issue due the business activities in natural resources. "Saudi Arabian oil revenue constitutes more than 70% of the nation's budget, and a $10 rise in the price of a barrel of oil results in a 14% rise in Gross Domestic Product" [6]. According to the OPEC's annual statistical report period for 2015, Saudi Arabia is the second largest crude oil producer after Russia, with burgeoning domestic consumptions of crude oil to meet energy demand (electricity) and public transportations estimated at about 9 million barrels a day, and the number increases during the summer [13].

Moreover, Saudi Arabia has no natural water resource thus Saudi Arabia is one of the driest countries on the world and on another hand Saudi Arabia on the top of countries whom consuming high natural water. "Despite the limitations of natural water resources in the Kingdom of Saudi Arabia, the Kingdom is one of the biggest water consumers worldwide" [1]. Saudi Arabia completely generates its electricity from unsustainable resource that makes damage on the water, land, climate and air [18]. Saudi Arabia placed in below the acceptable level for sustainable particularly in the housing sector [16]. However, "the government has already implemented a number of campaigns in order to increase people's awareness of the problem of water scarcity and the importance of its conservation in Saudi Arabia" [18].

The Kingdom's population is rapidly increasing, and the more it continues to increase the more waste it will be generate. The problem of harmful industrial companies and their waste is a result of inadequate policies. The Economic and Social Commission for Western Asia (CESCWA) debated that one major way to invest in sustainability in the region is to involve all stakeholders from government levels to private and community levels in educating the community about waste management, and creating laws policies and protocols for sustainability [17]. Sustainability can be enforced by law so that both public and private sectors will help to achieve a more sustainable future and livable country. The waste management business industry in the GCC has grown over the past few years and has turned into a billion-dollar industry. The GCC official members see the advantages of investing in the waste industry and turning it into renewable resources. "Renewable energy technologies can also offer additional benefits to the GCC states as alternatives to oil and natural gas, particularly in the sphere of environmental policy" [4].

The recycling industry provides substantial potential to generate and save energy, boost efficiency and capacity, decrease greenhouse gases and provide job opportunities. In recent years, governments in developed nations such as the United States have supported the development of waste management and the expansion of investment in recycling waste. Another country that has adopted a similar strategy to the US is Australia. In Adelaide, a city located in the southern part of the country, the local council has established an impressive strategy for managing its waste called "Zero Waste SA" [15].This strategy was legislated by the SA state government in 2004. The ZWSA has also resulted in positive outcomes since individuals, companies and industries are empowered to recycle and minimize the waste they produce. In terms of waste management, South Australia has a higher percentage of waste collection systems compared with other Australian cities. Hence, the Kingdom of Saudi Arabia should adopt similar initiatives taken from both South Australia's "Zero Waste" program and similar US practices and implement these measures within its large provinces [15, 17].

Therefore, these issues need to be addressed by involving and integrating all participants from government sectors, education and society in general. One of the most mechanisms that developed and developing nations have turned to for the past 10 years is education especially higher education as a tool for effecting a change toward sustainability. In 2010, the Australian government held a national conference about introducing sustainability in higher education in order to encourage participation and promoting sustainability as part of its curricula in all universities throughout Australia. "Education for sustainability is part of the process of building an informed, concerned and active civil society" [5].

The Kingdom of Saudi Arabia is highly interested in implementing a more environmentally sustainable platform in all their business practices and processes privately and publically in order to respond to the rising global concerns about sustainability awareness and the increasing expectations of stakeholders. "In Saudi Arabia, the government established the Saudi Energy Efficiency Center (SEEC) in 2010 to focus on reducing power through audits, load management, regulation and education" [12]. As the Kingdom is still in a developing state, with rapid industrialisation, urbanisation developments and the rapid growth of population, the result is an increase in the consumption of fossil fuels and wastage. The Kingdom has realised the need for improved waste management services, and has recently attracted waste companies to invest in the Kingdom to tackle waste issues. Its eagerness to embrace green practices and to catch up with the rest of developed nations has attracted attention from both national and international investors. This could bring the Kingdom another source of economic incentives via the implementation of recycling solutions, composting, energy via waste production and adopting green building. "Kuwait and Saudi Arabia have both launched Green Building Forums in 2012 to support the development of green building regulations" [10].

The opportunities and risks were identified and re-constructed using Visio Microsoft to depict the findings of opportunities in a clear way (see Figs. 1 and 2). As can be seen from the map, one factor was labelled by the researcher as 3Rs. This stands for "Reduce pollution", "Reduce emissions" and "Reduce health hazards" variables which were variables with the highest values. These factors are environmental burdens which were recognised by those students who responded to the survey. The researcher considered these outcomes as an opportunity to raise awareness of the need for sustainability development. Furthermore, depicted in the advantages map are identify namely: economic and business opportunities. In Fig. 2: Risk concept map, there are two main sets of factors which have the highest variables for risks. The first sets of factors are Compliance, Fraud and Security, while the second set of factors are Insurance, Competitions and new Regulations.

The literature review found that schools, colleges and universities have been involved in discovering the increased stresses being placed on the environment,

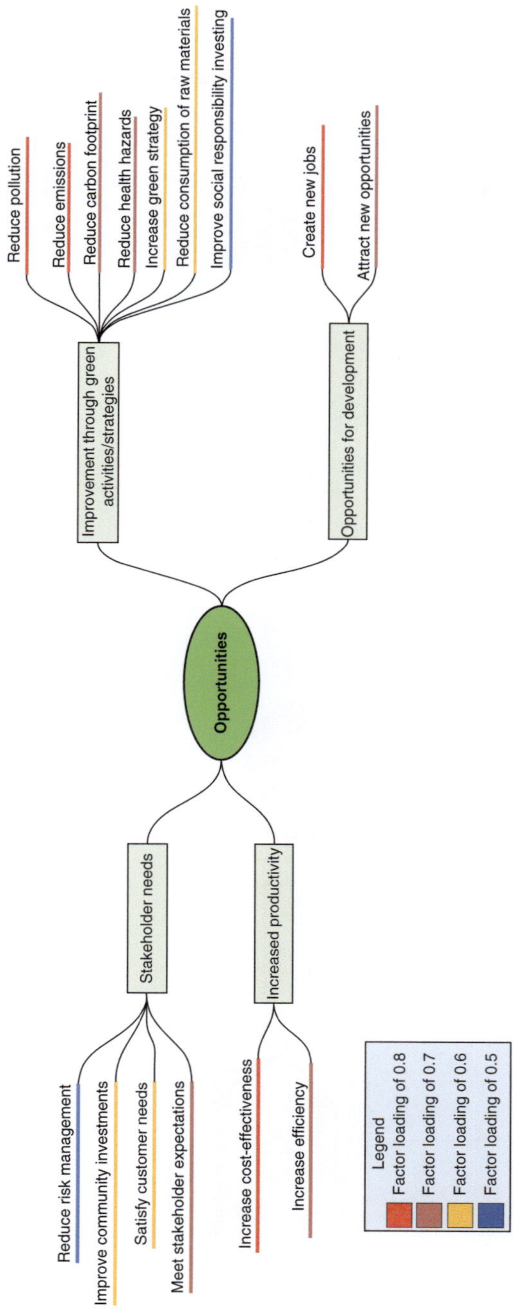


**Fig. 1** Opportunities related to sustainability awareness in Saudi Arabia (prepared by the authors)

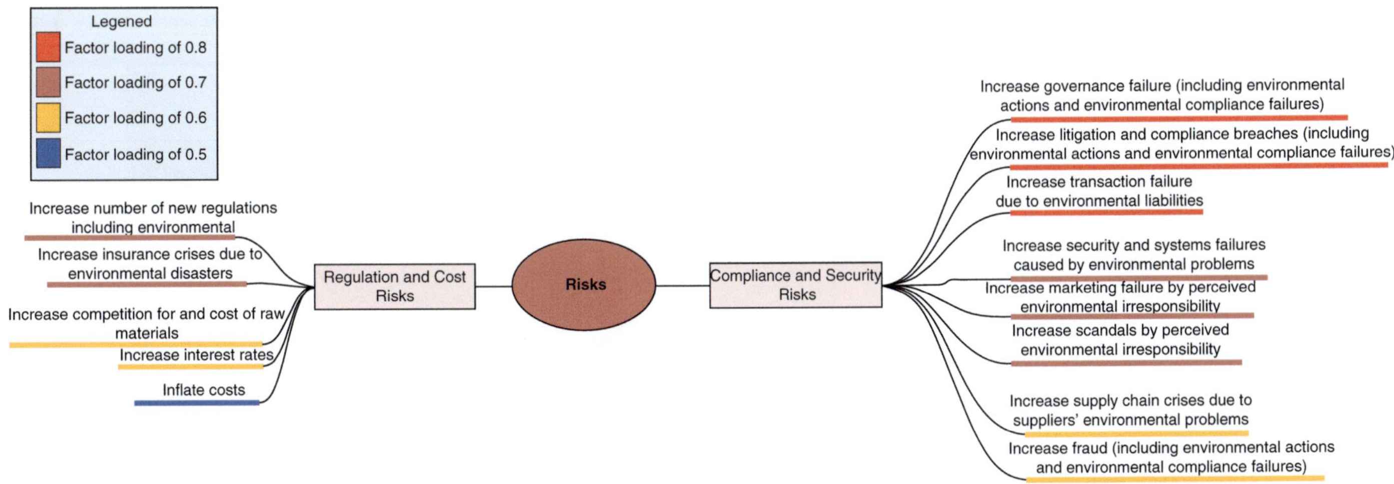


**Fig. 2** Risks related to sustainability awareness in Saudi Arabia (prepared by the authors)

economy and society which have amounted to a crisis facing both industrial and non-industrial nations. Institutions such as higher education ones have the ability to take initiatives to lead new generations on a new path into sustainable development and sustainability which is achievable. Yet, there is still opposition across many higher education institutions in developed and developing nations to creating or integrating sustainable development and sustainability initiatives in their teachings, research courses and curricula. Sustainable development and sustainability has become an essential concern throughout the world. Even though numerous advocates calling for changes to courses so that they incorporate sustainability education some argue that sustainability development is an unrealistic notion and is difficult to achieve.

## 3 Methodology

A Qualtrics designed online survey was the chosen method to collect data for this research via email, Facebook, Twitter and Whatsapp as collection instruments. There are numerous benefits to using an online survey for data collection purposes. Some of these are its ease of reach to and for participants and its favourable cost. It also offers features such as a variety of downloadable formats, lower error rate of responses and, for the researcher, the ability to analyse results with greater speed and convenience [8].

The survey design comprised two parts: firstly, a section dedicated to demographic and personal information via closed-ended questions, followed by scaled-response questions on sustainability on a Likert continuum and, secondly, the collection of views and opinions through the capture of anecdotal material. Social Science Statistics Software (SPSS) was then applied as an analytical tool together with those available in Qualtrics.

## 4 Data Analysis

The primary research question is "What are the opportunities and risks of sustainability awareness in Saudi Arabia education sector?".

On thousand eight hundred and twenty-one has respondents from Saudi Arabia participated in the survey: 52% of these were male and 48% were female. The age range of the participants was 18–65, with 62% of the participants drawn from the 18–24 age group. Of these respondents, 40% held bachelor degrees whilst 17% were at pre-university level and 15% were Master's degree holders.

Looking at respondents' use of technology, 47% reported that they spend up to 5 h per day on the computer, followed by 25% who spent from 5–10 h. In terms of usage of the Internet, it can be noted that most participants (43%) use it for up to 10 h with 30% in the category of 5–10 h.

In terms of exposure to and knowledge of sustainability and Green Information Technology, the highest number of respondents (335) reported that they were first introduced to these concepts via the Internet while 20% stated that they first encountered sustainability issues at high school. This was closely followed by respondents who first learned of sustainability and Green IT while in higher education (19%). While the majority of participants reported that they do not read the company sustainability report before buying a device (33%), it is encouraging to note that both "yes" and "maybe" responses were at an equal 29%. A large majority of takers of the survey (43%) thought that frequent swap-over of devices could damage the planet with 33% positively stating that they believe that environmental damage is caused by rapid obsolescence cycles. In regard to whether it is possible to change the mindset of designers and users of devices, the highest number of participants considered education as the best vehicle to promote change with similar percentages of people indicating that mindsets can be changed through the Internet and social networking (32.73% and 33.44% respectively).

In order to enhance current understanding of the opportunities and risks associated with sustainability awareness in Saudi Arabia, a conceptual model was developed based on the outcomes of the survey and the application of factor analysis to identify key factors:

Reliability testing on the data in relation to opportunities related to sustainability awareness in Saudi Arabia was in Table 1:

There were 24 items evaluated for consistency among the opportunities related to sustainability awareness in Saudi Arabia. From the results above, it is seen that the reliability statistics were highly consistent (Cronbach's Alpha = 0.919).

Similarly, reliability regarding risks associated with sustainability awareness in Saudi Arabia can be represented as in Table 2:

Thirteen components were evaluated to establish the reliability of the risks associated with sustainability awareness in Saudi Arabia with an extremely consistent outcome of a Cronbach's Alpha of 0.905.

**Table 1** Reliability statistics on the data in relation to opportunities

| Reliability statistics | |
|---|---|
| Cronbach's Alpha | No. of items |
| 0.919 | 24 |

**Table 2** Reliability statistics on the data in relation to risks

| Reliability statistics | |
|---|---|
| Cronbach's Alpha | No. of items |
| 0.905 | 13 |

Factor analysis, a data reduction technique via a group of factors, was used for reduction of the variables for opportunities and risks associated with sustainability awareness in Saudi Arabia.

## 4.1 *Opportunities Related to Sustainability Awareness in Saudi Arabia*

From the data analysis, the items relating to opportunity that belong to each of the components can be grouped and clearly labelled as in Table 3; and Fig. 1 presents the opportunities related to Sustainability Awareness in Saudi Arabia.

## 4.2 *Risks Related to Sustainability Awareness in Saudi Arabia*

In regard to risk-related items, the factors can be grouped and labelled according to the components to which they belong as shown in Table 4, and Fig. 2 presents the risks related to Sustainability Awareness in Saudi Arabia.

**Table 3** Group of factors relating to opportunity that belong to each of the components

| Variable name | Factor loading | Label |
|---|---|---|
| Reduce pollution | 0.852 | Improvement through green activities/ strategies |
| Reduce emissions | 0.827 | |
| Reduce carbon footprint | 0.756 | |
| Reduce health hazards | 0.752 | |
| Increase green strategy | 0.683 | |
| Reduce consumption of raw materials | 0.609 | |
| Improve social responsibility investing | 0.541 | |
| Meet stakeholder expectations | 0.700 | Stakeholder needs |
| Satisfy customer needs | 0.660 | |
| Improve community investments | 0.624 | |
| Reduce risk management | 0.561 | |
| Create new jobs | 0.825 | Opportunities for development |
| Attract new opportunities | 0.716 | |
| Increase cost-effectiveness | 0.867 | Increased productivity |
| Increase efficiency | 0.705 | |

**Table 4** Group of factors relating to risks that belong to each of the components

| Variable name | Factor loading | Label |
|---|---|---|
| Increase governance failure (including environmental actions and environmental compliance failures) | 0.832 | Compliance and Security Risks |
| Increase litigation and compliance breaches (including environmental actions and environmental compliance failures) | 0.828 | |
| Increase transaction failure due to environmental liabilities | 0.807 | |
| Increase security and systems failures caused by environmental problems | 0.788 | |
| Increase marketing failure by perceived environmental irresponsibility | 0.715 | |
| Increase scandals by perceived environmental irresponsibility | 0.706 | |
| Increase supply chain crises due to suppliers' environmental problems | 0.690 | |
| Increase fraud (including environmental actions and environmental compliance failures) | 0.647 | |
| Increase number of new regulations including environmental | 0.759 | Regulation and Cost Risks |
| Increase insurance crises due to environmental disasters | 0.739 | |
| Increase competition for and cost of raw materials | 0.686 | |
| Increase interest rates | 0.663 | |
| Inflate costs | 0.569 | |

## 5 Discussion of Findings and New Significance

As discussed previously in this article, Saudi Arabia is on the brink of a more enlightened attitude to sustainability issues although practical steps have yet to be vigorously embraced or implemented by the general populace in tandem with government policy and regulations [9]. Indeed, there is some evidence that both the government and Saudi citizens are not well versed in understanding green concepts applied to a variety of sectors, such as education and housing construction [3, 14]. Nonetheless, respondents to the survey indicated at least a basic level of understanding and consideration of key issues relevant to sustainability. For the factor "Improvement through green activities/strategies", reduction of pollution and emissions received the highest ratings, perhaps due to the general global recognisability of these concepts. In terms of "stakeholder needs", respondents believed that the major opportunity inherent in sustainability practices is the ability to better fulfil stakeholder expectations (0.700) and customer needs (0.660), recognizing that business practices will need to become aligned with sustainability initiatives to better serve the interests of increasingly sophisticated and knowledgeable consumers. Linked to this, survey participants highly ranked the creation of new job opportunities and the attraction of new business and organizational opportunities within the component "Opportunities for development" at 0.825 and 0.716 respectively. Moreover, for organisations, opportunity lies in increased cost effectiveness and efficiency in the factor "Increased productivity".

Despite the general awareness of the opportunities afforded by sustainability activities and strategies and their positive impact on stakeholders and organisations, respondents also pinpointed some obstacles or risks associated with the implementation of sustainability practices in Saudi Arabia. There were two factors in this area: "Compliance and Security Risks" and "Regulation and Cost Risks". The implication of these is that the greatest risk in terms of adopting sustainable behaviour is the escalation of regulations involved in enforcing compliance with sustainability actions (0.759). This leads to the potential for governmental lapses in administering and managing compliance (0.832), the potential for increased breaches of such regulations by organisations and individuals and possible follow-up litigation (0.828).

Looking at gender as a variable, there are commonalities in the top five factors identified by male and female respondents. Both male and female participants chose "reduce pollution" (4.10 out of 5) as the top factor. Other similarities were the selection of "reduce carbon footprint" and "reduce health hazards" which scored in the top five factors for both genders. Differences were in the female choice of "create new jobs" and "improve corporate social responsibility" whereas males chose "reduce emissions" and "reduce paper usage" as top five factors. It can be noted that most of the elements chosen by both genders belong to the category of "Improvement through green activities/strategies".

In terms of low-rated factors, both male and female participants chose similar items from the group of risks. The common elements were "Increase security and systems failures caused by environmental problems", "Increase transaction failure due to environmental liabilities" and "Increase governance failure (including environmental actions and environmental compliance failures". Differences were in the following items: "Increase fraud (including environmental actions and environmental compliance failures)" and "Increase insurance crises due to environmental disasters" chosen by males and "Increase litigation and compliance breaches (including environmental actions and environmental compliance failures)" and "Increase marketing failure by perceived environmental irresponsibility" selected by females in the bottom five factors. It can be theorized that these factors were chosen due to a lower level of awareness of risk factors and how these affect individuals and organisations.

The significance of this research is that it will seek to understand whether the sustainability concept exists amongst Saudi Arabian students and faculties throughout the general education and tertiary education sectors. A survey was distributed to students across different sectors, focusing specifically on Saudi Arabian students in tertiary institutions. However, the data gathered for this paper is expected to be limited to that specific area meaning further research is required. The significance will give educators, learners and academics an overview of the sustainability concept in Saudi Arabia's education system, particularly the universities. In addition, it will reflect the importance of sustainability as well as its embedment in education sectors to both academics in universities and students in public schools.

## 6 Conclusion

This study has looked at the current state of awareness of sustainability issues in Saudi Arabia, a country which has, to date, not yet achieved a high level of activeness and commitment in the area of sustainability. However, there is some indication that both governmental and individual attitudes towards sustainability are poised to change and that the sector will become invigorated in years to come. From the surveys conducted with Saudi respondents, it can be seen that there is a basic level of knowledge and awareness in regard to such sustainability fundamentals as pollution, emissions, health hazards and wastage of resources. Similarly, survey participants demonstrated an awareness of the business opportunities associated with applying green measures to organizational practices. In contrast, though, participants rated risk factors related to sustainability relatively poorly, possibly suggesting that public understandings of the risks are not as developed as they could be. Education, as provided by the higher education sector in Saudi Arabia, will play a vital role in assisting to create more nuanced understandings of sustainability concepts and activities in future.

## References

1. Al-Gahtani K, Alsulaihi I, El-Hawary M, Marzouk M (2016) Investigating sustainability parameters of administrative buildings in Saudi Arabia. Technol Forecast Soc Change 105:48–48. doi:10.1016/j.techfore.2016.01.016
2. Borgmann M (2016) Saudi Arabia's amazing new renewables target: 9.5GW by 2023. Available via Renew Economy http://reneweconomy.com.au/2016/potentially-game-changing-saudi-arabian-government-restructuring-bolsters-9-5-gw-renewable-energy-target-by-2023. Accessed 11 May 2016
3. Cortese AD, Hattan AS (2010) Research and solutions: education for sustainability as the mission of higher education. Sustain J Rec 3:48–52. doi:10.1089/SUS.2009.9802
4. El-katiri L, Husain M (2014) Prospects for renewable energy in GCC states: opportunities and the need for reform. Available via The Oxford Institute for Energy Studies http://www.oxfordenergy.org/wpcms/wp-content/uploads/2014/09/MEP-10.pdf
5. Fien J (2001) Education for sustainability: reorientating Australian schools for a sustainable future. Tela: Environment, Economy and Society. http://eev.vic.edu.au/wp-content/uploads/2015/09/tela08_education_-for_sustainability-Fein.pdf
6. Hashmi MA, Al-Habib M (2013) Sustainability and carbon management practices in the Kingdom of Saudi Arabia. J Environ Plan Manag 56:140–157. doi:10.1080/09640568.2012.654849
7. Hoekstra AY, Wiedmann TO (2014) Humanity's unsustainable environmental footprint. Science 344:1114–1117. doi:10.1126/science.1248365
8. Issa T (2013) Online survey: best practice. In: Isaias P, Nunes MB (eds) Information systems research and exploring social artifacts: approaches and methodologies. IGI Global, Hershey, pp 1–19

9. Karban A (2014) Developing a framework for neighborhood-level urban sustainability assessment in Saudi Arabia. The University of Texas at San Antonio, San Antonio
10. KPMG (2012) Future-proofing business in the GCC: opportunities for sustainable growth. http://www.kpmg.com/NL/nl/IssuesAndInsights/ArticlesPublications/Documents/PDF/High-Growth-Markets/Future-proofing-business-GCC.pdf
11. Kuhlman T, Farrington J (2010) What is sustainability? Sustainability 2:3436–3448. doi:10.3390/su2113436
12. Meltzer J, Hultman N, Langley C (2014) Low-carbon energy transitions in Qatar and the Gulf Cooperation Council Region. Available via The Brookings http://www.brookings.edu/~/media/Research/Files/Reports/2014/03/low-carbon-energy-transitions-qatar-meltzer-hultman/low-carbon-energy-transitions-qatar-meltzer-hultman-full.pdf
13. OPEC (2015) OPEC Annual Statistical Bulletin. Available via Vienna http://asb.opec.org/images/pdf/ASB2015.pdf
14. Saied Al Surf M, Trigunarsyah B, Susilawati C (2013) Saudi Arabia's sustainable housing limitations: the experts' views. Smart Sustain Built Environ 2:251–271. doi:10.1108/SASBE-04-2013-0022
15. Song Q, Li J, Zeng X (2015) Minimizing the increasing solid waste through zero waste strategy. J Clean Prod 104:199–210. doi:10.1016/j.jclepro.2014.08.027
16. Susilawati, C, Al Surf, M (2011) Challenges facing sustainable housing in Saudi Arabia: a current study showing the level of public awareness. In: The 17th Pacific Rim Real Estate Society Conference. pp. 1–12. Gold Coast, Australia: PRRES Conference. http://www.prres.net/papers/Susilawati_and_Al-Surf_Challenges_Facing_Sustainable_Housing_Saudi.pdf
17. Taher N, Hajjar B (2014) Energy and environment in Saudi Arabia: concerns and opportunities. Energy Environ Saudi Arab Concerns Oppor. doi:10.1007/978-3-319-02982-5
18. Taleb HM, Sharples S (2011) Developing sustainable residential buildings in Saudi Arabia: a case study. Appl Energy 88:383–391. doi:10.1016/j.apenergy.2010.07.029

# Sustainability Awareness: Colombia Perspective

**Juan Esteban Otalvaro Restrepo**

**Abstract** This study intends to provide a snapshot of the status of sustainability awareness in Colombia, its relationship with economic growth, poverty and ecology in Colombia. From the literature review those concepts and their interaction are explored, as well as the risk and opportunities of sustainability across business, society and governments. Together with the analysis of Colombia, the study will also explore the different concepts of sustainability and its evolution in the last decades. It was noticed that the concept of sustainable development was no more than a catch phrase, trying to capture the different global aspects of the environmental challenges, and the definition given by the World Commission on Environmental and Development has been the starting point for global agencies, governments and non-government organizations to establish government and business policies for environmental issues. One of the major contributions of the research is the recognition that, in addition to or in conjunction with ecological factors, there are social conditions that affect the ecological sustainability or unsustainability of the interaction between people and nature. Most significantly, people seem to be unaware that, given the rate of population growth, the subsequent increase in the consumption of resources cannot be sustained forever. From the literature review, it can be recognized that the number of scientific studies related to sustainability in Colombia is limited; most of the research so far has been done by government agencies and organizations. Nevertheless, previously studies have failed to identify the rating of sustainability awareness in the country, and the advantages and disadvantages that it can bring to organizations, government bodies and the public in general. This study was performed using a quantitative method and an online survey was distributed through email mainly among Colombian individuals living in Colombia. The analysis of the data produced three factors that can be associated with the advantages or opportunities of sustainability. The results also indicated that the same number of factors, including a major number

J.E.O. Restrepo (✉)
School of Information Systems, Curtin University, Perth, Western Australia, Australia
e-mail: j.otalvaro@postgrad.curtin.edu.au



T. Issa et al. (eds.), *Sustainability, Green IT and Education Strategies in the Twenty-first Century*, Green Energy and Technology, DOI 10.1007/978-3-319-57070-9_14

of variables, was found for the risks associated with sustainability in Colombia. The research data and results indicate the Colombian public's attitudes toward green products, green strategies, sustainability policies and corporate social responsibility.

## 1 Introduction

A proper understanding of the concept of sustainability and sustainable development emerged only in the '1980s, just after the publication of Our Common Future report issued by the UN-sponsored World Commission on Environmental and Development [46]. The concept of sustainable development was no more than a catch phrase, trying to capture the different global aspects of the environmental challenges [24, 36]. The definition given by the [46] has been the starting point for global agencies, governments and non-government organizations to establish government and business policies for environmental issues.

Since then, multiple initiatives have been taken by global agencies, governments, ONGs and the private sector with the intention of addressing various environmental challenges. The WCED emphasized that the industrial society was going to exceed most of the ecological limits within a matter of decades, if it continues to promote the kind of economic growth witnessed in the '60s and '70s [24]. Some of the issues presented in environmental policies are based on vague definitions; on the other hand, some authors use the wide definition of sustainability and sustainable development in reference to practices that are reputed to be more environmentally sound than other long-range planning strategies for society or the economy [14].

This research intends to provide valuable information about sustainability, sustainable development and its relationship with economic growth, poverty and ecology in Colombia. These concepts and their interaction will be explored in more detail further in the research. The simplest way to define *sustainable* is "that which can be maintained over time". In recent years, this word has been wrongly used to refer mainly to practices that are reputed to be more environmentally sound those others.

The book chapter research topic has been selected as today economics, ecological and social circumstances have grown to a point where the great needs for clean and equitable economic growth remains the biggest single difficulty presenting the biggest challenge to sustainable development [6]. Therefore, it follows that in the long term, a country's economic growth, trade expansion and environmental protection are goals that can only be reached in conjunction with one another.

From the literature consulted, it appears that even though the concepts of sustainability, sustainable growth and environment have been fairly used, there still confusion about their definitions, mainly because of two misconceptions [24]: firstly, environment and ecology have become pretty much synonymous, limiting the environmental focus to the natural environment. Secondly, there is a danger of abstraction due to the distorted notion of the word "environment".

The intention of this research is to determine how sustainability awareness in Colombia can bring advantages and disadvantages for sustainable development, economic growth, the poverty issue, and ecology policies for Colombia. In addition to the lack of a definitive explanation for environment and ecology, today sustainability is seen in terms of three dimensions: social, economic and environmental which are contained in the definition below which has been adopted by the United Nations: "Development is a multidimensional undertaking to achieve a higher quality of life for all people. Economic development, social development and environmental protection are interdependent and mutually reinforcing components of sustainable development" [20].

This chapter will employ an online survey to examine the Colombia perspective toward sustainability awareness and to identify new risks and opportunities toward sustainability in Colombia.

## 2 What Is Sustainability?

The term sustainability originally came from the field of forestry in Germany: first coined in 1713 by Hans Carl von Carlowitz of Saxony. His idea was not to fell more trees than could grow back. He spoke of nachhaltende Nutzung, a term that began to be used in works on forestry and was translated into English as "sustainable". From this economic standpoint, an independent discourse was developed within the scientific community. It began with a seminal work on sustainable development, the study on the Limits to Growth, written by Dennis Meadows and co-authors in 1972 for the Club of Rome [37].

According to the UN Commission on Economic Development, the definition given in the Brundtland Commission report titled Our Common Future, sustainability is defined as development that "meets the needs of the present without compromising the ability of the future generations to meet their own goals" [46]. Mebratu [24] has agreed that the WCED definition is acceptable in the sense that economic growth around the whole world is essential for improving the livelihood of poor people, for sustaining the growing population, and eventually for stabilizing the population.

Although some authors have criticized this definition [18, 21, 40], many others believe that it covers most aspects of sustainability. The critics emphasized the fact that it is often difficult to determine the future needs of the next generation since they may be different from the needs of people today. In addition, the way in which developed countries view the concept of needs could be completely different from the views of the developing countries [40]. For instance, the needs of a community in Norway may differ greatly from the needs of a community in Colombia.

Authors such as Heinberg [14] argued that the definition given by the Brundtland Report has omitted to explicitly note the unsustainability of the use of nonrenewable resources, and has generally disregarded the problem of population growth. Linked to the problem of population growth and the necessity of having an accounting or indexing scheme by which to measure sustainability, the concept of "ecological footprint" was implemented as the amount of land and water area a human population would hypothetically need to provide the resources required to support itself and to absorb its wastes.

An interesting fact implicit in the scheme is the recognition that for humanity to achieve sustainability, the total footprint must be less that the total amount of Earth's land and water [14]. Currently, the global footprint has been calculated at about 40% greater than what the planet can regenerate.

Kuhlman and Farrington [20] defined sustainability as a state of affairs where the sum of natural and man-made resources remains at least constant for the foreseeable future, in order that the well-being of future generations does not decline.

Sustainability has been defined as the requirement of our generation to manage resources in such a way that the average quality of life that we ensure ourselves can potentially be shared by all future generations; in other words, development is sustainable if it involves a non-decreasing average quality of life. Linked to sustainability definitions, concepts such as sustainable consumption have been established around finding workable solutions to social and environmental imbalances through more responsible behavior of humans, organizations and countries. In a similar way to sustainability, its objective is to ensure that the basic needs of the community are met, avoiding and reducing the excess consumption of materials and energy [10].

# 3 Sustainable Development

The simplest way to define *sustainable* is "that which can be maintained over time". In recent years, this word has been wrongly used to refer mainly to practices that are reputed to be more environmentally sound than others [14].

Sustainable development finally emerged on the global stage during the 1992 "Earth Summit" in Rio de Janeiro. The purpose of the summit was to work towards achieving a new balance between the use and the preservation of resources and nature's potentials. The concept of sustainable development is the result of the growing awareness of the global links between environmental problems, socioeconomics to do with poverty and inequality, and concerns about a healthy future for humanity [17].

The notion of sustainable development was a response to the failure of past growth models, which did not eradicate poverty globally or within countries' sustainable development has been broadly implemented. Before the rise of sustainability awareness, growth models were dominated by the economic policy that international prosperity and human well-being can be achieved by increasing global trade and industry.

Plessis [31] presents a conceptual model consisting of three interconnected sectors to describe the term "sustainable development". These interconnected sectors are the Economy, the Environment and the Society as Fig. 1. Within the model, there must be a reasonable level of balance of interactions between these sectors for the world to achieve sustainable development.

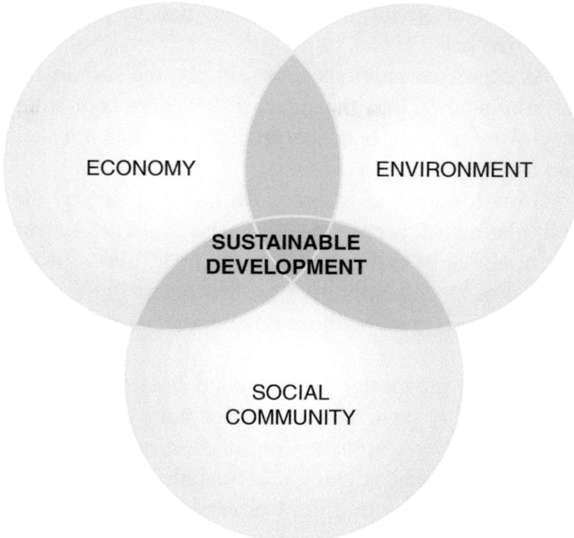


**Fig. 1** Interconnection between the Sectors of Sustainable Development. Adapted from Plessis [31] and prepared by the author

The majority of people in some cases use the term *sustainable development* together with ecologically sustainable or environmentally sound development. This unproven connection is in fact due to sustainability being interpreted as ecological sustainability [21] and the additional ecological sustainability concept is another objective of the sustainable development process. As a consequence, sustainable development is frequently reduced to mean a process of change that can be continued forever. One misinterpretation is that it is the process of material consumption; this assumption is either impossible or trivial as the amount of usable resources is not infinite.

One of the major contributions to the debate has been the recognition that in addition to or in conjunction with ecological factors (ecological conditions for ecological sustainability, patterns that determine environmental responses to human activities and human ability to use the environment), there are social conditions that affect the ecological sustainability or unsustainability of the people-nature interaction [21] theory that is aligned with that of Plessis [31] as mentioned previously.

## 4 Difference Between Sustainable Development and Sustainable Growth

Other authors explain the interaction of growth and sustainability as the continuing economic growth in industrial countries and the efforts of those countries to at some time reduce their contribution to global pollution, which in fact is only a compensation for the inevitable rise in pollution from growth in developing countries [6].

As expressed previously, put simply, the sustainability of something is its capacity for continuance into the future [6]. Where economic activity or more generally, a way of human life is concerned, this sustainability will depend on economic, social and ecological factors.

Growth means to increase naturally in size by the addition of material through assimilation of accretion. Whereas, *to develop* means to expand or realize the potentialities of bringing gradually to a fuller, greater, or better state. Growth can be seeing as a quantitative improvement or unfolding of potentialities, when an economy can grow without developing, or develop without growing or do both or either [5].

Usually, prosperity is associated only with economic growth. However, Fritz and Koch [9] have taken a wider perspective of prosperity based on the 2007 Sustainable Development Commission report. It considers prosperity according to three dimensions: Ecological Sustainability, Social Inclusion (individual well-being) and Quality of Life. Conversely, the definition of "de-growth" establishes a link between ecological sustainability, social equity and individual wellbeing in terms of an equitable downscaling of production that increases well-being and enhances ecological conditions at the local and global levels [9].

The 2005 World Summit on Social Development identified sustainable development goals as social development, economic development and environmental protection [42]. A more detailed analysis of Sustainable Goals will be given later.

Using a broader definition, Hopwood et al. [17] have summarized the ideas of sustainable development into five equity principles:

- Futurity – Intergenerational equity
- Social Justice – Intergenerational equity
- Trans-frontier Responsibility – Geographical equity
- Procedural Equity – People treated openly and fairly
- Inter-species Equity – Importance of biodiversity

These principles help to give clarity to the ideas of sustainable development and how human equity is linked to the environment. At the November 2015, the Paris Climate Change Conference organized by the UN was held. It is expected that the results and agreed terms of the conference will be focused on how the world will move to zero emissions at some time in the second half of the century. As an outcome, the conference will provide an essential framework for business and government so that the right decisions are made now on key energy and infrastructure issues [25]. Moreover, it will inspire inventors, engineers and businesses to meet this challenge and deliver.

Probably one of the most important outcomes as a result of a major shift toward sustainability awareness is the fact that society has moved from a world where everyone said it was someone else's problem, to one where everyone knows this can only be solved collectively [25], where the status-quo is not in confined to the usual world of business.

# 5 Economic Debate

The global goal of sustainability is to create the conditions for a sustainable and effective socio-economic development of the world, transforming the specific objectives of structural policy to define the needs of the economy in some form or other, be it goods, activities or services.

Beckerman's argument is that poor countries need economic growth to pull them out of poverty, while rich countries pursue economic growth because of the net benefits it brings [6]. Experts are calling for a new era of economic growth, where growth is forceful and at the same time socially and environmentally sustainable. Statistics show that between 1972 and 1986, the relationship between energy use and economic growth in industrial countries has undergone a significant change, moving from the broadly proportional relationship that it had previously.

The relationship between water and poverty has broadly been explored as a close relationship between a family's health and social dignity. Poverty can have negative effects on the management of water resources and services [44], as it can stop the development of livelihood opportunities in agricultural areas as well as existing water development investments such as sanitation facilities, water treatment and dams.

Poverty still remains at high levels in countries like China, Brazil and India, where strong economic growth has not guaranteed widespread social progress. In addition, many developing countries face the reality of a big gap between rich and poor people and between those who can and who cannot exploit new opportunities brought by economic growth. Therefore, access to good education, health care, electricity and safe water still remains exclusively for people who live in developed economies [44], conditions that are in conflict with sustainable development and equity principles.

In addition to having access to education, health care and basic benefits, serious poverty reduction will require population control and redistribution of wealth to minimize this particular inequality [5]. Some experts however, do not deny that fighting poverty will be very difficult without any growth.

The Water for a Sustainable World report by the United Nations Educational, Scientific and Cultural Organization [44] found that income disparity is the factor most likely to cause an impact on the global scale in the near future. Therefore, economic growth must focus on four areas:

- Enhancing equity, including the gender dimension
- Enabling more participation of citizens
- Confronting environmental pressures
- Managing demographic change

The United Nations Conference on Environment and Development (UNCED) (1992) stated that the need for clean, equitable economic growth remains the biggest single difficulty with the largest challenge in terms of sustainable development.

Business and industry face the challenging tasks of ensuring that such growth is possible [6]. The rapid economic growth around the world will apply environmental pressures to non-sustainable situations, and as a consequence will increase the pressure to reduce increasing poverty. On the other hand, the rapid increase and high demand for energy and non-renewable raw materials could significantly increase their cost and it will be the same for other goods.

Developing countries which cannot afford any reduction in per-capita resource use must concentrate their efforts on population control [5]. Countries that have high rates of per-capita resource usage frequently have low rates of demographic growth and consequently must aim their efforts more at consumption control rather than population control.

Another point raised in the economic debate as a result of the Global and Local Environmental Sustainability, Development and Growth report (2015) is that efforts to achieve sustainable development must be encouraged; moreover, the elimination of poverty must be a collaborative effort between developed countries and third world countries. According to the report, the collaboration must be based on three principles:

- Promoting green investment through strengthened and targeted finance development and other technical assistance;
- Promoting innovation, technology cooperation and knowledge transfer; and
- Facilitating trade in environmentally-sustainable goods and services to expand green markets and economic opportunities.

## 6 What Does Sustainability Mean for Business?

A great deal of attention has been given to the social and environmental dimensions of business activities in organizations which are linked to the triple bottom line: People (social), Planet [26] and Profit (economic) [7]. Organizations are increasing their accountability to stakeholders on environmental and social issues; this is evident as they are publishing their efforts and future strategies towards sustainable development on their executive reports. Figure 2 illustrates the evolution of industries' responses to sustainability.

However, Hoek [16] suggests that the view of sustainability as an additional burden has changed to become a potential source of competitive advantage. Other authors identify this as a value approach, whereby organizations integrate into their business strategy environmental activities such as green purchasing and ISO implementation as strategic initiatives [16, 39].

In addition to other authors, Ramirez [33] has identified that companies have recently adopted sustainability practices and policies faster than expected, as most of them believe sustainability is a core activity for marketing, logistics and human resources.

The business models used by many organizations increasingly seek to include environmental aspects of performance in line with Elkington [7] pivotal Triple

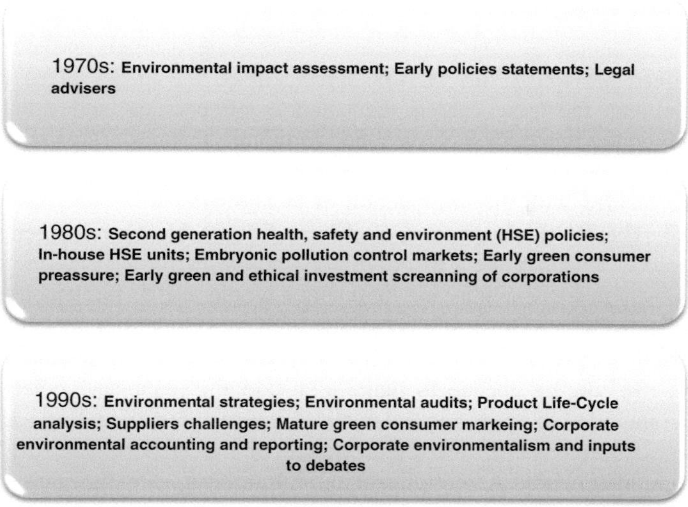


**Fig. 2** Organizations responses to sustainability. Adapted from Elkington [7]

Bottom Line (TBL) concept. Some management approaches argue that TBL incorporates profits, the planet and people, and that the framework goes beyond the traditional measures to include environmental and social dimensions. The TBL concept has found wide acceptance in firms, governments and non-government organizations. In several organizations, innovations are seen as a potential means of generating economic sustainability for social entrepreneurial initiatives [38] which focus primarily on social issues that have not previously been regarded as attractive commercial business opportunities.

Another positive result in terms of organizations is the implementation of an Environmental Management Strategy (EMS). It is a complete management tool that monitors and reports on the environmental performance of organizations and their systematic control of environmental issues [10]. EMS provides tools and principles designed to guide the allocation of resources, assignment of responsibilities, evaluation of practices, procedures and processes and environmental concerns that organizations need to integrate into operations.

Organization can be separated into two groups: the ones that have adopted sustainability as part of their core business, and those still lagging behind in its implementation. Investments in sustainability have rapidly increased over the few last years as it is becoming a source of competitive advantage [33]. With the adoption of sustainability policies within the framework of a better quality environment and corporate accountability, organizations are more likely to receive financial assistance from banks and investors. This gives to organizations the opportunity for cost reduction in energy consumption, and better market positioning where consumers are identified with sustainable products and services.

# 7 Current Progress Level of Sustainability

As highlighted previously, the lack of agreement on the theory of sustainability, does not affect the practical implementation of policies and environmental approaches required on both ecological and economic grounds. The aim would is to internationalize the policy on the environmental externalities and their reduction that can be tackled with the introduction of technologies to reduce environmental impacts [6]. The World Business Council for Sustainable Development [45] has develop a 2050 Vision project to provide a framework for a vision of the world well on the way to sustainability by 2050 which would lead to fundamental changes in business, governance structures, economic frameworks and human behavior.

The current progress of sustainability has established the things that must happen during the coming decade to make a sustainable planetary society possible [45]. These include incorporating the costs of externalities, including carbon, ecosystem services and water, into the structure of the marketplace; the WBCSD [45] report is also planning to duplicate agricultural output without increasing the amount of land or water used; halting deforestation and increasing yields from planted forests: keeping carbon emissions at the same labels as 2005 by 2050. This would be possible by shifting to low-carbon energy systems and improved demand-side energy efficiency, and providing universal access to low-carbon mobility.

In order to achieve these outcomes, more companies would have to join the Vision 2050 scheme. Business will lead market change by creating efficiencies, competitive advantage and meeting customer needs. This shift towards sustainability will create zero-waste cities, better ecosystems and livelihoods and offer opportunities for finance, information technology and partnerships around the globe [45].

The creation of similar initiatives as those of WBCSD, and the increase in the number of sustainable development intergovernmental organizations have created the challenge to define a measurement framework and then select relevant Sustainable Development indicators that are able to capture policy goals and ensure standardization [13]. The designed frameworks have centered on human and ecological systems and their relationship, whereas others have been elaborated based on theories of economics.

The tendency so far has been to design a framework that is independent of political priorities (based on a model of sustainable development processes and their interaction). Hak et al. [13] have identified this gap as an opportunity for improvement as the framework should be developed based on the strategic issues; moreover, it will need to support a policy cycle starting from a policy formulation (extend), policy legitimization, policy implementation, policy evaluation and policy adjustment.

Another contribution to the objectives of Sustainable Development has been the development as an opposition to the proposed framework within which the human activities are minimizing the long-term ability of the natural environment to provide goods and services, as well as decreasing human health and well-being.

A way to personalize the fundamental goals, objectives and means of sustainable development has been proposed by Lele [21] as follows:

- Helping the very poor, because they are left with no options but to destroy their environment
- The idea of self-reliant development – having constraints in natural resources
- Cost-effective development – using nontraditional economic criteria
- Health control – technology solutions, food self-reliance, clean water and shelter for all

All these efforts made by countries, non-governmental organizations, private organizations and the United Nations have contributed greatly to the public's awareness of the world's environmental issues and this has generated a cognizance at all levels to reduce the impact of economic activities on the environment. Some levels of success have been achieved in relation to sustainable development through awareness and well-considered strategies. However, this success can be achieved only with the integration of social and economic impact on the society.

## 8 Sustainability, Future and Technology

The future of sustainability calls for a new model of development focused on human well-being and the interface with the natural environment. It should consider the future within a social, environmental and economic framework, and challenging the costs and benefits of business models of growth against alternative more sustainable options. The Real World Coalition points out that the usual current business models are themselves a source of danger as they boost inequality and poverty, environmental degradation and world instability [3].

The futuristic approach to tackling global and local environmental issues will require a global multi-stakeholder partnership, as well as participation from communities to national governmental institutions. The OECD [29] suggests that a twin-track policy agenda of national and international action would help to achieve environmental objectives in all countries.

The twin-track agenda is a key to persuading governments and international organizations to introduce much-needed major reforms. The reforms should address the areas of technology, and science and information systems; moreover, modifications need to be made to the market and government policies [17]. Organizations such as the OECD have detailed high level priorities by means of which developed countries can help developing countries to achieve the reforms [29]:

- Promoting green investment by providing finance and other technical assistance for development.
- Promoting innovation, technology co-operation and knowledge transfer.
- Facilitating commercial agreements in environmentally-sustainable goods and services.

The transformation to sustainability goes beyond assessing and implementing current responses to global change and addressing shortfalls in development needs. It will consider more fundamental and innovative long-term transformations that are needed in order to move towards a sustainable future [37].

Those long-term transformations must secure a more sustainable future for developing countries. Some of the key actions to achieve this will be developed according to the guidelines below [29]:

- Sustainable management of natural assets which are vital to build economic growth and human well-being.
- More secure and sustainable livelihoods for people depending on agricultural land, fertile soil and fisheries.
- Poverty reduction through new policies that influence the equitable distribution of cost and benefits.
- New economic growth and job opportunities to support a green economy, technological innovation and new markets from goods and services.
- Avoid locking countries into fossil-fuel based energy and emission-intensive pathways.
- Promote better livelihoods through access to clean water, sanitation services and diverse energy supplies, regulating lower pollution and better use land use planning and disaster risk management.

The goal of sustainable development will require significant changes in the technologies and the infrastructure that supports them. It is commonly known that technology can benefit the environment, and one of the major accomplishments has been the dramatic increase in energy efficiency and change in energy from fossil fuel to renewable sources [17]. A much more comprehensive planning and sophisticated modeling than we are accustomed to will be needed. A world in which 50% of a new mix of energy requirements are now produced from biomass will call for major transformations in agriculture, in the energy supply, and consumption infrastructure [32].

The introduction of new technologies will offer market opportunities for business and they should evolve and take advantage of those changes for their internal economic benefit ad social benefit in addition to making a positive impact on the environment. The implementation of new technologies would fail if promoted in isolation. Green economists argue for government action to change the taxation schemes and subsidies in order to benefit employment and environmental initiatives that encourage business to change their production technology [35].

## 9 Advantages and Disadvantages of Sustainability

The literature review so far has hinted at the advantages of sustainability. It is no secret that the benefits of sustainability can be specifically identified. However, the topics introduced so far have been discussed from a world-wide humanity perspective.

From the business perspective, the implementation of sustainability can bring a wide range of benefits for organizations. Opportunities to reduce cost, preserve resources, increase efficiency, business differentiation, higher customer satisfaction, higher capital investment and others can be achievable by the implementation of sustainable organizational policies.

- Cost Reduction: This perhaps is one of the companies' objectives that attract more attention from management. With sustainability, organizations have access to tools and techniques to reduce product cost by using fewer resources (raw materials, people and power) and minimizing waste [47].
- Preserve Resources: This can be achieved together with cost reduction, as both outcomes have common origins. Organizations are more conscious nowadays of the threat of a shortage of non-renewable resources needed to manufacture their products [47]. Reduction of the amount of resources needed for manufacture can be achieved by redesigning products or patterns of production and consumption. It helps to reduce the disposal and handling costs as it avoids the cost of recycling, landfilling and combustion [10].
- Increased Efficiency: This can be looked at not only from a business point of view. Increased efficiency can be achieved across multiple fronts. Energy efficiency, for example, has become a mandatory outcome for companies and international organizations that want to achieve sustainable performance.
- Business Differentiation: As mentioned previously, Ramirez [33] has identified that companies have recently adopted sustainability practices and policies as most of them believe sustainability is a core activity that provides them with a competitive advantage.
- Greater Customer Satisfaction: The implementation of sustainability has encouraged a significant number of customers to purchase sustainable products [47], thereby sharing their level of commitment towards sustainability.
- Higher Capital Investment: Together with sustainability, the execution of social responsibility policies by organizations has become an attractive means of securing funds from investors and banks.

From a more global perspective, sustainability and sustainable development has seen countries set their targets on tackling environmental issues, eliminating poverty and improving human well-being. Other advantages as the reduction of the carbon footprint, reduction of pollution, increase of environmental awareness, and improvement of human rights.

- Reduction of Carbon Footprint: The emissions of carbon dioxide can be reduced with the implementation of global and national policies that aim for a sustainable industrial growth, automobile usage and use of landfill. The implementation of such policies will help the long run to reduce the impact on climate change.
- Reduction of Pollution: This would not only have a positive impact on climate change, but also will also reduce the risk of wildlife extinction as well as improving the well-being of humans and animals.

- Increase of Environmental Awareness: Widespread environmental awareness among people will differentiate our environmental behavior as it could be considered as the trigger for a more sensitive environmental behavior and promote naturally eco-friendly initiatives.
- Improvement of Human Rights: Together with poverty eradication, sustainable development goals also comprise a social policy of gender equity, minority group integration and equal rights for all humans without any discrimination.

Despite all the advantages already mentioned, sustainability implementation could also carry with it several disadvantages or risks. Some authors have explicitly stated that due to the vague and non-specific definition, sustainability is becoming a "plastic" word [27] that can mean almost anything, but in the whole world, people have agreed upon nothing.

From the implementation perspective, companies and governments may incur additional cost when implementing sustainability. It could also increase the number of fraud allegations, litigations and government failures when tackling environmental issues and social issues.

- Cost Inflation: This could be driven by the implementation of costly technologies, and the design of products and services that promote sustainable practices. In the majority of cases, those costs are transferred to the consumers, who must pay a higher price for eco-friendly products [12].
- Fraud Allegations and Litigations: These can arise in organizations or government bodies and the consequences can be seriously damaging. Usually, fraud allegations are made from an inadequate sustainable claim to obtain financial benefits or can be also brought about by from wrong auditing processes [2].
- Government Failure: This issue can have different explanations. Firstly, it can be seen as a government failure due to unsuccessful sustainable development policies. On the other hand, it can be seen as the failure of global initiatives to reduce environmental issues, reduce poverty and improve human well-being due to the conflict of interest between parties, conflicting ideas, and the lack of a common goals and instruments.

## 10 Sustainability in Developing Countries

Countries have found that collaborative initiatives involving companies, governments, inter-governmental organizations, investors, NGOs and communities are the key to achieve benefits for developing economies such as Colombia [23].

The list below of the developing countries was developed by UNFCC in 2009.

In China, which is one of the countries included in Table 1, sustainable development is still analyzed more according to the impact it will have on the economy not according to the three dimensions: the environment, the economy and the society. If an economic benefit of an environmental policy is recognized to be low, such policy will be downplayed without considering the trade-offs. This is still the major challenge to bring China on board as it has drifted away with economic growth.

**Table 1** Developing Countries based on UNFCC Classification (2009)

| | | | | |
|---|---|---|---|---|
| Afghanistan | Bangladesh | Burkina Faso | Comoros | Dominica |
| Albania | Barbados | Burundi | Congo | Dominican Republic |
| Algeria | Belize | Cambodia | Cook Islands | Ecuador |
| Angola | Benin | Cameroon | Costa Rica | Egypt |
| Antigua & Barbuda | Bhutan | Cape Verde | Cuba | El Salvador |
| Argentina | Bolivia | Central Africa Republic | Cyprus | Equatorial Guinea |
| Armenia | Bosnia & Herze-govina | Chad | Cote d'Ivoire | Ethiopia |
| Azerbaijan | Botswana | Chile | Democratic Peo-ple's Republic of Korea | Fiji |
| Bahamas | Brazil | China | Democratic Republic of the Congo | The former Yugoslav Republic of Macedonia |
| Bahrain | Brunei Darus-salam | Colombia | Djibouti | Togo |
| Gabon | Kyrgyzstan | Myanmar | Saint Lucia | Tonga |
| Gambia | Lao | Namibia | Saint Vincent and the Grenadines | Trinidad and Tobago |
| Georgia | Lebanon | Nauru | Samoa | Tunisia |
| Ghana | Lesotho | Nepal | San Marino | Turkmenistan |
| Grenada | Liberia | Nicaragua | Sao Tome and Principe | Tuvalu |
| Guatemala | Libyan | Niger | Saudi Arabia | Uganda |
| Guinea | Madagascar | Nigeria | Senegal | United Arab Emirates |
| Guinea-Bissau | Malawi | Niue | Serbia | United Republic of Tanzania |
| Guyana | Malaysia | Oman | Seychelles | Uruguay |
| Haiti | Maldives | Pakistan | Sierra Leone | Uzbekistan |
| Honduras | Mali | Palau | Singapore | Vanuatu |
| India | Malta | Panama | Solomon Islands | Venezuela |
| Indonesia | Marshall Islands | Papua New Guinea | South Africa | Viet Nam |
| Iran | Mauritania | Paraguay | Sri Lanka | Yemen |
| Israel | Mauritius | Peru | Sudan | Zambia |
| Jamaica | Mexico | Philippines | Suriname | Zimbabwe |
| Jordan | Micronesia | Qatar | Swaziland | |
| Kazakhstan | Mongolia | Republic of Korea | Syrian | |
| Kenya | Montenegro | Republic of Moldova | Tajikistan | |

On the other hand, developing countries which cannot afford any reduction in per-capita resource use will perforce have to concentrate their efforts on population control. Countries that have high rates of per-capita resource usage frequently have low rates of demographic growth and consequently must aim their efforts more at consumption control than population control [5].

In the last decade, Latin America has seen significant reductions in poverty, high economic growth and greater macroeconomic stability as well as the emergence of a stronger middle class [44]. Nevertheless, it remains the most unequal in the world. Most of the national policies are primarily focused on economic development and poverty reduction. The majority of Latin American countries base their economy on the exports of natural resources, which uses large quantities of water in production. This situation can decrease water availability as the effect on higher water demand due to population growth as well as economic growth, increases hydrological variability in many rivers as a result of climate change.

McPhail [28] identified that the economic growth and poverty reduction in developing countries were driven by three main factors:

- Reformed and more predictable mineral legislation
- Improved macroeconomic management
- Improvement on general governance polices

In South American countries such as Peru and Chile, included in Table 1, the fight against poverty measured by the revenue/GDP proportion has improved due to a better climate for investors as a result of better macroeconomics policies. In Chile, for example, poverty has fallen by over 40% [23]. However, a better improvement has been seen in mining regions, where government and mining companies had strong policies for employment generation, training, skills development and local procurement of goods and services.

In addition to poverty reduction, other mechanisms proposed by UNESCO [44] are intended to improve access to potable water. Reducing poverty through water management is a useful pro-poor framework for action, allowing for the introduction of inter-related issues of governance, water quality, access, livelihood opportunities. Access to water for household use is critical for a family's health and social dignity. Access to water for productive purposes like agriculture and family-run businesses is vital to realize livelihood opportunities, generate income and contribute to economic productivity. The UNESCO [44] report also states that one quarter of the global population live in developing countries with shortages of water due to weak government policies and a lack of proper infrastructure to transport the water from rivers and aquifers.

## 11 Sustainability in Colombia

Colombia has been one of the most dynamic economies around the world (2003–2013), with a constant improvement according to its development indicators. GDP has grown an average of 4.7% each year for the same period. The inflation rate has fallen below 4% for the last 5 years and foreign direct investment has multiplied by ten times [34]. Also, Colombia has one of the highest (51st) positions in terms of highest greenhouse gas (GHG) emissions per capita. Colombia also suffers extreme weather events that cause costly natural disasters.

In Colombia, poverty and environmental degradation is described from two points of view:

- Agricultural producers are forced by population growth to exploit fragile, marginal soils, resulting in soil degradation, while other authors link poverty in Colombia to the low productive potential of the resource.
- Labor productivity and per capita food production have been eroded to the point of subsistence, while population growth has become unsustainable [1].

Policy instruments have recently attracted attention as they can be an efficient mechanism to create an environment for harmonious development of production systems. Sustainable resources management can be achieved by the improvement of agricultural skills and new technologies that can boost the productivity of farms in Colombia [1].

The Sustainable Development Plan in Colombia was developed by the government only in 2002, as part of the National Development Plan for 2002–2006. The plan was intended to mainstream environmental issues and concerns into the national priorities such as: Democratic Security, economic growth, employment, social equity and public participation [26].

Each of the priorities was developed under the framework of public participation and communitarian state by the Colombian government and communities. One of the most critical problems in Colombia is the fact it has suffered a civil war for more than 50 years, since most of the financial resources are spent combating the guerrillas, making this probably the first item on the agenda to resolve. In order to provide democratic security, the National Development Plan focused on three points:

- Efficiency and transparency within the State
- Elimination of illicit crops
- International cooperation

The delivery of those initiatives was to be supported by regional development strategies, prioritization of forestry projects and environmental initiatives; and partnerships with communities and NGOs [26]. At the time of writing, the Colombian conflict was not over. However, negotiations between the main guerilla force and the government were held for about 1 year to try to put an end of the conflict. It is worth noting that as part of the Millennium Development Goals set up by the UN, the lack of employment opportunities has been a source of a civil and armed conflict in many countries. In the UN's (2011) annual report, it was announced that no conflict-prone, low-income country has achieved a single Millennium Development Goal [43]. Hence, the Colombian Sustainable Development Plan must realign it policies for a post-conflict era.

In addition to the National Development Plan for 2002-2006, the Colombian green growth framework is supported by three long-term policies [34]:

- Strategic policies — Including the National Development Plan which defines a crosscutting strategy of Green Growth.
- Operational Policies — Including Institutional Strategy for the Articulation of Policies and Actions Related to Climate Change which creates a National Climate Change committee as well as a Financial Committee.
- Specific incentives — Economic incentives for green investments, such as tax-related incentives and R&D incentives.

On the other hand, the achievement of national security must go hand-in-hand with sustainable economic growth and employment generation. For that purpose, the Colombian government designed an action plan based on five main strategies:

- Housing program to build 400,000 units.
- Improved Oil and Gas and Mining exploration and exploitation.
- Transportation Infrastructure projects.
- Biotechnology as an instrument for productivity.
- Environmental Sustainability.

The construction of 400,000 units in 4 years was considered to be a catalyst for economic growth. It was designed as a whole plan to provide subsidies for low income families, income tax benefits, build rural housing with the objective to employ local people and involve local communities during the construction period. Like many other developing countries, Colombia still has an ever-increasing number of slum-dwellers. Worldwide, more than 820 million urban residents live in slums, and the number will continue to rise in the near future [43]. The faster expansion of basic services is critical to improve the living standards of Colombia's growing urban population as well as the adoption of policies like the housing program that can create better employment opportunities in urban and rural areas.

The Colombian economy is highly dependent on oil and gas and coal exports. The oil and gas industry represents 50% of the total exports and 5% of GDP [34] so it made sense that an improved environmental licensing process was required.

It was planned to be a lean and faster process, with a reduction of 61% for the approval period [26]. It was also required that the licensing process was based on the best management practices within the sectors as well as being focused on strategic environmental assessment.

One of the most outstanding characteristics of Colombia is its biodiversity. It is considered the second richest country in terms of species in the whole world after Brazil [30]. In Colombia, there are more than 45,000 plant species, over 1721 bird species and about 358 mammal species. This means that, on average, 1 of 10 species of the world's fauna and flora inhabits Colombia [30]. With all these unique characteristics, Colombia has a great opportunity to establish biotechnology as one of its economic pillars. Within the National Development Plan, rules and policies concerning the sustainable use of biodiversity, access to genetic resources, biosecurity and proper rights were to provide guidelines for the future years.

One of the main achievements of having the National Development Plan was the inclusion of Environmental Sustainability as part of the national agenda. The plan was basically developed to support a national policy of sustainability. The policy is based on five main actions:

- Supply of natural resources for national production.
- Sustainability in National Production Sectors – Which was developed as a the guide for Strategic Environmental Assessment, the environmental licensing system as an instrument of control and the regulation of use water, air, hazardous waste, forest and biosafety.
- Conservation and sustainable use of natural resources.
- Generation of green employment and markets.
- Planning and efficient administration from the regional environmental authorities.

On the other hand, and in addition to its internal conflict, a combination of corruption, negligence and carelessness about people's needs, has delayed the implementation of the national sustainability plan, putting at risk social and environmental amelioration [30]. Developing countries similar to Colombia will incur large costs associated with the negative impact of changing weather patterns, water shortages and rising sea levels. The policies established in the Sustainable Development Plan are closely linked to the UN worldwide program for sustainable development. Global issues such as desertification, land degradation and drought continue to be the most pressing environmental challenges [43]. The UN [43] report states that land degradation affects almost 1.5 billion people globally, most of whom derive their income from agriculture and food production. In addition, the impact of land degradation and climate change would have stronger effects on poor countries, where so many people suffer hunger and live in poverty as they are more vulnerable and exposed. Some of the issues related to climate change that could affect Colombia are the shortage of water, food and land, and the increase of violence and armed conflict.

## 12 The Chosen Research Method and Research Question

This study aims to examine and understand the advantages and disadvantages of sustainability awareness in Colombia. For this purpose, an online survey was distributed to Colombian-only participants. The first part of the questionnaire contains a list of 15 questions to elicit demographic information; the second part consists of 37 summated-rating questions, using the Likert Scale rating where respondents are asked how strongly they agree or disagree with a series of statements related to the advantages and disadvantages of sustainability awareness in Colombia. This method helps to obtain a lot of data, usually in a statistical form, covering a significant sample [22]. The survey is usually a component of the deductive research approach and is commonly used in business and management research. What researchers find useful about the survey is that it allows the collection of real-world data about practices, processes and events during a specific time period. Moreover, surveys can cover research questions about people's beliefs, behavior and knowledge that can be transformed into measured variables. Surveys differ from experiments because they can sample many respondents who answer the same questions, enabling researchers to measure many variables and test multiple hypotheses [28]. For this particular research, a Web-based survey research was designed, conceptualizing and operationalizing the variables as questions. The Web-survey was distributed with a hyperlink via email and various social media networks.

The survey method provides both advantages and disadvantages. Firstly, the use of online survey is a popular way to collect the data, as they can be standardized and used on a large population without the need to invest a lot of resources and money. Surveys also allow researchers to make comparisons easily. In addition, questionnaires have the advantage that they are easy to explain and can be customized to facilitate the respondents' understanding of the objective and questions.

Other advantages of using the survey strategy is that data can be used to suggest possible reasons for particular relationships between variables and to produce models of those relationships [22]. Surveys also facilitate the use of samples, rather than conducting the research across the entire population, which enables the research to be carried out at low cost. However, the researcher must ensure that the sample is representative and the response rate is adequate for the purposes of the research. On the other hand, researchers can face several challenges when developing a survey strategy. The most common challenge is related to the construction of the questionnaire. In particular, the wording of questions is linked to the validity and reliability of the research and usually takes time, patience and creativity [28]. In order to achieve what can be considered a good questionnaire, questions must avoid [4, 41]:

- Jargon, slangs and abbreviations
- Ambiguity, confusion and vagueness
- Emotional language and prestige bias
- Double barreled questions
- Leading questions
- Questions beyond respondents capabilities
- False premises
- Asking about future intentions
- Double negatives
- Overlapping or unbalanced responses

The length of the survey is another challenge often faced by researchers; in essence, there is no set length, but usually researchers opt for long surveys as they are the most cost effective; however, the general public is usually comfortable with a 10–20 min questionnaire. If the questionnaire is likely to take longer to complete, then options like face-to-face interviews are commonly used. When opting for long surveys, researchers must ensure that the order and sequence of questions create minimal confusion and distraction of respondents. When planning the survey questionnaire, researches should ensure that the questions are clear and unambiguous, and the structure is neat and easy to follow [28].

The challenges mentioned above can be mitigated with a good questionnaire survey plan, as those can be controlled by the researcher and are linked to his/her capabilities. On the other hand, researchers face the challenge of obtaining responses from participants, but this seems to be beyond the control of the researcher, and is a significant issue when generalizing from the results. Similar to the length of the questionnaire, there is no such thing as an adequate response rate [28] as it depends on the population, tools being used, and the topic. However, researchers usually feel comfortable with a response rate higher that 90%. If the response rate is lower that 50%, it is considered as a poor response; for responses below 75%, researchers are hesitant to generalize the results as they can differ from results obtained if everyone had responded [36].

A Web-based survey was the type of questionnaire chosen for the research, which falls into the self-completed category. Online surveys are widely used by researchers who distribute the survey via emails, intranet or social media networks. Some of the advantages of using a Web-based survey is that it is by far the cheapest option, it can be done by an individual researcher [4] without much effort, and can be used as well when a big sample is required as its distribution is much simpler than other methods. Data manipulation easier to do as inputs are automatically captured on dedicated software that simplifies data cleaning and statistical analysis.

On the other hand, Web-surveys limit the sample to computer-literate individuals, that can be contacted via email or have access to Internet or intranet. As explained before, usually self-completed surveys can face a low response rate. Previous studies have indicated that the content of Web questionnaires as well how the surveys are presented are perhaps the two main reasons for a low response rate [8]. Who the sponsors are, what the topic is, and how long the survey takes to complete are some of the factors affecting the Web questionnaire content. How to present Web surveys, question writing, question ordering, and visual display of the Web questionnaire are also factors more related to the layout and presentation of the survey itself. However, it can be mitigated in this case if the communication is done via email and ensuring that it is sent to the right person. In addition to those constraints, potential respondents may encounter technical difficulties when trying to access the survey or accessing the link from mobile devices. Some other potential risks and errors when using a Web-based survey are that Web surveys have unique challenges related to bias resulting from coverage error, sampling error and measurement error [41]. A risk factor that can affect survey results is the fact that researchers cannot control the conditions under which the responses are made. Also, this type of survey does not offer any type of assistance in case the respondents have any questions about it. In addition, the time taken by participants to respond can vary and again is not under the researcher's control. As a consequence, the analysis of the data can be delayed and this can affect the researcher's milestones and timely completion of the research.

The research survey consists of a Web-based questionnaire that is completely voluntary and anonymous. It contains a group of 15 list-type questions used primarily for demographic analysis. It also contains a second section containing 37 summated-rating questions, using the Likert Scale rating where respondents are asked how strongly they agree or disagree with a series of statements related to the advantages and disadvantages of sustainability awareness in Colombia. The below table indicates the rating scale used for the survey, where 1 represents the lower rating as strongly disagree and 5 represents the maximum rating as strongly agree.

The survey has been designed to respect respondents' integrity as well as confidentiality. In any case, if the respondents feel uncomfortable in answering certain questions, they can feel free to disregard them. Participants will also have the freedom to withdraw at any time without prejudice or negative consequences, and do not need to provide a reason.

As mentioned previously, the information provided by the respondents through the survey will be held as strictly confidential by the Curtin Business School, and data collected will not be disclosed to any parties besides the researchers, unless required to do so by law. Finally, no published material will contain any information that can identify a respondent and his/her organization.

The people targeted for the research study are called the population, which for this particular case would be Colombian citizens; the population of the target population will not be segregated by age, gender, financial status or education level. However, as the size of the population is quite substantial, data cannot be collected for the entire population. Other factors such as budget and time constraints supported the decision regarding sampling for this research.

The sample size was selected using a volunteer sampling method called snowball sampling, due to the difficulties of identifying members of the desired population. However, once those cases were identified, they assist in identifying more cases and distribute the questionnaire to them. In order to reproduce the statistical analysis required to achieve the research objectives and establish the size of the sample, a 95% level of certainty was considered; therefore, if the sample was selected 100 times, at least 95 of these samples would be certain to represent the characteristics of the population [36].

As previously mentioned, the data collection process involved asking the respondents to complete the questionnaire via email, Facebook or Link ID. An introduction letter explaining the purpose of the survey was attached to the email in order to communicate the topic, objectives of the survey as well as the agreement pertaining to confidentiality.

Some of the challenges that arose during the data collection process include the risk of having a low response rate, which will affect the representation of the population for the research. In addition, as the research is totally voluntary and there are no economic incentives for its completion, it might also prevent the desired high response rate. The limitation of obtaining a high number of respondents is linked to four common causes: respondents have complete freedom to simply refuse to respond, ineligibility to respond, inability to respond (due to technical IT issues) and survey not answered on time. However, a strict follow-up process was conducted to mitigate the impact of a possible low response rate.

The American Association of Public Opinion Research [19] has defined four levels of non-response for returned questionnaires that can be reported as part of the research findings.

- Complete refusal: None of the questions are answered.
- Break-off: Less than 50% of the questions are answered.
- Partial response: From 50% to 80% of all questions is answered.
- Complete response: Over 80% of all questions are answered.

Once all data had been collected, it was analyzed using SPSS Version 2.1 which a tool is provided by Curtin Business School. The data analysis was initially done as an exploratory analysis by looking at individual variables, its components and how these are related to the advantages and disadvantages of sustainability awareness in Colombia. As mentioned previously in the chosen research method section, Qualtrics software was the tool used to collect the data through the Web-based survey. The surveys were distributed via email in most of the cases. However, other social media interfaces such as Facebook and LinkedIn were also used. It must be stressed that regardless of the manner of survey distribution, all respondents received the cover letter as well as the link enabling them to access the survey. A total of 158 surveys were collected. However, a significant number of surveys were invalid or incomplete, and therefore were removed from the data analysis.

The majority of respondents were male with a value of 63.875% of the total. Female respondents represented less than 50% of the total, with a total value of 35%. On the other hand, when evaluating the distribution of the respondents by age, the majority of them were aged between 31 and 35 years, which accounts for more than 35%. The second biggest group was people over 65 years at slightly over 20%. The rest of the distribution is depicted in Table 2 shows the participants' education level. From the online survey result, almost 60% of the participants had a Postgraduate Diploma or Master Degree. This higher number can be explained by the fact that the majority of participants are professionals aged between 25 and 35 years.

Furthermore, Table 3 confirmed that 60% spend on the computer per day more than 5–10 h, while 55% spend up to 5 h on the Internet and email, while 66% spend on the social networking per day less than an hour.

Furthermore, the online survey identified that 93% of the Colombian users are using the Internet to access their email; while 80% for working, and 63% for banking online (see Table 4)

Furthermore, the online survey identified the devices used to access the Internet. Eighty-five percent of respondents are using smartphones and 71.7% laptop (Table 5).

It came to our attention that Colombian users were first introduced to the concepts of sustainability and green information technology via higher education and internet with 50% respectively (see Table 6).

**Table 2** Online survey statistics Australia – prepared by the author

| Number and percentage of online survey | |
|---|---|
| Questionnaires distributed | 158 |
| Questionnaires returned | 105 |
| Response rate | 66.4% |
| *Gender* | *Response* |
| Male respondents | 102 |
| Female respondents | 56 |
| *Age* | *Response* |
| 17 years and under | 0 |
| 18–20 | 1 |
| 21–24 | 4 |
| 25–30 | 26 |
| 31–35 | 60 |
| 36–40 | 14 |
| 41–45 | 3 |
| 46–50 | 4 |
| 51–55 | 3 |
| 56–60 | 7 |
| 61–65 | 3 |
| Over 65 | 35 |
| *Qualifications* | *Response* |
| Primary Education | 0 |
| Higher Secondary /Pre-University | 8 |
| Professional Certificate | 18 |
| Diploma | 9 |
| Advanced/Higher/Graduate Diploma | 5 |
| Bachelor's Degree | 18 |
| Post Graduate Diploma | 36 |
| Master's Degree | 37 |
| PhD | 2 |
| Others | 0 |

**Table 3** Technology use by Australian users – prepared by the author

| Answer | Hours spend on the computer per day | Hours spend on the Internet per day | Hours spend on the email per day | Hours spend on the Social Networking per day |
|---|---|---|---|---|
| | Response | Response | Response | Response |
| Less than an hour | 5 | 15 | 37 | 86 |
| Up to five hours | 26 | 72 | 71 | 42 |
| Five to ten hours | 78 | 31 | 17 | 2 |
| Ten to twenty hours | 19 | 12 | 4 | 0 |
| Over twenty hours | 2 | 0 | 0 | 0 |

**Table 4** Internet usage by Australian users – prepared by author

| Answer | Response |
|---|---|
| Email | 121 |
| Playing games | 6 |
| Studying | 74 |
| Working | 104 |
| Shopping online | 59 |
| Chatting | 49 |
| Researching hobbies | 31 |
| Banking online | 88 |
| Buying goods or services | 42 |
| Buying stocks or investing online | 14 |
| Researching travel information or making reservations | 57 |
| Others – Please specify | 9 |

**Table 5** Devices usage by Colombian users – prepared by author

| Answer | Response |
|---|---|
| PC | 48 |
| Desktop | 17 |
| Laptop | 94 |
| Netbook | 4 |
| PDAs | 0 |
| Workstation | 4 |
| Tablet | 59 |
| Smartphone | 112 |
| Others - please specify | 2 |

**Table 6** First introduced to the concepts of sustainability and green information technology by Colombian users – prepared by author

| Answer | Response |
|---|---|
| School | 20 |
| Higher education | 64 |
| Internet | 64 |
| Books | 20 |
| Magazine | 25 |
| News media | 23 |
| Conferences | 32 |
| Others – Please Specify | 6 |

The online survey examined the companies, which were associated with Colombian users' devices. Apple and Google are the leaders at 41.8% and 31% respectively (see Table 7).

Furthermore, the survey asked Colombian users whether they read the sustainability report of the manufacturer before buying a device. The survey shows that 58% do not read the report; 16% responded "maybe"; and 13% read the report. This outcome indicates that users should take more responsibility for their actions,

**Table 7** Colombian users devices – prepared by author

| Answer | Response |
|---|---|
| Apple | 54 |
| Google | 40 |
| Dell | 11 |
| IBM | 7 |
| Others – please specify | 17 |

**Table 8** Colombian users reading the sustainability report of the company before buying a device – prepared by author

| Answer | Response |
|---|---|
| Yes | 17 |
| No | 74 |
| Maybe | 21 |
| Not at all | 15 |

**Table 9** Colombian users changing their device – prepared by author

| Answer | Response |
|---|---|
| Every six months | 2 |
| Every 12 months | 13 |
| Every 18 months | 21 |
| Every 24 months | 30 |
| Every 30 months | 16 |
| Every 36 months | 19 |
| Every 42 months | 21 |
| Other – please specify | 6 |

**Table 10** Colombian users "Why do you change your device" – prepared by author

| Answer | Response |
|---|---|
| Size | 19 |
| Speed | 57 |
| Functionality | 85 |
| Keeping with technology | 37 |
| Others – please specify | 13 |

and awareness of their responsibility to the planet needs to be raised via education and training (see Table 8).

In addition, Colombian users change their device after 24–42 months with percentages ranging from 23% to 16% respectively (see Table 9).

When asked by the survey "Why do you change your device?" the majority of participants (67%) indicated that they changed because of the functionalities offered by the new device; 29% want to keep up with technology, and 45% want more speed (see Table 10).

Additionally, the online survey sought to determine the Colombian users' attitudes to their moral responsibilities toward the planet by asking whether changing their devices frequently will cause damage to our planet. Table 11 indicated that

**Table 11** Colombian users "changing device frequently will cause damage to our planet" – prepared by author

| Answer | Response |
|---|---|
| Yes | 78 |
| No | 13 |
| Maybe | 35 |
| Not at all | 1 |

**Table 12** Colombian users: "can we change the mindset of designers and users regarding sustainability" – prepared by author

| Answer | Response |
|---|---|
| Training | 45 |
| Education | 95 |
| Awareness | 52 |
| Workshop | 30 |
| Internet | 63 |
| T.V. | 33 |
| Social Networking | 65 |
| Others - please specify | 2 |

27% responded "Maybe" while 61% showed their awareness that changing devices frequently would cause damage to our planet.

The online survey examined Colombian users' recommendations of ways to change the mindset of designers and users toward sustainability. The survey concluded that via education and awareness (74% and 51% respectively) designers and users could change their mindset and attitude (see Table 12).

## 13 Data Analysis and Discussion

### *13.1 Reliability Statistics for Opportunities and Risks*

As opportunities and risks of sustainability were measured using Likert-type scales, it is imperative to calculate and report Cronbach's Alpha coefficient for internal consistency reliability for any scales or subscales one may be using. The Cronbach's Alpha reliability coefficient normally ranges between 0 and 1. However, authors are not in agreement regarding what can be considered as the lowest Cronbach's Alpha coefficient value that signifies reliable and acceptable data. However, for the purposes of this study, the values below will be considered as measures of reliability [11]:

- _ > 0.9 – Excellent
- _ > 0.8 – Good
- _ > 0.7 – Acceptable
- _ > 0.6 – Questionable
- _ > 0.5 – Poor
- _ < 0.5 – Unacceptable

Table 13 Cronbach's Alpha for opportunities – prepared by the author

| Cronbach's Alpha | N of items |
|---|---|
| 0.852 | 24 |

Table 14 Cronbach's Alpha for risk – prepared by the author

| Cronbach's Alpha | N of items |
|---|---|
| 0.944 | 13 |

A total of 24 items were considered as the opportunities or positive effects of sustainability. Table shows the results for the reliability test. As explained previously, a value of 0.852 can be considered as good (see Table 13).

In addition, 13 items were considered as risks of sustainability. After performing Cronbach's Alpha with a result of 0.944, the reliability of the risks or negative impact of sustainability can be also considered as very good for this particular study (see Table 14).

## 13.2 Factor Analysis for Opportunities

The principal objective of the factor analysis is to satisfy the researcher's need to prove the relationship between variables that apparently are closely connected and establish a set of factors that supports those links. Consequently, with factor analysis, a small number of factors can be derived from a large number of variables which can explain the observed variance in the larger number of variables.

The first step to performing the factor analysis is to perform the KMO and Bartlett's Test. For the purposes of this study, the KMO values would be assumed to be the following [15]:

- Recommend 0.5 as minimum
- Values between 0.7 and 0.8 are acceptable
- Values above 0.9 are excellent

Table 15 present a KMO value higher than 0.5 and a Sig value equal to 0.000; therefore, factor analysis is suitable for this research study; furthermore, the same table present the Bartlett's test which is presented well from this study.

From the communalities in Table 16, it can be seen that at least 14 variables have an extraction value lower than 0.5 and were removed using seven iterations. The next step when calculating factors analysis is the rotated component matrix as shown in Table 17. The significance of these results is that from the respondents' answers the opportunities or benefits of sustainability awareness in Colombia can be consolidated into three factors or groups. The items highlighted in yellow in show those groups.

- Factor 1 – Three variables were linked to the improvement of human well-being and social responsibility in Colombia (see Table 18).

**Table 15** KMO Bartlett's test for opportunities and risks – prepared by the author

| | | |
|---|---|---|
| Kaiser-Meyer-Olkin Measure of Sampling Adequacy. | | 0.712 |
| Bartlett's Test of Sphericity | Approx. Chi-Square | 1058.647 |
| | Df | 276 |
| | Sig. | 0.000 |
| Kaiser-Meyer-Olkin Measure of Sampling Adequacy. | | 0.883 |
| Bartlett's Test of Sphericity | Approx. Chi-Square | 1110.771 |
| | Df | 78 |
| | Sig. | 0.000 |

**Table 16** Communalities for opportunities – prepared by the author

| Communalities | | |
|---|---|---|
| | Initial | Extraction |
| Reduce energy and water usage | 1.000 | 0.542 |
| Reduce paper usage | 1.000 | 0.425 |
| Reduce consumption of raw materials | 1.000 | 0.342 |
| Increase efficiency | 1.000 | 0.523 |
| Increase cost-effectiveness | 1.000 | 0.569 |
| Enhance reputation | 1.000 | 0.294 |
| Improve corporate social responsibility | 1.000 | 0.428 |
| Reduce risk management | 1.000 | 0.356 |
| Differentiate business | 1.000 | 0.255 |
| Increase green strategy | 1.000 | 0.475 |
| Improve community investments | 1.000 | 0.444 |
| Improve human rights | 1.000 | 0.515 |
| Improve social responsibility investing | 1.000 | 0.545 |
| Increase triple bottom line – People, Planet and Profit | 1.000 | 0.434 |
| Increase productivity | 1.000 | 0.669 |
| Attract quality employees | 1.000 | 0.444 |
| Satisfy customer needs | 1.000 | 0.678 |
| Meets take holder expectations | 1.000 | 0.300 |
| Attract new opportunities | 1.000 | 0.288 |
| Create new jobs | 1.000 | 0.337 |
| Reduce carbon footprint | 1.000 | 0.578 |
| Reduce emissions | 1.000 | 0.594 |
| Reduce pollution | 1.000 | 0.645 |
| Reduce health hazards | 1.000 | 0.439 |

Extraction Method: Principal Component Analysis.

- Factor 2 – Four variables were associated to the reduction of environmental negative effects in Colombia (see Table 18).
- Factor 3 – Three variables were connected to the improvement of organizations performance in Colombia (see Table 18).

**Table 17** Rotated component matrix for opportunities – prepared by the author

| Rotated component matrix[a] | | | |
|---|---|---|---|
| | Component | | |
| | 1 | 2 | 3 |
| Improve social responsibility investing | 0.721 | 0.155 | |
| Improve human rights | 0.695 | 0.171 | |
| Improve community investments | 0.656 | 0.107 | |
| Increase green strategy | 0.651 | 0.204 | −0.101 |
| Improve corporate social responsibility | 0.629 | | −0.160 |
| Increase triple bottom line – People, Planet and Profit | 0.605 | 0.258 | |
| Reduce risk management | 0.523 | | 0.286 |
| Create new jobs | 0.517 | 0.101 | 0.244 |
| Attract quality employees | 0.514 | | 0.424 |
| Reduce paper usage | 0.481 | 0.439 | |
| Attract new opportunities | 0.465 | | 0.267 |
| Differentiate businesses | 0.390 | −0.150 | 0.283 |
| Reduce pollution | | 0.795 | |
| Reduce emissions | | 0.769 | |
| Reduce carbon footprint | | 0.757 | |
| Reduce energy and water usage | 0.219 | 0.702 | |
| Reduce consumption of raw materials | 0.329 | 0.482 | |
| Reduce health hazards | 0.383 | 0.449 | 0.300 |
| Increase productivity | | 0.247 | 0.780 |
| Satisfy customer needs | 0.162 | −0.242 | 0.770 |
| Increase cost-effectiveness | −0.186 | 0.217 | 0.698 |
| Increase efficiency | | 0.341 | 0.637 |
| Meets take holder expectations | | −0.198 | 0.506 |
| Enhance reputation | 0.284 | | 0.461 |

**Table 18** Factor 3 for opportunities – prepared by the author

| Variable name | Factor loading | Label |
|---|---|---|
| Improve social responsibility investing | 0.721 | Improvement of human well-being and social responsibility |
| Improve human rights | 0.695 | |
| Improve community investments | 0.656 | |
| Reduce pollution | 0.795 | Reduce environment negative effects |
| Reduce emissions | 0.769 | |
| Reduce carbon footprint | 0.757 | |
| Reduce energy and water usage | 0.702 | |
| Increase productivity | 0.780 | Improvement of organizations performance |
| Satisfy customer needs | 0.770 | |
| Increase cost-effectiveness | 0.698 | |

## 13.3 Factor Analysis for Risks

Using the same approach as that used for the opportunities, a factor analysis was done for the risks associated with sustainability. The analysis had the same level of rigor as that for the opportunities; therefore, the results of the analysis were populated without going into detail about the steps of the process. Table 19 shows the communalities after a factor analysis was performed for the variables identified as risks of sustainability in Colombia. For this opportunity, there was no need to perform further iterations as all the variables had an extraction value higher than 0.5.

The significance of the results shown in Table 20 is that from the respondents' answers the risks associated with sustainability awareness in Colombia can be consolidated in three groups.

- Factor 1 – Six variables can be associated with the growth of fraud and governance failure applying sustainability in Colombia (refer to Table 20).
- Factor 2 – Three variables can be linked with the intensification of supply chain issues in Colombia (see Table 21).
- Factor 3 – Three variables can be related to the increment of general cost and financial costs (see Tables 20 and 21).

**Table 19** Communalities for risk – prepared by the author

| Communalities | | |
|---|---|---|
| | Initial | Extraction |
| Increase fraud (including environmental actions and environmental compliance failures) | 1.000 | 0.853 |
| Increase litigation and compliance breaches (including environmental actions and environmental compliance failures) | 1.000 | 0.873 |
| Increase governance failure (including environmental actions and environmental compliance failures) | 1.000 | 0.845 |
| Increase transaction failure due to environmental liabilities | 1.000 | 0.829 |
| Increase security and systems failures caused by environmental problems | 1.000 | 0.774 |
| Increase scandals by perceived environmental irresponsibility | 1.000 | 0.835 |
| Increase marketing failure by perceived environmental irresponsibility | 1.000 | 0.701 |
| Increase supply chain crises due to suppliers' environmental Problems | 1.000 | 0.827 |
| Inflate costs | 1.000 | 0.727 |
| Increase interest rates | 1.000 | 0.888 |
| Increase insurance crises due to environmental disasters | 1.000 | 0.712 |
| Increase number of new regulations including environmental | 1.000 | 0.731 |
| Increase competition for and cost of raw materials | 1.000 | 0.819 |

Extraction Method: Principal Component Analysis.

**Table 20** Rotated component matrix for risks – prepared by the author

| Rotated component matrix[a] | | | |
|---|---|---|---|
| | Component | | |
| | 1 | 2 | 3 |
| Increase litigation and compliance breaches (including environmental actions and environmental compliance failures) | 0.889 | 0.222 | 0.184 |
| Increase fraud (including environmental actions and environmental compliance failures) | 0.867 | 0.119 | 0.294 |
| Increase transaction failure due to environmental liabilities | 0.865 | 0.114 | 0.258 |
| Increase governance failure (including environmental actions and environmental compliance failures) | 0.858 | 0.210 | 0.253 |
| Increase security and systems failures caused by environmental problems | 0.782 | 0.234 | 0.328 |
| Increase scandals by perceived environmental irresponsibility | 0.706 | 0.580 | |
| Increase marketing failure by perceived environmental irresponsibility | 0688 | 0.440 | 0.182 |
| Increase supply chain crises due to suppliers' environmental problems | 0.447 | 0.789 | |
| Increase competition for and cost of raw materials | 0.110 | 0.766 | 0.469 |
| Increase number of new regulations including environmental | 0.148 | 0.741 | 0.399 |
| Increase interest rates | 0.305 | 0.118 | 0.883 |
| Iniate costs | 0.196 | 0.440 | 0.704 |
| Increase insurance crises due to environmental disasters | 0.439 | 0.296 | 0.657 |

Extraction Method: Principal component Analysis
[a]Rotation converged in 7 iterations.

The use of statistical tool, SPSS, has facilitated the analysis of the data collected on the survey. The data analysis process began by confirming the validity of the data. This was done using the Cronbach's Alpha for both the opportunities and risks variables that were identified. Both results were highly acceptable as values obtained were higher than 0.8. After performing the Factor Analysis for the opportunities, three new factors were obtained as a result. Factor 1 is related to the improvement of human well-being and social responsibility in Colombia; factor 2 is linked to the reduction of environmental negative effects in Colombia; and; factor 3 is connected to the improvement of organizations' performance in Colombia. The Factor Analysis for the risk variables produced three new factors. Factor 1 is associated with the increase of fraud and governance failure to implement sustainability in Colombia; factor 2 is linked with the intensification of supply chain issues in Colombia; and lastly, factor 3 is related to the increase of general and financial costs.

**Table 21** Factor 3 for risks – prepared by the author

| Variable name | Factor loading | Label |
|---|---|---|
| Increase litigation and compliance breaches (including environmental actions and environmental compliance failures) | 0.889 | Growth of fraud and Governance failure |
| Increase fraud (including environmental actions and environmental compliance failures) | 0.867 | |
| Increase transaction failure due to environmental liabilities | 0.865 | |
| Increase governance failure (including environmental actions and environmental compliance failures) | 0.858 | |
| Increase security and systems failures caused by environmental problems | 0.782 | |
| Increase scandals by perceived environmental irresponsibility | 0.706 | |
| Variable name | Factor loading | Label |
| Increase supply chain crises due to suppliers' environmental problems | 0.789 | Intensification of supply chain issues |
| Increase competition for and cost of raw materials | 0.766 | |
| Increase number of new regulations including environmental | 0.741 | |
| Variable name | Factor loading | Label |
| Increase interest rates | 0.883 | Increment of general cost and financial costs. |
| Inflate costs | 0.704 | |
| Increase insurance crises due to environmental disasters | 0.657 | |

## 14 Conclusion from the Analysis

After analyzing the data results, it can be said that the research clearly indicated the advantages and disadvantages of sustainability awareness in Colombia. Figs. 3 and 4 illustrate the advantages and disadvantages of sustainability awareness in Colombia after factor analysis had been performed.

The three factors generated for the advantages or opportunities are totally aligned with the real meaning of sustainability explored in the literature review. The fact that one of the factors was the improvement of human well-being and social responsibility in Colombia shows that the levels of sustainability awareness in Colombia are not linked solely to the environment issues. Moreover, it covers the extensive role of sustainability and sustainable growth in terms of the economy, society and environment. On the other hand, the factors generated for the disadvantages and risks associated with sustainability, are related to the possibility of fraud and governance failure. This can be a general feeling in Colombia as

| | | |
|---|---|---|
| **Advantages** | **Improvement of human well-being and social responsibility** | Improve social responsibility investing |
| | | Improve human rights |
| | | Improve community investments |
| | **Reduce environment negative effects** | Reduce pollution |
| | | Reduce emissions |
| | | Reduce carbon footprint |
| | | Reduce energy and water usage |
| | **Improvement of organizations performance** | Increase productivity |
| | | Satisfy customer needs |
| | | Increase cost-effectiveness |

**Fig. 3** Advantages of sustainability awareness in Colombia, prepared by the author

| | | |
|---|---|---|
| **Disadvantages** | **Growth of fraud and governance failure** | Increase fraud |
| | | Increase litigation and compliance breaches |
| | | Increase transaction failure |
| | | Increase governance failure |
| | | Increase security and systems failures |
| | | Increase scandals environmental irresponsibility |
| | **Intensification of supply chain issues** | Increase competition for and cost of raw materials |
| | | Increase supply chain crises environmental problems |
| | | Increase number of new regulations |
| | **Increment of general cost and financial costs** | Increase interest rates |
| | | Inflate costs |
| | | Increase insurance crises |

**Fig. 4** Disadvantages of sustainability awareness in Colombia, prepared by the author

factors such as poverty, violence and the civil war can negatively influence the activities of the government and other agencies. It should be interesting to explore in more detail why the implementation of sustainability is considered to have a negative impact on the supply chain and create additional costs for organizations.

## 15 Limitations of the Study

The findings of the research should be interpreted within the context of the following limitations:

- **Number of participants:** Although participants were drawn mostly from the central region of Colombia, they do not represent a holistic view of the country. This of course, represents a restriction within the target participant and it reflects the limited time available to conduct the research. Further researches should not be limited to only a specific area in the country.
- **Sample:** A total of 154 responses were collected. However, the majority of the responses were collected from residents in Antioquia. Due to the time constraint, the sample could not be bigger and the surveys could not be delivered to different regions within the country. Consequently, the results of the survey should not be considered as a representation of all the people from Colombia.
- **Research strategy:** Whilst the Web-based method has many advantages, the limitations relating to the exploratory nature of the research include the generally non-scientific nature of the approach or data interpretation. In addition, most of the surveys were not fully completed and many respondents did not provide any further comments about the research although a specific section of the questionnaire gave them the opportunity to do so.

## 16 Future Research

It would be worthwhile to extend this type of study to other regions in Colombia. Due to the time constraint and the lack of a proper distribution method of the survey, the collection of data was limited to a specific region. This study has the potential to make a broader impact if it could be done with local resources that can obtain a bigger sample from other regions in the country. A different direction for future researches could also include an investigation into the different levels of education as this could determine the issues that are faced by the educational sector in Colombia in terms of introducing sustainability awareness to students at an early stage. Using the same approach, future studies can also explore the levels of sustainability awareness according to type of industry. As mentioned in the study, Colombia's economy depends greatly on the exploration of non-renewable resources such as oil and gas and coal. However, those industries are not the largest employers in the country, since a substantial amount of Colombians income is from small-crop agriculture.

## 17 Recommendations

One recommendation taken from the data results is that the adoption of sustainability practices must be seen as a means of improving human well-being and social responsibility in Colombia. However, as indicated by the literature review, governments and international organizations must design a framework to eradicate poverty, bring equity, social justice and sustainable growth in the country. In relation to the first factor for the opportunities, the framework must be also designed to promote better livelihoods through access to clean water, sanitation services and diverse energy supplies, regulating lower pollution, better planning of land use, and disaster risk management. In addition to the social improvement required, the use of new technologies is critical to achieve a reduction of negative environmental effects in Colombia. The promotion of green investment through development finance and other technical assistance is crucial, as well as other initiatives such as innovation, technology co-operation and knowledge transfer. Free trade agreements between the countries must also facilitate new commercial agreements in environmentally-sustainable goods and services. The two previous recommendations would apply on a macro country level. However, some other recommendations emerging from the study can be taken on board by companies and organizations. Organizations are increasing their accountability to stakeholders regarding environmental and social issues. Moreover, future company strategies should closely monitor and include sustainable development. The study has recognized that the implementation of sustainability brings improvement to organizations' performance. However, according to the study, one of the factors that can be improved and might require further investigation is the fact that sustainability appears to be seen as an additional cost for Colombian organizations.

## 18 Conclusion

It could be noted that, generally speaking, the amount of scientific research related to sustainability in Colombia is scarce. To date, most of the research has been conducted by organizations or government bodies. There is an important amount data mainly related to environmental health in Colombia. Nevertheless, previously studies have not determined the extent to which there is awareness of sustainability in the country, and the advantages and disadvantages that it can bring to organizations, government and the general public. The research study intended to ascertain the status of sustainability awareness levels in Colombia and the advantages and disadvantages that the general public identified. The findings from the data analysis could encourage the implementation of sustainability policies that help Colombian government to increase public awareness. It could be also used as a first measurement of the current situation, which will trigger a contingency plan for schools and the education sector to realize the importance of sustainability awareness. More specifically, the research data and results would also provide a good understanding of the Colombian public's attitude to green products, green strategies, sustainability policies and corporate social responsibility.

## References


1. Agudelo C, Rivera B, Tapasco J, Estrada R (2003) Designing policies to reduce rural poverty and environmental degradation in a hillside zone of the Colombian Andes. World Dev 31(11):1921–1931
2. Ariwa E, Omoneye OO, Mauri JL (2012) Green communication and corporate sustainability of computer aided audit techniques and fraud detection. In: Lecture notes in electrical engineering. Springer, Dordrecht, the Netherlands, pp 843–862
3. Christie I, Warburton D (2001) From here to sustainability. Earthscan, London
4. Couper MP, Traugott MW, Lamias MJ (2001) Web survey and administration. Public Opin Q 65:230–253
5. Daly HE (1990) Toward some operational principles of sustainable development. Ecol Econ 2:1–6
6. Ekins P (1993) 'Limits to growth' and 'sustainable development': grappling with ecological realities. Ecol Econ 8:269–288
7. Elkington J (1994) Towards the sustainable corporation: win-win-win business strategies for sustainable development. Calif Manag Rev 36(2):90–99
8. Fan W, Yan Z (2010) Factors affecting response rates of the web survey: a systematic review. Comput Hum Behav 26:132–139
9. Fritz M, Koch M (2014) Potentials for prosperity without growth: ecological sustainability, social inclusion and the quality of life in 38 countries. Ecol Econ 108:191–199
10. Glavic P, Lukman R (2007) Review of sustainability terms and their definitions. J Clean Prod 15:1875–1885
11. Gliem J, Gliem R (2003) Calculating, interpreting, and reporting Cronbach's alpha reliability coefficient for Likert-type scales. In: Midwest Research-to-Practice Conference in Adult, Continuing, and Community Education, Columbus, Ohio, pp 82–88
12. Gudmundsson E, Wessells CR (2000) Ecolabeling seafood for sustainable production: implications for fisheries management. Mar Resour Econ 15:97–113
13. Hak T, Janouskova S, Moldan B (2016) Sustainable develpment goals: a need for relevant indicators. Ecol Indicat 60:565–573
14. Heinberg R (2010) What is sustainability? In: The post carbon reader: managing the 21st century's sustainability crisis. Watershed Media, Healdsburg
15. Hill B (2011) The sequential Kaiser-Meyer-Olkin procedure as an alternative for determining the number of factors in common-factor analysis: a Monte Carlo simulation. Degree of Doctor of Philosophy Oklahoma State University, Oklahoma
16. Hoek RIV (1999) From reversed logistics to green supply chains. Supply Chain Manag 4:129–135
17. Hopwood B, Mellor M, O'Brien G (2005) Sustainable development: mapping different approaches. Sustainable Cities Research Institute. Sustain Dev 13(1):38–52. doi:10.1002/sd.244
18. Jabareen Y (2008) A new conceptual framework for sustainable development. Environ Dev Sustain 10:179–192
19. Kaplowitz MD, Hadlock TD, Levine R (2004) A comparison of web and mail survey response rates. Pub Opin Q 68(1):94–101. doi:10.1093/poq/nfh006
20. Kuhlman T, Farrington J (2010) What is sustainability? Sustainability 2:3436–3448
21. Lele SM (1991) Sustainable development: a critical review. World Dev 19(6):607–621

22. McNeill P, Chapman S (2005) Research methods. Taylor and Francis, London
23. McPhail K (2008) Contributing to sustainable development through multi-stakeholder processes: practical steps to avoid the 'resource curse'. Corp Gov 8(4):4471–4481
24. Mebratu D (1998) Sustainability and sustainable development: historical and conceptual review. Environ Impact Assess Rev 18:493–520
25. Miliban E (2015) Yes, the Paris Climate Change Conference can save the planet. *The Guardian*. https://www.theguardian.com/commentisfree/2015/nov/22/paris-climate-change-conference-zero-emissionsplanet. Accessed 1 June 2017
26. Ministry CE (2002) Colombia's sustainable development reforms: mainstreaming environment towards growth and employment. Bogota. http://www.eisourcebook.org/cms/Colombia%20Sustainable%20Development%20Reform.pdf. Accessed 1 June 2017
27. Mitcham C (1995) The concept of sustainable develpment: its origins and ambivalence. Technology Soc 17(3):311–326
28. Neuman WL (1999) Social research methods: qualitative and quantitative approaches, 4th edn. Pearson Education Limited, Needham Heights
29. OECD (2015) Global and local environmental sustainability, development and growth. OECD and Post-2015 Reflections
30. Olivero-Verbel J (2011) Colombia: environmental health issues. University of Cartagena, Cartagena
31. Plessis CD (2000) Cities and sustainability: sustaining our cultural heritage. In: Brandon P, Lombardi P, Perera S (eds) Cities and sustainability: sustaining our cultural heritage. The Millennium Conference (2000), Kandalama, Srilanka
32. Raleigh CB (1999) Mission statement of the Center for a Sustainable Future. The Center for a Sustainable Future, Honolulu
33. Ramirez GA (2012) Sustainable development: paradoxes, misunderstandings and learning organizations. Learn Organization 19(1):58–76
34. Ramirez JM, Bravo GA, Botero JL (2015) Aligning Colombia's financial system with sustainable development. International Finance Corporation, Washington
35. Roodman D (1997) Getting the signals right: tax reform to protect the environment and the economy. Worldwatch Institute, Washington
36. Saunders M, Lewis P, Thornhill A (2012) Research methods for business students, 6th edn. Pearson Education Limited, Harlow
37. Schafer T (2013) Sustainability – some thoughts on the term and its meaning. Available via Aluminiportal Deutschland. https://www.alumniportal-deutschland.org/en/global-goals/2030-agenda/sustainability/. Accessed 1 June 2017
38. Seelos C, Mari J (2005) Social entrepreneurship: creating new business models to serve the poor. Bus Horizons 48(3):241–246
39. Srivastava SK (2007) Green supply chain: a state of the art literature review. Int J Manag Rev 9(1):53–80
40. Taylor J (2002) Sustainable development: a dubious solution in search of a problem. Policy Analysis 449:1–49
41. Umbach PD (2004) Web survey: best practices. New Dir Institutional Res 121:23–28
42. UN (2005) 2005 World Summit Outcome. vol Resolution A/60/1. New York
43. UN (2011) Accelerating progress towards the millennium development goals: options for the sustained and inclusive growth and issues for advancing the United Nations Development Agenda Beyond 2015. Annual Report of the Secretary-General
44. UNESCO (2015) Water for a sustainable world. Paris
45. WBCSD (2010) Vision 2015 – The new agenda for business. Conches, Geneva
46. WCED (1987) Our common future. Oxford University Press, London
47. Weybrecht G (2013) The sustainable MBA: a business guide to sustainability. Wiley, Chichester

# Sustainability Awareness in the Brazilian Higher Education

**Erika Kuroda**

**Abstract** This exploratory study examines Brazil's awareness of sustainability in the higher education, and examines the advantages, disadvantages and risks associated with the adoption of sustainability. From the literature reviewed in this study, it is evident that the concept of development and growth has shifted from individual factors to one increasingly concerned about sustainable development that particularly includes social welfare. The Marrakech Climate Change Conference held in November 2016 reiterated that our climate is an alarming situation worldwide and, now more than never; it calls for an urgent duty to react. It is believed that in order to move forward with the sustainable development, an immediate commitment and actions must be foster from 2017 to 2020 to benefit and support the 2030 Agenda for Sustainable Development. Hence, sustainability has become an essential concept to ensure the future of humanity. What is more, in a world scenario that is rapidly changing, the resources and the ecosystem on which we depend play a crucial role and ought to be comprehensively viewed so that sustainability can be assessed and enhanced in order to ensure a better future. At first, sustainability may not seem to be an easy task, and in fact, sustainable development might be one of the most difficult challenges of our time. This is partly because sustainability is also susceptible to overwhelming risks and threats to business and society overall. Consequently, this exploratory research study has examined sustainability approaches by highlighting the positive aspects of sustainability in the Brazilian higher education as well as the possible risks that addressing sustainability in Brazil might produce. This study also briefly reviews sustainability approaches used in the economic, environmental and social spheres including the strategic and management role as well as the sustainability role in developing countries with an emphasis on Brazil. Through a quantitative method analysis combined with an online survey questionnaire, this study identified three positive factors related to sustainability awareness in the Brazilian higher education: enhancement of business assets, the 4 Rs (reduction of emissions, pollutions,

E. Kuroda (✉)
School of Information Systems, Curtin University, Perth, Western Australia, Australia
e-mail: erika.kuroda@postgrad.curtin.edu.au



T. Issa et al. (eds.), *Sustainability, Green IT and Education Strategies in the Twenty-first Century*, Green Energy and Technology, DOI 10.1007/978-3-319-57070-9_15

carbon footprint and health hazards), and the reduced consumption of raw material. Subsequently, three risks to sustainability in the Brazilian higher education were also revealed: increase of management failure, increase of costs, compliances and regulations, and increase of procedures and competition.

## 1 Introduction

A number of studies have been conducted into the dilemma presented by the concept of sustainability. The ever-growing number of studies that are promoting sustainability and reinforcing the idea of sustainable development that can meet the needs of the present without compromising the ability of future generations to satisfy their own needs, has rapidly engaged the world's concern [6]. Moreover, this growing importance also shows the urgency of drawing attention to this matter.

Given this scenario and the lack of understanding of the sustainability concept, Merriman et al. [33] posit that companies, businesses and entities from both the private and public sectors, are increasingly being held responsible for environmental hazards and depletion of natural resources, which place at risk the future of mankind. This accountability is part because of the rapid changes in products provided by technology aligned with the world's demand and the need for products and services, as well as the population growth. These factors have negative implications for the planet and its people, and to sustain our world, sustainable development is required [43]. Consequently, in order to overcome the sustainability issue and address this dilemma, there is a vital need to engage all the stakeholders so that sustainable development can be accomplished. In terms of Brazil, before sustainability development can be introduced and promoted, it is necessary to determine how people perceive and understand the concept.

According to several scholars, Brazil is a promising nation, having one of the largest and fastest growing economy in the world [38]. Notwithstanding, Brazil's economy relies mainly on the production and commercialisation of agricultural commodities entailing the consumption of a considerable amount of limited natural resources as well as non-renewable resources [22]. Therefore, the link between the need to remain competitive and maintain a growing economy together with the rise of sustainability awareness plays a crucial role in sustainability development in Brazil. Thus, this exploratory research study examines sustainability awareness and its relevance to the social, cultural and economic environment in Brazil and on the global scale. Furthermore, this study is intended to increase sustainability awareness in the Brazilian higher education and encourage positive Brazilian attitudes towards sustainability development. In doing so, the positive aspects of sustainability development and the risks associated with addressing sustainability in Brazil will be further analysed.

## 2 Background of Research

### *2.1 What is Sustainability?*

The term 'sustainability' first emerged in the literature shortly after the 1970s [34]. A few years later, the term was related to sustainable development when it appeared in a report by Brundtland in 1987 in Our Common Future, published by the World Commission on Environment and Development (WCED) [4, 2]. The concept of sustainable development was widely disseminated by the WCED and defined as a process whereby the people's current needs can be met without interfering with and compromising the needs of future generations [4]. The safeguarding of resources is crucial to the long-term prosperity of society.

According to Borim-de-Souza et al. [4] sustainability concerns the relationship between non-renewable natural resources, climate changes and ecologic development, and the preservation and maintenance of its systems. This relationship comprises environmental, economic and social dimensions. Taking the same viewpoint, David and Tobin [8] made an analogy that connected sustainability to Maslow's hierarchy of needs, with the authors suggesting that sustainability is perceived as a balance between the satisfaction of an individual's basic needs, the viability of the ecosystem, and economic growth.

Additionally, Pater and Cristea [34] believe that the sustainability concept has been developed and clustered into two concepts. The first functional concept is related to sustainable development and encompasses the notion of a conscious development. In other words, it is the efficient usage of natural resources, thereby ensuring the needs of the future generation [35]. The second sustainability concept suggested by Pater and Cristea [34] stems from the first and brings the idea of a systemic approach that takes the dynamical system approach to the context of a continuous future. In other words, it is a conscious development followed by the concept of a continuous future.

Pater and Cristea [34] state that one of the dilemmas faced by the twenty-first century is to constantly innovate not only to develop efficiency of time, resources and products but also to accomplish sustainable progress in every possible manner. Expanding on the same viewpoint, Pribeanu and Toader [35] conclude that achieving sustainable progress is vital since human beings have a tremendous influence on the environment. Nowadays, this has been recognised as a crucial issue for society, far more so than in the past. Furthermore, the increase of human activities, along with the negative impacts of economic growth, has degraded the global environment [4].

Borim-de-Souza et al. [4] believe that sustainable development and sustainability are more than a social phenomenon; their characteristics and manifestation in different parts of the world have given rise to various research questions. It is clear that there is a need to combine economic development with an ecological approach to further establish sustainability around the world [35]. Given that there is a limit to our planet's natural resources, sustainable development is fundamental to society and business and should be viewed holistically.

Due to the growing importance of sustainability and its multifaceted dimensions across the world, a number of authors have tried to conceptualise and unify approaches to this concept [37]. Thomas et al. [41] concluded that even though sustainability seems to be well-established in some parts of the world, it is a critical research area that has some limitation in scope. Moreover, Little et al. [29] also pointed out that this limitation stems from the difficulty of quantifying and explaining the sustainability concept. Notwithstanding this ongoing discussion, an overview of sustainability approaches shows that many authors pursue similar goals which can be mainly categorised under three areas: environmental, economic and social sustainability [29]. It is clear in the literature that each of the pillars (environmental, economic and social) needs to be addressed simultaneously in order for sustainability to be established.

## 2.2 The History of Sustainability

The word 'sustainability' was not always a buzzword as it is in this millennium; nowadays, it is receiving renewed interest worldwide [31]. More importantly, sustainability has become an important fundamental strategy of organisations in the twenty-first century [20]. For Kuhlman and Farrington [26], however, the sustainability concept has always been a concern for society in terms of the well-being of future generations and the replacement of resources. For instance, the first people to worry about the possible extinction of their prey were hunters, since hunting was a fundamental means of survival [26]. Hence, one can conclude that the concern with preserving now for the future has always been there; it has now attracted greater attention worldwide and is known as sustainable development. Therefore, Kuhlman and Farrington [26] emphasize that the relationship between humankind and nature has always been a consideration.

Despite this relationship between humanity and environment, it is evident that for the past five decades there have been changes in many areas with the development of industries, increase of life expectancy, decline in infant mortality, and an exponential increase of food production and consumption [31, 32]. While on the one hand these changes and developments have been beneficial for mankind, on the other hand, certain activities have been directly and negatively affecting the economic, social and environmental dimensions, gradually bringing an awareness of sustainable development and sustainability, both of which are now critical modern issues [31].

Sustainability received global public attention after the Brundtland Report which revealed that many natural resources that are vital for the survival of humankind would be depleted within the next generation or so [26]. After the Brundtland Report was made public, the concept of sustainability was widely disseminated, as was the recognition and concern regarding this issue [26]. The Brundtland Report aroused public concern about how to prevent this dire future. Kuhlman and Farrington [26] argued that the Brundtland committee believed that sustainable development was the only way to reconcile human needs with the limited natural resources and the doom that the rapid exhaustion of these resources could lead to.

Even though the term 'sustainability' gained worldwide attention after the WCED Brundtland Report, Thomas Robert Malthus is considered to be the first economist to predict, in 1978, the hazards of environmental exploitation caused by the scarcity of resources [32]. It is believed that this was the first time that the concept of sustainability emerged – not necessarily using the word 'sustainability'. Thomas Malthus stated that there was a limit amount of environmental resources; later, his view was considered the antecedent of sustainable development [32].

In addition, Mebratu [32] emphasises that the history of sustainability can has three major milestones. The first was the UN Conference on Environment held in Stockholm in 1972. This was followed in 1987 by the WCED (World Commission on Environment and Development) also known as the Brundtland Report and finally, the Post-WCED that had its starting point in 1992.

The UN Conference on Human Environment recognised the importance of tackling the environmental issue [32]. At about the same time, a group of scientists – known as Club Rome – produced an important report on the state of the natural environment which stated that if society continues to promote economic growth as it was, the exploitation of the natural environment would lead to devastation. Therefore, from this time the idea of addressing the environment and development became recognised globally as it was acknowledged that these two matters should be addressed together and immediately [32].

A few years later, in 1987, following the same conceptual insight, the WCED Report (Brundtland Report) brought the sustainable development statement to light [32]. The Report's main concept is that development should meet the needs of the present while not compromising the needs of future generations [32].

Finally, the Post-WCED became known as the UN Conference on Environment and Development (UNCED), also called the Rio Conference or the Earth Summit [32]. Mebratu [32] explains that in the UNCED Conference, each one of the member countries was required to prepare a national report explaining the current environmental situation, and to produce an action plan promoting sustainable development [32]. This process would indicate the extent to which each country was committed to adopting the sustainability concept [32]. Moreover, the UNCED created significant documents and declarations, one of the most important of which was the Rio Declaration and Agenda 21. In particular, the sustainable management of climate change, human impacts on the environment, and biodiversity issues, were highlighted and showed to the world that those were issues that needed to be considered globally, nationally and locally [32].

Nowadays, the term 'sustainability' refers to enhancing economic, social and environmental aspects in the long-term. Moreover, the approach to sustainability is crucial to companies if they are to remain competitive over the long term [20]. For that reason, Kuhlman and Farrington [26] declare that environmental concerns are important now and were important then, but regardless of its environmental importance, the authors suggest that the value of sustainability lies in the necessity of safeguarding the future for our children. That is why it is extremely important to be aware of and understand sustainability.

Moreover, at present, there are a number of campaigns, regulations, technologies and innovations that bring to the agenda three types (environmental, economic and social) of sustainability that ought to be the fundamental pillars of progress in order to create a better world and ensure sustainable development.

## 2.3 What Sustainability Means for Business

The integration of sustainability into everyday activities has become vital, strategic and crucial for any sort of business nowadays according to Weybrecht [45]. Dyllick and Muff [16] assert that from the organisational perspective, embedding sustainability throughout the entire organisation and managing risks and opportunities along with sustainability brings to companies a better performance at the operational as well as strategic level. The concept of sustainability has also become popular in regards to employees applying sustainable strategies to their work, which has become increasingly evident and some of them build a business case for sustainability development [45]. Bonn and Fisher [3] say that in order for organisations to become sustainable, sustainability needs to be integrated from the beginning into every project on an ongoing basis.

Weybrecht [45] asserts that for business enterprises, sustainable development means adopting business strategies and activities that satisfy the organisation's goals as well as the requirements of stakeholders. In addition, sustainability in business also means meeting all the needs while at the same time protecting, sustaining and enhancing the human being and the limited natural resources that will be needed in the future. Additionally, an organisation's vision needs to be aligned with strategic decision-making as well as the organisational culture in a holistic way so that all stakeholders are taken into consideration [3].

Despite the importance for business of adopting sustainability principles, some organisations seem to find it difficult and challenging to make their business sustainable [3].

Weybrecht [45] believes that the reasons for integrating sustainability practices into business strategies may vary from company to company; however, it can be assumed that some arguments are relevant and should be considered if organisations are to adopt sustainability as part of their business practices. Among the many reasons for organisations wanting to establish a sustainable business, this literature review focuses on several determinant reasons: financial aspects, cost reduction, environmental – preservation of resources, risk, enhancement of business reputation and increased corporate social responsibility. These determinants can be better seen in the following concept map (see Fig. 1).

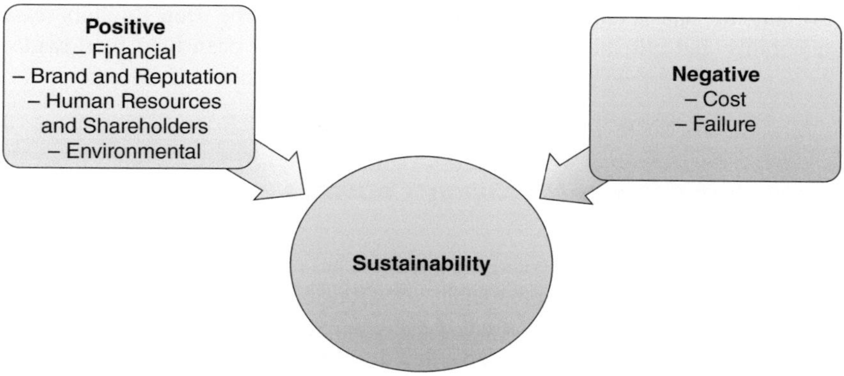


**Fig. 1** Adopting sustainability in business: concept map

## *2.4 Current Progress Level of Sustainability*

One of the greatest challenges that most organisations face in today's world includes the difficulties in prioritising issues, promoting strategies, regulations and policies, and measuring performance [3]. Not only are organisations concerned about the future of humankind; society also plays a decisive role since there are pressures from governments to adopt sustainable practices [27]. Because of that, sustainability has become a global concern characterised by new policies, strategies, new products and services to a degree that has affected the horizons for future investments, research and development [37].

Ratiu and Anderson [37] believe that the sustainability concept has been transformed from a notion to be a context open to anyone's interpretation over the years. Given this broad idea of sustainability and its multifaceted concepts, the sustainability issue has given rise to a debate regarding climate change as a major environmental problem that is currently faced by society [44]. Climate change is believed to be due to environmental exploitation that results in the degradation of nature thus causing changes in the natural pattern of the earth such as changes to the environment, weather alterations and increased greenhouse gases emissions, as a result of human activities [44]. In other words, human activities have impacted on the environment, resulting in changes to weather patterns.

The climate change issue has led to several debates and conferences and more importantly brought this issue to the attention of business [46]. One of these conferences is the Kyoto Protocol signed in 1997 in which countries were required to sign an agreement to reduce gas emissions by 6%. Along with this Protocol, the Conference of the Parties (COPs) takes place regularly in order to track and discuss current progress concerning gas emissions [46, 18]. Therefore, those debates and conferences demonstrate the world awareness and attention given to issues such as sustainable development, the future of sustainability and the safeguarding of resources for the next generation.

To sum up, the actual progress of sustainability can be seen through several sustainability and climate change conferences that have been held worldwide in order to monitor climate change and sustainable development effectively.

## *2.5 Sustainability in Developing Countries*

Punnett [36] asserts that the developing countries are changing the world of business and what is more, there is a remarkable yet strong interaction between developing and developed countries in the current world that experts have never seen before. In this interaction, it is possible to observe that developed countries are adapting to the environment of developing countries and the reverse is also occurring [36]. Punnett [36] describes developing countries as countries where the economic resources and the level of industrialisation are little when compared to countries where these activities are greater. According to Punnett [36], developing countries represented 80% of the world in 2013. More importantly, The UN's [42] Prospect for the World Economy state that the developing economies are expected to contribute more to the global growth than developed countries, that is, the emerging economies all together are expected to growth 4.3% in 2016 and 4.8% in 2017 against a slowdown of developed countries as a rebound to 5.6% growth in both 2016 and 2017, is expected. The developing countries are also believed to mitigate global risks, stimulate global investment and demand of products and services [42].

As a result of changes in the global economy and the attention given to emerging economies, the dominance of some countries in the global context has also changed substantially over the years [45]. That is, although the big economies are still leaders in the global economy, many other players have appeared such as the developing countries and their emerging markets [45]. The globalisation phenomenon has given developing countries more opportunities and growth, and hence it has become important to address sustainability in developing and emerging countries. In other words, developing countries and emerging countries are playing an important role in the global scene; thus, they are vital players in global sustainability development and environmental impacts. Eweje [17] states that sustainable development in developing countries, as well as CSR, are both increasing. On the other hand, at the same time that discussions concerning developing economies are increasing, Lourenço and Branco [30] report that there are few studies available on sustainability in developing countries.

Given this statement, there is an urgent need to address sustainability in developing countries. Eweje [17] believes that the developing business economy plays a crucial role in reducing environmental exploitation, poverty, and social disparity, and is likely to improve society's welfare. What is more, Lourenço and Branco [30] demonstrated that when a comparison between developed countries and emerging countries is made some similarities can be traced such as the fact that companies listed in the stock market receive more funding from the capital market than the ones not listed. For developing countries, this is a crucial factor considering the greater difficulty that companies located in emerging economies face in raising funds [30]. Moreover, Lourenço and Branco [30] see that in developing countries, enterprises that have their business listed in the stock market show signs of maintaining corporate governance practices and make an effort to address social and environmental matters.

Along with the actual sustainability situation in developing countries, it is important to also consider the role of corporate social responsibilities in emerging economies [17]. Eweje [17] believes that CSR should be addressed in developing countries because they do not have good governance. This results in social and economic issues such as underdevelopment, poverty, uncertainty and corruption; therefore, multinational companies and large corporations are required to do more for their country and community [17].

In other words, in some developing countries, CSR meets its responsibilities to stakeholders at the same time as it addresses social issues [17]. That is, companies are expected to contribute to the economy of the developing countries as well as the growth and improvements of social issues and environmental concerns. Therefore, CSR and sustainability initiatives are needed to change methods of production and labour relationships as well as to reduce environmental impacts and investments in infrastructure, thereby benefiting the local communities in developing countries [17]. What is more, in some developing countries, CSR is mandatory and organisations are required to present reports on CSR and sustainability practices demonstrating that their company initiatives will benefit society [17].

If on the one hand the sustainability awareness is increasing in developing countries, on the other hand – and stressing the actual scenario – Dyllick and Muff [16] believe that a sustainability approach in developing countries has a long way to go. The authors assert that sustainability is likely to occur only if the development of products and services helps to counteract the common problems faced by developing countries, such as hunger, poverty, human rights, poor public health, water and energy problems [16]. That is, the fundamental common issues faced by developing countries need to be taken into consideration in order to achieve sustainable development.

## *2.6 Sustainability in Brazil*

As stated previously, emerging economies are currently attracting a lot of attention in the global market due to the globalised economy. Developing economies such as Brazil are benefitting from the interconnections provided by the global scenario [28]. In Brazil for instance, it is clear that companies are striving to expand their operations to include developed countries such as the United States and Europe in order to increase their market share [28]. However, to thrive in competitive global markets such as those in the United States and Europe, Brazilian companies have been facing challenges regarding sustainability practices since developed economies embraced and adopted sustainable regulations in the financial, environmental and social pillars some time ago [28].

Jabbour et al. [23] state that when talking about emerging countries and sustainability practices, Brazil is at the forefront of discussions given its abundance of natural resources and its extensive bio-diversity. Gavronski et al. [21] point out that Brazil has the fourth-ranked largest number of companies credited with ISO 14001. According to Gavronski et al. [21], this ranking indicates that these companies are concerned about their environmental performance, and value the reputation that an ISO (international organisation for standardisation) gives them. Further, it indicates that they are willing to take the path to sustainable development.

Regardless of the Brazilian business attitude towards sustainability, Barbosa and Veloso [1] suggest that the Brazilian individual's awareness of sustainability is gradually yet slowly increasing. However, the authors declare that the Brazilians' daily behaviours in domestic life, everyday practices and consumption habits are still considerably unsustainable. What is more, the authors emphasise that in Brazil there is a misuse of water, electricity and an increase of both consumption and waste. One of the reasons for such behaviour is pointed out by Barbosa and Veloso [1] through an ethnographic research where it is made clear that Brazil faces a cultural dilemma. The everyday non-sustainable actions are embodied in Brazilian's cultural assumptions such as concerns regarding domestic cleaning, personal hygiene, how they do their dishes and so forth, all of which are detrimental to the environment and therefore should be highlighted in the Brazilian sustainability agenda [1].

This view about sustainability awareness in Brazil was also reinforced by Duarte [15] in an exploratory study about the barriers associated with sustainability that are perceived by Brazilian employees working in a sustainable company. According to Duarte [15], the lack of clarity and understanding of the sustainability concept is widely evident in Brazilian respondents. Moreover, the author believes that the lack of understanding of sustainability is also due to issues relating to the lack of good public education and therefore the indifferent attitude towards sustainable practices [15].

On the other hand, to counterbalance these Brazilian aspects of sustainability, Castanheira et al. [5] highlight the emerging alternative of bioenergy in Brazil which supports environmental sustainability practices. According to Castanheira et al. [5], the environmental sustainability of biodiesel and bioethanol (biofuel) in Brazil has grown substantially in the past few years and it is believed to be a sustainable alternative to generating fossil fuel as it reduces greenhouse effects and climate change consequences. Hence, the alternative means used to generate fuel, as well as the reduction of pollutant gas emissions, demonstrate Brazil's effort to adopt a sustainable approach.

Duarte [15] asserts that Brazil is currently living a paradox. On the one hand, the country is recognised as a leader in sustainability activities with increasing attention being given to social and environmental responsibilities [15]. Moreover, Duarte [15] states that Brazil is the third-ranked country in the world to present the highest number of sustainability reports and policies as well as making tremendous investments in social and environmental projects. However, a review of Duarte [15]'s article clearly shows that the opinions expressed by his survey respondents are not what one would expect from people living in a country that is a leader in sustainability practices. Hence, there appear to be barriers to sustainability awareness in Brazil.

## 2.7 *Sustainability and the Future*

An increasing trend is that organisations are responding to the sustainability and climate change issues by adopting strategies such as the publication of triple bottom line reports and increasing their level of CSR. As stated previously, it is evident from the literature review that many organisations are attempting to promote sustainable practices and integrate sustainability in their operations [15]. However, according to Dentchev et al. [9], the sustainable business model poses a big challenge since organisations need to address sustainability promptly and at the same time maintain or increase their profits, which ultimately is the reason for their existence.

It is clear for Dyllick and Muff [16] that a further step toward future sustainability is to recognise that sustainability requires a response to social and environmental issues at the same time as it addresses the economic concerns of business. What is more, the value creation of sustainability goes beyond economic performance and includes social and environmental values. Given the current situation and despite the increasing efforts to adopt sustainable practices, Duarte [15] asserts that our society is far from being sustainable since there are still some barriers to sustainability.

Taking the same viewpoint, in their recent paper, Dias et al. [10] state that our current world cannot be considered sustainable. This is due to the obvious depletion of natural resources, the fact that millions of people are still undernourished and/or do not have access to clean water and adequate sanitation [10, 16]. The literature review demonstrates that there is a huge discrepancy between the theoretical approach to sustainability and the reality [16].

Swart et al. [40] expand on this view by drawing on the fact that the population of the world is expected to double in this century and issues such as poverty, hunger and health conditions call for an urgent transition to a better and more desirable level of sustainability [40]. However, prior to this transition Swart et al. [40] assert that there is a need to understand and acknowledge sustainability problems, and this understanding has to be global. In other words, the value of sustainability needs to be created and promulgated to stakeholders worldwide [16].

Additionally, Lahsen [27] outlines an international research program called Future Earth as a platform to support the transition to a sustainable future. The Future Earth research for global sustainability program stresses that the future of sustainability will consist of better socioeconomic equality, better environmental practices and a prosperous society. To achieve this, Lahsen [27] believes that action-oriented moves to integrate social, environmental and economic aspects need to be made to transform the sustainability agenda into actuality. What is more, the author suggests that a bottom-up movement is needed to structure reform that is appropriate for the current situation so that sustainability can be achieved.

According to Dyllick and Muff [16], on the other hand, the challenges to achieving a sustainable world are enormous and require a change in the society mind-set and a change in the overall rules of business. In other words, the authors suggest that the challenges towards a sustainable structure start from issues such as public wealth, income disparity, unemployment rate, climate change, education to energy efficiency and renewable resources to name a few [16].

Dias et al. [10] believe that the current situation calls for modifying methods in all areas, not only the business environment. An awakening to critical issues, as well as the way that reality is perceived by human beings, needs to be addressed, so that a more sustainable future becomes possible. Bonn and Fisher [3] reiterate that sustainability is a crucial driver of innovation and needs to be integrated at all levels within organisations and society in general. Finally, Kelley and Nahser [24] claim that a sustainable future is likely to eventuate when a change in mind-set occurs so that a social transformation can take place globally and locally.

## 3 Research Methods and Questions

For Creswell [7], the research method is a process that involves data collection, analysis, and interpretation of the results. That is, through the collection of quantitative data – usually via surveys and interviews – and qualitative data – primarily exploratory research – researchers can obtain the answer to the research question as well as obtain a better understanding of the collected data [7].

For the purposes of this study, it has been chosen a combination of quantitative and qualitative inquiry. The quantitative method is used in this study to measure and analyse the primary data drawn from the sample of the web-survey. The primary data was collected via an online survey questionnaire. The qualitative research method was used to analyse the secondary data collected through peer-reviewed journal articles, books and relevant sources in the existing literature.

The chosen research method is based on online surveys questionnaire. Web-based surveys have attracted much attention in recent years since the advent of the Internet, providing a convenient way to disseminate surveys effectively and efficiently [47]. Undoubtedly, the Internet provides a reliable, efficient and less time consuming (economic) means of collecting data through the use of various types of tools.

Online surveys have several advantages, two of which concern time and resources – both of which are reduced, making data gathering process much easier [14]. Taking the same approach, Wright [47] also pointed out that an online survey can reach individuals in distant locations, giving access to a unique population. Moreover, it is convenient as the results of the data collection can be automated reducing time, cost and effort. Furthermore, Dillman et al. [12] suggested that the use of web surveys can increase the response rates.

Although online surveys have many advantages over the traditional surveys methods, there are also some disadvantages associated with online survey research. Researchers should consider, for example, that the survey questions may be susceptible to bias since individuals are required to read and understand the questions by themselves before completing and returning the questionnaire to the researcher.

Therefore, sampling issues are likely to occur in terms of eliciting accurate demographic and characteristics information as well as accuracy in size of the target population and response rates [47]. Additionally, the innovative way of developing a survey provided by a web survey tool can also result in some respondents making survey errors [11]. That is, the level of sophistication of some surveys can make it impossible for some respondents to receive and complete the survey.

Subsequently, Wright [47] state that the disadvantages when pursuing an online survey are mostly based on the fact that researchers ought to face uncertainty regarding the validity of the collected data and sampling issues. What is more, the author reinforces that problems with the survey design, implementation, and evaluations can also be an issue when using an online survey.

Moreover, Dillman et al. [13] and Smyth et al. [39] pointed out that the uncertainty that researchers might face when conducting a web-based survey can occur given the fact that some sectors of the population have no access to the Internet (low coverage), or might have access to the Internet but lack Internet skills. Dillman et al. [13] also assert that the sector of the population that for various reasons cannot participate in web-based surveys may be crucial to the validity of some researches since the composition of this population mainly includes older households with less education and lower incomes.

The primary data collection method used for this research consisted of a number of questions ranked on a Likert scale and distributed through the use of the Internet. The use of the Internet enabled the questionnaires to be sent through social media such as Facebook and LinkedIn, mobile applications such as WhatsApp as well as emails. The survey distribution had nearly 2 months of circulation and resulted in a total of 250 questionnaires being returned.

As a result, a sample of 250 people was used to determine the negative and positive aspects of sustainability awareness in Brazil, and to examine the factors that influence sustainability awareness in Brazil.

In this study, the data is analysed using a quantitative approach. The IBM SPSS Statistics version 24 (Statistical Package for the Social Science) designed by IBM is used to analyse the collected data. Moreover, Factor Analysis and Factor Scores are used to find patterns in the data as well as new factors related to the positive and negative aspects of sustainability in Brazil. In addition, the data analysis is also based on Cronbach's Alpha reliability coefficient to ensure that the data analysed is reliable and that unreliable data is not used when drawing conclusions.

The purpose of this research is to explore the negative and positive aspects of sustainability awareness in the Brazilian higher education as well as the advantages and risks associated with addressing sustainability. Therefore, the primary research question of this study is:

> What are the factors that influence the adoption of sustainability awareness in Brazil in terms of opportunities and challenges?

Following this, the objective of this study is to identify, through an exploratory study, the advantages and risks associated with the adoption of sustainability in

Brazil. By achieving this research objective, it will be possible to understand the extent to which Brazilian people are aware of sustainability, as well as its negative and positive aspects of the adoption of sustainability.

## 4 Results

This research study used both quantitative and qualitative research methods. The quantitative method is used to analyse and quantify primary data collected from the web-based survey designed by Qualtrics software, and further input into SPSS for a statistical analysis. The qualitative method is used to analyse the secondary data such as the relevant literature, journal articles, books and other sources.

The web-based survey was distributed mainly through the use of social media such as Facebook and LinkedIn, as well as through emails and WhatsApp messages. The online survey questionnaire comprised questions regarding sustainability awareness, positive sustainability outcomes and negative aspects of sustainability. In addition, the survey also collected basic demographic information about respondents as well as their opinions about the benefits and risks of pursuing sustainability in Brazil. The survey results were collected from a total number of 250 survey respondents. Of these, 125 were discarded because they were incomplete due to questions being omitted by respondents and/or missing data; therefore, they could not be considered for the analysis. Hence, a total of 125 valid questionnaires were used for the purpose of analysis.

The demographic section comprises questions regarding gender, age, job title, field of study and education in order to determine factors that may influence respondents' answers. This enables the researcher to classify respondents and allows the statistical analysis of data. Therefore, through the demographic questions it is possible to cluster the gathered information based on common variables. In addition, this section was intended to determine the number of hours that respondents usually spend on the computer, on the Internet, on the Internet for social networking, and time spent for emails. The purposes of using the Internet and types of devices used to access the Internet are also discussed. Following, questions regarding sustainability concepts, sustainability awareness as well as advantages and disadvantages of a sustainability approach in Brazil are also examined.

The gender of the research respondents demonstrates that females comprised the majority (62%) of respondents. The male portion is responsible for 38% of the sample. The total number of participants who answered this question was 228 where 141 are female and 87 are male participants (see Fig. 2).

The majority of the respondents were aged between 25 and 30 years, accounting for 33.91% of the total number of respondents. The second biggest group comprising 27.04%; was aged between 31 and 35 years 11.16% were aged between 21 and 24 years; 10.30% were in the 36–40 years age group; followed by an equal number of participants aged in the group of 18–20 and 36–40 and, finally

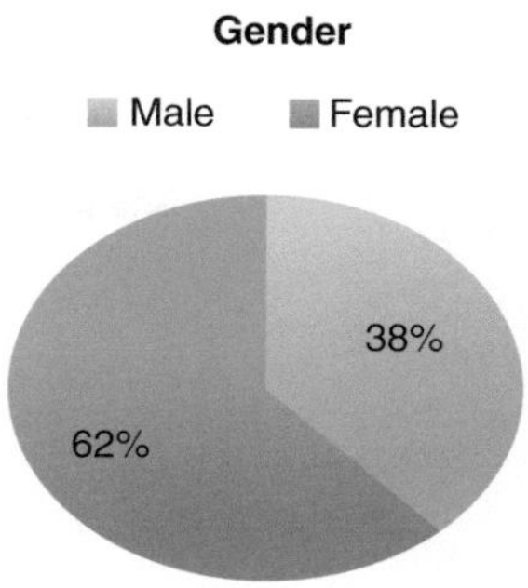


**Fig. 2** Gender of survey responses

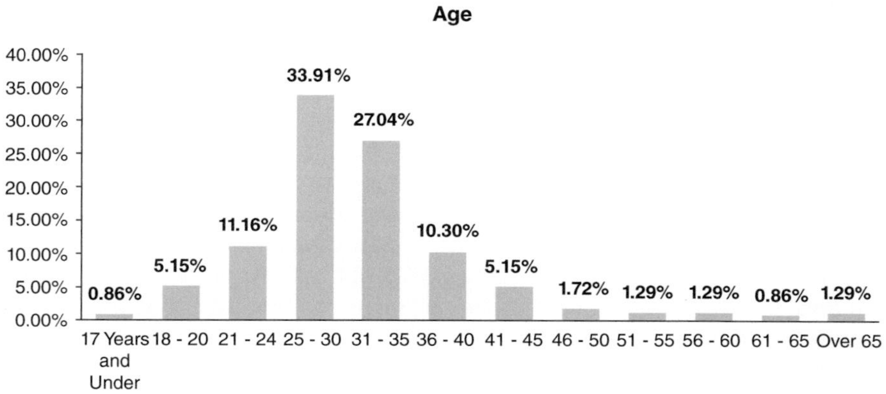


**Fig. 3** Age of survey responses

a minority of respondents were in groups aged 17 years and under, 46–50 years old, 51–55 years old, 56–60 years old and over 65 years old. As can be observed, the majority of the participants were aged between 21 and 40 years (see Fig. 3).

Two hundred and twenty-two respondents answered the question pertaining to job title. The research sample has shown that respondents were from different areas and held different positions as follows: Students, Interns, Professors, Teachers, Managers from different areas within the organisations, Directors from various areas of the business, Company's Vice-President, Business Associates, Veterinary, Dog Trainer, Geographers, Geologists, Geophysicist, Biologists, Architects, Artist, Chefs, Engineers, Dentist, Journalists, Pharmacists, Massage Therapist, Physiotherapist, Accountants, Hairdressers, Designers, Digital Marketers, Publicists, Consultants from various areas of the business, Business Coordinators, Business Supervisors, Business Analysts, Business Assistants, Entrepreneurs, Medical Doctor, Software Developer, Lawyers, Travel Agent, Hospitality Staff, Mountain Guide, Mailman, Drivers, Nurses, Retired, Unemployed and Stay-at-home mother.

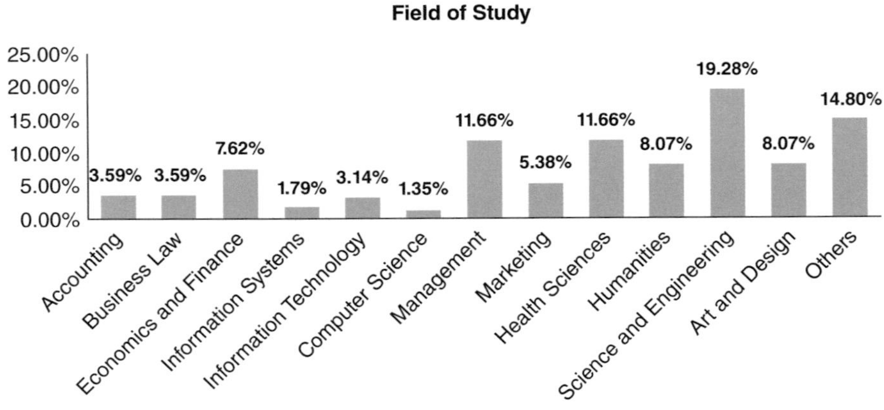


**Fig. 4** Field of study survey responses

The majority of the survey respondents have a Science and Engineering background (19.28%). The same number related to Business and Management (11.66%) and Health Sciences (11.66%), then Humanities (8.07%), Arts and Design (8.07%), Economics and Finance (7.62%), Marketing (5.38%), an equal amount of backgrounds in Accounting (3.59%) and Business Law (3.59%), Information Technology (3.14%) and a small number had backgrounds in Information Systems (1.79%) and Computer Science (1.35%). In addition, the 'others' option was selected by some respondents who did not know where their field of study belonged to any of the groups stated in the question. That is, some respondents who chose the 'others' option had backgrounds in Business Administration, Engineering, Law Film and Television. Apart from that, the 'others' option consisted of respondents that had backgrounds in the following fields of study: Philosophy, Architecture, Construction, Environmental Science, Tourism, Hospitality, Health and Safety, Education, Medicine, Fashion, Social and Public Communication, Physical Education, Hydrogeology, and Teaching and Education (see Fig. 4).

The education status statistics shows that 31.84% of the respondents hold a Bachelor's Degree, followed by 21.97% with a Post Graduate Diploma, 12.11% are Graduates and 11.66% have a Master's Degree. Very few (4.93%) participants hold a PhD and a few participants fall between the Primary Education and Diploma levels. Furthermore, the 'others' option was selected by respondents who did not know the category to which their studies belonged (see Fig. 5).

From the survey results, it is evident that the majority of respondents, 37.50% of the sample, spend from 5 to 10 h per day on computers. 31.25% of the respondents, on the other hand, spend up to 5 h; 15.18% spend less than an hour; 14.29% spend 10–20 h on the computer per day and 1.79% of respondents spend over 20 h on the computer per day (see Fig. 6).

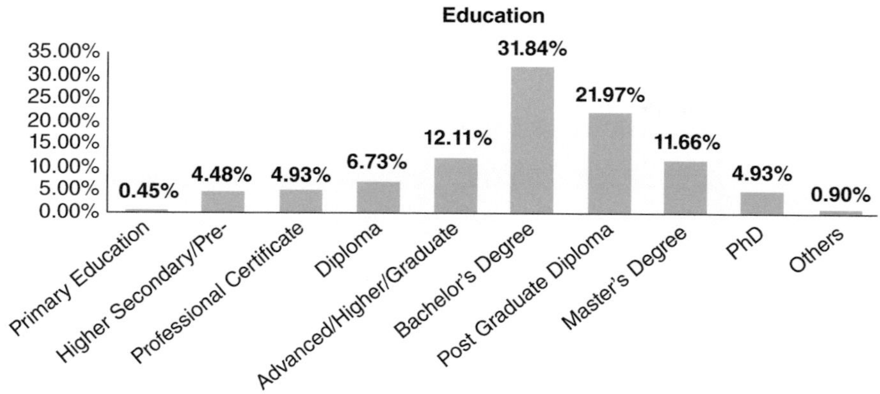


**Fig. 5** Education survey responses

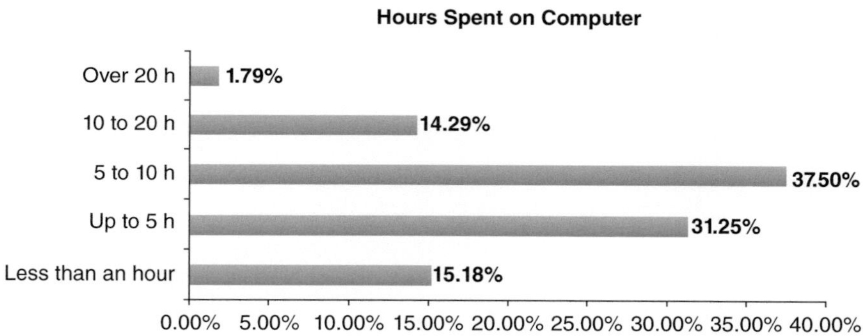


**Fig. 6** Hours spent on computers survey responses

Considering the survey results regarding the average hours spent on the Internet per day, it is possible to conclude that the vast majority (41.96%) of respondents spend up to 5 h on the Internet. 36.16% of the sample, on the other hand, spend between 5 and 10 h on the Internet followed by 13.84% of the participants who spend 10–20 h on the Internet. Few respondents (5.80%) spend less than an hour on the Internet and 2.23% of the sample stated that they spend over 20 h per day on the Internet (see Fig. 7).

Per day, 49.10% of the survey respondents spend up to 5 h on the Internet for social networking. Another 41.44% of the respondents spend less than an hour on the Internet for social networking followed by 5.86% that spend 5–10 h, 3.25% that spend 10–20 h and a very small number (0.45%) spend over 20 h on the Internet for social networking (see Fig. 8).

According to the survey responses, over half (56.32%) of the respondents spend less than an hour per day on the Internet for email. This is followed by

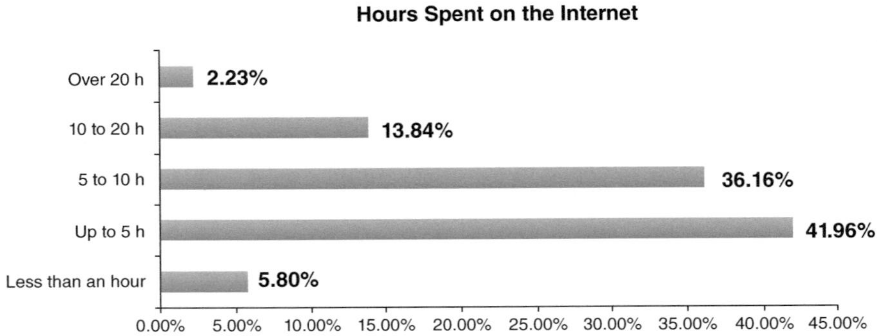


**Fig. 7** Hours spent on the Internet survey responses

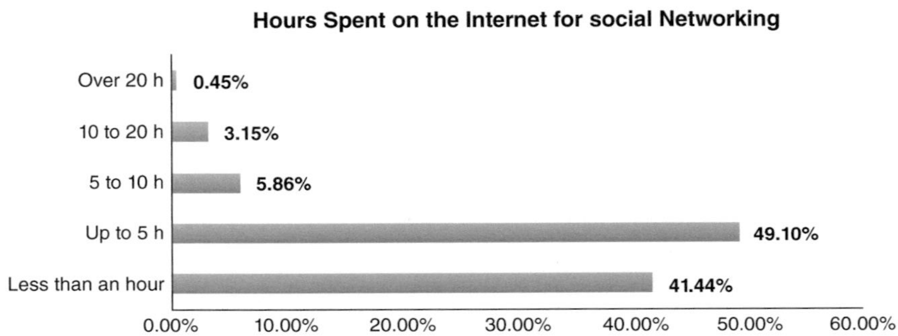


**Fig. 8** Hours spent on the Internet for social networking survey responses

32.43% of the respondents who spend up to 5 h on the Internet. 9.46% of the survey respondents spend 5–10 hours checking emails and 1.80% spend 10–20 on the Internet for emails. None of the respondents spends over 20 h on the Internet for email (see Fig. 9).

The purpose of the use of the Internet question allowed participants to choose more than one option if this applied to them.

From the results, it is possible to cluster six specific groups that consisted of 50% or more of respondents who use the Internet for the purpose of email, working, studying, online banking, chatting and researching travel information or making reservation. Consequently, 84.68%, or in other words, the vast majority of the total of survey respondents use the Internet for email. The second largest group comprises 78.38% of respondents who use the Internet for working purposes followed by the third largest group of 73.42% who use the Internet for studying. The fourth largest group comprises 63.51% of the respondents who use the Internet for online banking and another 54.50% of the respondents use the Internet for chatting, comprising the fifth largest group. Moreover, 50.90% of respondents claimed to use the Internet for researching travel information or making reservations. 46.40% of the survey responses demonstrate that shopping online is also one of

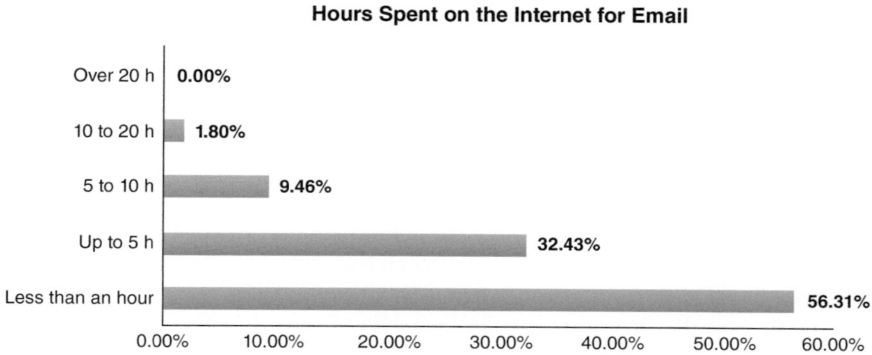


**Fig. 9** Hours spent on the Internet for email survey responses

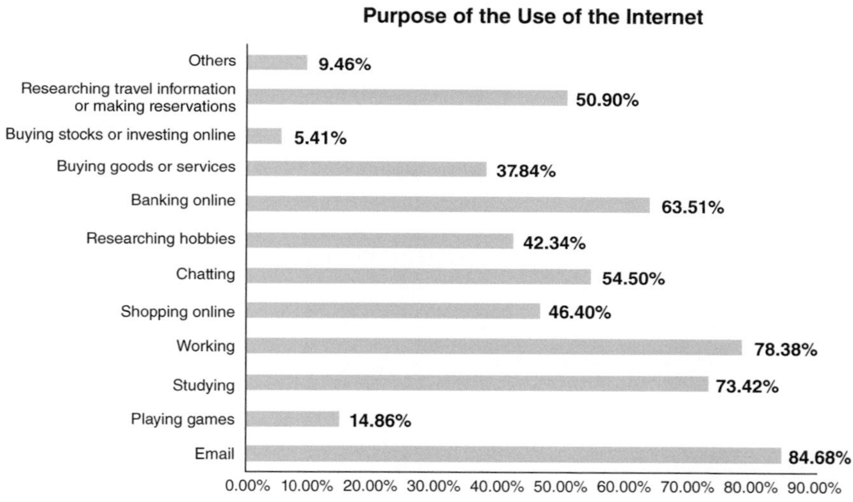


**Fig. 10** Purpose of the use of the Internet survey responses

the reasons for using the Internet, accounting for 46.40% of the total sample. Moreover, the Internet is also used by respondents for researching hobbies (42.34%), buying goods or services (37.84%), playing games (14.86%) and buying stocks or investing online (5.41%).

The ‘others’ option included reasons such as: watching TV series/shows and movies/Netflix, watching videos and tutorials, watching sports games, reading books, reading the news and watching the stock exchange (see Fig. 10).

From the chart below, it is possible to conclude that the vast majority (92.79%) of the survey responses suggest that smartphones are widely used by respondents to access the Internet. 72.52% of the respondents use laptops to access the Internet, followed by 28.38% who use tablets, 22.97% use PCs, 21.17% a Desktop, 6.76% a Netbook, 4.95% a Workstation and 0.90% PDAS.

In addition, the 'others' option drew responses about the use of Smart TV although respondents might have been confused about the classification of their devices (see Fig. 11).

From the pie chart below, it can be said that the majority of respondents were first introduced to the concept of sustainability via the Internet, comprising the highest percentage of 26% of the respondents. The second largest group is school with 21% followed by higher education 16%, news and media 14%, magazine 9%, books 7% and conferences 5%. The 'others' option refers to work environment, family and friends. Furthermore, two respondents state that they have never been introduced to the concept of sustainability (see Fig. 12).

According to the survey, the majority (46%) of respondents are Apple's users followed by Google (32%), Dell 11% and IBM 1% users. In addition, the 'others'

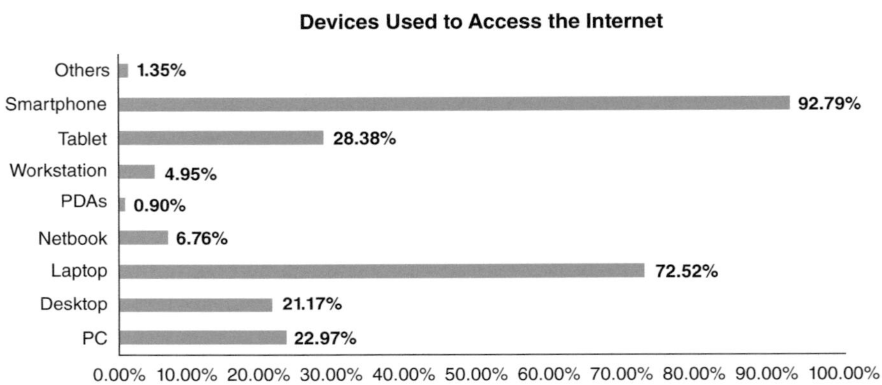


**Fig. 11** Devices used to access the Internet survey responses

**Fig. 12** First introduction to the concepts of sustainability survey responses

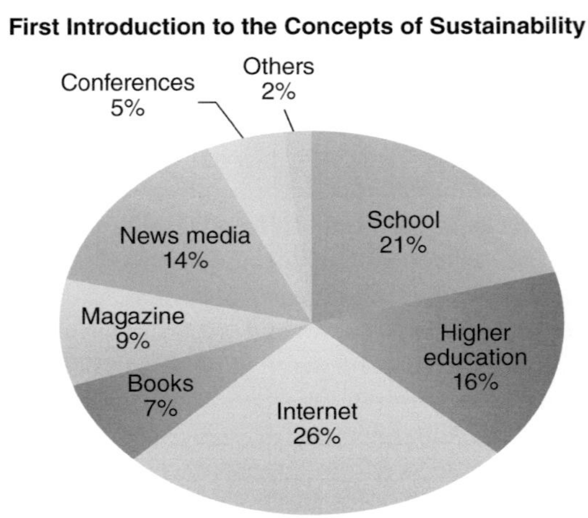

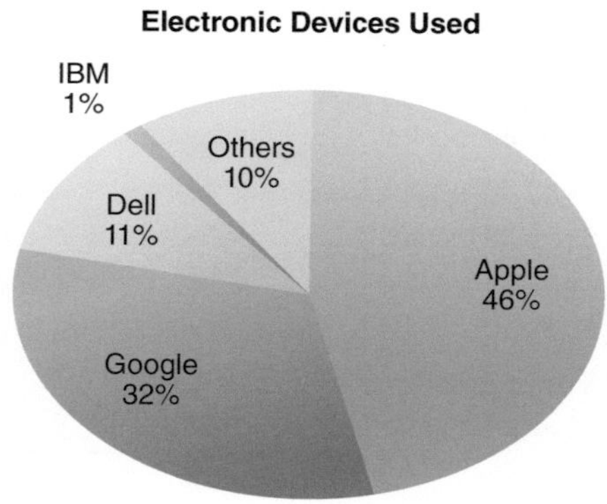


**Fig. 13** Electronic devices used survey responses

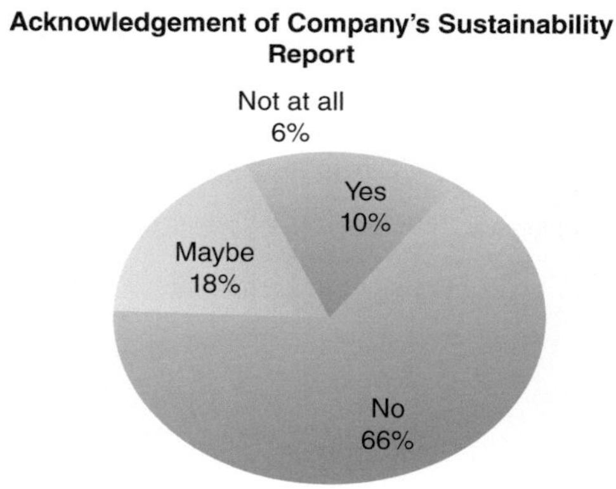


**Fig. 14** Acknowledgement of company's sustainability report survey responses

option was chosen by respondents who use brands such as Samsung (most of the responses), Lenovo, Asus, HP, Motorola, Nokia, and Itautec which is a Brazilian computer company (see Fig. 13).

It is clear that more than half (66%) of the respondents do not check the company's sustainability report before buying a new device. 18% of the respondents said 'maybe', 10% read the company's sustainability report and 6% do not read it at all (see Fig. 14).

The bar chart below shows that 25.91% of the respondents change their devices every 2 years (24 months), followed by 16.82% of the sample that changes their devices every 36 months. 14.09% of the respondents pointed out through 'other' option that they change their devices only when the device breaks down or suddenly stops working and cannot be fixed. The same number of respondents (10.91%) said that they change their devices every 42 and 30 months. This was followed by 10% of respondents who change devices every 18 months, 7.73% every 12 months and another 3.64% of the respondents change devices every 6 months (see Fig. 15).

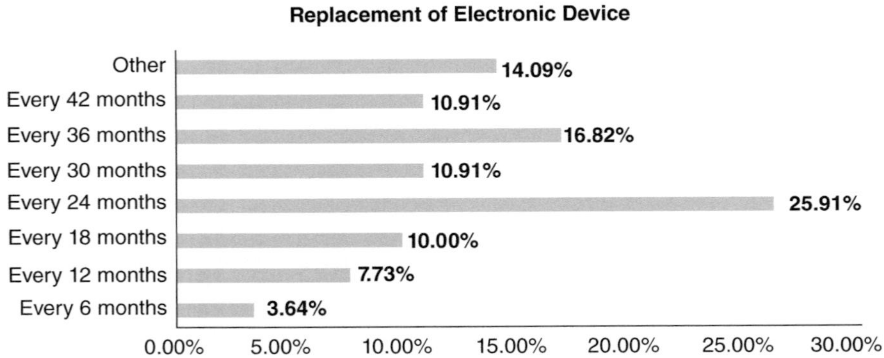


**Fig. 15** Replacement of electronic devices survey responses

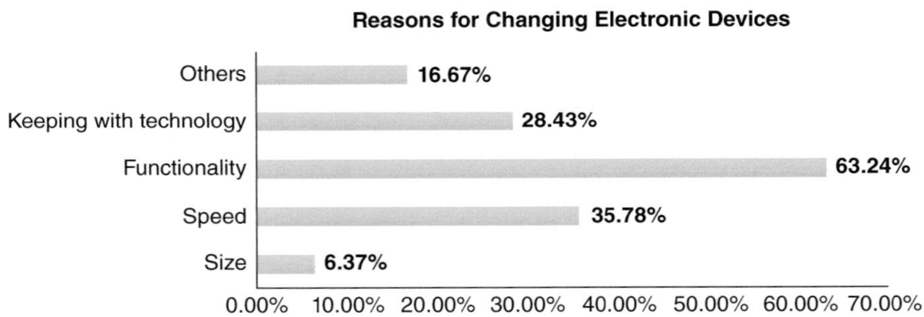


**Fig. 16** Reasons for changing electronic devices survey responses

Respondents were allowed to choose more than one option for this following question.

From the following chart, it can be seen that the majority (63.24%) of respondents change their electronic devices for reasons of functionality. 35.78% of respondents change devices because of speed, 28.43% because they want to keep up with technology, and 6.37% because of the size of the electronic device. For the 'other' option, 16.67% of the respondents reported that they change their devices only when the devices are stolen (Brazilian pick pocketing issues), or the devices are broken, malfunctioning, lost, have poor battery life, or when the contract agreement with telecommunication company expires see (Fig. 16).

Results show that the majority (69%) of respondents stated that the constant changing of devices will damage our planet. This number is followed by 24% of the respondents who think that perhaps the frequent changes of devices will damage our planet. However, 5% believe that the changing of devices will not damage our planet and 2% think that it will not produce any damage at all (see Fig. 17).

According to the results below, the vast majority (82.27%) of respondents believe that it is through education that a sustainability mind-set can be embedded. 57.14% of the respondents think that the Internet can be a useful tool for changing people's mind-set toward sustainable practices, followed by 55.67% of the respondents who believe that this occurs through social networking and 51.72% of respondents believe that this can happen through awareness. In addition, 32.02% believe that it can happen through the use of TV and training, and another 23.65% believes in workshops.

The 'others' option produced comments from respondents such as 'need to require companies to act sustainable', 'the industry should present better products' and 'the not being sustainable problem consist in a business problem where companies want to make a profit and therefore will manipulate buyers' (see Fig. 18).

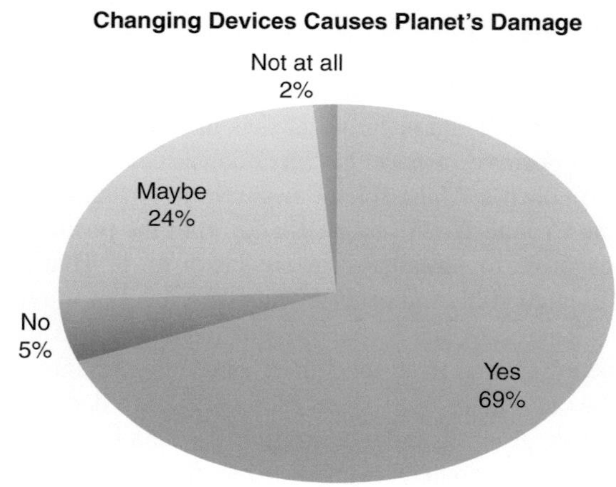


**Fig. 17** Changing devices causes planet's damage survey responses

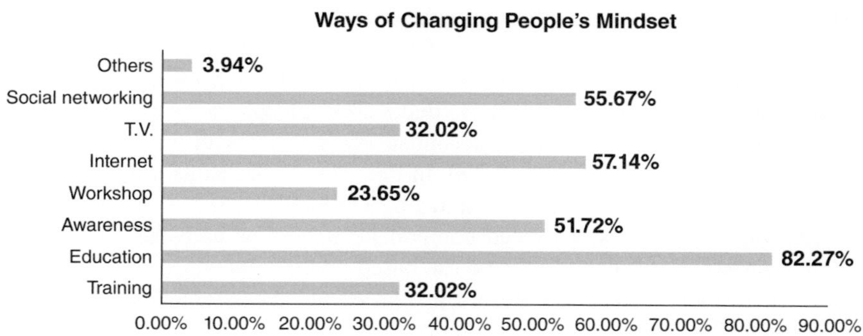


**Fig. 18** Ways of changing people's mind-set survey responses

### 4.1 Positive and Negative Aspects Associate With the Adoption of Sustainability

As stated previously, this study aimed to identify the positive and negative aspects of adopting sustainability in Brazil. Therefore, the web-based survey questionnaire had 24 items regarding the positive aspects of addressing sustainability in Brazil. Moreover, a total of 13 questions were asked in order to determine the perceived risks associated with the adoption of sustainability in Brazil. The respondents indicated their level of agreement with sustainability issues by means of a Likert Scale.

Several survey responses could not be considered because some answers were incomplete or omitted. Hence, with SPSS software, a listwise deletion was used to exclude cases not valid, and Cronbach's Alpha test (test consistency reliability) was run to test the level of validity and accuracy of the results.

Regarding the positive aspects of sustainability in Brazil, from a total of 125 respondents, 13 questionnaires were discarded because of missing data, leaving 112 useful surveys while the negative aspects associate with sustainability from the total of 125 cases, 8 were excluded and 117 were valid.

The level of validity and accuracy of both results, positive and negative aspects, indicate that this research study has excellent reliability in terms of the analysis of the positive and negative aspects of sustainability in Brazil. The Cronbach's Alpha for positive aspect is very consistent with an accuracy of 0.917 and the Cronbach's Alpha for the negative to sustainability is equal to 0.940. Equally so, the Cronbach's Alpha results for both the positive aspects of sustainability and the negative to sustainability are close to 1. That is, the reliability of the present research study indicates an excellent level of internal consistency of responses.

Therefore, the factor analysis technique was used to cluster and summarise the data gathered from the survey based on variances. Factor analysis in research studies is widely used when there is a large data set consisting of several variables. When grouped and categorised, the data can be better understood and interpreted by the researcher [48]. Subsequently, the variables that have high correlation are also known as factors [25].

To check whether the studied sample is satisfactory, the Kaiser-Meyer-Olkin (KMO) – that measure of sampling adequacy value – and Bartlett's Test of sphericity – when is less than 0.05 of the significance indicates that the factor analysis is useful and adequate – were run. The result for the positive aspects of sustainability in Brazil is equal to 0.843 with a Bartlett's Test of sphericity value is 0.000. Subsequently, the KMO for risks to sustainability is 0.909 with the Bartlett's Test of sphericity value again equal to 0.000. In this case, both results indicates that in this case the sample size is superb as stated by Field [19].

Based on these results, a rotated component matrix for both the positive and negative aspects of sustainability were run based on three factors.

The three highest variables present in the second factor are the reduction of emissions (0.900), pollutions (0.868) and carbon footprint (0.800). Hence, it can be concluded that the majority of Brazilian respondents acknowledge that by addressing sustainability awareness, this will reduce Brazilian's emissions, pollution and carbon footprint.

In addition, the two highest variables present in the first factor are increase productivity (0.742) and reduce risk management (0.723) what indicates that the respondents believe that sustainability awareness is likely to increase productivity and reduce risk management. Subsequently, the third factor demonstrates that the reduction of energy and water usage (0.791), as well as the reduction of paper usage (0.778), can also play an important role in sustainability awareness in Brazil (see Table 1).

From the rotated component matrix results listed above, it is possible to combine and cluster the various variables into groups.

**Table 1** Rotated factor matrix for positive aspects to sustainability

| Rotated component matrix[a] | | | |
|---|---|---|---|
| | Component | | |
| | 1 | 2 | 3 |
| Increase productivity | 0.742 | | 0.124 |
| Reduce risk management | 0.723 | 0.154 | 0.176 |
| Attract quality employees | 0.682 | 0.181 | 0.105 |
| Satisfy customer needs | 0.653 | | |
| Meet stakeholder expectations | 0.630 | | |
| Increase efficiency | 0.622 | 0.113 | 0.234 |
| Improve community investments | 0.618 | 0.467 | |
| Improve human rights | 0.588 | 0.457 | |
| Attract new opportunities | 0.568 | 0.216 | 0.151 |
| Improve social responsibility investing | 0.553 | 0.312 | 0.232 |
| Create new jobs | 0.548 | 0.388 | |
| Increase cost-effectiveness | 0.546 | | 0.434 |
| Differentiate businesses | 0.523 | 0.175 | 0.220 |
| Increase triple bottom line – People, Planet and Profit | 0.450 | 0.181 | 0.307 |
| Reduce emissions | 0.121 | 0.900 | 0.199 |
| Reduce pollution | 0.130 | 0.868 | 0.210 |
| Reduce carbon footprint | | 0.800 | 0.325 |
| Reduce health hazards | 0.217 | 0.772 | 0.140 |
| Increase green strategy | 0.229 | 0.614 | 0.186 |
| Improve corporate social responsibility | 0.195 | 0.444 | 0.440 |
| Reduce energy and water usage | | 0.248 | 0.791 |
| Reduce paper usage | | 0.210 | 0.778 |
| Enhance reputation | 0.412 | 0.141 | 0.470 |
| Reduce consumption of raw materials | 0.242 | 0.193 | 0.440 |

Extraction Method: Principal Component Analysis.
Rotation Method: Varimax with Kaiser Normalization.
[a]Rotation converged in 6 iterations.

**Table 2** Factor number 1 for positive to sustainability

| Variables | Factor Loading | Positive Factor Label 1 |
|---|---|---|
| Increase productivity | 0.742 | Enhancement of Business Asset |
| Reduce risk management | 0.723 | |
| Attract quality employees | 0.682 | |
| Satisfy customer needs | 0.653 | |
| Meet stakeholder expectations | 0.630 | |
| Increase efficiency | 0.622 | |
| Improve community investments | 0618 | |

**Table 3** Factor number 2 for positive to sustainability

| Variables | Factor Loading | Positive Factor Label 2 |
|---|---|---|
| Reduce emissions | 0.900 | 4 Rs |
| Reduce pollution | 0.868 | |
| Reduce carbon footprint | 0.800 | |
| Reduce health hazards | 0.772 | |
| Increase green strategy | 0.614 | |

**Table 4** Factor number 3 for opportunities to sustainability

| Variable | Factor Loading | Positive Factor Label 3 |
|---|---|---|
| Reduce energy and water usuage | 0.791 | Reduce Raw Material Consumption |
| Reduce paper usage | 0.778 | |

Factor 1 consists of the following variables: increase productivity; reduce risk management; attract quality employees; satisfy customer needs, meet stakeholder expectations; increase efficiency and improve community investments. These variables have been named as enhancement of business assets since they all are related to improving business advantages by addressing sustainability awareness (see Table 2).

Factor 2 comprises the reduction of emissions; pollution; carbon footprint and health hazards, and an increase of green strategies. All variables presented can be grouped as reduction. Therefore, factor 2 is related to the 4Rs (see Table 3).

Factor 3, on the other hand, comprises the reduction of energy, water and paper usage. Hence, these variables receive the label of reduction of raw material consumption (see Table 4).

The positive aspects of sustainability awareness in Brazil are depicted in the following figure (see Fig. 19).

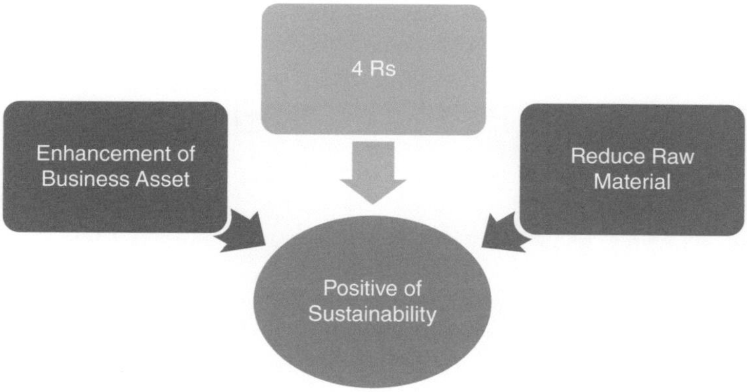


**Fig. 19** Positive aspects of sustainability awareness in Brazil

The rotated factor matrix for risks to sustainability awareness in Brazil was also based on the components of three factors.

The first factor comprises five variables while the second factor has six variables. The third factor, however, has two variables.

The highest variable is presented in the first factor as increase marketing failure by perceived environmental irresponsibility (0.873). The second highest variable is present in the third factor and comprises the increasing number of new regulations including environmental (0.869). The third highest variable is also present in the first factor and comprises the increase in scandals due to perceived environmental irresponsibility (0.864). Therefore, the highest variables are present in the first and third factor analysis. Thus, it can be said that the majority of Brazilian respondents believe that the risks to sustainability awareness in Brazil comprise the increase of marketing failure, increase in the number of new regulations and scandals associated with environmental negligence.

In addition, the two highest variables presented in the second factor are the increase in the interest rates (0.743) and the increase in costs (0.721). Hence, this indicates that the Brazilian respondents believe that the risks of addressing sustainability awareness in Brazil are based on the increase of rates and costs (see Table 5).

From the rotated component matrix demonstrated in the previous section about the risks to sustainability awareness in Brazil, three factors comprising risks to sustainability can be conceptualised.

Factor 1 consists of the following variables: increase of marketing failure resulting from perceived environmental irresponsibility; increase of scandals due to perceived environmental irresponsibility, increase of security and systems failures caused by environmental problems and increase of supply chain crises due to suppliers' environmental problems (see Table 6).

Hence, these variables have been named as increase in management failure since the variables presented assume that risks are due to managerial failure.

**Table 5** Rotated factor matrix for risks to sustainability

| Rotated component matrix[a] | Component | | |
|---|---|---|---|
| | 1 | 2 | 3 |
| Increase marketing failure by perceived environmental irresponsibility | 0.873 | 0.188 | 0.134 |
| Increase scandals by perceived environmental irresponsibility | 0.864 | 0.204 | 0.219 |
| Increase security and systems failures caused by environmental problems | 0.712 | 0.451 | 0.159 |
| Increase supply chain crises due to suppliers' environmental problems | 0.690 | 0.327 | 0.383 |
| Increase Interest rates | 0.283 | 0.743 | 0.341 |
| Inflate costs | | 0.721 | 0.276 |
| Increase litigation and compliance breaches (including environmental actions and environmental compliance failures) | 0.561 | 0.716 | |
| Increase governance failure (including environmental actions and Environmental compliance failures) | 0.560 | 0.688 | 0.124 |
| Increase fraud (including environmental actions and environmental compliance failures) | 0.583 | 0.670 | |
| Increase transaction failure due to environmental liabilities | 0.613 | 0.652 | |
| Increase insurance crises due to environmental disasters | 0.446 | 0.598 | 0.309 |
| Increase number of new regulations including environmental | 0.190 | | 0.869 |
| Increase competition for and cost of raw materials | 0.138 | 0.418 | 0.731 |

Extraction method: Principal Component Analysis.
Rotation Method: Varimax with Kaiser Normalization.
[a]Rotation converged in 8 iterations

**Table 6** Factor number 1 for risks to sustainability

| Variables | Factor Loading | Risks Factor Label 1 |
|---|---|---|
| Increase in marketing failure by perceived environmental irresponsibility | 0.873 | Increase in Management Failure |
| Increase in scandals by perceived environmental irresponsibility | 0.864 | |
| Increase in security and systems failures caused by environmental problems | 0.712 | |
| Increase in supply chain crises due to suppliers' environmental problems | 0.690 | |

Factor 2 relates to the increase of rates and costs, increase of litigation and compliance, increase government failure and fraud, increase of environmental actions and environmental compliance failures as well as increase of transaction failures due to environmental liabilities. Therefore, factor 2 is labelled as increase in costs, compliances and regulations (see Table 7).

**Table 7** Factor number 2 for risks to sustainability

| Variables | Factor Loading | Risks Factor Label 2 |
|---|---|---|
| Increased interest rates | 0.743 | Increased Costs, Compliances and Regulations |
| Inflated costs | 0.721 | |
| Increased litigation and compliance breaches (including environmental actions and environmental compliance failures) | 0.716 | |
| Increased governance failure (including environmental actions and environmental compliance failure) | 0.688 | |
| Increased fraud (Including environmental actions and environmental compliance failures) | 0.670 | |
| Increased transaction failures due to environmental liabilities | 0.652 | |

**Table 8** Factor number 3 for risks to sustainability

| Variables | Factor Loading | Risks Factor Label 3 |
|---|---|---|
| Increased number of new regulations including environmental | 0.869 | Increased Procedures and Competition |
| Increased competition for and cost of raw materials | 0.731 | |

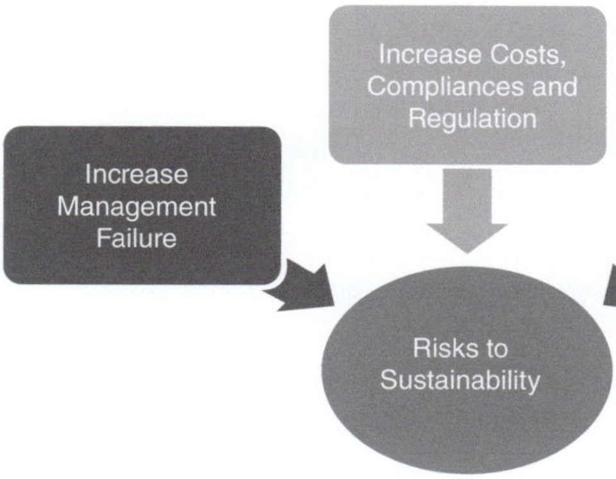


**Fig. 20** Risks to sustainability awareness in Brazil

Factor 3 pertains to the increase of new regulations including those for environment protection and an increase of the competition for and cost of raw material. For this reason, the third factor is labelled as increased procedures and competition (see Table 8).

Therefore, the risks to sustainability awareness in Brazil are depicted below (see Fig. 20).

## 5 Discussion and New Significance

Sustainability has attracted a lot of attention during the past few years. It is now a buzzword not only on the United Nations agenda, but also in every society and in every organisation. This is in part because of the direct relationship between sustainability and the environment. Therefore, concepts such as the Triple Bottom Line (people, planet and profit) and the three sustainability pillars (social, economic and environmental) play a crucial role in today's world. Following this line of reasoning, sustainability and sustainable development play – or should play – an essential role in safeguarding a decent future for the next generation. In other words, it is about striking a balance between economic growth and sustainable development. Moreover, when aligned, these two factors can foster an ongoing development that is significant for all society. In such a manner, scholars believe that the interconnection between growth, development and sustainability is due to human activities, the natural system and the environmental impacts of individual behaviours, and this link has become clearer over time.

As a result, several research studies have been undertaken on the subject of sustainability in Brazil, but little significant attention was given to the positive aspects of sustainability in Brazil and the risks of sustainability in Brazil. In other words, hitherto little has been revealed about the extent of sustainability awareness in Brazil.

This particular study, therefore, intends to ascertain the awareness of Brazilians regarding the positive and negative (risks) aspects of sustainability.

Brazil is believed to be an important driver in the global economy given that it has the fastest growing economy. Brazil's natural resources and commodity export trades make it necessary to study sustainability awareness in Brazil. Hence, this study's findings are likely to assist Brazilian society as a whole to become aware of sustainability, thereby assisting all Brazilian stakeholders, all sorts of businesses, and government bodies, to make better decisions.

The research objective of this study is:

> To identify, through an exploratory study, the advantages and risks associated with the adoption of sustainability in Brazil.

To achieve this objective, an extensive literature review was conducted of reliable sources, and a web-based survey was designed and distributed to Brazilian people. The results of the data analysis revealed three positive aspects of sustainability in Brazil as well as three risks of addressing sustainability in Brazil.

The three positive aspects to sustainability in Brazil are: enhancement of business assets; 4Rs, and reduced consumption of raw materials. On the other hand, sustainability in Brazil can pose three risks such as the increase of management failure; increase in costs, compliance, regulations and procedures, and increased competition which are likely to focus business attention on the mitigation of those factors. In saying this, the awareness of sustainability needs to be better understood and strengthened in society and business in general.

The research question was successfully answered. That is, the aim of this study was to identify the positive aspects of sustainability in Brazil and the possible risks associated with sustainability in Brazil. Three major factors for each of the two variables (positive aspects and negative/risks) were generated through the analysis of the web-based survey.

The sampling adequacy tests used for this study were the KMO and Bartlett's test which were applied to verify the validity of the survey results. The values obtained were satisfactory being 0.843 for the positive aspect of sustainability in Brazil and 0.909 for risks to sustainability in Brazil. Thus, both sample sizes were adequate, useful and indicated an outstanding sample analysis.

In addition, this study discovered that the majority of Brazilians spend about 5 hours browsing the Internet, using the computer as well as surfing on the Internet for social networking. What is more, Brazilian people are likely to prefer the use of devices such as smartphones and laptops over other electronic devices available in the market.

It is also asserted by this study that the Internet plays a substantial role in disseminating the sustainability concept. It appeared that most Brazilians first encountered the concept of sustainability through the Internet and academic life. Another significant finding from this study is that most Brazilians do not read a company's sustainability report prior to buying electronic devices. Moreover, most Brazilian people replace their devices every 2 years due to functionality, speed and the desire to keep up with technology. However, more than half of the respondents believe that the constant changing of electronic device will jeopardise our planet.

This study has shown that 82.27% of the Brazilian respondents agree that it is through education that a new mind-set towards sustainable development can be created and encouraged. Additionally, factors such as the Internet, social networks and awareness are also believed by the Brazilians to be effective and efficient means of changing people's mind-sets.

## 6 Limitation

This exploratory study had several limitations. Three important factors limited the scope of this study. First, this study had time constraints. The survey was distributed and circulated for approximately 2 months and took about 10 min to complete.

Second, several incomplete surveys were returned. That is, several respondents were unable to complete the whole survey. Some of the participants claimed that some questions were not very clear, perhaps due to their lack of proficiency in the English language or other language barriers. Moreover, some of them complained that the online survey was difficult to access from smartphones, especially in regards to the last part of the questionnaire that seemed to not have the right fit on the mobile screen.

Third, the sample size also appeared to be one of the limitations. Although this study comprised 250 respondents in total, the results of this survey cannot be considered to represent the awareness of all Brazilian people towards sustainability and its positive and negative aspects.

Another important factor regarding the sample size is that the majority of the survey respondents were in 25–30 years age group which means that the findings mainly represent this particular group. Albeit, this particular research study consisted of a wide-ranging group of participants regarding age, field of study and job position.

## 7 Future Research

This exploratory study identified significant factors regarding sustainability awareness in Brazil. The positive aspects of sustainability, as well as its associated risks, were revealed through a comprehensive online research based on the quantitative method. Therefore, further research on sustainability awareness in other countries could be undertaken as a comparative study so that the findings can be generalised. In addition, this study methodology has mainly used a quantitative method, whereas a mixed-methods approach could be taken by addressing qualitative data gathering such as interviews.

## 8 Recommendations

As mentioned earlier, this exploratory study revealed three positive factors and three risks associated with sustainability awareness in Brazil. Based on these findings, the researcher strongly believes that any business or organisation, and society in general, should consider encouraging an awareness of sustainability and promote sustainability so that a better future can be assured.

Clearly, caution should be exercised to eliminate and mitigate possible risks that sustainability can pose to business and society. Thus, sustainability awareness can on the one hand increase the advantages of pursuing sustainable development, and on the other hand mitigate risks by developing a contingency plan to anticipate and deal with those risks. The researcher's recommendations are as follows.

- The concept of sustainability should be broadcast. Our planet is evidently undergoing some drastic changes. Therefore, people ought to be provided with an awareness of and knowledge about the sustainability concept and sustainable development. People ought to be made aware of the impact that human activities have on the environment, ecosystem and on the development of mankind in general. Therefore, the education system has a crucial role to play in raising sustainability-aware young people who will soon be future leaders. Sustainability should be included in every school's curriculum so that a proper mind-set can be established. Education and knowledge are powerful tools for building a better future.

- The threat that the world faces should be emphasised and discussed more often through the various communication channels. Marketing and publicity combined with the dissemination of knowledge will result in more people becoming involved in the sustainable development cause.
- Business of any kind should be less greedy. Today's greed and selfishness is senseless, particularly in view of the fact that they will jeopardise the future. The effects of today's irresponsibility will still be felt after the current generation has disappeared. Hence, future generations must be given due consideration right now in the present.
- Organisations should encourage employees to embrace sustainable development. Hence, organisations should provide training and establish policies and goals to ensure the wholehearted commitment of their employees to a common vision and work that is aligned with company objectives. Therefore, CSR will become more evident and persuasive.
- Finally, the government and public organisations should be leading by example and simultaneously encouraging people to be good citizens.

## 9 Conclusion

This exploratory study reviewed existing literature on sustainability and sustainable development as well accentuated the sustainability scenario in developing countries with an emphasis in the positive and negative sustainability aspects in Brazil. To reiterate, the path to achieving sustainability requires the preservation of the environment comprising the conscious use of limited natural resources, economic growth, including companies' generation of profit, and well-being rights to live in a pleasant world. Therefore, the idea of safeguarding the planet for the next generation is at the core of sustainable development.

As pointed out in the beginning of this study, the ever-growing number of studies that are promoting sustainability and reinforcing the idea of sustainable development that can meet the needs of the present without compromising the ability of future generations to satisfy their own needs, has rapidly engaged the world's concern. Thus, the lack of understanding of the sustainability concept might jeopardise the future of mankind.

The findings of this study revealed responses to survey questions pertaining to respondents' awareness of sustainability, the channel(s) through which they became aware of this concept, the positive aspects as well as the risks associated with sustainability in Brazil. The demographic results demonstrated the distribution of the respondents in terms of gender, age, job title, field of study and education. Then, three positive aspects of sustainability and three risks associated with sustainability awareness in Brazil. The positive aspects of sustainability are the enhancement of business assets, 4Rs, and reduced consumption of raw materials. The risks of sustainability awareness in Brazil are the increase of management failure, the increase of costs, compliance and regulations as well as the increase in procedures and competition.

It is also asserted by this study that the Internet plays a substantial role in disseminating the sustainability concept. It appeared that most Brazilians first encountered the concept of sustainability through the Internet and academic life. Another significant finding from this study is that most Brazilians do not read a company's sustainability report prior to buying electronic devices. Moreover, most Brazilian people replace their devices every 2 years due to functionality, speed and the desire to keep up with technology. However, more than half of the respondents believe that the constant changing of electronic device will jeopardise our planet.

This study has shown that 82.27% of the Brazilian respondents agree that it is through education that a new mind-set towards sustainable development can be created and encouraged. Additionally, factors such as the Internet, social networks and awareness are also believed by the Brazilians to be effective and efficient means of changing people's mind-sets.

Although the findings indicated that even though sustainability awareness in Brazil presents some risks, there is an urgent need to address sustainable development because the advantages of sustainability awareness in Brazil are greater than any potential risks. The results clearly show that Brazilians are aware of and concerned about the sustainability dilemma and they believe that education can exert a major influence in changing people's mind-set regarding a sustainable world.

This exploratory study had several limitations. Three important factors limited the scope of this study. That is, this study had time constraints, several incomplete surveys were returned, and although this study comprised 250 respondents in total, the results of this survey cannot be considered to represent the awareness of all Brazilian people towards sustainability and its positive and negative aspects.

Further research on sustainability awareness in other countries could be undertaken as a comparative study so that the findings can be generalised and extended. In addition, this study methodology has mainly used a quantitative method, whereas a mixed-methods approach could be taken by addressing qualitative data gathering such as interviews.

## References

1. Barbosa L, Veloso L (2014) Consumption, domestic life and sustainability in Brazil. J Clean Prod 63:166–172. doi:10.1016/j.jclepro.2013.09.020
2. Blowfield M (2013) Business and sustainability. Oxford University Press, Oxford
3. Bonn I, Fisher J (2011) Sustainability: the missing ingredient in strategy. J Bus Strategy 32(1): 5–14. doi:http://dx.doi.org/10.1108/02756661111100274
4. Borim-de-Souza R, Balbinot Z, Travis EF, Munck L, Takahashi ARW (2015) Sustainable development and sustainability as study objects for comparative management theory. Cross Cult Manag 22(2):201–235. doi:10.1108/CCM-02-2013-0027
5. Castanheira ÉG, Grisoli R, Freire F, Pecora V, Coelho ST (2014) Environmental sustainability of biodiesel in Brazil. Energy Policy 65:680–691. doi:10.1016/j.enpol.2013.09.062

6. Constanza R, Fioramonti L, Kubiszewski I (2016) The UN sustainable development goals and the dynamics of well-being. Front Ecol Environ 14(2):59. doi:10.1002/fee.1231
7. Creswell JW (2014) Research design: qualitative, quantitative and mixed methods approaches. Sage Publications Inc., California
8. David RB, Tobin GA (2016) Addressing the sustainability of Austin, Texas, water policy. Pap Appl Geogr 2(1):96–104. doi:10.1080/23754931.2015.1107617
9. Dentchev N, Baumgartner R, Dieleman H, Jóhannsdóttir L, Jonker J, Nyberg T, Rauter R, Rosano M, Snihur Y, Tang X, Hoof BV (2016) Embrancing the variety of sustainable business models: social entrepreneurship, corporate intrapreneurship, creativity, innovation, and other apprаches to sustainability challenges. J Clean Prod 113:1–4. doi:10.1016/j.jclepro.2015.10.130
10. Dias MAdP, Vianna JNdS, Felby C (2016) Sustainability in the prospective scenarios methods: a case study of scenarios for biodiesel industry in Brazil, for 2030. Futures 82:1–14. doi:10.1016/j.futures.2016.06.005
11. Dillman DA (2007) Mail and Internet surveys: the tailored design method. John Wiley & Sons, Inc., New York
12. Dillman DA, Phelps G, Tortora R, Swift K, Kohrell J, Berck J, Messer BL (2009) Response rate and measurement differences in mixed-mode surveys using mail, telephone, interactive voice response (IVR) and the Internet. Soc Sci Res 38(1):1–18. doi:10.1016/j.ssresearch.2008.03.007
13. Dillman DA, Reips U-D, Matzat U (2010) Advice in surveying the general public over the Internet. Int J Internet Sci 5(1):1–4
14. Dolowitz D, Buckler S, Sweeney F (2008) Researching online: Palgrave study skills. Palgrave Macmillan, New York
15. Duarte FdP (2015) Barriers to sustainability: an exploratory study on perspectives from Brazilian Organization. Sustain Dev 23(6):425–434. doi:10.1002/sd.1603
16. Dyllick T, Muff K (2016) Classifying the meaning of sustanable business: introducing a typology from business-as-usual to true business sustainability. Organization Environ 29(2):156–174. doi:10.1177/1086026615575176
17. Eweje G (2014) Introduction: trends in corporate social responsibility and sustainability in emerging economies. In: Corporate social responsibility and sustainability: emerging trends in developing economies. Emerald Group Publishing Limited, USA, vol 8. pp 3–17. doi:10.1108/S2043-905920140000008001
18. Ferry L, Eckersley P (2016) Debate: climate change and 21st Conference of the Parties (COP21) Paris – addressing the 'ultimate' wicked issue. Public Money Manag 36(4):241–242. doi:10.1080/09540962.2016.1163021
19. Field A (2009) Discovering statistics using SPSS. SAGE, London
20. Galpin T, Whittington JL, Bell G (2015) Is your sustainability strategy sustainable? Creating a culture of sustainability. Corp Govern 15(1):1–17. doi:10.1108/CG-01-2013-0004
21. Gavronski I, Ferrer G, Paiva EL (2008) ISO 14001 Certification in Brazil: motivations and benefits. J Clean Prod 16(1):87–94. doi:10.1016/j.jclepro.2006.11.002
22. Henders S, Persson UM, Kastner T (2015) Trading forests: land-use change and carbon emissions embodied in production and exports of forest-risk commodities. Environ Res Lett 10(12):1–13
23. Jabbour CJC, Jugend D, Jabbour ABLdS, Gunasekaran A, Latan H (2015) Green product development and performance of Brazilian firms: measuring the role of human and technical aspects. J Clean Prod 87:442–251. doi:10.1016/j.jclepro.2014.09.036
24. Kelley S, Nahser R (2014) Developing sustainable strategies: foundations, method, and pedagogy. J Bus Ethics 123(4):631–644. doi:10.1007/s10551-013-2014-6
25. Kootstra G (2004) Exploratory factor analysis: theory and application. pp 1–15. https://www.scribd.com/document/103738250/Exploratory-Factor-Analysis-Kootstra-04. Accessed 1 June 2017

26. Kuhlman T, Farrington J (2010) What is sustainability? Sustainability 2(11):3436–3448. doi:10.3390/su2113436
27. Lahsen M (2016) Toward a sustainable future earth. Sci Technology Hum Values 41(5): 876–898. doi:10.1177/0162243916639728
28. Laquimia MB, Eweje G (2014) Collaborative governance toward sustainability: a global challenge on Brazil perspective. In: Corporate social responsibility and sustainability: emerging trends in developing economies. Emerald Group Publishing Limited, USA, vol 8. pp 371–413. doi:10.1108/S2043-905920140000008018
29. Little JC, Hester ET, Carey CC (2016) Assessing and enhancing environmental sustainability: a conceptual review. Environ Sci Technology 50(13):6830–6845. doi:10.1021/acs.est.6b00298
30. Lourenço IC, Branco MC (2013) Determinants of corporate sustainability performance in emerging markets: Brazilian case. J Clean Prod 57:134–141. doi:10.1016/j.jclepro.2013.06.013
31. Lozano R (2008) Envisioning sustainability three-dimensionally. J Clean Prod 16(17): 1838–1846. doi:10.1016/j.jclepro.2008.02.008
32. Mebratu D (1998) Sustainability and sustainable development: historical and conceptual review. Environ Impact Assess Rev 18(6):493–520. doi:10.1016/S0195-9255(98)00019-5
33. Merriman KK, Sen S, Felo AJ, Litzky BE (2016) Employees and sustainability: the role of incentives. J Managerial Psychol 31(4):820–836. doi:10.1108/JMP-09-2014-0285
34. Pater LR, Cristea SL (2016) Systemic definitions of sustainability, durability and longevity. Procedia Soc Behav Sci 221(7):362–371. doi:10.1016/j.sbspro.2016.05.126
35. Pribeanu G, Toader C-S (2016) The success in business in the context of sustainable development. Agric Manag 18(2):99–102
36. Punnett BJ (2013) Management: a developing country perspective. Taylor and Francis, New York
37. Ratiu C, Anderson BB (2015) The multiple identities of sustainability. World J Sci Technology Sustain Dev 12(3):194–205. doi:10.1108/WJSTSD-05-2015-0022
38. Sivramkrishna S (2016) Cracks in BRICS: a sectoral financial balance analysis and implications for macroeconomic policy. Theor Appl Econ 23(3):53–78
39. Smyth JD, Dillman DA, Christian LM, O'Neill AC (2010) Using the Internet to survey small towns and communities: limitations and possibilities in the early 21st century. Am Behav Sci 53(9):1423–1448. doi:10.1177/0002764210361695
40. Swart RJ, Raskin P, Robinson J (2004) The problem of the future: sustainability science and scenario analysis. Global Environ Change 14(2):137–146. doi:10.1016/j.gloenvcha.2003.10.002
41. Thomas RW, Fugate BS, Robinson JL, Tasçioglu M (2016) The impact of environmental and social sustainability practices on sourcing behavior. Int J Phys Distribution Logist Manag 46(5):469–491. doi:10.1108/IJPDLM-02-2015-0041
42. UN (2016) World economic situation and prospects 2016. http://www.un.org/en/development/desa/policy/wesp/wesp_current/2016wesp_ch1_en.pdf. Accessed 20 Nov 2016
43. Vergragt PJ, Dendler L, Jong Md, Matus K (2016) Transitions to sustainable consumption and production in cities. J Clean Prod 134:1–12. doi:10.1016/j.jclepro.2016.05.050
44. Vlassopoulos CA (2012) Competing definition of climate change and the post-Kyoto negotiations. Int J Clim Change Strategies Manag 4(1):104–118. doi:10.1108/17568691211200245
45. Weybrecht G (2010) The sustainable MBA: the manager's guide to green business. Wiley, Chichester
46. Wittneben BBF, Kiyar D (2009) Climate change basics for managers. Manag Decis 47(7): 1122–1132. doi:10.1108/00251740910978331
47. Wright KB (2005) Researching Internet-based populations: advantages and disadvantages of online survey research, online questionnaires, authoring softwares packages, and web survey services. J Comput Mediat Commun 10(3). doi:10.1111/j.1083-6101.2005.tb00259.x
48. Yong G, Pearce S (2013) A beginners guide to factor analysis: focusing on exploratory factor analysis. Tutor Quantitative Methods Psychol 9(2):79–94. doi:10.20982/tqmp.09.2.p079

# Sustainability and Green Supply Chain Awareness in Nigerian Organizations

**Ogubuike Chukwuemeka Inmpey**

**Abstract** The concept of organizational growth has experienced a paradigm shift from the sole objective of profit maximization to the overall notion of social welfare. Consequently, organizations strategies, policies, and operations now include environmental considerations. All these considerations come under the umbrella of sustainability. The objective of this chapter is to analyze factors that determine the attitude of Nigerian organizations towards sustainability in terms of its Advantages and disadvantages. The author collected primary and secondary data, and carried out quantitative as well as qualitative analysis of these data. The instrument used for primary data collection was a survey questionnaire that consisted of 24 items that measured the Advantages of sustainability and 13 items that measured the challenges of sustainability. The web link for the survey instrument was distributed to Nigerian contacts on social media platforms (email, Facebook, WhatsApp, LinkedIn). The research findings revealed two key factors for the Advantages of sustainability: (1) increased operational and social productivity, and (2) increased adherence to environmental standards. Two key factors for the disadvantages to sustainability were also revealed: (1) increased risk of management failure, and (2) increased regulatory and compliance costs. Future studies may expand on the findings of the current study to identify additional factors and diverse perspectives that may be found for other countries. Based on the research findings, it is recommended that organizations and governments focus their sustainability initiatives on the fulfillment of two broad goal categories that will enable them to achieve most of the desirable advantages of sustainability. These goals are: (1) increasing operational and social productivity, and (2) increasing adherence to environmental standards. The disadvantages associated with the adoption of sustainability initiatives can be managed by developing risk management plans that mitigates against risks associated with increased management failure and potential increase in sustainability-related regulatory and compliance costs. The major limitation of the study is the difficulty of generalizing the research findings globally given

O.C. Inmpey (✉)
School of Information Systems, Curtin University, Perth, Western Australia, Australia
e-mail: o.inmpey@postgrad.curtin.edu.au



T. Issa et al. (eds.), *Sustainability, Green IT and Education Strategies in the Twenty-first Century*, Green Energy and Technology, DOI 10.1007/978-3-319-57070-9_16

the fact that the focus country, Nigeria, is a developing country with an economy and circumstances that are not typical of most other countries in the world. However, the findings can be of benefit to other developing countries in Africa that share similarities with Nigeria. Other limitations are the limited study sample size, the survey time-frame and financial constraints.

# 1 Introduction

Corporations are increasingly being held responsible for the environmental and societal devastations resulting from the effect of their business activities. This has resulted to increased research focus on the sustainability debate, and increased attention to sustainability issues in recent corporate literature [45]. Despite this increasing interest barriers to the adoption of sustainability still exist. The complexity of the sustainability concept and the different sustainability assessment methods plays a crucial role in the existence of these barriers. Sustainability assessment may be regarded as a range of processes encompassing broad aims for integrating the concepts of sustainability concepts into decision making. It is also regarded as a process by which the implications of sustainability initiatives are measured. Furthermore, sustainability assessment is also described as a process aiming to operationalize concept of sustainability and making it a guide for decision-making. It helps in the early identification of the future impacts of any sustainability action plan. Various tools have been designed for assessing sustainability. However, there is a lack of consensus in sustainability assessment literature on the effective use of these tools, the choice of tool, and the effective use of methodologies [17]. Sustainability as well as sustainable development have been contested concepts, and interpretations vary in the academic literature.

The concept of sustainability and sustainable development and their related issues need to be integrated with the functions of management, education, and research. The question of measuring sustainability focuses on the question of measuring development as a whole [22]. Development that was once viewed only as economic growth, needed to be adapted to include considerations for environmental preservation.

Sustainability is regarded as "the next mega trend" for public and private sector organizations. The harmful effects of continuing urbanization on the environment make sustainability initiatives of particular importance in the contemporary context. The growing importance of sustainability across the world during the past two decades has resulted in the development of sustainability measures such as sustainability indicators, sustainability ratings, and sustainability indices. Sustainability measures may be utilized in different contexts. There are ratings available even for measuring the sustainability performance of countries such as Robeco and EIRIS [57]. Other measurement systems focus on assessing the sustainability performance of organizations. These measurements reflect the importance of the sustainability initiatives. Harmful gaseous emissions (popularly called Greenhouse Gases) are mostly result from energy consumption during the business operations. Hence, organizations are uniquely positioned to serve as key stakeholders in efforts to adopt sustainability initiatives.

The philosophical and historical background that led to the idea of sustainability may be elaborated by explaining the theories of the basic nature of historical change. The two historical change perspectives emphasize that history is either progressive or cyclical. Hence, the aim of any progress or development is either away from an imperfect past or toward a perfect future [48].

During the last 10 years, there has been a significant transformation in the intellectual discourse related to environment-development. The question that was focused on in earlier studies was, “Do development and environmental concerns contradict each other?” However, now the researchers are focusing on the question, “How can sustainable development be achieved?” [39]. The notion of Sustainable Development has gained huge prominence. The term is being used by development planners, international aid agencies, and environmental activists.

The terms “sustainable growth” and “sustainable development” have gained prominence in the environmental domain but there is a distinction between the two terms. Growth implies increasing in size in a natural way through adding material, done by way of accretion or assimilation. Whereas, development implies realizing or expanding the potentialities and brings improvements from the current state to a greater, fuller, or better state. Hence, it may be asserted that growth implies increase in physical scale and refers to quantitative increase. On the other hand, development implies unleashing potentialities and refers to qualitative improvement [19]. An economy may have growth but no development. Also, an economy may be developing without any growth. In other cases, there may be presence or absence of both phenomena.

The notion of Green Supply Chain Management emerged from the literatures of supply chain management as well as environment management. The addition of the green aspect to supply chain management implies analyzing the relationships and impact between the natural environment and the supply chain.

The chapter topic was selected due to the increasing importance and relevance of sustainability and energy-efficient products. The environmental issues have become an area of concern for all governments due to the rise of pollution, increase in carbon emissions, and the changes and uncertainties in the weather and climatic conditions [47].

All environments in the world can be classified broadly into the natural or built environment. Human beings have little or no control over the natural environment e.g. the changes and cycle of seasons, rains, and extreme weathers. However the built environment is a man-made habitat over which human beings have full control over [24]. In this context, this chapter intends to focus on the disadvantages and advantages of a sustainable environment that promote the safety and protection of the natural environment.

The environment determines the form and character of the existence of people/organisms it surrounds, the way people live, and the development of society in general. Consequently, societal progress, the individuals that make up the society and the environment are closely linked, and can pose threat to each other. Human activity, if not checked, can potentially damage the environment and adversely affect the survival of present and future generations of human, plant and animal life in the society.

Issues related to environmental sustainability is considered as one of the top priority world problems, and is included the Millennium Development Goals (MDGs) set out in the United Nations Millennium Declaration [68]. As a result of this, many organizations from different sectors are investing on research and adopting strategies/programs aimed at the world's natural resources so that present and future generations can benefit from it. There is also concerted effort by first world countries and international donor agencies in the provision sustainability related technical support and aid to developing countries.

Throughout life, the human body is exposed to constant attack from a wide range of environmental factors which impacts directly on the life, health and life expectancy of people. In urban environments, there are five main groups of factors that affect human health: social, biological, living environment, production factors, and lifestyle. Lifestyle factors have the greatest impact on the health of urban residents and include such things as smoking, and recreational air pollutants. The main sources of environmental pollution are large industrial plants, thermal and nuclear power plants, motor transportation etc. Power plants and waste dust contain silicon dioxide, heavy metals, arsenic, mercury, vanadium, lead, which are particularly harmful to the human body [9].

The Information Technology sector, due to rapid advancements in technologies, has become a major contributor to environmental degradation. Technology devices are increasing the carbon emissions. Also, when people purchase new information technology products, they discard the old products in a non-environmental-friendly manner. Any flaws in the supply chain of the information technology can pose serious threats to the environment. Hence, it has become crucial to study their impacts and recommend improvements for a better and greener supply chain of information technology products.

The researcher aims to identify the factors that influence the attitude of Nigerian organizations towards the concept of sustainability in terms of its Advantages and disadvantages. Nigeria is a country in the African continent. Hence, the findings of this study will be beneficial not only for the policy makers in Nigeria, but in general, to the whole African community. The government institutions, civil society organizations as well as the private sector will benefit from this research. This is because the issue centers on climatic changes, energy efficiency, green environment, and reducing carbon footprints – all of which have become an area of focus in every field and in all parts of the world. The researcher has selected only one country in Africa, so that the study could be completed within the stipulated time period and a comprehensive analysis could be carried out through focused and dedicated work. The theoretical framework of this research is based on the theories of sustainability and green environment. The study will add to the theoretical foundations established in past studies. It will suggest several guiding principles particularly for academics, PhD and Master level students, and researchers to assist them in their endeavors related to environmental sustainability and green supply chain. The study will add to the theoretical foundations established in past studies. It will suggest several guiding principles particularly for academics, PhD and Master level students, and researchers to assist them in their endeavors related to environmental sustainability and green supply chain.

# 2 The Concept of Sustainability and Green Supply Chain Management

This section presents background information on the research topic and briefly reviews key literature that relates to sustainability and green supply chain. It discusses the concepts of sustainability, green supply chain, and the triple bottom line. The history of sustainability, its importance for business and academic environments, and current progress of sustainability in developing countries (with Nigeria as the focus country) is also discussed. Finally, the Advantages of sustainability, and barriers to sustainability, as well as future of sustainability are summarized.

## *2.1 Sustainability and Green Supply Chain Management*

The scope and definition of green supply chain management varies and depends on the objectives of the researcher. Some of these include: the integration of green supply chains to green purchasing [62]; a closed loop supply chain with forward and reverse flow of material, information and money [23, 60, 74]; environmental perspective definition [7]; and the sustainability viewpoint [42]. Despite their differences, the fundamental theme in these definitions/scope is that green supply chain management is the inclusion of environmental impact considerations in the professional practice of supply chain management, so as to reduce and possibly eliminate any negative impact supply chain management activities may have on the environment. Hence, green supply chain management can then be generally defined as a methodology that tries to incorporate considerations for environmental sustainability issue into the practice of supply chain management, from the time the product idea is conceived down through its design, raw material procurement, manufacturing, warehousing, sales, transportation, delivery and final end-of-life management [31].

The concept of green supply chain management integrates environmental sustainability thinking into the domain of supply chain management. It focuses on eliminating or minimizing wastages that contain hazardous emissions, hazardous chemical, solid and energy waste throughout the supply chain. Green supply chain management includes material selection and sourcing, manufacturing process, product design, delivery to the final customer, reverse logistics, and product's end-of-life management. In this regard, green supply chain management plays a crucial role in determining the overall environmental impacts of the supply chain activities and contributes to the enhancement of sustainability performance. With the increase in competition in the 1990s, there was an enhanced awareness of green practices and the use of environmental-friendly products. This trend triggered the need for organizations to be socially and ethically responsible in their supply chains. In early 1995, the concept of green supply chain management had generated substantial interest in studies by academics [14]. Taking all these factors into consideration, businesses began to develop strategies respond proactively to emerging environmental legislation and requirements. The importance of an integrated approach for green supply chain management that could address different aspects of sustainability development have been stressed [66] (Fig. 1).

A multi-objective optimization model for green supply chain network design attempts to achieve a compromise between overall network cost and environment influence [70]. However, if the emission per facility is to be considered, then there are no considerations for demand uncertainty. Routing studies aim to optimize travel distance, the volume of greenhouse emissions, the amount of fuel consumed, travel times and all associated costs [8].

Designing sustainable supply chains under the emissions trading scheme gave rise to mixed-integer linear-programming-based framework for sustainable supply chain design [13]. This model looked at Life Cycle Assessment (LCA) principles and the material balance constraints that usually exist at each node in the supply chain. It considers limit on carbon emission and determines the number of carbon credits to be bought and sold.

A supply chain network design aimed at reducing carbon emissions via network design, analyzes the relationship between carbon-di-oxide emissions and

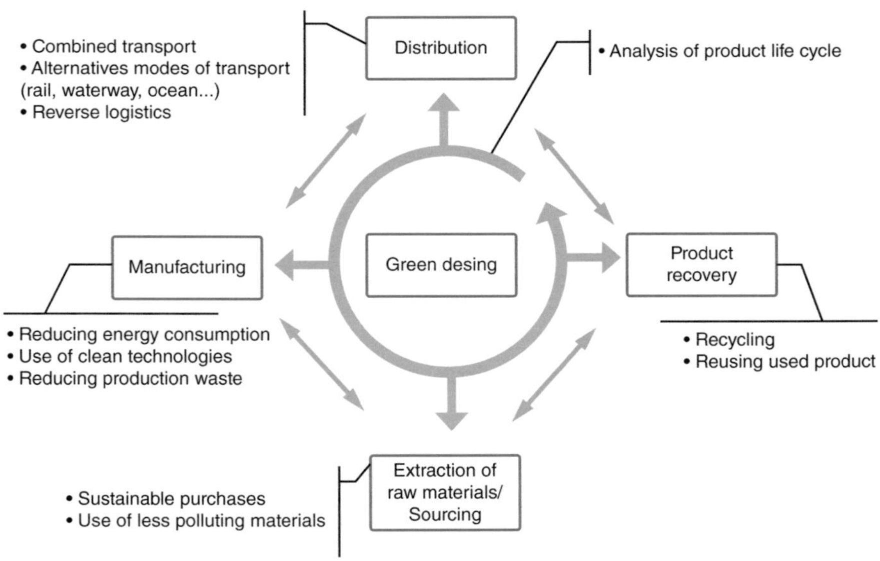


**Fig. 1** Different approaches for green supply chain management [66]

vehicle weight using a concave function that creates a concave minimization problem. Lagrangian relaxation is used to breakdown the problem into a capacitated facility location problem with single sourcing [26]. This makes it a concave knapsack problem that can be solved by the application of Lagrangian relaxation to a concave mixed integer programming model.

Studies have also analyzed the carbon emission per product produced [1, 20, 49, 54]. They consider the periodic carbon emission constraint, rolling carbon emission constraint, cumulative carbon emission constraint, and global carbon emission constraint. Green supply chain network optimization, the compromise between environmental and economic considerations has also been studied. They establish the interplay between emissions, supply chain volume based cost and the energy mix. The results of this showed that optimizing the energy mix reduces supply chain carbon dioxide emissions by 30% with no resultant increase in the variable cost of production [65].

## 2.2 *History of Sustainability*

By the late 1960s, the environmental movement gained prominence because of rising consumer/public concerns about environmental degradation. Subsequently, the Environmental Protection Agency was established, and mandated with the formulation of standards that will regulate all manufacturing activities on a product's supply chain [50]. These developments led to the realization of the need to apply environmentally-sound decisions in the context of Supply chain management, shifting the paradigm from reactive approach to proactive approach.

Businesses are constantly pressured to align the strategic goals, internal policies and practices with emerging environmental legislations and sustainability initiatives. Despite different industrial conceptions of "sustainable business practice", the underlying theme of achieving the goals of profitability while maintaining accountability for the social/environmental well-being of the animate and inanimate specimen of their operating environment remain constant This underlying theme is the key tenets of sustainability, and has always being referred to as Corporate Social Responsibility (CSR) [32].

Currently, the application of data mining in measuring sustainable business practice usually occurs from a linear, Newtonian perspective that seeks to find a cause and effect relationship aimed at the generation of profits from sustainable business ventures [59]. Studies indicate that one of the main issues faced by corporations in overcoming this single-purpose data analysis of financial reporting is the ability to determine what to report and how to report it [12, 15, 56]. However, these factors are dependent on the mission of each organization, thereby making it difficult to develop a standardized approach.

Green Supply Chain Management evolved as a reaction to the emerging need for environmental compliance, and reflected an integration into the movement of environmental thinking. Green Supply Chain Management gained importance and was driven by diminishing natural resources, environmental degradation, and increase in the levels of pollution. In essence, Green Supply Chain Management is based on different environmental criteria that interlink with the various stages of the supply chain that both interact with and stem from all suppliers along the chain [50].

In attempting to adopt the green supply chain concept, there are overlapping and complementary components for ensuring environmental neutrality of the manufacturing and delivery system. Green supply chain management originated from the literature of Environmental Management and supply chain management. With the incorporation of a green component, the relationships and influences between natural environment and Supply Chain Management are intertwined [50]. This is depicted in Fig. 2 below showing that green operations focus specifically on reducing the ecological burdens.

Triple Bottom Line (TBL) term was first introduced in 1994 by John Elkington. Triple Bottom Line (TBL) concept refers to an accounting framework that focuses on three key elements: financial (profit), social (people), and environmental (planet) [34]. Many organizations have started adopting the TBL framework for analyzing their performance in a broader framework in order to create greater business value. An example of TBL could be a social enterprise that functions as a non-profit organization working in the recycling business. The enterprise earns income by offering employment opportunities to handicapped people. These handicapped people would not otherwise have been employed to earn a living. The organization earns a profit from the work of these handicapped people (financial element), which is ploughed back into the community as the organization is a non-profit organization. The social element in this case is the meaningful employment of disadvantaged citizens. Another social element is the reduction in the disability costs and welfare costs to the society. The environmental element in this case is the recycling that is being done. This serves as a measure of their commitment to corporate social

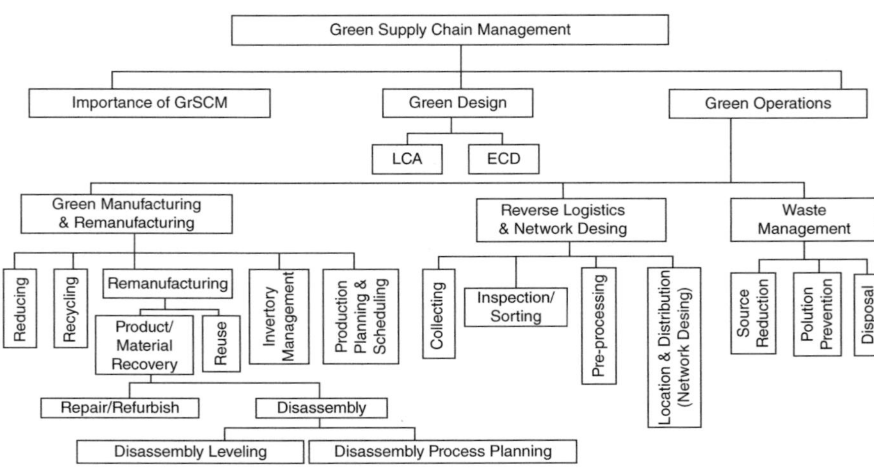


**Fig. 2** Problem context classification in supply chain [50]

responsibility (CSR) and demonstrates transparent reporting about the physical impact of their business activities on the people and the planet.

TBL does not have a common unit of measure. This is because it comprises three elements: people, planet and profit. Profits may be measured in dollars; however, it is hard to quantitatively determine the social capital and the environmental health. One of the disadvantages facing TBL implementation has been to find a common unit of measurement.

It has been argued that there is substantial literature available in relation to the appropriate measures for incorporating sustainability at both state and national levels [4]. However, there is a lack of established models, theories, and frameworks in this respect. Eventually, the availability of data drives the calculations of TBL. The traditional measures of sustainability include economic, environmental and social measures. Economic measures consider issues related to finance like individual income, employment rates, number of firms and revenue by sector, etc. that contributed to the gross product. Environmental measures include the rate of environmental pollution resulting from different pollutants, level of environmental nutrients, energy utilization, hazardous/toxic waste management and the extent of deforestation. Social measures look at issues like poverty, crime level, average death and life expectancy, population demographics etc. [34].

The triple bottom line has become a fundamental principle of business sustainability. Traditionally, businesses have been focusing on increasing profit, irrespective of how these profits have been earned. However, to succeed in the evolving drive toward sustainability, organizations need to utilize five identified levers for generating promising and profitable ventures through the execution of sustainability-based strategies. Two out of these five identified levers focus on organizational management relying on the employees to carry out their sustainability strategies. The remaining three trains the employees to recognize the advantages of sustainability strategies and make a commitment to the initiatives/ programs that will execute the strategies [40]. The triple bottom line modifies the focus of the organizations so that they focus not only on profits, but on the generation of profits in a manner that is beneficial for the people as well as the planet.

## 3 Sustainability Significance for Business and Education

Embracing business practices that are environmentally sustainable provides a number of significant advantages for businesses and education. These advantages are summarized below.

1. It fosters innovation through constant review of process to identify new solutions.
2. Employees are employees are more invested in the business culture and have fulfilling job experience.
3. The organization builds a culture of accountability across the organization.

4. It contributes to the attainment of the Millennium Development Goals (MDG): poverty and hunger eradication, improved standards of education/healthcare, gender equality, sustainable economic growth, reduction in climate change activities etc. [73].
5. It improves organizational eco-friendly reputation and attracts customer loyalty [27].
6. It forms a critical part of organizational risk management strategy and improves the success rate of risk management process [64].

It is therefore incumbent upon an organization to realize the importance of sustainable practices for more reliable products, better corporate culture, and greater long-term profitability.

A recent study that analyzed the current progress level of Green Supply Chain Management practices among the largest corporations of the world, provided insights into the extent to which large corporations are engaging in sustainable practices in their supply chain and the type of activities undertaken to accomplish this objective. It also revealed that customers expect an organization to assure ecologically and socially sound production, and modern Green Supply Chain Management practices focus on prevention of pollution and reduction of operational cost [72]. There are also positive changes in the corporate sector in the adoption and evolution of proactive sustainable approaches [10]. One study has found that over 66% of organizations studied have developed a sustainability report in one form or another [43]. This indicates the level of commitment of large corporations and the importance that they are placing on sustainability. There were also a few reports that mentioned terms such as "green procurement", "supplier audit", and "sustainable supply chain". The study also found that firms who has understood relevance sustainability and the green supply chain are aiming to build a management measurement system to track their sustainability performance level. These include the installation of sustainability key process indicators, development of an environmental management system, and making it mandatory for the suppliers to meet sustainability requirements.

The conventional wisdom that economic growth always leads to human progress has been challenged by the presence of modern anomalies such as global climate change, ecological destruction, continuous economic crisis, resource depletion, and social inequality. In light of recent economic crises and tensions over resources and climate, warnings about the present economic paradigm have become even more pronounced. The need for a paradigm shift is one of the most popular themes in the economic sustainability literature. However, what passes for a paradigm shift in this continually mutating political economy varies from one theorist to another. For example, while one argues for abandoning the economic ideology that regards the growth of gross domestic product (GDP) as an absolute value, proponents of the increasingly popular Green Growth model of economic development – a model that presents itself as a new paradigm – still frames economic growth as essential for achieving human well-being. Initiatives that fall under the category of Green Growth commonly attempt to equate economic growth with environmental protection, although they have been formulated under a variety of names, modes, and contexts. Green Growth (sustainability) initiatives have been founded on a belief that environmental protection measures will make economic growth sustainable, and thus help save the planet from ecological collapse [6].

Regardless of how one defines it, the preponderance of arguments in favor of a paradigm shift reflects the seriousness of the economic maladies that have arisen in the contemporary world. Conventional wisdom has hitherto believed that technological progress would solve the side effects of economic development. However, the diversity and severity of ecological catastrophes across the globe, have shown that technological solutions offer limited prospects for widespread success in face of the constant emergence of new ecological problems and the persistence of severe and escalating ones [44].

Another serious problem is the ever-growing chasm between the rich and the poor. This phenomenon has not been restricted to the jurisdiction of any one country. A recent report showed that almost 50% of the world's total wealth is controlled by 1% of its population and that the combined wealth of the 85 wealthiest individuals equals the wealth of the poorest half of the world's population [29]. Even in the wealthiest country, the U.S., inequality among working-age people has risen by 25% since 1980 [29].

## 4 Sustainability in Developing Countries

In 2015, a UN Conference on Trade and Development (UNCTAD) focused on three fundamental questions related to sustainability: "What would it cost to become sustainable?", "Do we have the money?" and "How can we mobilize it?" These follow-on conferences were intended to find inter-governmental agreement as they related to three core elements for transitioning to sustainable development [36]. At the first conference in Addis Ababa, a package of investment and finance policies and initiatives was assembled specifically for developing countries. It was hoped that these initiatives would help developing countries understand the importance of the sustainability agenda, and as a consequence take actions necessary to adopt them. The second conference developed a framework of global Sustainable Development Goals (SDGs). The conference also set the targets in relation to these goals for the period 2015–2030.The third conference sought to reach an agreement on a legally binding climate accord, and address the issues of dangerous climate change. According to the World Investment Report, developing countries are getting more than half of the total flows as they relate to sustainable development. However, it revealed a gap of approximately $2.5tn annually between what is being received by the developing countries and what is needed by them for transitioning to sustainable development [36].

A fundamental concept in management science is sustainable development. Sustainable development is a concept managers use to satisfy the current world's survival needs without negatively affecting the ability of up-coming generations to satisfy their survival own needs. The concept of sustainable development refers to all types of socio-economic development models that concentrate on making sure that a balance exists social, economic, ecological, and natural resources. Societies can achieve sustainable economic development by focusing on and supporting the development of small and medium scale enterprises. The results from the analysis of available data indicate that these types of businesses are a vital part the socio-economic development of developing countries. Societies (developing countries) can achieve sustainable economic development by supporting small and medium enterprises [63].

Managers use the concept of sustainable development to find solutions to ecological crises created by unchecked industrial consumption of natural resources, and by so doing tackle the unabated environmental degradation of the environment and with the aim of protecting the environment. The expansion of the sustainable development concept includes the quality of all aspects life. Sometimes, sustainability refers to management ability to respond to changes in the operational environment. In this context, management must look at the bottom line in terms of the business effect on the economy, society and environment [71].

For scholars the basis for sustainability stems from the idea that human activities are dependent on natural resources. Hence the quality of life is essentially defined by health, social security and economic stability. Small business managers may realize substantial advantages through the incorporation of sustainability practices as part of their general business strategy. The United Nations (UN) recommend sustainable development to all their member countries as a tool for socio-economic development. For developing countries, one important part of sustainable development is economic growth in which small businesses can play a pivotal role in achieving sustainable economic development. Managers use sustainable development programs to create jobs and stimulate entrepreneurship. Sustainable economic development is the insurance to businesses spanning from appropriate use of resources to the efficient implementation of management practices. To conform to sustainable development, managers must ensure that sustainable financial situations exist as a result of improving asset structure, social development, and dynamic teams in external environments. Societies can achieve sustainable economic development by supporting small and medium enterprises [63].

Small business managers help to create competitive market environments that supports innovation, create employment opportunities and helps bridge the gap between the poor and the rich through the creation of the middle class. By focusing on development of small and medium scale enterprises, developing countries are better placed to achieve their sustainability initiatives.

Lagos State is regarded as the Nigerian economic hub. It is gives a reasonable idea of Nigeria's situation in terms of sustainability. Construction generally has an irreversible and significant effect on the environment. Resultant effects like environmental pollution, extensive use of environmental resources, and high-energy consumption are some of the factors to consider in the Lagos supply chain. Some scholars have argued that the supply chain management analysis in Nigeria has attracted little attention, particularly in the construction sector. The supply chain management awareness is low and an awareness campaign is proposed for sensitizing the construction industry's stakeholders [53].

The reduction in environmental degradation by means of supply chains is expected to positively improve the environment. The economy of Nigeria is heavily dependent on petroleum. Hence, the environmental wastes from the petroleum sector are at an elevated level particularly from the supply chain. It have been asserted that the petroleum supply chain downstream stakeholders may adopt green practices if they can be assured of commitment and trust from the major partners [67]. The major partners in the context of Nigeria include the Nigerian National Petroleum Corporation (NNPC) and its affiliates. The affiliates include Pipelines and Products Marketing Company (PPMC), refining companies, and independent petroleum marketers. The study emphasized that if all organizations in Nigeria set individual targets of reducing carbon emission, the downstream partners in the supply chain such as the small size petroleum retail outlets and filling stations are more likely to adopt their environmental policies [67].

In September 2000, the MDG included the MDG7 which aims at making sure that global environmental sustainability is achieved. The MDGS are people oriented goals that are measureable and expected to be achieved within a given timeframe. Despite the universality of the MDG goal, countries were expected to adapt them to their local peculiarities and resource constraints. The Nigeria situation with respect to the adoption and implementation of the issues related to MDG7 is so critical that the United Nations Development Programme (UNDP) observed that the country's natural resources is under serious threat. Recognizing this problem, and the need to strive toward reversal of this threat, the Nigeria government set up mechanisms at national and state levels for achieving the MDGs by 2015 [2].

Nigeria is a country located in the western parts of the African continent. It is found along the coast of the Gulf of Guinea and covers a total area of 923,768 square km. The total length of the state border is 4047 km in the west – with Benin (773 km) to the north – with the Niger (1,497 km) to the north-east – with Chad (87 km) to the east – with Cameroon (1,690 km); the coast line is 853 km long. The highest point of the country is a mountain located in the state of Taraba close to the Nigeria-Cameroon border. Nigeria is one of the most populous countries of Africa and the eighth largest producer of oil in the world, but its success has been weakened in modern times by political instability, religious and ethnic conflict, an ailing economy, and rampant official corruption [28].

Rivers Niger and Benue divide the country into two parts, tropical rain forests are located in the southern part of coastal plains, and in the north is dominated by low plateaus [5]. Most of the country is coastal plains formed mainly by river sediments. In the west, the plains along the coast are sandy braid chain, which are connected with each other and the Gulf of Guinea. To the north, the country's coastal plain area becomes a low plateau – a plateau Yoruba Plateau west of the river Niger and the Udi Plateau to the east. Next is the Northern Plateau, the height of which varies from 400–600 m to over 1,000 m high at the most central part of the plateau – Jos Plateau, whose highest point is Mount Cheret (1,735 m). In the northwest, the North Plateau becomes the Sokoto plain, and in the north-east, the Borno plain.

Oil-rich Nigeria has a history of political and religious instability, high poverty levels, and poor infrastructure which is compounded by successive corrupt and inefficient governments. The economy has remained undiversified with a total dependence on revenue generated from the oil sector. Recent governments have begun the process of economic reformation and are encouraging private sector involvement in infrastructural development.

Present-day Nigeria is a recent creation. Nigeria has a rich, multifaceted culture consisting of the traditions of more than 250 ethnic groups [51]. The country has produced writers, musicians, dance troupes, and athletes whose professional reputation has spread beyond the country's borders and whose talents have contributed to world culture.

Having attained independence for more than 50 years, the country is one of the wealthiest, most culturally diverse African nations. Nigeria's natural landscapes range from deserts to tropical rain forests. Nigeria is the most populous African country, with a population of almost 186 million. Most Nigerians live in rural areas, but increasing numbers of people now see the cities as their means to a better life. Nigerians belong to many ethnic groups and speak different languages, mainly Hausa, Yoruba, and Ibo, or Igbo. English is commonly used and is the official language.

Nigeria is bordered by the Gulf of Guinea in the south, Republic of Benin in the west, Cameroun on the east and Niger and Chad in the north. There are two main rivers that divide the country into three. These rivers are River Niger and River Benue. Both rivers meet at countries center Lokoja, before flowing down south into the Niger Delta. Spanning 150 miles (241 km) from north to south and 200 miles (322 km) on the coast, the Niger Delta covers 14,000 square miles (36,000 square km). Inside it, the Niger River branches out into hundreds of shifting channels. The Niger Delta was formed by soil deposits brought by the river and contains huge amounts of crude oil and natural gas reserves have been discovered in the delta. The channels in the delta are shallow and difficult to navigate, apart from the Escravos River. Big seafaring ships regularly navigate the Escravos, which provides access to the ports of Burutu, Koko, Sapele, and Warri. Nigeria's coastline is sheltered from the open sea by a narrow belt of coastal swamps, lagoons, and creeks. North of the coast, the mangrove swamps give way to tropical rain forest, and the terrain gets drier farther toward the north as the rain forest gives way to savanna and finally to the desolation of the desert in the extreme north.

Nigeria has a predominant tropical climate with two main seasons called rainy season and dry season. A third minor season exists during the dry season. The seasons depend on the movement of two air masses, which flow over the country. These are a dry air mass that flows over the Sahara Desert via the north and a moist air mass that flow over the South Atlantic Ocean via the south. These air masses create the appropriate weather conditions when they meet over the country, and depending on the time of the year, they create either the dry, wet or harmatan weather condition [52].

The two main rivers supply the country with a constant supply of water all year round. These rivers are also used for the purposes of irrigation, fishing, electricity production and transportation.

## 5 Advantages and Barriers of Sustainability and Green Supply Chain

In 2010, BearingPoint carried out a research based on the incorporation of green practice into the supply chain. The organization had also conducted a similar survey in 2008. Hence, this survey also allowed a comparison of the findings of the 2010 survey with the findings of the 2008 survey. According to the survey, the motivations for implementing green actions in organizations included improvement to the of brand image, the commitment decision by the board of executives, environmental protection laws, opportunities for process and product, reduction in overall operating cost, level of competitive pressure, the drive for new market acquisitions and pressure from opinion leaders. Among these motivators, the brand image improvement was the biggest motivator and the pressure of opinion leaders was the least powerful motivator [55].

Based on the survey, a concept map of the advantages of sustainability and green supply chain management is given in Fig. 3 below:

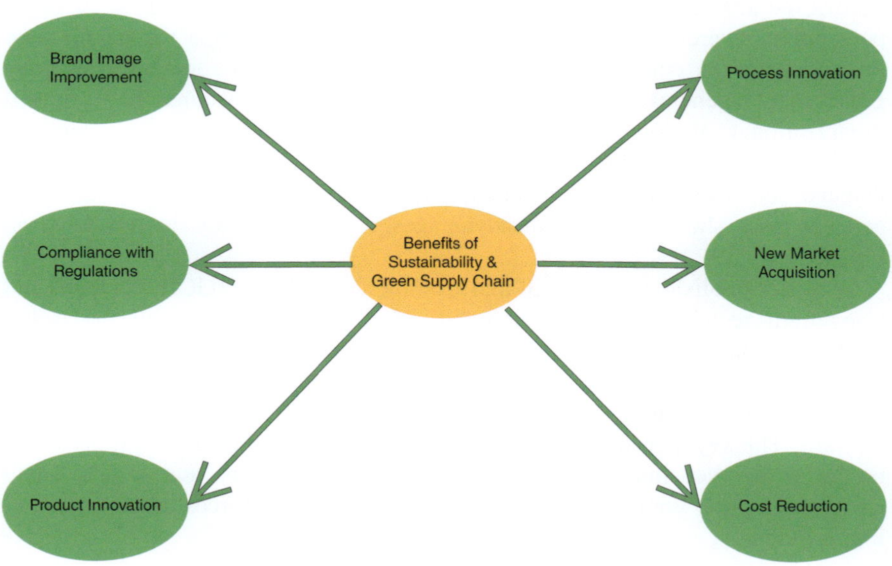


**Fig. 3** Concept map of the advantages of sustainability – prepared by the author

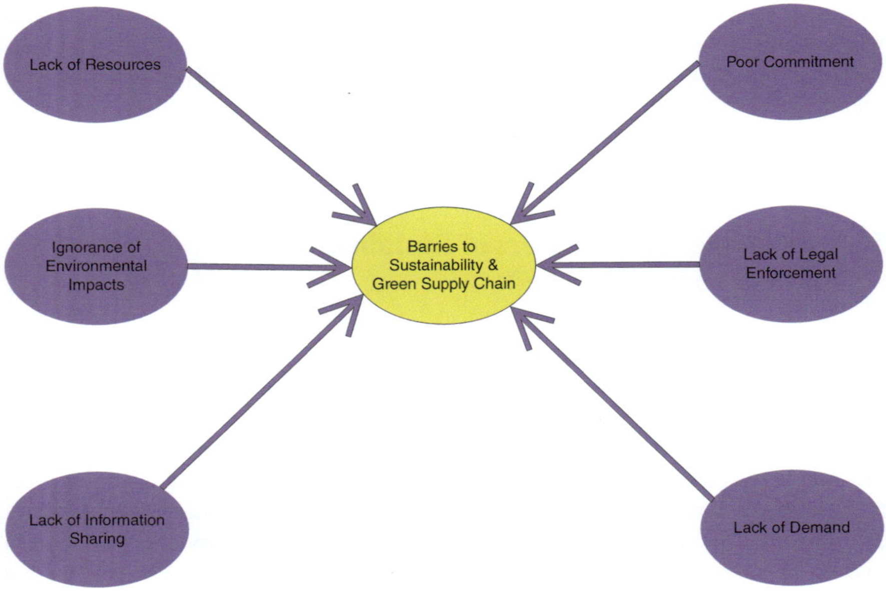


**Fig. 4** Concept map of barriers to sustainability – prepared by the author

A survey of Nigerian construction firms identified nine barriers that mitigate against the attainment of green supply chain management goals. These included poor public awareness, poor knowledge of environmental impact issues, lack of top management commitment, weak enforcement of environmental laws by government agencies, poor resource base, absence of sustainability oriented organizational mission and vison, absence of markets for recyclable materials, poor sharing of information between supply chain partners, and lack of societal demand for sustainable supply chain practices [53]. Based on the survey, a concept map showing these disadvantages to sustainability is presented in Fig. 4.

A report published by The Global Commerce Initiative (GCI) and Intel and Capgemini recommends that improved collaboration between all the stakeholders of the value chain are essential for achieving a more effective and efficient value chain for meeting the consumer needs. The report focuses on the future of the supply chain and the critical role of collaboration in moving forward. The key step in defining suitable scenarios for the future supply chain is the comprehension of the relevant trends and forces that would make an impact in the coming years [30]. These trends and forces are depicted in Fig. 5 below:

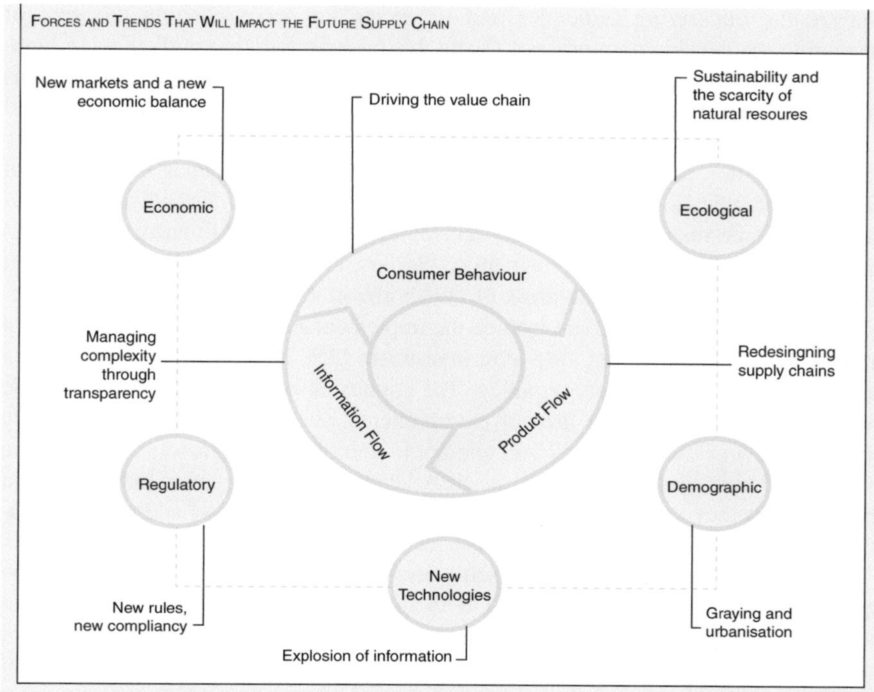


**Fig. 5** Forces and trends for the future supply chain [30]

# 6 Research Method and Question

The researcher has formulated the following primary research question for the study: "What are the factors that influence the attitude of Nigerian organizations towards sustainability in terms of advantages and barriers?"

It is worth noting that in the year 2000, the Millennium Development Goals (MDGs) were established, and were adopted by 189 member countries of the United Nations. These member countries also include Nigeria. These goals aim to expedite key developmental issues related to Nigeria and other developing countries [3]. They focus on raising the standard of living, ensuring the provision of basic life-sustaining goods, and expanding the range of social and economic choices available to people world-wide. One of these goals is to achieve environmental sustainability by investing and engaging with countries and organizations to achieve worldwide partnership that will drive sustainable development. Taking these goals into consideration, it is important for the economy of Nigeria to analyze the underlying principles and prospects for accomplishing the goal of sustaining the development process. Given this background, this study aims to highlight the factors that play a crucial role in how Nigerian citizens and organizations view the Advantages and disadvantages related to the concept of sustainability.

The key aim and objective of the study have been derived from the primary research question as follows: "To identify the factors that influence the attitude of Nigerian organizations towards sustainability in terms of its advantages and barriers". Three main barriers to the development of sustainable energy in Nigeria has been cited. These are legal/regulatory barriers, cost/pricing barriers, and market performance barriers. Possible areas of collaboration and focus that can help stakeholders overcome these barriers include the implementation of a legal and regulatory framework that could attract long-term investment [25]. Nigeria needs to establish a coherent and predictable energy policy. The country also needs to encourage public and private initiatives in the area of environmental sustainability and green supply chain. In view of this, this study will attempt to identify critical factors that influence attitudes towards sustainability and its associated Advantages and disadvantages in the hope that they can be factored into creating opportunities and addressing disadvantages in relation to environmental sustainability and green supply chain.

The survey process is a very useful one for many reasons. It can easily be applied to any type of research and it facilitates data analysis based on mathematical and statistical laws. This unique quality greatly reduces the possibility of making hasty conclusions from the research data. It is also a cost effect method of having access to large populations simultaneously. A research survey instrument can be distributed to the target population in various ways that helps ensure answers to the research questions come from geographically dispersed population. Depending on the research budget and time considerations, the survey can be limited scope without watering down the value of its findings. The survey does, however, have some disadvantages as a research method. These disadvantages are mostly related to the propensity of error to find its way into the survey process and invalidate the survey results. For example, respondents who have no available answer for a question may put down answers that they feel is expected of the researcher. However, the possibility of errors and their negative impact can be guarded against through meticulous planning, design and execution of the research. This improves the reliability and validity of the research results [46]. This study uses the method of survey research for the collection of primary data. The researcher aims to capitalize on the Advantages of survey research and aims to minimize the disadvantages as much as possible.

Although little empirical evidence exists, many researchers believe Web-based surveys offer several advantages [58]. The advantages include the following. The first benefit is lower costs. Web-based surveys save costs related to mailing, phone calls, and transcribing data. However, the development of the system may result in higher start-up costs. The empirical results regarding cost savings are mixed. The second benefit is time savings. Data is electronically entered into a database. The research process is not subject to postal delays. In addition, data cleansing, transcription, and validation are performed electronically upon data entry. The third benefit is enriched survey experience. The survey can be enhanced through images, video, different interaction opportunities (checkboxes, rank order, text boxes), multiple typefaces, and colors. The fourth benefit is that the Internet has the ability to reach many subjects that may be more difficult to reach through mail or phone. The fifth benefit is better item completion rate: empirical research has shown that respondents to Web surveys are more likely to complete all the questions.

Several disadvantages have also been mentioned for Web-based survey [11]. The disadvantages include the following. The first disadvantage is related to response rates. Historically, Web-based surveys suffer from low response rates compared to other forms of surveys. In a study of IT professionals in China, 5,000 individuals were contacted, 37% visited the Web site, and only 5.8% submitted the survey [21]. The response rate problem can be mitigated by sending a postal mail reminder to the surveyed population. The second disadvantage is bias due to sampling errors. Web-based surveys target those who have access to the Internet and are computer literate. This target may bias the survey results. The third disadvantage is privacy concerns. Some individuals will not respond to Web-based surveys because of the fear that they can be tracked. On the other hand, some respondents believe that the Internet is more private than paper or the phone. The fourth disadvantage relates to technical issues. Several studies have cited technical problems, such as a down Web server. The fifth disadvantage is sugarcoating and clipping. Poorly designed surveys frequently produce implausibly favorable results or compress the range of scores (i.e., clipping). Sometimes favorable results are caused by confidentiality concerns (e.g., using a Web survey on a company Web server to provide feedback on management). Some believe that sugarcoating and clipping are common in Web-based surveys [61]. In this study, the researcher uses Web-based method for survey research. The researcher endeavors to benefit from its advantages and also avoid the disadvantages as much as possible.

The research design selected for this study is the survey questionnaire. Before filling out the survey questionnaire, the researcher collects the demographic information about the participants through a series of questions (numeric code is added with each option to facilitate the coding of responses in SPSS). See Appendix 1 for the survey questionnaire.

The research participants responded on a five-point Likert scale ranging from Strongly Disagree, Disagree, Neutral, Agree to Strongly Agree. The researcher coded these responses as 1, 2, 3, 4, and 5 respectively. The researcher analyzed the merits and demerits of adopting sustainability and green supply chain on the basis of the strengths and weaknesses of the responses.

The sample population for the study consisted of those organizations that have their head offices or main offices located in Nigeria. The researcher selected a sample of 183 organizational representatives from these organizations to analyze the factors that are influencing the attitude of Nigerian organizations towards sustainability and green supply chain in terms of advantages and disadvantages. The sampling strategy used was judgment sampling. The researcher selected those participants who were very familiar with the research topic as per the judgment of the researcher and who were willing to participate in the study.

The researcher collected primary data through a Web-based survey. Participants completed the survey on the Web. The researcher allowed one-week time for questionnaire completion. All participants completed and returned the questionnaire within the stipulated time frame.

Data was analyzed through descriptive statistics. The researcher analyzed the responses to each question through frequency distribution tables. Data was also plotted in the form of bar charts. Secondary data was analyzed in the form of thematic analysis and key themes were identified for the study.

Data is considered to be valid if is representative of the target population and can lead to it valid when they represent what they purport to represent, and can lead to significant conclusions. Reliable data can easily be reproduced with consistent results irrespective the time, and the researcher that conducts the survey process. It is important to point out that the notions of validity and reliability do not have the same emphasis, and hence to some degree differ in meaning when used in in quantitative and qualitative data.

Reliability and validity are vital constituents in the development of a research design. Reliability deals with the data consistency and revolves around the degree to which the survey instrument produces the same results irrespective of the number of times it is used. Three categories of reliability exist. These are inter-observer reliability, test-retest reliability, and internal consistency reliability. Validity, however explores the measurement rate and accuracy of the survey instrument in measuring the variable it was intended to measure. Validity also helps the researcher decide if the survey instrument serves its intended purpose. Three type of validity exist. These are content validity, criterion validity, and construct validity.

Survey instruments must be reliable and valid for researchers to have assurance in the quality of the research data, and the subsequent analysis of the data. The questions that measure a particular variable should produce the same response and be applicable to the variable that they measure. Measurement errors decreases when there is high level of data validity and reliability. Note also that it is possible for a survey instrument to possess only one of these quality i.e. it can be valid buy not reliable or vice versa. For instance, if a research methods instructor included items on an exam that were not covered in class lectures or reading materials, the items could demonstrate reliability through consistency of response even as they fail to constitute a valid assessment of learning material covered in the course – assuming that the purpose of giving the exam is to measure the extent to which students learned about the topics covered in the course [33].

The researcher ensured the validity of the data collection instrument through content validity. It was ensured that all questions of the questionnaire address the designed purpose of analyzing the attitudes of Nigerian organizations towards sustainability and green supply chain. The researcher evaluated the reliability of the instrument by calculating Cronbach's Alpha. This statistic calculates the internal consistency of responses and the value of 0.7 and above is considered reliable. The calculations showed reliable results for this study.

# 7 Results

The survey questionnaire elicited demographic information from the respondents to enable statistical analysis and grouping based on these variables. The respondents were also asked about the number of hours spent on the computer and the number of hours spent using the Internet/emails. The survey is divided into three sections: demographics and background, advantages, and risks. The results of the questions on demographics and background are presented in the charts below.

The gender distribution of the survey respondents is presented in Fig. 6. Gender distribution shows that majority of the research participants were male (63.86%). This is because the researcher could not obtain informed consent from an equal number of female participants, and it was difficult to find female participants in organizations that have their head offices or main offices located in Nigeria, and that have awareness about sustainability and green supply chain in terms of advantages and disadvantages.

The age distribution of the survey respondents is presented in Fig. 7. Age distribution shows that the highest number of participants were in the age bracket of 36–40 years (29%), followed by the age bracket of 31–35 years (24%). This is a good indication for the data because the responses required participants to have some experience regarding the attitude of Nigerian organizations towards sustainability, and their green supply chain awareness.

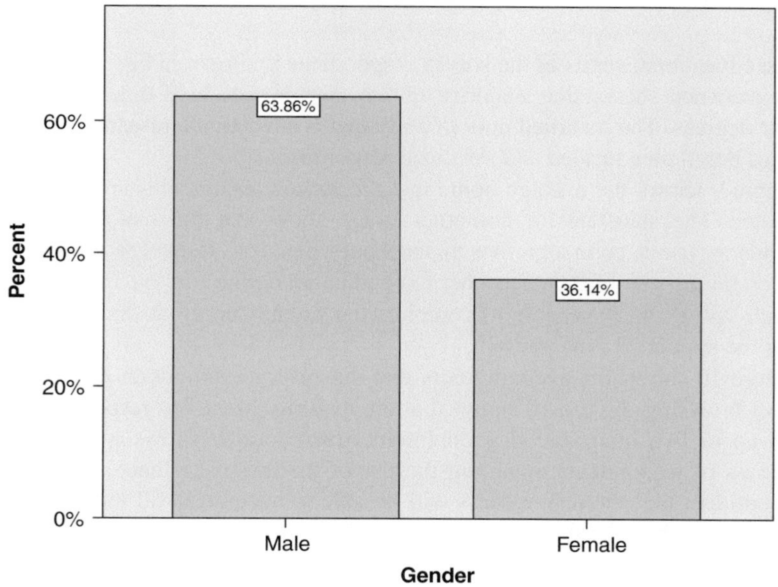


**Fig. 6** Gender distribution of survey respondents

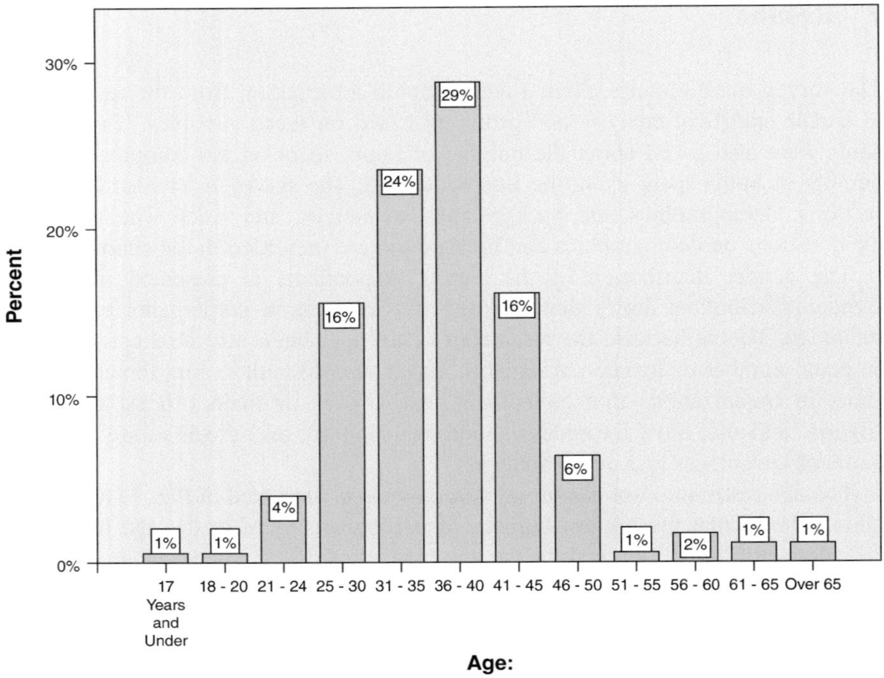


**Fig. 7** Age distribution of survey respondents

The educational status of the survey respondents is shown in Fig. 8. Educational status statistics shows that majority of the respondents held either Bachelor or Master degrees. The qualified pool of participants enabled the researcher to acquire findings based on educated and informed responses.

Figure 9 shows the average hours that the survey respondents spend daily on a computer. The statistics for computer usage show that the majority of survey respondents use a computer five to ten hours per day. Hence, it was a prudent decision on the part of the researcher to conduct an online survey. The respondents are tech-savvy and were able to complete the online survey at their convenience within the stipulated time period.

Figure 10 shows the average hours that the survey respondents spend daily on Internet browsing. Figure 10 shows that the majority of survey respondents (41%) spend up to five hours per day on Internet browsing. It shows a good level of awareness of respondents regarding the use of the Internet. Hence, the researcher was confident that the respondents will be able to accurately fill the survey questionnaire over the Internet.

Figure 11 shows the job titles of the survey respondents. It is evident that respondents belonged to a variety of professions that enabled diversity and variety of perceptions. The highest number of respondents belongs to the profession of

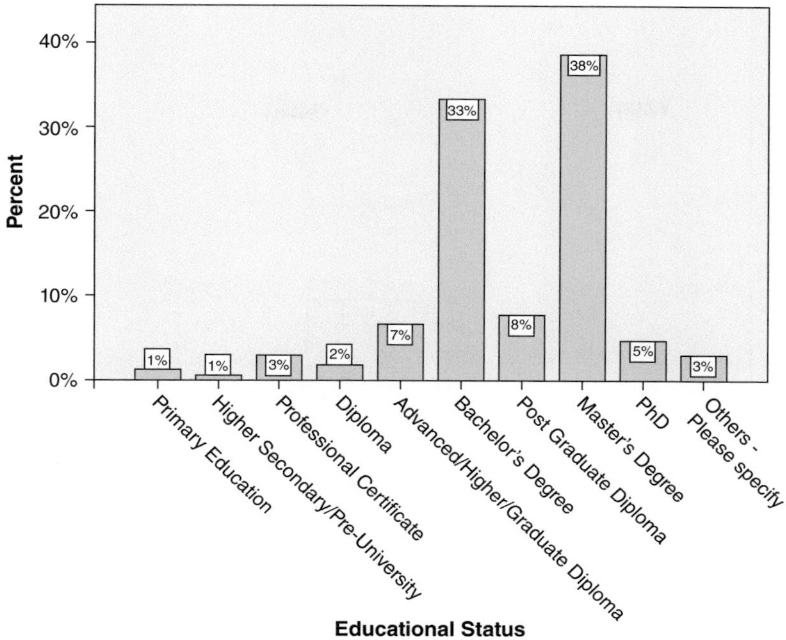


**Fig. 8** Educational status of survey respondents

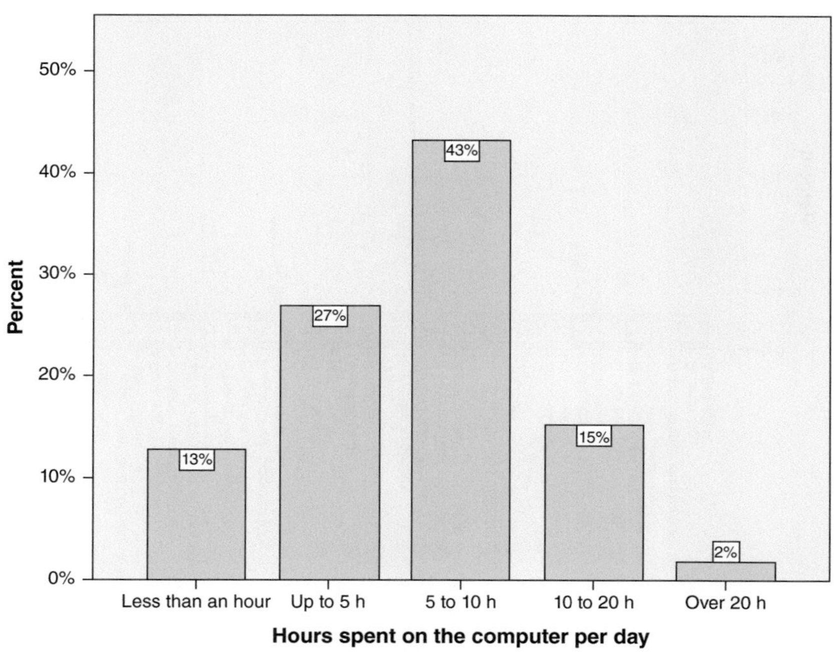


**Fig. 9** Computer usage of survey respondents

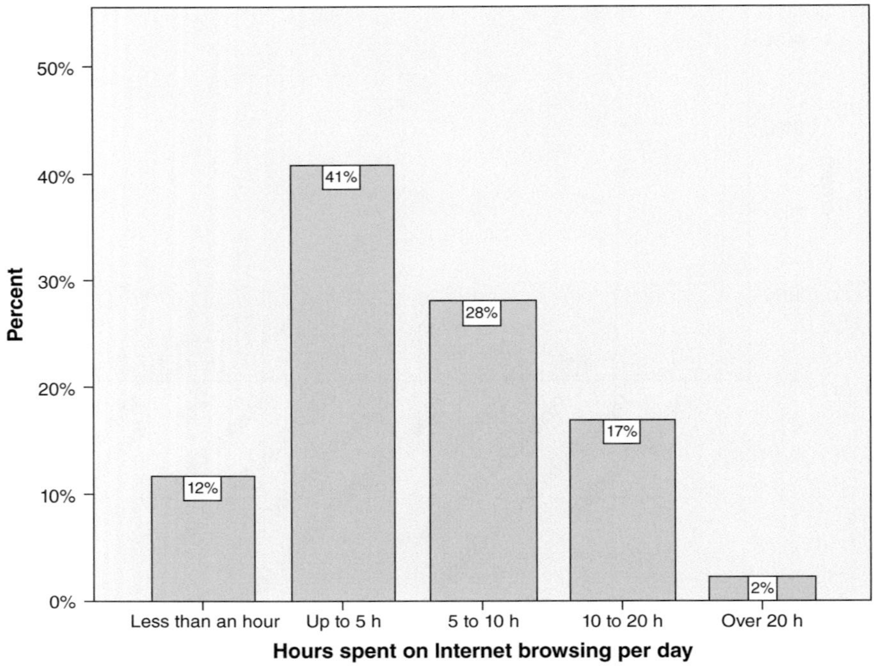


**Fig. 10** Internet usage of survey respondents

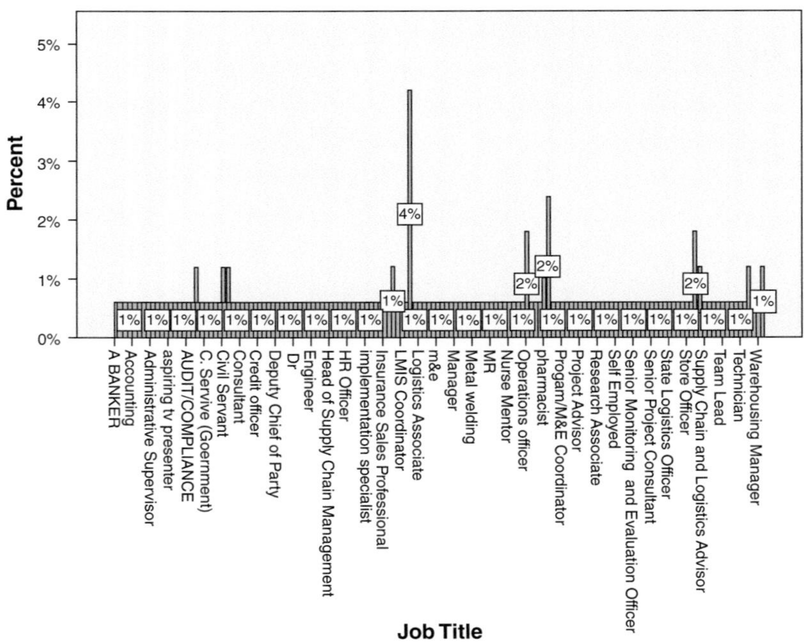


**Fig. 11** Job title of survey respondents

logistics; this is highly desirable as the topic of the study relates to sustainability and green supply chain management.

Figure 12 shows the main fields of study represented by the survey respondents. The research participants belonged to different areas of study including accounting, economics and finance, information technology, management, health sciences, science and engineering. This enabled the researcher to obtain responses from people with different academic backgrounds for the factors that influence the attitude of Nigerian organizations towards sustainability.

Figure 13 shows the number of hours that survey respondents spend on the Internet for social networking. It is obvious that a most of the respondents were making use of social networking sites. This result was important for the researcher as the questionnaire was distributed online, and the researcher also made use of social networking sites for this purpose.

Figure 14 shows the number of hours that survey respondents spend on the Internet for email purposes. It is apparent that the most of the respondents were frequent email users. This result was important for the researcher because the survey was distributed online. If the respondents were not frequent email users, it would have adversely affected the response rate.

Figure 15 shows the purposes for which survey respondents use the Internet. It is evident that respondents were using the Internet for a variety of purposes. These also included email with a percentage of 18. Hence, based on this fact, the

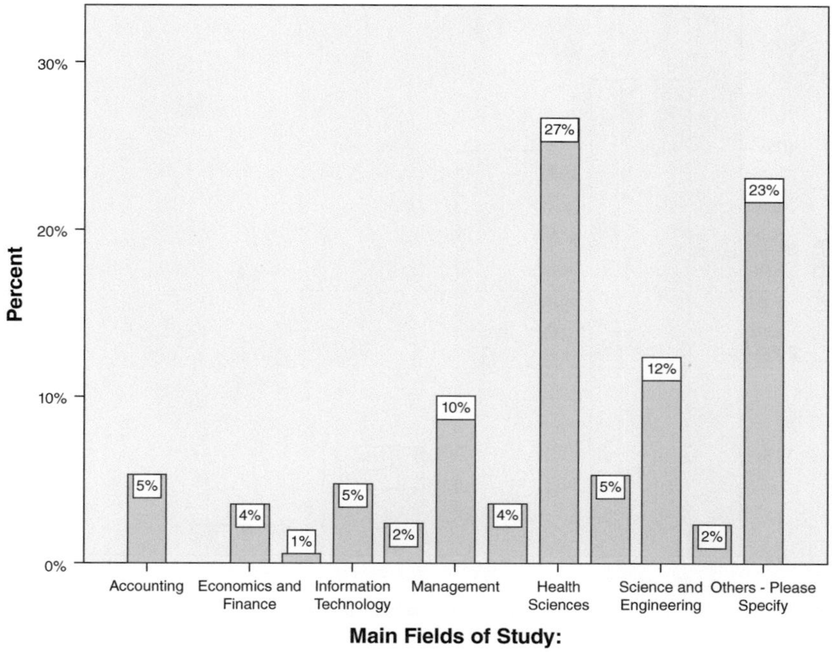


**Fig. 12** Main fields of study of survey respondents

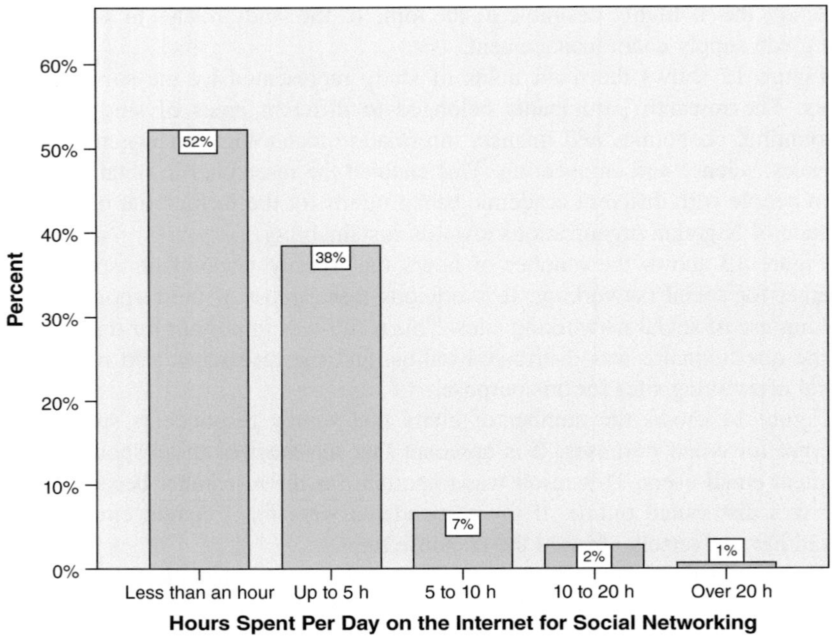


**Fig. 13** Hours spent on Internet for social networking

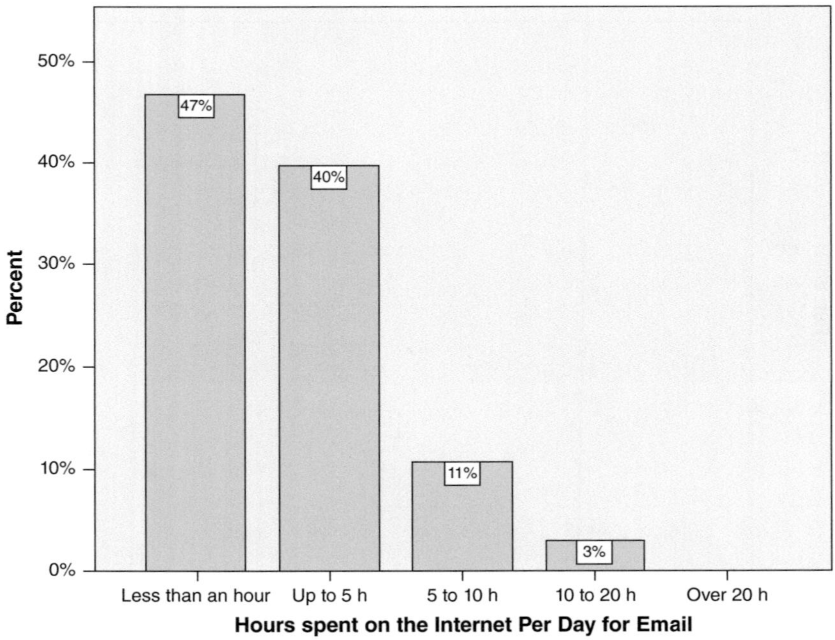


**Fig. 14** Hours spent on Internet for email

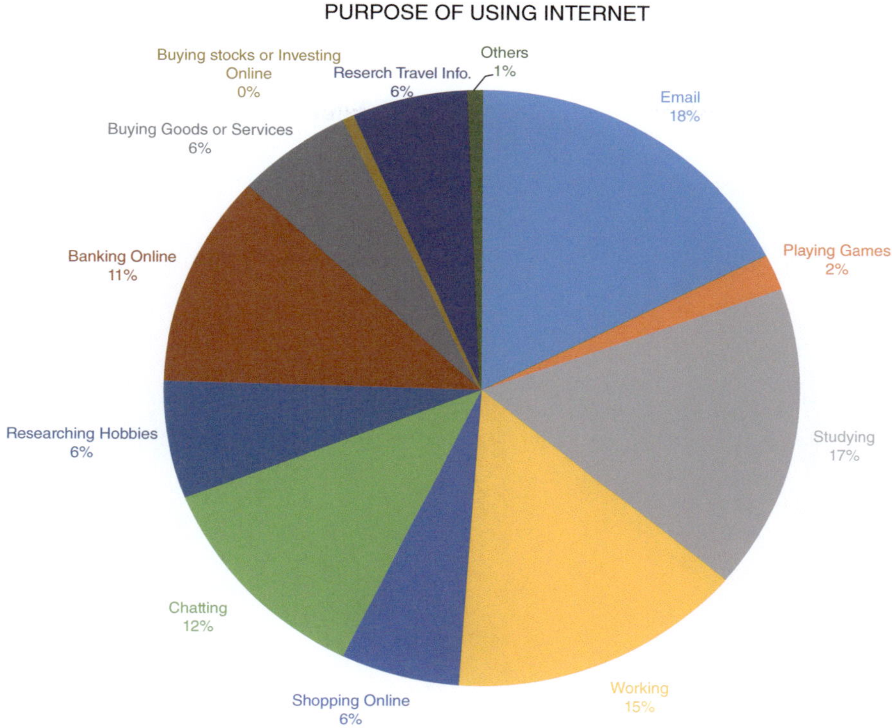


**Fig. 15** Purpose of using Internet for survey respondents

researcher assumed that the response rate to the questionnaire would be good because of the interest of the respondents in using the Internet.

Figure 16 shows the devices which survey respondents use to access the Internet. It is obvious that most of the respondents use desktop computers, laptop, and smartphone. Hence, the disadvantages associated with the use of online surveys were minimal due to the tech-savvy nature of the participants.

Figure 17 shows how the respondents first became acquainted with the concept of sustainability. Figure 17 shows that most of the respondents had a good level of understanding of the concept of sustainability, making them competent as research participants for the study. Their knowledge and understanding of the concept had developed not just through higher education, but also through media, Internet, books, conferences, and magazines.

The survey questionnaire contained 24 items related to the advantages (advantages) of sustainability. The respondents were requested to indicate their level of agreement regarding these advantages. Agreement would imply that the adoption of sustainability and green supply chain by businesses will provide these advantages to the organization or individual.

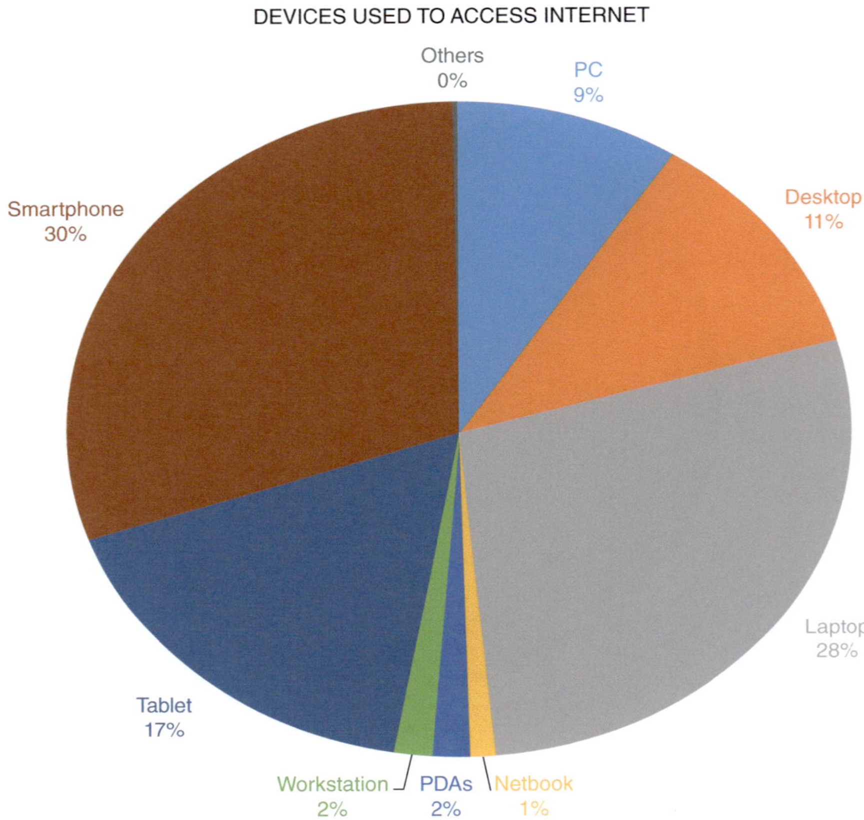


**Fig. 16** Devices used to access Internet

The researcher analyzed the responses for these 24 items for internal consistency through Cronbach's Alpha analysis. The researcher cleaned the data before doing the reliability analysis. This was accomplished through a listwise deletion of cases. In listwise deletion, a case is dropped from an analysis when it has a missing value for at least one of the 24 items considered [38]. The reliability analysis was run only on cases that contained values for all 24 items. Based on these criteria, of the 183 cases, 89 cases were considered as valid for analysis and 94 cases were excluded as shown in Table 1.

Table 2 shows that Cronbach's Alpha is 0.895 that indicates a high level of internal consistency in the responses for the 24 items measuring advantages of sustainability. The minimum acceptable value of Cronbach's Alpha is 0.5. A value of 0.8 and higher is considered good; whereas a value of 0.9 and higher is considered excellent [41]. Based on this classification, the above value of Cronbach's Alpha is regarded as good. The survey questionnaire contained 13 items related to the disadvantages (barriers) to sustainability. The respondents were requested to

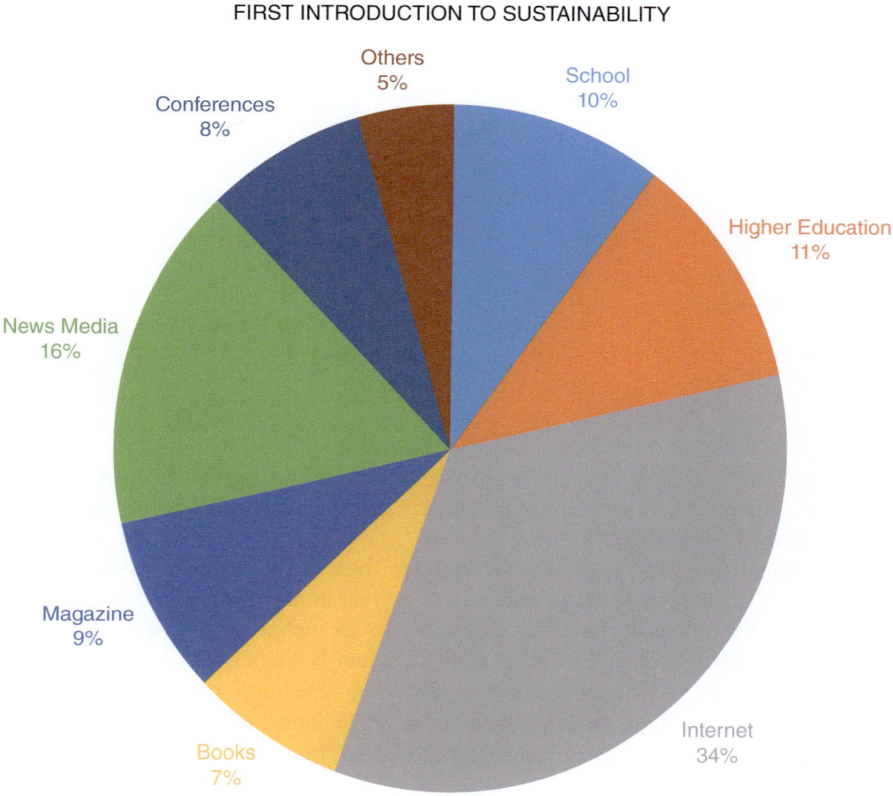


**Fig. 17** First introduction to concept of sustainability

**Table 1** Listwise deletion based on 24 benefit items

| Case processing summary | | | |
|---|---|---|---|
| | | N | % |
| Cases | Valid | 89 | 48.6 |
| | Excluded[a] | 94 | 51.4 |
| | Total | 183 | 100.0 |

[a]Listwise deletion based on all variables in the procedure.

**Table 2** Reliability statistics for advantages of sustainability

| Reliability statistics | | |
|---|---|---|
| Cronbach's Alpha | Cronbach's Alpha based on standardized items | No of Items |
| 0.895 | 0.899 | 24 |

**Table 3** Listwise deletion based on 13 barrier items

| Case processing summary | | | |
|---|---|---|---|
| | | N | % |
| Cases | Valid | 95 | 51.9 |
| | Excluded[a] | 88 | 48.1 |
| | Total | 183 | 100.0 |

[a]Listwise deletion based on all variables in the procedure.

**Table 4** Reliability statistics for barriers to sustainability

| Reliability statistics | | |
|---|---|---|
| Cronbach's Alpha | Cronbach's Alpha based on standardized items | N of items |
| 0.898 | 0.898 | 13 |

indicate their level of agreement regarding these barriers. Agreement would imply that the adoption of sustainability and green supply chain practices in businesses is hindered by these barriers.

The researcher analyzed the responses for these 13 items for internal consistency using Cronbach's Alpha analysis. The researcher cleaned the data before conducting the reliability analysis through listwise deletion of cases. In listwise deletion, a case is excluded from an analysis when it contains a missing value for at least one of the 13 items considered [38]. The reliability analysis was applied on only those cases that contained values for all 13 items. Based on these criteria, of the 183 cases, 95 cases were considered to be valid and 88 cases were excluded, as shown in Table 3.

The result of Cronbach's Alpha for the barriers to sustainability is shown in Table 4.

Table 4 shows that Cronbach's Alpha is 0.898 that indicates a high level of internal consistency in the responses for the 13 items measuring the barriers to sustainability. A value of Cronbach's Alpha that is greater than 0.5 is considered as acceptable and the value greater than 0.8 indicates a good level of internal consistency [18]. Based on this classification, the value of Cronbach's Alpha in the above table shows a good level of internal consistency.

The results of Cronbach's Alpha for both advantages of and barriers to sustainability showed a value closer to 1. This indicates a good level of internal consistency in responses for the scales measuring positive effects as well as negative effects.

The researcher cleaned the data before conducting the factor analysis by excluding the cases that did not have values for all 24 items related to the positive effects of sustainability. Moreover, the researcher opted for fixed number of factors and entered the value 2 in the Factor Analysis extraction dialog box.

The first step in this analysis was to execute KMO and Bartlett's Test for obtaining the sampling adequacy value and Chi-Square significance value. The results are shown in Table 5.

**Table 5** KMO Bartlett's Test for advantages

| KMO and Bartlett's Test | | |
|---|---|---|
| Kaiser-Meyer-Olkin Measure of Sampling Adequacy | | 0.816 |
| Bartlett's Test of Sphericity | Approx. Chi-Square | 1150.334 |
| | *df* | 276 |
| | Sig. | 0.000 |

Table 5 shows that sampling adequacy value for the items related to advantages is 0.816, which is greater than 0.6. In addition, the Bartlett's Test of Sphericity shows the significance value of less than 0.05. Hence, it was concluded that the factor analysis is suitable for the purpose of this study. Factor analysis assists in establishing the notion that the sets of questionnaire items (measuring advantages and disadvantages of sustainability) are measuring the same underlying factor, and hence the items can be combined together to arrive at a more reliable measure of that factor [16].

The second step in the factor analysis is to evaluate the table of communalities. In communalities, the items having extraction values below 0.5 are removed from the analysis, while the items having extraction values greater than or equal to 0.5 are retained. The communalities table is shown in the Table 6.

The results of the above table show 13 items with values less than 0.5. Hence, these items are removed from the list used for measuring the advantages of sustainability. It is important to note the relative size of communalities. They indicate the strength of the linear association among component and variables. Communalities indicate how well the indicators can be combined for the identification of different components [35]. The third step is to evaluate the Total Variance Explained table. Factors with an Eigenvalue of 1.0 or more are retained according to Kaiser's criterion; factors with a value less than 1.0 are removed. The Total Variance Explained is shown in Table 7.

Table 7 shows that two factors of total variance explain the cumulative percentage of advantages with 44.86% which is acceptable to clarify the variance of the extracted factors. Eigenvalues refer to the variance accounted with respect to the number of items worth of variance. Furthermore the percentage of variance shows covariation between items by each factor before and after rotation [37].

The fourth step is to evaluate the table of rotated component matrix. The sample size 183 used for this research study has a factor loading of 0.750, which is deemed a significant statistical value to interpret the factors. The rotation method used was Varimax with Kaiser Normalization. Table 8 shows the rotated component matrix which items load on which components after rotation. Hence, it gives a clear depiction of the principal components [69].

The final step is the grouping and labelling of the factors that belong to each component. Factor 1 contains those items that result in enhancement of the organization's operational and social productivity. This factor has been labelled as Increased Operational and Social Productivity. Factor 2 contains those items that

**Table 6** Communalities for advantages

| Communalities | | |
|---|---|---|
| | Initial | Extraction |
| Reduce energy and water usage | 1.000 | 0.306 |
| Reduce paper usage | 1.000 | 0.322 |
| Reduce consumption of raw materials | 1.000 | 0.250 |
| Increase efficiency | 1.000 | 0.458 |
| Increase cost-effectiveness | 1.000 | 0.324 |
| Enhance reputation | 1.000 | 0.333 |
| Improve corporate social responsibility | 1.000 | 0.500 |
| Reduce risk management | 1.000 | 0.074 |
| Differentiate businesses | 1.000 | 0.207 |
| Increase green strategy | 1.000 | 0.421 |
| Improve community investments | 1.000 | 0.576 |
| Improve human rights | 1.000 | 0.544 |
| Improve social responsibility investing | 1.000 | 0.537 |
| Increase triple bottom line – People, Planet and Profit | 1.000 | 0.424 |
| Increase productivity | 1.000 | 0.686 |
| Attract quality employees | 1.000 | 0.563 |
| Satisfy customer needs | 1.000 | 0.400 |
| Meet stakeholder expectations | 1.000 | 0.260 |
| Attract new opportunities | 1.000 | 0.510 |
| Create new jobs | 1.000 | 0.477 |
| Reduce carbon footprint | 1.000 | 0.640 |
| Reduce emissions | 1.000 | 0.658 |
| Reduce pollution | 1.000 | 0.731 |
| Reduce health hazards | 1.000 | 0.566 |

Extraction method: principal component analysis

**Table 7** Variance for advantages

| Total variance explained | | | | | | | | | |
|---|---|---|---|---|---|---|---|---|---|
| Component | Initial eigenvalues | | | Extraction sums of squared loadings | | | Rotation sums of squared loadings | | |
| | Total | % of variance | Cumulative % | Total | % of variance | Cumulative % | Total | % of variance | Cumulative % |
| 1 | 7.685 | 32.020 | 32.020 | 7.685 | 32.020 | 32.020 | 6.378 | 26.576 | 26.576 |
| 2 | 3.083 | 12.844 | 44.864 | 3.083 | 12.844 | 44.864 | 4.389 | 18.288 | 44.864 |

Extraction method: Principal Component Analysis.

**Table 8** Rotated component matrix for advantages

| Rotated component matrix[a] | Component | |
|---|---|---|
| | 1 | 2 |
| Increase productivity | 0.826 | |
| Attract quality employees | 0.750 | |
| Improve community investments | 0.706 | 0.279 |
| Improve social responsibility investing | 0.702 | 0.211 |
| Improve human rights | 0.697 | 0.240 |
| Improve corporate social responsibility | 0.682 | 0.185 |
| Attract new opportunities | 0.677 | 0.226 |
| Increase efficiency | 0.672 | |
| Create new jobs | 0.650 | 0.233 |
| Satisfy customer needs | 0.628 | |
| Increase triple bottom line – People, Planet and Profit | 0.557 | 0.337 |
| Enhance reputation | 0.543 | 0.196 |
| Meet stakeholder expectations | 0.510 | |
| Increase cost-effectiveness | 0.444 | 0.356 |
| Differentiate businesses | 0.325 | 0.318 |
| Reduce risk management | 0.200 | −0.185 |
| Reduce pollution | | 0.854 |
| Reduce emissions | | 0.807 |
| Reduce carbon footprint | | 0.799 |
| Reduce health hazards | 0.108 | 0.744 |
| Reduce energy and water usage | | 0.548 |
| Reduce paper usage | 0.214 | 0.525 |
| Increase green strategy | 0.404 | 0.508 |
| Reduce consumption of raw materials | 0.144 | 0.479 |

Extraction method: Principal Component Analysis. Rotation method: Varimax with Kaiser Normalization.
[a]Rotation converged in three iterations.

reduce the negative effect the business operations have on the environment. This factor has been labelled as Increased Adherence to Environmental Standards. The grouping of factors is shown in Tables 9 and 10.

The researcher cleaned the data before conducting the factor analysis. This was accomplished by excluding the cases that do not have values for all 13 items related to the disadvantages associated with the adoption of sustainability.

The first step in this analysis was to execute KMO and Bartlett's Test for obtaining the sampling adequacy value and Chi-Square significance value. The results are shown in Table 11.

**Table 9** Factor 1 for advantages

| Advantages of sustainability | Factor loading | Label |
|---|---|---|
| Increase productivity | 0.826 | Increased Operational and Social Productivity |
| Attract quality employees | 0.750 | |
| Improve community investments | 0.706 | |
| Improve social responsibility investing | 0.702 | |
| Improve human rights | 0.697 | |
| Improve corporate social responsibility | 0.682 | |
| Attract new opportunities | 0.677 | |

**Table 10** Factor 2 for Advantages

| Advantages of sustainability | Factor loading | Label |
|---|---|---|
| Reduce pollution | 0.854 | Increased Adherence to Environmental Standards |
| Reduce emissions | 0.807 | |
| Reduce carbon footprint | 0.799 | |
| Reduce health hazards | 0.744 | |

**Table 11** KMO Bartlett's Test for disadvantages

| KMO and Bartlett's Test | | |
|---|---|---|
| Kaiser-Meyer-Olkin Measure of Sampling Adequacy | | 0.854 |
| Bartlett's Test of Sphericity | Approx. Chi-Square | 709.574 |
| | df | 78 |
| | Sig. | 0.000 |

Table 11 shows that sampling adequacy value for the items related to disadvantages is 0.854, which is greater than 0.6. In addition, the Bartlett's Test of Sphericity shows the significance value of less than 0.05. Hence, it was concluded that the factor analysis is suitable for the purpose of this study [16]. The second step in the factor analysis is to evaluate the table of communalities. In communalities, the items having extraction value below 0.5 are removed from the analysis, while the items having extraction value greater than or equal to 0.5 are retained. The communalities table is shown in the Table 12 below:

The results of the above table show three items with values less than 0.5. Hence, these items are removed from the list used for measuring disadvantages associated with sustainability. Communalities indicate how well the indicators can be combined for the identification of different components [35].

The third step is to evaluate the Total Variance Explained table. Factors with an Eigenvalue of 1.0 or more are retained according to Kaiser's criterion; factors with a value less than 1.0 are removed. The Total Variance Explained is shown in the Table 13 below:

**Table 12** Communalities for disadvantages

| Communalities | | |
|---|---|---|
| | Initial | Extraction |
| Increase fraud (including environmental actions and environmental compliance failures) | 1.000 | 0.493 |
| Increase litigation and compliance breaches (including environmental actions and environmental compliance failures) | 1.000 | 0.481 |
| Increase governance failure (including environmental actions and environmental compliance failures) | 1.000 | 0.765 |
| Increase transaction failure due to environmental liabilities | 1.000 | 0.818 |
| Increase security and systems failures caused by environmental problems | 1.000 | 0.670 |
| Increase scandals by perceived environmental irresponsibility | 1.000 | 0.721 |
| Increase marketing failure by perceived environmental irresponsibility | 1.000 | 0.658 |
| Increase supply chain crises due to suppliers' environmental problems | 1.000 | 0.570 |
| Inflate costs | 1.000 | 0.631 |
| Increase interest rates | 1.000 | 0.612 |
| Increase insurance crises due to environmental disasters | 1.000 | 0.643 |
| Increase number of new regulations including environmental | 1.000 | 0.248 |
| Increase competition for and cost of raw materials | 1.000 | 0.652 |

Extraction method: Principal Component Analysis.

**Table 13** Variance for disadvantages

| Total variance explained | | | | | | | | | |
|---|---|---|---|---|---|---|---|---|---|
| Component | Initial eigenvalues | | | Extraction sums of squared loadings | | | Rotation sums of squared loadings | | |
| | Total | % of variance | Cumulative % | Total | % of variance | Cumulative % | Total | % of variance | Cumulative % |
| 1 | 6.077 | 46.749 | 46.749 | 6.077 | 46.749 | 46.749 | 5.051 | 38.851 | 38.851 |
| 2 | 1.885 | 14.501 | 61.250 | 1.885 | 14.501 | 61.250 | 2.912 | 22.400 | 61.250 |

Extraction method: Principal Component Analysis.

Table 13 shows that three factors of total variance explain the cumulative percentage of risks with 61.25% which is acceptable to explain the variance of the extracted factors. The fourth step is to evaluate the table of rotated component matrix. The sample size (183) used for this research study has a factor loading of 0.785, which is a significant statistical value to interpret the factors. The rotation method used was Varimax with Kaiser Normalization. As shown in Table 14, the rotation converged in three iterations.

The final step is the grouping and labelling of the factors that belong to each component. Factor 1 contains those items that increase the risk of environmental management failure in terms of transactions, governance, security, systems, marketing, scandals, and supply chain crises. This factor has been labelled as Increased Risk of Management Failure. Factor 2 contains those items that increases the costs

**Table 14** Rotated component matrix for disadvantages

| Rotated component matrix[a] | | |
|---|---|---|
| | Component | |
| | 1 | 2 |
| Increase transaction failure due to environmental liabilities | 0.889 | 0.168 |
| Increase governance failure (including environmental actions and environmental compliance failures) | 0.874 | 0.014 |
| Increase scandals by perceived environmental irresponsibility | 0.832 | 0.171 |
| Increase security and systems failures caused by environmental problems | 0.807 | 0.135 |
| Increase marketing failure by perceived environmental irresponsibility | 0.757 | 0.292 |
| Increase litigation and compliance breaches (including environmental actions and environmental compliance failures) | 0.664 | 0.200 |
| Increase supply chain crises due to suppliers' environmental problems | 0.663 | 0.360 |
| Increase fraud (including environmental actions and environmental compliance failures) | 0.658 | 0.246 |
| Increase competition for and cost of raw materials | −0.001 | 0.808 |
| Inflate costs | 0.104 | 0.788 |
| Increase interest rates | 0.295 | 0.725 |
| Increase insurance crises due to environmental disasters | 0.378 | 0.707 |
| Increase number of new regulations including environmental | 0.158 | 0.472 |

Extraction method: Principal Component Analysis. Rotation method: Varimax with Kaiser Normalization.
[a]Rotation converged in three iterations.

**Table 15** Factor 1 for disadvantages

| Barriers to sustainability | Factor loading | Label |
|---|---|---|
| Increase transaction failure due to environmental liabilities | 0.889 | Increased Risk of Management Failure |
| Increase governance failure (including environmental actions and environmental compliance failures) | 0.874 | |
| Increase scandals by perceived environmental irresponsibility | 0.832 | |
| Increase security and systems failures caused by environmental problems | 0.807 | |
| Increase marketing failure by perceived environmental irresponsibility | 0.757 | |
| Increase supply chain crises due to suppliers' environmental problems | 0.663 | |

associated with complying with environmental related regulations and laws. These items include increase in the competition for and cost of raw materials, increased inflation costs, increased interest rates, and increased insurance crisis resulting from environmental disasters. This factor has been labelled as Increased Regulatory and Compliance Costs. The grouping of factors is shown in Tables 14 and 15.

**Table 16** Factor 2 for disadvantages

| Barriers to sustainability | Factor loading | Label |
|---|---|---|
| Increase competition for and cost of raw materials | 0.808 | Increased Regulatory and Compliance Costs |
| Inflate costs | 0.788 | |
| Increase interest rates | 0.725 | |
| Increase insurance crises due to environmental disasters | 0.707 | |

# 8 Discussion and the New Significance

This study analyzed sustainability and green supply chain awareness in Nigerian organizations. The research question of the study was as follows: "What are the factors that influence the attitude of Nigerian organizations towards sustainability in terms of its advantages and disadvantages?" The study was based on the collection of both primary and secondary data. In the review of the literature, the major themes analyzed included the concept of sustainability and green supply chain, history of sustainability, triple bottom line, significance of sustainability for business/education, current progress level of sustainability, sustainability in developing countries, and sustainability in Nigeria. The review also focused on the advantages and barriers to sustainability based on the survey data.

The researcher selected the quantitative research method for primary data analysis and the qualitative research method for secondary data analysis. The instrument for primary data collection was a survey questionnaire. The questionnaire contained 24 items related to the positive effects of sustainability and 13 items related to the disadvantages of sustainability. The survey was a Web-based survey and it was distributed online using Qualtrics software. The study sample consisted of employees working in organizations that have their head offices or main offices located in Nigeria. The sample size of the study was 183. The analysis of the survey focused on the demographic results, reliability statistics, and factor analysis. The reliability statistics showed an acceptable value of Cronbach's alpha for the items related to the scale of positive effects as well as the disadvantages of sustainability. The factor analysis showed two major factors for advantages of sustainability. These included increased operational and social productivity, and increased adherence to environmental standards. The factor analysis also showed two major factors for disadvantages of sustainability. These included increased risk of management failure, and increased regulatory and compliance costs. The sections below briefly summarize the research findings in terms of the advantages of sustainability and the disadvantages to the implementation of sustainability initiatives by governments and organizations. The initial concept maps presented for these in Chap. 3 were also revised to incorporate the new factors that arose from the factor analysis of the research data. The new concept maps for the advantages of, and barriers to, sustainability are presented in Figs. 18 and 19 below.

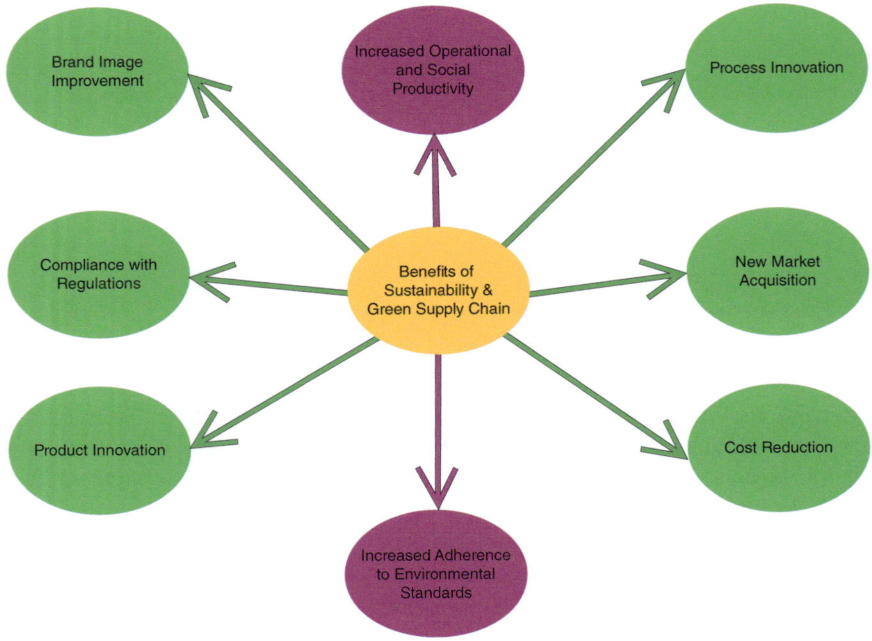


**Fig. 18** Revised concept map of the advantages of sustainability

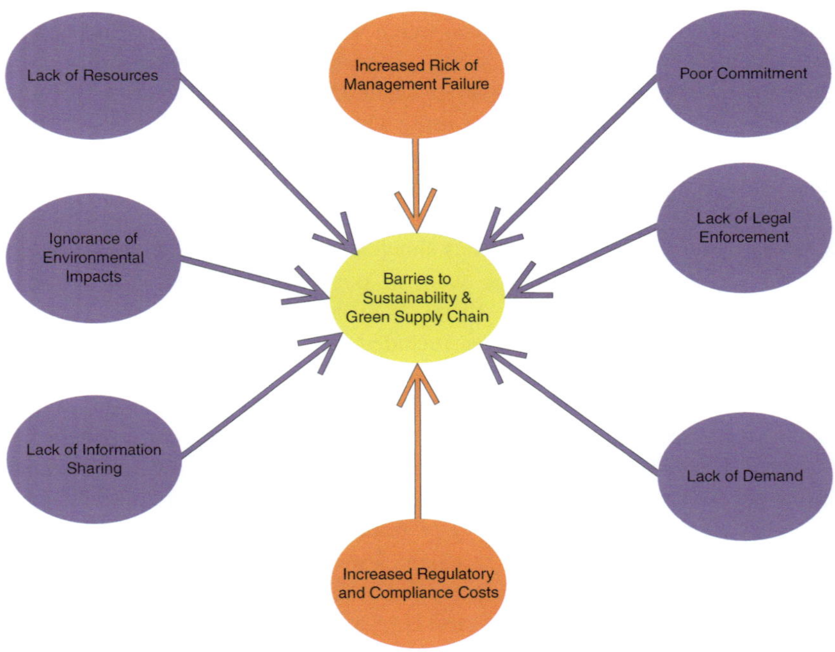


**Fig. 19** Revised concept map of the barriers to sustainability

The first benefit of the adoption and implementation of sustainability initiatives consists of organizational efforts that have the ability to improve the organization's overall productivity and simultaneously enhance their respect and acceptance in the society that they operate in. There were seven items that contribute to this benefit and they were collectively named "Increased Operational and Social Productivity".

The second benefit consists of those items that help reduce the negative effect that business operations have on the environment and its ecological components. This benefit helps in the reduction of environmental negatives like carbon footprints, gaseous emissions, pollutions, and health hazards. These advantages were collectively named "Increased Adherence to Environmental Standards". Figure 18 present a new version of concept map for the advantages of sustainability which has been revised to reflect these new advantages.

The new disadvantages to the implementation of sustainability initiatives were also extracted from the factor analysis of the survey data.

The first barrier consists of the group of factors that the potential of increasing the risk of system failure as a result of poor management practices. They include environmentally-based problems relating to transaction problems, governance failure, security and system issues, management scandals, market share loss, and supply chain crisis. These factors were collectively named "Increased Risk of Management Failure" and based on the author's opinion, they suggest that a more environmentally aware management model may help mitigate the emergence of these issues.

The second group of factors that were identified as potential disadvantages to the adoption of sustainability initiatives are those factors that increases the cost of complying with environmental sustainability laws. They include: increase in raw material competition and cost, increase in inflation costs, increase interest rates, and increase in environmentally related insurance crisis. These factors were considered together under the heading "Increased Regulatory and Compliance Costs".

Figure 19 present a new version of concept map for the disadvantages to sustainability which has been revised to reflect these new disadvantages.

## 9 Limitations, Future Research and Recommendations

This study had several limitations. Firstly, it focused only on the Nigerian organizations. Although, the researcher collected both primary and secondary data for the study, and the findings of the primary data were compared with the analysis of the secondary data, yet the generalizability of the findings may be difficult for other parts of the world due to the specific context and circumstances of Nigeria. Since Nigeria is an African country, the researcher contends that the findings are particularly beneficial in the African context and are useful for other parts of the world in general. The second limitation of the study is the sample size. Although the sample of 183 may be considered a good sample size, the canvas and scope of sustainability analysis could have been broadened by conducting the study with a larger sample size. This would have elicited more views, factors, and perceptions in terms of the advantages and disadvantages related to sustainability.

## 10 Conclusion from the Analysis – Primary and Secondary Data

The key research question of the study was: "What are the factors that influence the attitude of Nigerian organizations towards sustainability in terms of advantages and disadvantages?" Based on the analysis of primary and secondary data, the researcher identified two main factors for advantages and three main factors for disadvantages of sustainability. The factors for advantages included increased operational and social productivity, and increased adherence to environmental standards. Increased operational and social productivity results in the organization improving their overall productivity through the attraction of quality employees and new opportunities. They are also able to enhance their corporate social image by investing more in the community, protecting human rights, and improving on their social responsibility investments. Increased adherence to environmental standards enables a reduction of the carbon footprints, gaseous emissions, pollutions, and health hazards. The factors for disadvantages included increased risk of management failure, and increased regulatory and compliance costs. Risk of management failure may increase due to: transaction failures due to environmental liabilities, governance failure, security and system issues, scandals resulting from environmental irresponsibility, marketing failure, and supply chain crises. Increased regulatory and compliance costs can result from increased competition and cost of raw materials, increased inflation costs, increased interest rates, and increased insurance crises.

Prior to now, the use of sustainability in business has been viewed as independent initiatives that belong to the realms of strategic business management. This has presented a barrier to the successful implementation of the sustainability initiatives in business, since it is in direct opposition to the current trend of operationalization of sustainability practices at every level of organizational structure. Contemporary literature highlights the importance of entrenching sustainable business practices in all the business functional areas, and makes it the responsibility of every employee in the business. There should be sustained efforts to continuously monitor and evaluate the performance level of current business sustainability practice with the view of adapting to current trends.

The triple bottom line modifies the focus of organizations so that they focus not only on profits, but on the generation of profits in a manner that is beneficial for the people as well as the planet.

Sustainability is an area of concern for all businesses today, irrespective of their place or size in the marketplace. Businesses should realize the importance of sustainable practices for more reliable products, better corporate culture, and greater long-term profitability.

The conventional wisdom that economic growth always leads to human progress has been challenged by the presence of modern anomalies such as global climate change, ecological destruction, continuous economic crisis, resource depletion, and social inequality. Warnings about the future of the growth-driven economy have been given by numerous economic philosophers and international bodies.

The reduction in environmental degradation by means of supply chains is expected to affect the environment to a large extent. The economy of Nigeria is heavily dependent on petroleum. Hence, the environmental wastes from the petroleum sector are at an elevated level particularly from the supply chain. Stakeholders in the Nigerian petroleum industry supply chain may adopt green practices if they could be assured of commitment and trust from the major partners. Additionally, if all organizations in Nigeria set individual targets of reducing carbon emission, the downstream partners in the supply chain, such as the small size petroleum retail outlets and filling stations, may more readily adopt their environmental policies [67].

The Nigeria situation with respect to the adoption and implementation of the issues related to MDG7 is so critical that the United Nations Development Programme (UNDP) observed that the country's natural resources is under serious threat. Recognizing this problem, and the need to strive toward reversal of this threat, the Nigeria government set up mechanisms at national and state levels for achieving the MDGs by 2015 [2].

In 2010, BearingPoint carried out a study linked to green supply chain. The organization had also conducted a similar survey in 2008. Hence, this survey also compared the findings of the 2010 survey with the findings of the 2008 survey. According to the survey, the motivations for implementing green actions in organizations included improvement to the of brand image, the commitment decision by the board of executives, environmental protection laws, opportunities for process and product, reduction in overall operating cost, level of competitive pressure, the drive for new market acquisitions and pressure from opinion leaders. Among these motivators, enhanced brand image was the biggest motivator and the pressure of opinion leaders was the least powerful motivator.

Nine barriers that mitigate against the attainment of green supply chain management in Nigeria has also been identified. These included poor public awareness, poor knowledge of environmental impact issues, lack of top management commitment, weak enforcement of environmental laws by government agencies, poor resource base, absence of sustainability oriented organizational mission and vison, absence of markets for recyclable materials, poor sharing of information between supply chain partners, and lack of societal demand for sustainable supply chain practices [53].

A recent global commerce initiative report has also recommended that improved collaboration between all the stakeholders in the value chain is essential for achieving a more effective and efficient value chain for meeting consumer needs. The report focuses on the future of the supply chain and the critical role of collaboration in helping organizations to move forward. The key step in defining suitable scenarios for the future supply chain is acquiring an understanding of the relevant trends and forces that would make an impact in the coming years [30].

## 11 Future Research and Recommendations

This study identified important factors that influence the attitude of Nigerian organizations towards sustainability in terms of its advantages and disadvantages. The study identified three important factors in terms of advantages and three important factors in terms of disadvantages. Future studies may expand on the findings of the current study and explore further factors influencing organizational attitude towards sustainability. This study was based on the attitude of Nigerian organizations. Future studies may be conducted with a broader scope such as the whole African context or the global context. Future studies may also be conducted with a sample size larger than this study to incorporate additional perspectives and diverse opinions.

This study identified two crucial factors that can provide new opportunities and advantages for organizations that adopt sustainability initiatives. Organizations should focus their sustainability initiatives on the fulfillment of two broad goals: increasing operational and social productivity, and increasing adherence to environmental standards. The goal of increasing operational and social productivity should focus on adopting practices that will attract investment opportunities and quality employees that can be motivated to efficiently utilize scarce resources. Concerted efforts should be channeled towards improving/enhancing corporate reputation, community investments, human rights conditions, and social responsibility investments. The goal of increasing adherence to environmental standards should be divided into the objectives of reducing the carbon footprints, gaseous emissions, pollutions, and environmental health hazards.

The study also identified two crucial factors that may become disadvantages and disadvantages for organizations in the context of sustainability initiatives. Organizations should take these factors into consideration and develop risk assessment measures and contingency planning to address these disadvantages and disadvantages. The two disadvantages identified in this study are: increased risk of management failure, and increased regulatory and compliance costs. For the increased risk of management failure, organizations should take into account that sustainability initiatives may lead to increased transaction failures due to environmental liabilities, increased governance failure, increased demand on security and system integrity, increased scandals resulting from environmental irresponsibility, increased marketing failure, and increased supply chain crises. For the increased regulatory and compliance costs, organizations should be aware that sustainability initiatives may increase competition and cost of raw materials, inflation costs, interest rates, and environmentally related insurance crises.

## 12 Conclusion

The concept of sustainability is a next global mega trend that has fueled a paradigm shift towards the inclusion of short-term and long-term environmental considerations in organizational strategies, policies, and operations. Green supply chain management involves the addition of the sustainability considerations in supply chain management, and implies analyzing the relationships and impact between the natural environment and the supply chain. The growth of sustainability and its adoption by government and organization is associated with advantages, and disadvantages that may facilitate or hinder successful implementation of sustainability programs. Available secondary data provides information on the possible advantages and disadvantages associated with the adoption of sustainability initiatives. A 2010 BearingPoint survey revealed six advantages of adopting sustainability initiatives in the supply chain. These advantages are brand image improvement, compliance with regulation, product innovation, process innovation, new market acquisition, and cost reduction. Among these identified advantages, brand image improvement was the biggest motivator. A survey in Nigeria identified some challenging factors to the implementation of sustainability initiatives the supply chain. These disadvantages are lack of resources, ignorance of environmental impacts, lack of information sharing, poor commitment, lack of legal enforcement, and lack of demand. The research findings revealed two new key factors each for the advantages of sustainability, and the disadvantages to sustainability that adds to the existing primary data. The new key factors for the advantages of sustainability are: (1) increased operational and social productivity, and (2) increased adherence to environmental standards. The new key factors for the disadvantages to sustainability are: (1) increased risk of management failure, and (2) increased regulatory and compliance costs. Based on the research findings, it is recommended that organizations and governments focus their sustainability initiatives on the fulfillment of three broad goal categories that will enable them to achieve most of the desirable advantages of sustainability, while mitigating the associated disadvantages. These goals are: (1) increasing operational and social productivity, (2) increasing adherence to environmental standards, and (3) developing risk management plans that mitigates against risks associated with increased management failure and potential increase in sustainability-related regulatory and compliance costs.

## References

1. Absi N, Dauzère-Pérès S, Kedad-Sidhoum S, Penz B, Rapine C (2013) Lot sizing with carbon emission constraints. Eur J Operational Res 227(1):55–61
2. Adeagbo A (2013) Achieving environmental sustainability (MDG7) in Nigeria: progress so far, challenges and prospects. Acad J Interdiscip Stud 2(6):47
3. Adejumo Akintoye V, Adejumo Opeyemi O (2014) Prospects for achieving sustainable development through the millennium development goals in Nigeria. Eur J Sustain Dev 3(1):33

4. Ahi P, Searcy C (2015) Assessing sustainability in the supply chain: a triple bottom line approach. Appl Math Model 39(10):2882–2896
5. Andersen I, Golitzen Katherin G (2005) The Niger river basin: a vision for sustainable management. World Bank Publications. http://documents.worldbank.org/curated/en/187941468097179065/pdf/345180PAPER0NR1Basin01OFFICIAL0USE1.pdf. Accessed 1 June 2017
6. Barbier E (2010) How is the global green new deal going? Nature 464(7290):832–833
7. Beamon Benita M (2005) Environmental and sustainability ethics in supply chain management. Sci Eng Ethics 11(2):221–234
8. Bektaş T, Laporte G (2011) The pollution-routing problem. Transp Res Part B Methodol 45(8):1232–1250
9. Bilgen S (2014) Structure and environmental impact of global energy consumption. Renew Sustain Energy Rev 38:890–902
10. Boons F, Lüdeke-Freund F (2013) Business models for sustainable innovation: state-of-the-art and steps towards a research agenda. J Clean Prod 45:9–19
11. Bourbonnais P, Morency C (2013) Web-based travel survey: a demo. In: Transport survey methods: best practice for decision making. pp 207–223
12. Brau James C, Sutton Ninon K, Hatch Nile W (2010) Dual-track versus single-track sell-outs: an empirical analysis of competing harvest strategies. J Bus Ventur 25(4): 389–402
13. Chaabane A, Ramudhin A, Paquet M (2012) Design of sustainable supply chains under the emission trading scheme. Int J Prod Econ 135(1):37–49
14. Chin Thoo A, Tat Huam H, Sulaiman Z (2015) Green supply chain management, environmental collaboration and sustainability performance. Procedia CIRP 26:695–699
15. Clark Bruce H (2001) A summary of thinking on measuring the value of marketing. J Target Meas Analysis Mark 9(4):357–369
16. Comrey Andrew L, Lee Howard B (2013) A first course in factor analysis. Psychology Press, UK
17. Coteur I, Marchand F, Debruyne L, Dalemans F, Lauwers L (2016) A framework for guiding sustainability assessment and on-farm strategic decision making. Environ Impact Assess Rev 60:16–23
18. Curtis E, Drennan J (2013) Quantitative health research: issues and methods: issues and methods. McGraw-Hill Education, Berkshire, England
19. Daly Herman E (1990) Toward some operational principles of sustainable development. Ecol Econ 2(1):1–6. doi:10.1016/0921-8009(90)90010-R
20. Davis Steven J, Caldeira K (2010) Consumption-based accounting of CO2 emissions. Proc Natl Acad Sci U S A 107(12):5687–5692
21. Davison RM, Li Y, Kam CSP (2008) Web-based data collection in China. Handb Res Information Manag Glob Landsc 7:1584
22. De Lima Rafael G, Lins Hoyêdo N, Pfitscher Elisete D, Garcia J, Suni A, De Andrade José Baltazar Salgueirinho O, Delle Finati Caroline R (2016) A sustainability evaluation framework for science and technology institutes: an international comparative analysis. J Clean Prod 125:145–158
23. Dyckhoff H, Lackes R, Reese J (2013) Supply chain management and reverse logistics. Springer Science & Business Media, Springer-Verlag, Berlin Heidelberg
24. Edgerton E, Romice O, Thwaites K (2014) Bridging the boundaries: human experience in the natural and built environment and implications for research, policy, and practice, vol 5. Hogrefe Publishing, Boston
25. Edomah N (2016) On the path to sustainability: key issues on Nigeria's sustainable energy development. Energy Rep 2:28–34
26. Elhedhli S, Merrick R (2012) Green supply chain network design to reduce carbon emissions. Transp Res Part D Trans Environ 17(5):370–379
27. Engardio P, Capell K, Carey J, Hall K (2007) Beyond the green corporation: imagine a world in which eco-friendly and socially responsible practices actually help a company's bottom line. It's closer than you think. Bus Week 4019:50

28. Falola T, Heaton MM (2008) A history of Nigeria. Cambridge University Press. http://assets.cambridge.org/97805218/62943/frontmatter/9780521862943_frontmatter.pdf. Accessed 1 June 2017
29. Fuentes-Nieva R, Galasso N (2014) Working for the few: Political capture and economic inequality, vol 178. Oxfam. https://www.oxfam.org/sites/www.oxfam.org/files/file_attachments/bp-working-for-few-political-captureeconomic-inequality-200114-en_3.pdf. Accessed 1 June 2017
30. Gci Capgemini (2008) The future supply chain 2016. Serving consumers in a sustainable way. Global Commerce Initiative and Capgemini. http://supplychainmagazine.fr/TOUTE-INFO/ETUDES/GCI_Capgemini-SC2016.pdf. Accessed 1 June 2017
31. Gnoni MG, Elia V, Lettera G (2011) A strategic quantitative approach for sustainable energy production from biomass. Int J Sustain Eng 4(2):127–135. doi:10.1080/19397038.2010.544420
32. Grant Robert M (2016) Contemporary strategy analysis: text and cases edition. John Wiley & Sons, UK
33. Hair Joseph F (2015) Essentials of business research methods, 2nd edn. ME Sharpe, Routledge, UK
34. Hall Tanya J (2011) The triple bottom line: what is it and how does it work? Indiana Bus Rev 86(1):4
35. Henry C (2003) Microfinance poverty assessment tool, vol 255. World Bank Publications. https://www.cgap.org/sites/default/files/CGAP-Technical-Guide-Microfinance-Poverty-Assessment-Tool-Sep-2003.pdf. Accessed 1 June 2017
36. Hohnen P (2014) $2.5tn shortfall for sustainable development in developing countries. *The Guardian*, 27 Oct 2014. https://www.theguardian.com/sustainable-business/2014/oct/27/25-trillion-shortfall-developingcountries-sustainable-development. Accessed 1 June 2017
37. Ibm (2014) Exploratory factor analysis and principal components analysis. IBM 74
38. Ibm ( 2016) Pairwise vs. listwise deletion: what are they and when should I use them? IBM
39. Lélé Sharachchandra M (1991) Sustainable development: a critical review. World Dev 19(6):607–621. doi:10.1016/0305-750X(91)90197-P
40. Lenssen G, Tyson S, Pickard S, Bevan D, Lacy P, Arnott J, Lowitt E (2009) The challenge of integrating sustainability into talent and organization strategies: investing in the knowledge, skills and attitudes to achieve high performance. Corp Govern 9(4):484–494
41. Liebenson C (2007) Rehabilitation of the spine: a practitioner's manual. Lippincott Williams & Wilkins, Philadelphia
42. Linton Jonathan D, Klassen R, Jayaraman V (2007) Sustainable supply chains: an introduction. J Operations Manag 25(6):1075–1082
43. Llp Ernest Young, Citizenship Boston College Center for Corporate (2013) Value of sustainability reporting: a study by EY and Boston College Center for Corporate Citizenship. http://www.ey.com/us/en/services/specialty-services/climate-change-and-sustainability-services/value-ofsustainability-reporting. Accessed 1 June 2017
44. Lorenz K, Crutzen PJ, Lal R, Töpfer K (2012) Atmospheric chemistry and climate in the anthropocene. In: Recarbonization of the Biosphere. Springer, the Netherlands. pp 41–58
45. Lozano R, Nummert B, Ceulemans K (2016) Elucidating the relationship between sustainability reporting and organisational change management for sustainability. J Clean Prod 125:168–188
46. Marsden Peter V, Wright James D (2010) Handbook of survey research. Emerald Group Publishing, Bingley
47. Mcmichael Anthony J (2013) Globalization, climate change, and human health. N Engl J Med 369(1):96
48. Mitcham Carl (1995) The concept of sustainable development: its origins and ambivalence. Technol Soc 17(3):311–326. doi:10.1016/0160-791X(95)00008-F
49. Nässén J, Holmberg J, Wadeskog A, Nyman M (2007) Direct and indirect energy use and carbon emissions in the production phase of buildings: an input–output analysis. Energy 32(9):1593–1602
50. Nelson D, Marsillac E, Rao S (2012) Antecedents and evolution of the green supply chain. J Operations Supply Chain Manag 1(1):29–43

51. Obono O (2003) Cultural diversity and population policy in Nigeria. Popul Devel Rev 29(1):103–111
52. Odjugo PA (2010) Regional evidence of climate change in Nigeria. J Geogr Regional Plan 3(6):142–150
53. Ojo E, Mbowa C, Akinlabi ET (2014) Barriers in implementing green supply chain management in construction industry. In: International Conference on Industrial Engineering and Operations Management Bali, Indonesia, January 7–9, 2014, pp 1974–1981. http://ieomsociety.org/ieom2014/pdfs/432.pdf. Accessed 1 June 2017
54. Peters Glen P, Hertwich Edgar G (2008) CO2 embodied in international trade with implications for global climate policy. Envir Sci Technol 42(5):1401–1407
55. Point Bearing (2010) Green supply chain: from awareness to action. Bearing Point Management and Technology Consultants. https://www.bearingpoint.com/files/TAP-SC_EN.pdf. Accessed 1 June 2017
56. Roberts N, Thompson F (2006) "Netcentric" organization. Public Administration Rev 66(4):619–622
57. Rogmans T, Ghunaim M (2016) A framework for evaluating sustainability indicators in the real estate industry. Ecol Indic 66:603–611
58. Sauermann H, Roach M (2013) Increasing web survey response rates in innovation research: an experimental study of static and dynamic contact design features. Res Policy 42(1):273–286
59. Savitz A (2012) The triple bottom line: how today's best-run companies are achieving economic, social and environmental success – and how you can too. John Wiley & Sons, San Francisco, California
60. Spengler T, Stölting W, Ploog M (2004) Recovery Planning in Closed Loop Supply Chains: An Activity Analysis Based Approach. In: Supply chain management and reverse logistics. Springer, Berlin Heidelberg. pp 61–89
61. Spitz G, Niles FL, Adler TJ (2006) Web-based survey techniques, vol 69. Transportation Research Board
62. Srivastava Samir K (2007) Green supply-chain management: a state-of-the-art literature review. Int J Manag Rev 9(1):53–80
63. Taranenco L (2013) Concept of sustainable development of small and medium enterprises in Moldova. Knowl Horizons Econ 5(2):126–129
64. Teuscher P, Grüninger B, Ferdinand N (2006) Risk management in sustainable supply chain management (SSCM): lessons learnt from the case of GMO-free soybeans. Corp Soc Responsib Environ Manag 13(1):1–10. doi:10.1002/csr.81
65. Tognetti A, Grosse-Ruyken PT, Wagner SM (2015) Green supply chain network optimization and the trade-off between environmental and economic objectives. Int J Prod Econ 170:385–392. doi:10.1016/j.ijpe.2015.05.012
66. Tonape S, Owk M (2013) An overview, trends and future mapping of green supply chain management – perspectives in India. J Supply Chain Manag Syst 2(3):48–56
67. Ukpabi DC, Ikaba YV, Enyindah CW, Idatoru AR (2015) Impact of relationship marketing on green supply chain management (GSCM): an empirical study in the oil and gas industry in Port Harcourt. IOSR J Bus Manag 17(3):70–75. doi:10.6084/M9.FIGSHARE.1348608.V1
68. United Nations. (2013) Millennium Development Goals and Beyond 2015. United Nations
69. Unt (2016) Principal Components Analysis in SPSS. University of North Texas
70. Wang F, Lai X, Shi N (2011) A multi-objective optimization for green supply chain network design. Decis Support Syst 51(2):262–269
71. Webb Kimberly J, Hodge Thomas G, Thompson James H (2012) Small business sustainability: what is the CPA's role? Int J Bus Soc Sci 3(12):1
72. Wu J, Dunn S, Forman H (2012) A study on green supply chain management practices among large global corporations. J Supply Chain Operations Manag 10(1):182–194
73. Zaccai E (2012) Over two decades in pursuit of sustainable development: Influence, transformations, limits. Environ Dev 1(1):79–90
74. Zhu Q, Sarkis J (2006) An inter-sectoral comparison of green supply chain management in China: drivers and practices. J Clean Prod 14(5):472–486

# Analysis of University e-Environment in Selected European Countries – IRNet Case Study

**Peter Svec, Martin Drlik, Tatiana Noskova, Tatiana Pavlova, Olga Yakovleva and Michal Munk**

**Abstract** The chapter aims at finding out how students from selected universities in different European countries use e-environment and e-resources at their universities. Students' opinion was gathered using a survey. The questionnaire consisted of 14 questions and created the basis for the evaluation of a university e-environment. It was assumed that participants of e-environment (academic teachers, students, administration) are involved in activities with the six different benefits. Almost 900 responses from five different countries joined in the project IRNet under European Union Seventh Framework program (Poland, Czech Republic, Slovakia, Ukraine and Russian Federation) were collected. The chapter presents the analysis of the answers with the result that students require a broad Internet connection and electronic versions of study material, but do not want to publish own achievements in a university e-environment. Half of the students require printed publications and students do not widely use scientific resources when searching for learning materials.

P. Svec (✉) · M. Drlik · M. Munk
Constantine the Philosopher University in Nitra, Nitra, Slovakia
e-mail: psvec@ukf.sk

M. Drlik
e-mail: mdrlik@ukf.sk

T. Noskova · T. Pavlova · O. Yakovleva
Herzen State Pedagogical University of Russia, Saint Petersburg, Russia
e-mail: noskovatn@gmail.com

T. Pavlova
e-mail: pavtatbor@gmail.com

O. Yakovleva
e-mail: o.yakovleva.home@gmail.com

M. Munk
e-mail: mmunk@ukf.sk



T. Issa et al. (eds.), *Sustainability, Green IT and Education Strategies in the Twenty-first Century*, Green Energy and Technology, DOI 10.1007/978-3-319-57070-9_17

# 1 Introduction

Universities in Europe took a long and complicated approach to e-learning systems implementation. There can be seen that two common approaches have been implemented; bottom-up and top-down [10, 11]. Successful implementations are those who combined these two approaches. Most universities started to use commercial e-learning solutions such as Blackboard or WebCT. Now most of them have shifted to use open source Moodle. In fact, it does not matter what virtual learning environment they use. E-learning became the natural method in education, and now it is time to look at how current technologies influence the use of e-learning. Many researchers try to assess the e-learning form and its effectiveness not just as the technical solution but also from the didactic view. There are many open questions, for example, the impact of social networks and cloud computing on standard e-learning systems. After so many years, it is common, that many teachers use the virtual learning system just as a repository for educational material. Researchers and teachers try to find new ways to motivate students to use the e-learning system in the same way as they use social networks.

One of the common features of modern education in different areas of training is the presence of positions that define the ability to apply e-learning and distance learning technologies in the educational process [14]. An important aspect of a university graduate is an ability to work in a professional environment with a variety of available information resources, ICT tools, and networking opportunities. Information and communication technologies are considered today as natural tools for education and professional activities. They quickly progress and are widely used by young generations in different areas of their activities [24].

Contemporary educational standards specify the goals and results of training, including a wide range of graduate and professional competencies and objectives that a student should be ready to achieve [18].

Regardless of professional activities students are expected to implement, information and communication technologies (ICT) have a high potential to achieve educational outcomes, improve the efficiency of network forms of educational process organization [28]. To realize this potential, it is necessary to form a system of targeted information and communication educational opportunities, taking into account the benefits of a modern e-environment.

An electronic environment of a modern university should create opportunities for the development of the twenty-first-century competencies, for the implementation of a lifelong learning strategy. In this chapter, the research implemented in the Work Package 3 of the IRNet project is described and the data obtained in several universities, the participants of the IRNet project, analysed. These universities are the following: Herzen State Pedagogical University of Russia, St. Petersburg (HSPU), The University of Silesia in Katowice, Poland (US), Borys Grinchenko Kyiv University, Ukraine (BGKU) and Constantine the Philosopher University in Nitra (UKF), Slovakia. The main aim of the research was to identify the ways in which students use ICT tools in their universities' e-environments and to outline the possible directions for improving educational interactions in network learning communities.

Among the aims of the survey was to determine if students understand the opportunities and educational benefits of e-environments: expansion of space-time coordinates, personalization of educational activities, individual requests, an increase in the degree of educational openness. The data can be not only the basis for determining the readiness of students for self-guided work and activity in an e-environment but can also help to identify ways to improve a university e-environment overall.

However, besides the analysis of individual questions, we would sometimes like to find different groups of respondents that answered the questionnaire similarly. In this case, we can apply several other knowledge-mining methods were applied, which can help to find statistically significant results.

In order to understand better the specifics of the students interviewed, universities participating in this study are briefly described in section two of this chapter. Section *Related work* offers current literature review. Main parts of the chapter are in Section 4 *The relationship between e-learning and environment* and Section 5 *Analysis of the answers*.

## 2 Participating Universities

Universities participating in European Union Seventh Framework Program IRNet (http://irnet.us.edu.pl/) were chosen. These Universities are similar in their size and their educational and research interests. All mentioned universities are focused on the development of a modern university e-environment and take into account different goals and requirements of its participants. In the following section, these universities are described in more detail.

### *2.1 Herzen State Pedagogical University of Russia, Russia*

Herzen State Pedagogical University of Russia (HSPU) is one of the oldest teacher-training institutions in the country. Today the University comprises 5 institutes, 21 faculties, and 3 affiliations. Alongside the training of highly qualified teachers for universities and secondary schools, HSPU offers a broad range of non-teaching specializations, allowing it to fit into the category of traditional universities. The university admits students to 57 educational programs integrated into 10 groups (32 Bachelor's programs, 2 Specialist programs and 23 Master's programs). The university offers programs in Arts & Humanities, Engineering & Technology, Life Sciences, Natural Sciences, and Social Sciences. The latest achievements in the spheres of education management and economics, juvenile justice, etc. form the foundation of the training programs for psychologists, economists, lawyers, managers, IT specialists and quality control managers. Still, the teacher-training programs remain the priority for HSPU.

The implementation of ICT in education is one of the HSPU priorities. The LMS Moodle is functioning at HSPU, and Adobe Connect has been recently introduced to support distant learning, e.g. for the university affiliations. There is a number of open electronic educational resources (e.g. open training courses, and public resources in the electronic repository). A single database accumulates information on teachers' and students' activities and research achievements. With the help of this database, an individual plan of each teacher's work is formed, which reflects the planned and the resulting performance of various types of professional activity (teaching, methodical, scientific, educational work). In addition, the data is collected and processed at the level of each division, resulting in a monthly ranking of university departments. In addition, information tools to form an individual educational student's trajectory are designed at HSPU. These tools are the electronic atlas (information about educational services), the electronic directory (individual educational routes), and the electronic guide (link to the labour market), allowing the student to navigate the available educational opportunities.

## *2.2 Constantine the Philosopher University in Nitra, Slovakia*

Constantine the Philosopher University in Nitra (UKF) has been formed as a modern European university where the dominant role is played by science, spiritual quality, high professionalism, pedagogical mastery, open and vivid communication both within the local environment and within the world.

The university is the fourth largest university in Slovakia and consists of five faculties. The university prepares young people for a wide range of professions, such as preschool, primary, secondary and university teachers in various fields, experts in cultural institutions, political scientists, marketers, managers of tourism, interpreters, translators, catechists, journalists, anthropologists, archaeologists, historians, museologists, physicists, specialists in natural sciences, psychologists, gemologists, medical staff, social workers and many others. The teacher study programmes are a double-major combination of 27 subjects at the undergraduate and graduate level. The university offers 5 single-major teacher study programmes at the undergraduate and graduate level, 105 scientific and professional study programmes and 31 scientific doctoral programmes.

UKF is gradually developing blended as well as distance form of e-learning. The main and sole LMS at the university is Moodle since 2008. Every university building is covered by Wi-Fi signal, and every student has access to the high-speed internet. The University also offers access to scientific databases through the services of the University Library. The university e-environment offers many other e-services for students, not just the access to the study materials and e-learning platform but also the whole management of the study process.

## 2.3 University of Silesia, Poland

The University of Silesia in Katowice (US) was established in 1968 and now, with 12 faculties and several interdisciplinary schools and centres, over 35,000 students, educated at bachelor, master, and doctoral levels and over 2,000 academic staff is one of the largest in Poland. The university, within the framework of its own activity in the of area study, research, science, innovation, cooperation, national and international projects has launched various initiatives in the use of e-learning for LLL. These initiatives include Distance Learning Centre of the University of Silesia, Project 'University as a Partner of the Knowledge Economy' UPGOW, University Television.

One of the first and the most actively functioning of distance learning platforms in the University of Silesia is the platform of Faculty of Ethnology and Sciences of Education in Cieszyn. The faculty e-learning platform contains a lot of interesting courses supporting learning. Its key objectives are to:

- provide support for teaching programme courses, run in the full-time and part-time mode (hybrid learning)
- prepare future teachers to take advantage of distance learning – to use e-learning in the own profession and to perform the role of a tutor
- provide assistance with scientific research and pedagogical experiments carried out by department staff, graduate students as well as post-graduate students
- foster international cooperation, in particular, through international projects
- provide non-formal and informal distance education in ICT area, and other subjects area for future and current teachers, other [23].

## 2.4 Borys Grinchenko Kyiv University, Ukraine

The University history began in 1874 with the establishment of pedagogical courses for teacher training in Kyiv. In 1919, Borys the Grinchenko Pedagogical Institute began operating, to be later reorganised into Kyiv Pedagogical Courses named after Borys Grinchenko. In 2002, it was reorganised into Kyiv Municipal Pedagogical University named after Borys Grinchenko, and it started offering degrees in teaching. Now, there are five institutes, over 8,000 students, educated at bachelor, master, and doctoral levels.

The university is actively implementing innovative educational technology, based on e-learning. That is why in the university there is a research laboratory of information education. This laboratory works on issues of creating e-learning environment of the university and guarantees free access to its students and teachers. LMS Moodle is used to implement e-learning and for the cooperation of all members of the educational process. Also, e-learning courses are stored there. All e-learning courses are certified by the approved quality criteria [23].

## 2.5 University of Ostrava, Czech Republic

The University of Ostrava was founded in 1991, but its origins can be traced back to 1953. The University is focusing on the study and research of humanities and social sciences, helping to redress a historical imbalance in a region traditionally dominated by heavy industry and technical studies. The University of Ostrava currently has six faculties: the Faculty of Social Studies, the Faculty of Fine Arts, the Faculty of Arts, the Faculty of Medicine, the Pedagogical Faculty, and the Faculty of Science. The University offers a wide spectrum of degrees at Bachelor's, Master's and doctoral levels, in addition to lifelong learning programs. The University of Ostrava incorporates two independent research institutes.

# 3 Related Work

It is important to analyse and systematize the main benefits of university e-environments that can be used by students during training. Such benefits can be evaluated from different perspectives: improvement of educational services quality, formation, and development of competencies for the knowledge society, formation of graduates' competitiveness [25, 36]. At the same time, the achievements of educational outcomes are directly related to the strategic and systematic use of ICT tools by students in the university e-environment.

An e-environment of a university can be defined as a set of information and communication technologies, which can contribute to better students' educational outcomes.

As a rule, there is no flexible model for building an e-environment for the implementation of blended learning, which takes into account; on the one hand, the rapid development of information technology, and the other – especially today's students belonging to other than teachers' generation [23].

It is proposed to divide indicators for an e-environment and ICT competencies evaluation into four groups [25]:

- indicators of university electronic scientific and educational environment (internal, external),
- indicators of e-learning development level,
- indicators of student' competencies,
- indicators of teacher's competencies.

Several examples of e-environments of universities are provided in Morze et al. [23].

Comparative analysis of the relation between an e-environment of several EU and Non-EU universities and indicators of students' competencies was described in Morze et al. [25]. Other groups of indicators of an e-environment development at universities accompanied by the IRNet project are summarized by Pavlova [34] and Prudencia Gutiérrez [14].

Different aspects of the VLE implementation into the e-environment of the university are described by Drlík and Skalka [10]. Negatives and positives of using e-courses for various groups of students are described by Capay [4]. The use of the web as an effective environment for science and technology education is described by Bilek et al. [3].

Some works tried to assess the use of an e-environment from the students' point of view and focused on the perception of the VLEs and their impact on the development of ICT competencies in the environment of the university. Costa et al. [7] examined the Moodle of the University of Aveiro (UA) in Portugal through a content analysis, complemented with a non-structured interview carried out with the responsible for the platform at the UA. Moodle' tools used by students were analysed afterwards. In this survey printed questionnaires were used and assessed in three areas: participant characteristics, general use of Moodle and specific Moodle tools. Students were asked about the general purposes for which they used the Internet in the learning context. 'E-mail' and 'Search' were the most purposes mentioned by the respondents. The 'Social Networks' were not mentioned in this context. Most students use the Moodle just for downloading materials; only a few of them use Moodle for communicating with the teachers.

The tools used at the UA can be grouped considering the level of importance assigned to them. 'News' and 'Assignments' were considered the most important, followed by 'Quiz/Survey', 'Questionnaires', 'Forums' and 'Wikis UA', with an intermediate level of importance. Finally, 'Chats', 'Blogs UA' and 'Video-conference' were considered less important.

The use of an e-learning environment at the La Laguna University in Spain was evaluated by Barberan et al. [1] by combining quantitative and qualitative observations. As a common quantitative method, La Laguna University supplies a yearly questionnaire to both teachers and students about the use of the virtual campus, and they monitor the growth of Moodle's resources used on campus. As a qualitative method, the direct observation of teachers and the triangle between assessors and teaching staff is considered. The number of teachers who use e-learning grows each year. However, teachers use the system mainly as a repository, then as the way to submit and receive submissions and discussion forum as the last. The main problem with the use of e-learning is the low level of teaching staff training.

The applicability of different types of resources and activity modules in the e-learning courses and the worthiness of their usage and compare outcomes of data analyses is described by Capay [5, 6].

The use of VLEs and Social Networks in the educational process of students and teachers of National Research Tomsk State University, one of the leading universities in Russia, is evaluated by Mozhaeva et al. [26]. The Google Form questionnaire with 68 various questions was used, and questions were divided into three separate groups. In the first group were participants' general information questions. The second group of questions tried to reveal the relation between students and teachers to various ways of e-learning organization and to compare it to the traditional face-to-face education. The third group of questions was directed to the definition of use frequency of eight different learning methods in LMS and social networks. A majority of the respondents considers graphical application interface is more convenient for social networks. Moreover, the efficiency, frequency, informational content, interaction, individual approach, cooperation, and emotionality are higher in the social networks than in LMS. However, factors disturbing learning process in Moodle are less visible than in social networks. Results of the research show that the LMS does not provide comparable pedagogical conditions for communication like social networks.

Possibilities of using heuristics, interactivity and feedback in electronic learning materials are described by Klocokova [17]. Experimental evaluation of the efficiency of heuristics approach into learning is presented in this research as well as detailed analysis of the user log-on data, on which it is better to understand the rate of using of heuristics elements together with other course activities in an electronic learning environment.

Advantages of e-learning at the university are discussed by Talebian et al. [38]. Main advantages are time and place of access, equity, enhancing group collaboration, direct access to resources, enhancing the international dimension of educational services and determination of the progress. Disadvantages are the absence of a teacher, access to unsupportive information, limited feedback from students or unsuitableness in practical courses.

There are also several other works, that tried to assess the impact of some LMSs, mainly Moodle on other internet services, like e-mail [2], SecondLife [39], Facebook [9]. More of them stated that LMS and other ICT Tools significantly improve students' learning outcomes. Moreover, they emphasize the importance of the continual development of e-environment of the university using ICT tools, which should be in line with the expectations and everyday habits of the students.

Klimova and Poulova [12] in a larger survey focused on the development of students' productive language skills, asked students from the Czech Republic and Kazakhstan questions about their opinions on e-learning. It was found that students prefer face-to-face learning.

Students, being the main educational services consumers, have been the subject of study many times. With the wide development of e-learning, the active building of university e-environments, students' behaviour – motivation, expectations, and demands for the e-environment have become of special interest.

Speaking of the motivation, there is no doubt today that students' e-learning efficiency correlates with their motivation [15]. For example, among motivation factors for e-environments researchers note problem-solving approach, highlighting the goals that may be achieved [27]. Students' expectations have also become the subject of study [32].

However, if an e-environment is considered, which is not merely e-learning, a wider range of issues must be addressed – some aspects of a university e-environment that may become means of attracting students. In this particular study, these aspects are the following. Firstly, massive open online courses provided by the university attract students. Not by chance, Coursera, the international MOOC platform, became the Webby Awards (international professional Internet award) winner in 2014. Secondly, an interesting content of a university website with a user-friendly interface and current information, together with the university social network fully correspond to the "Y" generation, or the millennials requests [16]. Thirdly, a comprehensive presentation of teachers' and students' achievements on the university website, along with the information on successful university graduates and their achievements is important, because 'in online learning faculties and tutors are symbols of the institution' [22]. Fourthly, cooperative education and experiential learning have become highly demanded. Learning in partnership, cooperative learning, are based on the constructivist approach. The leading role is played by a sense of belonging to the joint venture [13] widely used the potential of joint information activities and distributed networking. For those universities that specialise for the humanitarian sphere (e.g., teachers, psychologists, and social pedagogues), cooperation with schools, nursery schools, educational centres, and various companies forms the basis for professional practice. In addition, university cooperation in social activities and cultural life (volunteering, charity concerts, and exhibitions) may be related to extra-curricular activities of students [30].

## 4 The Relationship Between e-Learning and Environment

A dynamic development of ICT, as well as changes in the requirements of the educational outcomes, represent the factors, which determine the transformation of the educational environment of a modern university. Changing and improving are not only the components of the information space but also its content and the approaches to the quality of educational services. In many ways, the transformation of information processes related to the use of e-learning and distance learning technologies which potentially offer divergent ways of interaction. It can be talked about some specific benefits, which are essentially manifested in e-environment of a university.

The above mention work package was dedicated to the analyses and evaluation of the ICT level, e-learning and intercultural developments in every participating country and to the elaboration of the conceptual framework for a joint research project based on lasting collaboration with the project participants. It was assumed that participants of e-environment (academic teachers, students, administration) are involved in activities with the following benefits:

- Benefit 1: Increase of scientific and educational processes comfort, focus on lifelong learning goals, expansion of space-time coordinates.
- Benefit 2: Personalization of educational activities, individual request in e-learning.
- Benefit 3: Formation of new scientific and educational relations, cooperation, intercultural competence.
- Benefit 4: Empowerment of self-realization in educational and professional activities, support of initiatives.
- Benefit 5: Increase of the openness degree of the scientific and educational environment, expanding the influence of the university to the external cultural environment; positioning in the research and education community.

- Benefit 6: Enhancing self-organizational effects that support the sustainable development of the educational environment of the university and its participants.
- It is needed to analyse in more detail the benefits listed above. The relevance of such analysis is due to the fact that the development of a university information infrastructure requires constant material costs. Therefore, there is a need to track if students recognize these innovative benefits.

The most obvious benefits are associated with the increase of scientific and educational process comfort, focus on lifelong learning goals; the personalization of educational activities, the individual request of e-learning. It is important to emphasize that a variety of information and communication tools allow not only learning in a convenient mode but also showing to a greater extent the educational activity and independence, increasingly individualizing the educational process. The new task of education is to focus on lifelong learning, which involves the possibility for every person to fulfil own potential, in all ages, regardless of place and time, using all possible ways and means of cooperation [19]. This effect determines the expansion of educational and professional goals range. For students, these are the aspects of new competencies acquisition required for educational, and further professional tasks based on the new skills for the information environment, in accordance with the individual educational needs and requirements. For teachers, this is the focus on flexible educational support of students' independent activities in diverse media saturated space. An example of comfort indicators is an opportunity for an individual educational route [8]. These ideas are developed in the concept of a personal learning environment of a student [35].

The next benefit can be described as the formation of new scientific and educational relations, cooperation, intercultural competence. This is mainly communication benefit. Networking, distributed group work, which has become a common type of professional activities in many areas, also requires specific aspects of preparedness and qualification. Effective means of training to acquire required skills is to introduce into the educational process networking and cooperation, which are exactly the dominant technologies of e-learning [33]. It is necessary to take into account that the nature of the educational interactions should correspond to modern communication behaviour of young people in the network space [30]. Network communication is a means of self-empowerment in educational and professional activities; identifying and supporting educational initiatives. This is extremely important, both for students and for young professionals; so they get more opportunities to express themselves, to show their achievements and ideas. Network projects, competitions, scientific and educational communities are an integral part of modern educational and professional environment; an important means for formation of graduates' competitiveness.

It is important to cooperate with university social partners. For example, in Europe since 2010, a study of the effectiveness and research collaboration of countries, regions, and researchers (universities, research institutes) has been realized. In particular, in addition to such indicators as the number of publications, citations, impact factor, also compared such indicators as a number of joint (inter-institutional) publications [20]. In Russia, network association of universities is also actively developing. This means that a number of universities join efforts in creating new educational programs, support unified scientific electronic resource databases. We can see similar effort in many universities. The focus of study and the role of the university is the cooperation with the commercial companies in research and development.

The succeeding benefit is the increase of the openness degree of the scientific and educational environment, expanding the influence of the university to the external cultural environment; positioning of the actors in the research and education community. Interaction in an open information environment ensures that students have the opportunity to act first in quasi-professional conditions, and then enter the professional community to make full use of its resources, to position their educational, professional and scientific activities. For example, the Law 'On Education in the Russian Federation' determines the information transparency and public accountability of educational institutions as the basic principles of state policy in the sphere of education [21].

Thus, a modern e-learning environment of a university is a space of multiple-choice options. However, students and teachers can use variably this environment, as well as external educational resources. It is, therefore, important to assess the level of awareness and implementation of the available opportunities for students and teachers in the learning process, to see the real effects of the degree of manifestation of the e-learning environment, transformed by the introduction of a variety of information and communication technologies [29].

Manifestation of qualitative results of e-learning and ICT in education depends on a system of conditions:

- the degree of e-learning environment development (electronic space – electronic resources and information technology; interactions while solving scientific and educational problems);
- the level of competencies for major e-learning environment participants (faculty members, students, staff responsible for e-learning management).
- An e-environment of a university requires various vectors of interactions – interactions between students and teachers within certain learning courses, interactions between colleagues while developing interdisciplinary training, exchange and transfer of experience to other educational institutions, etc. [31]. That is why an e-environment of a university can be considered at three levels:
- micro level (achieving educational objectives at the level of discipline, e-course);
- meso level (solution of scientific and educational problems in the corporate environment of the university (interdisciplinary communication, cooperation, exchange of experience);
- macro level (achieving scientific and educational objectives by integrating into the external scientific and educational environment).

The complete vision of e-environment is presented in Fig. 1.

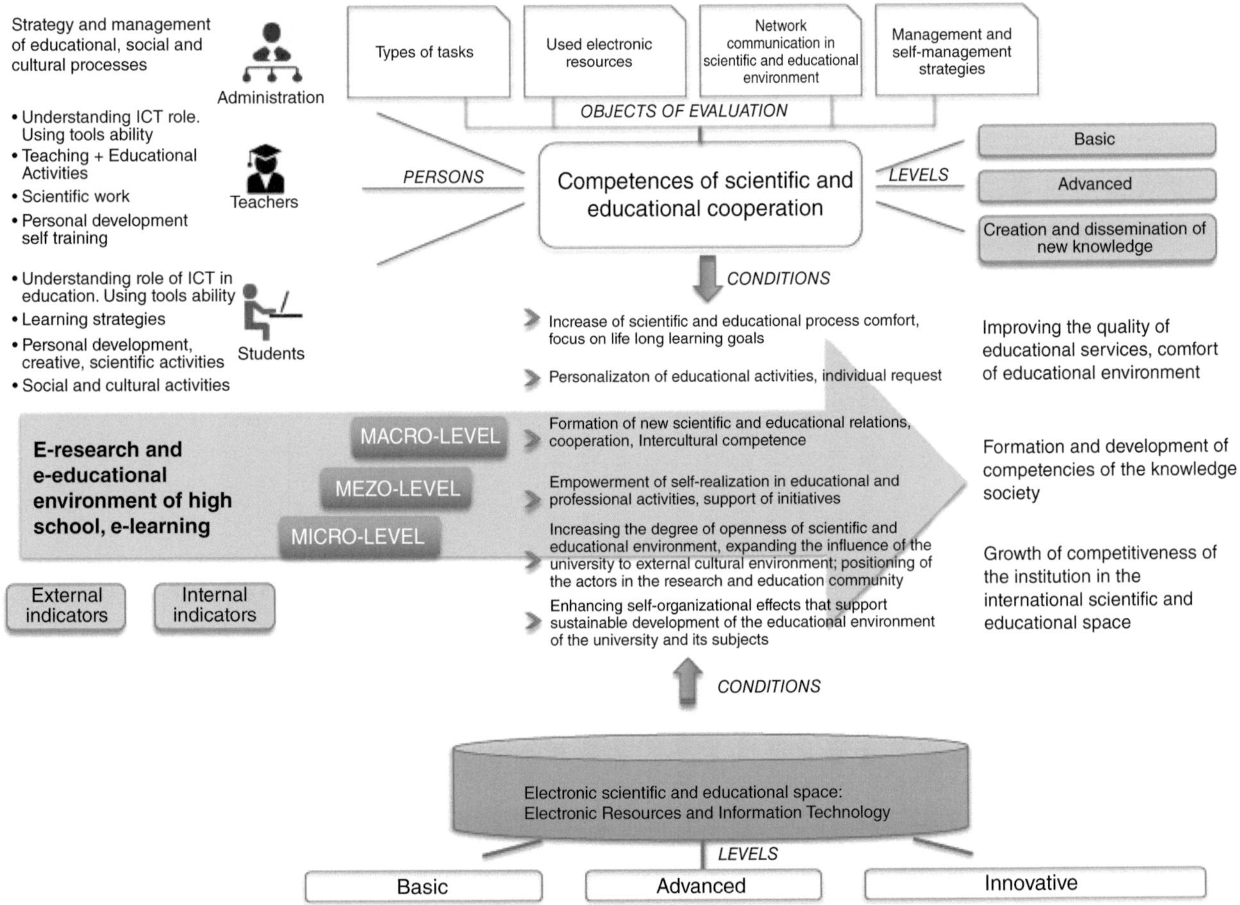


**Fig. 1** e-Environment: ICT tools, students' competencies, and learning benefits

In order to efficiently and effectively use all the opportunities of e-environment, all the environment participants should possess a certain level of e-learning and ICT competencies. That is why these competencies can be considered at three ascending levels: basic level, advanced level, and innovative level. Various activities in an e-environment make participants both form and manifest their competence, meeting the long-term requirements of training a specialist for the knowledge society.

E-learning and ICT competences can be indicated by:

- objectives of different types of activities;
- electronic scientific and educational resources;
- network communication in the scientific and educational environment;
- management strategies of scientific and educational activities in the information environment of the university.

The e-learning development and ICT contribute to the quality of educational services; the development of knowledge society competencies and the increase of the competitiveness of an institution in the world scientific and educational space. To achieve these following conditions must comply:

- an electronic space no less than advanced level should be created in an educational institution;
- a constant improvement e-learning and ICT competences should be performed; the preferred level of competencies is advanced;
- changing requirements, determining the competitiveness of an institution in the world scientific and educational space are considered at every level of the e-learning environment.
- Indicators for an e-learning environment and e-learning competencies can be divided into four sections:
- indicators of university electronic scientific and educational environment (internal, external);
- indicators of e-learning development level;
- indicators of student's competencies;
- indicators of teacher's competencies.

## 5 Analysis of the Answers

### *5.1 Research Method and Main Research Aim*

In order to collect and analyse students' responses and to summarize their vision of university e-environment, a questionnaire was elaborated. The questionnaire consisted of 14 questions, which can be integrated into six groups, corresponding to the key e-environment benefits. Consequently, evaluating university e-environment, we assumed that participants of an e-environment (academic teachers, students, administration) are involved in activities with the following benefits.

Benefit 1: Increase of scientific and educational processes comfort, focus on lifelong learning goals, expansion of space-time coordinates. (Q1, Q2, Q3). The corresponding questions, firstly, invited students to select the leading purpose to use e-environment of their university (for example, such resources as electronic libraries, distance learning platform Moodle, e-courses, university website, scientific databases of university subscriptions). Among the leading purposes were suggested the following: the use only required by a teacher and the use to ease tasks performance. Simultaneously students could choose reasons for non-use of e-resources: not understanding the rules or usefulness and finding other alternative Internet resources (e.g., MOOCs). In addition, respondents were asked to choose two main reasons influencing the effectiveness of the e-environment use: getting acquainted with the possibilities and usage rules; having an opportunity to perform tasks at own pace, anytime and anywhere; finding more diverse electronic resources. Finally, in this section students were suggested to select the most important indicators of e-environment comfort: availability of Wi-Fi access points; opportunity to use own gadgets; availability of electronic educational resources in different formats; university website with the relevant information for students and comfortable navigation; availability of distance support for disciplines; availability of a prompt feedback from a teacher.

Benefit 2: Personalization of educational activities, individual request in e-learning (Q4, Q5). The section included two questions. The first one revealed students' opinion on whether or not should teachers consider students' educational requests, interests, and needs while creating electronic resources in an educational environment (presentations, websites, tests, video lectures, etc.). The second question focused on what types of additional electronic educational services would students like to receive at their university: studying foreign languages; acquiring an additional profession; learning about start-ups and own business or any other types.

Benefit 3: Formation of new scientific and educational relations, cooperation, intercultural competence (Q6, Q7, Q8, Q9). Exploring cooperation, students were firstly asked to evaluate the need for it in solving educational problems: do teachers set these objectives; do students themselves strive to cooperate and ask teachers to offer such tasks; do students need such competences for being successful. Secondly, students commented on their use of social services for collaboration and teamwork: whether or not teachers offer such activities; do students in principle recognize the use of social services for collaboration. Thirdly, respondents were asked to specify the main reason for participation in virtual communities of students: getting additional cultural knowledge; helping with studies; finding new friends; demonstrating own personal experience, knowledge, and achievements; organizing own events.

Benefit 4: Empowerment of self-realization in educational and professional activities, support of initiatives (Q10, Q11). The section studied the possible reasons motivating students to demonstrate their academic, artistic, sporting activities in the university e-environment: the opportunity to present achievements to others; the opportunity to be noticed by a potential employer; prerequisite of studying a particular discipline; general interest; own status upgrade; the opportunity to make new friends. Students also could choose the options 'I do not want to show myself and my achievements to others because I have nothing to show' and 'I do not want to show myself and my achievements to others, because I am not interested in it'. Moreover, in this section students chose informational resources most often used when doing assignments: search engines; printed publications; electronic scientific databases from your university library subscription; digital libraries on the Internet; open storages of electronic educational resources; video channels; file sharing; webinars; podcasts.

Benefit 5: Increase of the openness degree of the scientific and educational environment, expanding the influence of the university to the external cultural environment; positioning in the research and education community (Q12). The focus of the section was on the key features of university e-environment that motivate students to choose the particular university to study. Among the features were named the following: MOOCs, provided by the university; an attractive university web site with user-friendly interface and up-to-date information; own university social network; clear presentation of the university students' and teachers' achievements on the web site; information about successful university graduates and their achievements; university social partners; university participation in social activities and cultural life.

Benefit 6: Enhancing self-organizational effects that support the sustainable development of the educational environment of the university and its participants (Q13, Q14). From students' prospective self-organization concerned selecting one of several main learning strategies: external monitoring by teachers; self-guided independent and systematic study; following peers' example. Furthermore, students estimated the possible benefits of ICT tools (electronic diaries, organizers, calendars, reminders, etc.) for planning educational and extracurricular activities – whether these tools produce any effect or not.

Each of the participating universities reviewed and accepted the questionnaire and created a version in their native language to facilitate students' answers and avoid the ambiguity of understanding. The supporting platform for this questionnaire was the Google Form tool as it offers simple and anonymous questionnaire delivery except the US, Poland that used the Lime Survey.

Some questions of the questionnaire were multiple choices, others not. For the purpose of further analysis using methods mentioned above, the transformation of all answers into true/false question was needed. For that reason, sub questions by this transformation were created, e.g. the Q3 became Q3-1, Q3-2, … Q3-6.

While preparing to conduct the survey, the possible disadvantages of this research method were taken into account, i.e., the inability to control the process of filling in the questionnaire, which can lead to the lack of independence of the respondent replies. In order to minimize all possible risks, each university had to collect at least 100 responses from different categories of students (first-year students, bachelors, and masters). This aim was accomplished, and it was collected 868 responses together (132 from Ukraine, 125 from Russia, 171 from Czech Republic, 340 from Slovakia and 100 from Poland).

Hereby, the main research aim was to identify the ways students use the ICT tools in the e-environments of their universities and to outline possible ways for improving educational interactions in students' network learning communities. In other words, of particular interest was to study how students perceive the ongoing transformations of the e-environment and are willing to participate in them.

The data provided only the basis for determining the readiness of students for self-guided learning and identifying the preferred activities in the e-environment. Moreover, it can also be helpful in identifying possible ways to improve the e-environment of the participating universities. In fact, for obtaining high-quality educational outcomes, universities need to improve both corporate strategies of e-environment development with the implementation of e-learning and educational competence of students in the extended information and communication space. For the students, these competencies are closely related with their information culture and the awareness of an individual request for the use of a variety of educational opportunities available in the electronic space.

In this chapter are some selected questions analysed using Cochran Q-test to test statistically significant differences between elements of questions. Tukey test [37] was used for multiple comparison and identification of homogenous groups. When it is talked about all answers from all universities, it is stated as 'Total' in all following tables. All questions and sub-questions are dichotomous ones, where 'zero' represent negative answer and 'one' positive one. The percentage of positive and negative answers is stated and also the 'Sum' of positive answers.

## 5.2 *Indicators of Comfort of the Electronic Environment*

The main application of this question is the use in self-development, self-realization, research and scientific activities. We assessed the availability of Wi-Fi access points (Q3-1), the opportunity to use own gadgets (Q3-2), the availability of electronic educational resources in different formats e.g. video, audio, hypertext (Q3-3), relevant information for students and comfortable navigation of the University website (Q3-4), the availability of distance support for disciplines e.g. e-tasks, e-journals, e-course (Q3-5) and the availability of a fast feedback from a teacher (Q3-6).

What can be seen from Table 1 is that answers of students from all countries form three homogenous groups. The first group contains questions with no statistical difference between questions Q3-5, Q3-3, Q3-6 and Q3-2. It means that these students put the similar weight to all these options. These students want e-courses, e-study materials, require fast feedback from the teacher and want to use own gadgets. The second group of answers creates a group where students want to have e-study materials, also require fast feedback from the

**Table 1** Homogenous groups identification according to question Q3 for all universities

| Total | Sum | Percent 0's | Percent 1's | 1 | 2 | 3 |
|---|---|---|---|---|---|---|
| Q3-5 | 181.00 | 79.15 | 20.85 | **** | | |
| Q3-3 | 193.00 | 77.76 | 22.24 | **** | **** | |
| Q3-6 | 199.00 | 77.07 | 22.93 | **** | **** | |
| Q3-2 | 222.00 | 74.42 | 25.58 | **** | **** | |
| Q3-4 | 235.00 | 72.93 | 27.07 | | **** | |
| Q3-1 | 389.00 | 55.18 | 44.82 | | | **** |

$Q = 180.57$, $df = 5$, $p < 0.00$.
Q – testing statistic, df – degree of freedom, p – significance value, 1 – the first homogenous group, … 3 – the third homogenous group, Sum – number of positive answers = 1's.

**Table 2** Answers to question Q3-1 (WiFi) at each university

| Q3-1 | Sum | Percent 0's | Percent 1's | |
|---|---|---|---|---|
| UKF | 71.00 | 79.12 | 20.88 | $Q = 15.12941$, $df = 5$, $p < 0.009825$ |
| OU | 57.00 | 66.67 | 33.33 | $Q = 54.08772$, $df = 5$, $p < 0.000000$ |
| US | 47.00 | 53.00 | 47.00 | $Q = 68.84000$, $df = 5$, $p < 0.000000$ |
| BGKU | 115.00 | 12.88 | 87.12 | $Q = 102.9956$, $df = 5$, $p < 0.000000$ |
| HSPU | 99.00 | 20.80 | 79.20 | $Q = 60.40161$, $df = 5$, $p < 0.000000$ |

Q – testing statistic, df – degree of freedom, p – significance value.

teacher, also want to use own gadgets, but they also want relevant information on the University website. The third homogenous group is the also the biggest one. There is statistically significant difference between the question Q3-1 (WiFi) and all other questions.

When a closer look is taken at the results in each University (Tables 2 and 3), the Wi-Fi question creates a homogenous group in the case of HSPU (Russia), BGKU (Ukraine) and US (Poland). At these Universities students' responses to Wi-Fi question represent the strong statistically difference between Q3-1 and all other questions. They also have a higher percentage of answers to this question, e.g. at the BGKU 87% of students selected the option Q3-1, at HSPU 79% of students. In the case of US, this number was just 47% of students, at OU and UKF even worse.

When this opinion of students in EU countries (Slovakia, Czech Republic, Poland) with non-EU countries (Ukraine, Russia) is compared, it can be seen that students in non-EU countries require Wi-Fi connection more than their colleagues in EU. There can be two reasons for this opinion. The first is the broad and cheap access to the mobile internet in EU countries. For that reason, the students do not think about the Wi-Fi anymore. The second reason can be the availability of Wi-Fi everywhere.

**Table 3** Homogenous groups identification according to question Q3 for US, BGKU, and HSPU

| US | Percent 1's | 1 | 2 | BGKU | Percent 1's | 1 | 2 | 3 | 4 | HSPU | Percent 1's | 1 | 2 |
|---|---|---|---|---|---|---|---|---|---|---|---|---|---|
| Q3-5 | 8.00 | *** | | Q3-5 | 31.06 | | | *** | | Q3-3 | 31.20 | *** | |
| Q3-6 | 8.00 | *** | | Q3-4 | 48.48 | *** | | | | Q3-4 | 43.20 | *** | |
| Q3-2 | 10.00 | *** | | Q3-6 | 49.24 | *** | | | | Q3-5 | 44.80 | *** | |
| Q3-3 | 11.00 | *** | | Q3-3 | 52.27 | *** | *** | | | Q3-2 | 47.20 | *** | |
| Q3-4 | 16.00 | *** | | Q3-2 | 65.15 | | *** | | | Q3-6 | 48.80 | *** | |
| Q3-1 | 47.00 | | *** | Q3-1 | 87.12 | | | | *** | Q3-1 | 79.20 | | *** |

## 5.3 Participation in Virtual Communities

Students were asked to specify the main reason for their participation in virtual communities of students (scientific, artistic, sports ones, etc.) in social networks or other Internet services. Six reasons why students want to participate in virtual communities were selected. They can get additional cultural knowledge, learn about events, broadening outlook, and adopt cultural experience (Q9-1). Virtual communities can help with studies by sharing knowledge or by the opportunity to ask for help (Q9-2). Students can find new friends in virtual communities (Q9-3), or they want just to demonstrate own personal experience, knowledge, and achievements (Q9-4). Virtual communities are often used for organizing own events (Q9-5). It was also given the option that students do not like to participate in such communities (Q9-6).

In Table 4 can be seen four homogenous groups of answers. These groups are not overlapping, so there is a statistically significant difference between each homogenous group. The biggest group (the fourth group) consists of answers that use virtual communities in the way that help them with a study by sharing knowledge. The second largest group (the second group) consists of answers which use virtual communities to get additional cultural knowledge or learn about events. The last single question group is Q9-6 (students do not participate in virtual communities). The rest of their options in questionnaire create the first homogenous group (there is no statistically significant difference among these answers). Students put similar weight to all these options in this group, and this group is the smallest one. For example, just 72 students selected the Q9-5 option (organizing own events). Using virtual communities is worth when propagating events and to share the knowledge.

Students from Ukraine and Russia to Slovak, Czech and Poland are compared in Table 5. As most students chosen Q9-1 and Q9-2, those questions were selected for analysis.

Students tend to use virtual communities in the way that helps them with studies or as the opportunity to ask for help (Q9-2).

**Table 4** Homogenous groups identification according to question Q9 for all universities

| Total | Sum | Percent 0's | Percent 1's | 1 | 2 | 3 | 4 |
|---|---|---|---|---|---|---|---|
| Q9-5 | 72.00 | 91.69 | 8.31 | **** | | | |
| Q9-3 | 80.00 | 90.76 | 9.24 | **** | | | |
| Q9-4 | 94.00 | 89.15 | 10.85 | **** | | | |
| Q9-6 | 153.00 | 82.33 | 17.67 | | **** | | |
| Q9-1 | 300.00 | 65.36 | 34.64 | | | **** | |
| Q9-2 | 402.00 | 53.58 | 46.42 | | | | **** |

$Q = 592.51$, $df = 5$, $p < 0.00$.

Q – test, df – degree of freedom, p – significance value, 1 – the first homogenous group, … 4 – the fourth homogenous group.

**Table 5** Answers to Q9-1 and Q9-2 from all universities

| | Q9-1 | | | Q9-2 | | |
|---|---|---|---|---|---|---|
| | Sum | Percent 0's | Percent 1's | Sum | Percent 0's | Percent 1's |
| UKF | 92.00 | 72.94 | 27.06 | 175.00 | 48.53 | 51.47 |
| OU | 60.00 | 64.50 | 35.50 | 105.00 | 37.87 | 62.13 |
| US | 31.00 | 69.00 | 31.00 | 18.00 | 82.00 | 18.00 |
| BGKU | 64.00 | 51.52 | 48.48 | 62.00 | 53.03 | 46.97 |
| HSPU | 53.00 | 57.60 | 42.40 | 42.00 | 66.40 | 33.60 |

## *5.4 Reasons Motivating to Demonstrate Own Achievements in the University e-Environment*

Students were asked about their motivation to publish in the university e-environment (website, social networks, etc.| their results of academic, artistic or sporting activities. The main application is the use in self-development, self-realization, and research. Following reasons motivating students were selected:

- opportunity to present me and my achievements to others (Q10-1),
- opportunity to be noticed by a potential employer (Q10-2),
- prerequisite for studying a particular discipline (Q10-3),
- general interest (Q10-4),
- own status upgrade (Q10-5),
- opportunity to make new friends (Q10-6).

Some students do not want to show themselves and their achievements to others, because they have nothing to show (Q10-7) or because they are not interested in it (Q10-8).

It can be seen that the presentation of own achievements in the university e-environment is not a very popular theme for students (Table 6). The highest percentage for question Q10-1 is just 25%. All other questions have a lower percentage of answers. Despite this fact, students' answers form three separate homogenous groups. The largest group consists of answers to Q10-1, Q10-2 and Q10-3. There is no statistically significant difference between 'opportunity to present myself and my achievements to others', 'opportunity to be noticed by a potential employer' and 'general interest'. A group of those three answers does not overlap with questions Q10-8 and Q10-7, which are negative for us (students do not show anything). Students use social networks, but it seems that only in their private life. They do not want to put status upgrades into their university e-environment. University environment should focus on motivating students not to hesitate to present own achievements. They should know that information in a university e-environment is valuable and trustworthy for a potential employer.

**Table 6** Homogenous groups identification according to question Q10 for all universities

| Total | Sum | Percent 0's | Percent 1's | 1 | 2 | 3 |
|---|---|---|---|---|---|---|
| Q10-5 | 49.00 | 94.34 | 5.66 | **** | | |
| Q10-7 | 58.00 | 93.30 | 6.70 | **** | | |
| Q10-3 | 69.00 | 92.03 | 7.97 | **** | | |
| Q10-6 | 78.00 | 90.99 | 9.01 | **** | | |
| Q10-8 | 170.00 | 80.37 | 19.63 | | **** | |
| Q10-4 | 203.00 | 76.56 | 23.44 | | **** | **** |
| Q10-2 | 207.00 | 76.10 | 23.90 | | **** | **** |
| Q10-1 | 220.00 | 74.60 | 25.40 | | | **** |

$Q = 323.17$, $df = 7$, $p < 0.00$.
Q – test, df – degree of freedom, p – significance value, 1 – the first homogenous group, … 3 – the third homogenous group.

## 5.5 *Informational Resources During Study*

Students were also asked what resources they use most often when doing assignments, doing research, preparing reports, etc. Students were able to choose from the following items:

- Q11-1: Search engines (Google, Yandex, etc.) - Search by keywords
- Q11-2: Printed publications (books, journals, guidelines, etc.)
- Q11-3: Electronic scientific databases from your university library subscription (databases of electronic journals, full-text electronic resources, etc.)
- Q11-4: Digital libraries in the Internet
- Q11-5: Open storages of electronic educational resources (institutional repository, WIKI)
- Q11-6: Video channels (YouTube)
- Q11-7: File sharing, torrents
- Q11-8: Webinars, podcasts

Informational resources are most valuable for studying any discipline. Knowing the source where students search for these resources is valuable for the teacher. As seen in Table 7 most students (nearly 86%) use search engines when looking for informational resources. This is strange because all universities offer access to scientific databases for all students. There is no statistical difference between questions Q11-3 (scientific e-databases), Q11-6 (YouTube) and Q11-5 (e-resources) and less than 30% of students use these resources. Students more often try to search relevant information using a search engine and often find not relevant one instead of using highly relevant source, e.g. a scientific database. It is surprising that more than half of students use printed publication (Q11-2). Even when students are well equipped with e-gadgets, when it comes to learning, they prefer paper copies of the books. What students do not discover yet are podcasts

**Table 7** Homogenous groups identification according to question Q11 for all universities

| Total | Sum | Percent 0's | Percent 1's | 1 | 2 | 3 | 4 | 5 | 6 |
|---|---|---|---|---|---|---|---|---|---|
| Q11-8 | 48.00 | 94.47 | 5.53 | | **** | | | | |
| Q11-7 | 138.00 | 84.10 | 15.90 | | | **** | | | |
| Q11-3 | 241.00 | 72.24 | 27.76 | **** | | | | | |
| Q11-6 | 255.00 | 70.62 | 29.38 | **** | | | | | |
| Q11-5 | 261.00 | 69.93 | 30.07 | **** | | | | | |
| Q11-4 | 345.00 | 60.25 | 39.75 | | | | **** | | |
| Q11-2 | 446.00 | 48.62 | 51.38 | | | | | **** | |
| Q11-1 | 745.00 | 14.17 | 85.83 | | | | | | **** |

Q = 1685.37, df = 7, $p < 0.00$.
Q – test, df – degree of freedom, p – significance value, 1 – the first homogenous group, … 6 – the sixth homogenous group.

and webinars. Just 6% of students know about this form of education. Many students do not want to widen their horizon beyond what is needed by the teacher.

## 5.6 *Elements of E-environment Which Imply Student Decision to Study*

Question 12 focuses on elements of an e-environment of a university, which imply student decision to study the selected field. Responses to this question can show the importance of some factors for the recruitments of new students, which factors most influence student decision to study some field at the university. The elements in this question were as follows:

- Massive open online courses provided by the university (Q12-1)
- The interesting content of the university website with a user-friendly interface and current information (Q12-2)
- An own social network of the university (Q12-3)
- A comprehensive presentation of the achievements of university teachers on the university website (personal pages, awards, publications, etc.). (Q12-4)
- A comprehensive presentation of the university students on the university website (awards, publications, etc.). (Q12-5)
- Information on successful university graduates and their achievements. (Q12-6)
- University cooperation with schools, nursery schools, educational centres, companies, etc.). (Q12-7)
- University cooperation in social activities and cultural life (volunteering, charity concerts, exhibitions, etc.). (Q12-8)

In Table 8 can be seen statistically significant differences for non-overlapping groups of columns. Respondents for the twelfth question form five groups

**Table 8** Homogenous groups identification according to question Q12 for all universities

| Total | Sum | Percent 1's | Percent 0's | 1 | 2 | 3 | 4 | 5 |
|---|---|---|---|---|---|---|---|---|
| Q12-3 | 181.00 | 21.37 | 78.63 | **** | | | | |
| Q12-5 | 197.00 | 23.26 | 76.74 | **** | | | | |
| Q12-4 | 225.00 | 26.56 | 73.44 | **** | **** | | | |
| Q12-6 | 253.00 | 29.87 | 70.13 | | **** | **** | | |
| Q12-1 | 262.00 | 30.93 | 69.07 | | **** | **** | | |
| Q12-8 | 305.00 | 36.01 | 63.99 | | | **** | **** | |
| Q12-7 | 350.00 | 41.32 | 58.68 | | | | **** | **** |
| Q12-2 | 393.00 | 46.40 | 53.60 | | | | | **** |

Q = 221.84, df = 7, p = 0.00.
Q – test, df – degree of freedom, p – significance value, 1 – the first homogenous group, … 5 – the fifth homogenous group.

based on their answers. The lowest weight for choosing the field of study has the group of elements Q12-3, Q12-4, Q12-5 (social network of the university and presentation of achievements of teachers and students). The highest weight has a group of elements Q12-2 and Q12-7 (content of university website and cooperation with other subjects). There is also a statistically significant difference between these two groups. The results for each university were also analysed. The results showed different views about the importance of elements in each country.

The UKF had the highest number of answers among other universities. Student responses form four groups of answers. These answers for question Q12-3, Q12-4, Q12-5 and Q12-8 form the group with the lowest impact on choosing the study at the university. The answers for Q12-7 and Q12-2 form the group with the highest impact for choosing the study at the university. Results at UKF almost copy the overall results. What is interesting, that students from the Czech Republic (Table 9) form very similar groups. Students from the University of Ostrava put the highest weight on the cooperation with schools, nursery schools, educational centres, companies (Q12-7). There are statistically significant differences between the Q12-7 question and all other questions.

When a closer look at results from Ukraine (Table 10) is taken, the highest weight is on the answers for Q12-2 (attractive university website) together with Q12-8 (university cooperation in social activities and cultural life). Answers for question Q12 form just two homogenous groups. The second group consists of the rest of Q12 subquestions, but the weight is lower. The highest weight is on Q12-8 also in Russia. In both countries, the students put a high focus on university cooperation in social activities and cultural life. In Russia, Q12-6 (successful university graduates) and Q12-8 (university cooperation in social activities and cultural life) form one homogenous group and the rest of question another group with the mean value similar to Ukraine.

**Table 9** Homogenous groups identification according to question Q12 for Slovak and Czech university

| UKF | Percent 1's | 1 | 2 | 3 | 4 | OU | Percent 1's | 1 | 2 | 3 | 4 |
|---|---|---|---|---|---|---|---|---|---|---|---|
| Q12-3 | 26.76 | **** | | | | Q12-3 | 16.05 | **** | | | |
| Q12-4 | 30.59 | **** | **** | | | Q12-5 | 17.90 | **** | | | |
| Q12-5 | 31.76 | **** | **** | | | Q12-4 | 19.14 | **** | **** | | |
| Q12-8 | 32.35 | **** | **** | | | Q12-6 | 21.60 | **** | **** | | |
| Q12-6 | 37.94 | | **** | **** | | Q12-1 | 24.69 | **** | **** | | |
| Q12-1 | 43.24 | | | **** | | Q12-8 | 32.10 | | **** | **** | |
| Q12-7 | 47.06 | | | **** | **** | Q12-2 | 45.68 | | | **** | |
| Q12-2 | 54.12 | | | | **** | Q12-7 | 67.90 | | | | **** |

**Table 10** Homogenous groups identification according to question Q12 for all Ukraine and Russia university

| BGKU | Percent 1's | 1 | 2 | HSPU | Percent 1's | 1 | 2 |
|---|---|---|---|---|---|---|---|
| Q12-6 | 19.17 | **** | | Q12-1 | 16.00 | **** | |
| Q12-5 | 20.83 | **** | | Q12-5 | 18.40 | **** | |
| Q12-3 | 23.33 | **** | | Q12-3 | 23.20 | **** | |
| Q12-1 | 27.50 | **** | | Q12-7 | 28.80 | **** | |
| Q12-7 | 29.17 | **** | | Q12-4 | 32.80 | **** | |
| Q12-4 | 33.33 | **** | | Q12-2 | 38.40 | **** | |
| Q12-8 | 50.83 | | **** | Q12-6 | 40.00 | | **** |
| Q12-2 | 61.67 | | **** | Q12-8 | 55.20 | | **** |

## 6 Discussion and New Significance

The questionnaire on five similar Universities in five different countries was taken with nearly 900 responses from students who provided their opinion about a university e-environment. When processing the data, not only exploratory techniques were used, but based on selected items of the questionnaire, profiles of universities were created. The Cochrane Q test was used to test differences between dichotomous variables of selected questionnaire items with distinguishing each university. Cochrane Q test is an extension of the McNemar test for dependent samples that provides a method for testing for differences between three or more dichotomous variables.

After the rejection of the global null hypothesis using multiple comparisons, significant differences, and homogeneous groups were identified. For this purpose, parametric (Tukey HSD test) and nonparametric methods were used. As the results were equally, they were considered to be robust. The use of advanced method of analysis gave this research new significance and it was possible to create a profile for all participating universities and find groups of respondents that answered the questionnaire similarly.

It is obvious that all universities participating in the described research realize their e-environment development strategies. However, the new significance of the survey lies in the student-centered approach. It means that students are considered as the main consumers of educational services, so their request to an e-environment is of a high value. On one hand, the obtained data can be helpful to critically analyse and find the answer to the question, does the existing e-environment and provided facilities match students' vision of an 'ideal' university. On the other hand, the results help to reveal weak points in students' understanding of all e-environment benefits that are already available for them.

One of the main research limitations can be related with the sample of respondents taken only from the Eastern European universities and Non-EU universities. However, this aspect can be one of the future research directions, because IRNet consortium includes also several Western and Southern European universities as well as one Australian university. In the future, data from Portugal, Spain, Netherland and Australia will be collected and analysed so it will be possible to better understand differences in e-environment strategies, implementation and opinion of students.

## 7 Conclusion

Following statements about a university e-environment can be concluded based on the answers of students.

Students require a Wi-Fi connection, but it seems that not for study purposes. Students from Ukraine and Russia require a Wi-Fi connection more than their colleagues from Slovakia, Czech Republic, and Poland, which more often use mobile technologies for internet access (mobile providers offer a cheap mobile internet access for students). Students want to e-study, but they do not require e-materials from the university enough. Even when the university offers access to e-documents, through access to digital libraries or scientific databases students do not use them and try to find relevant information using web search engines. It is surprising that more than half of students require printed publications for their studies.

There is a question that is difficult to answer unequivocally: to what extent teachers should take into account the information and educational needs of students, their interests, while creating electronic educational resources (presentations, websites, tests, videos, lectures, etc.). On the one hand, students can use the proposed resources in accordance with their needs. On the other hand, teachers have to take into account the information and educational needs of students, their interests, and provide resources adapted to individual request, especially when it comes to inclusive education. It is important to note that a modern e-learning environment allows variation of electronic resources. Furthermore, additional electronic educational services can be provided in such environment. For example, the study of foreign languages, additional profession, business start-ups, etc. Obtaining additional educational services ensures the readiness of the graduates for professional careers in the dynamically changing conditions.

Students actively use external information resources and less actively are turning to university resources. This allows to specifically recommend to acquaint students with these opportunities offered by each university participating in the IRNet project.

E-learning environment ICT tools allow to enrich the educational process with visual representation of educational results, scientific, university artistic and sports activities and achievements of particular students. Thus, students have the opportunity to show themselves and their achievements to others: potential employers, teachers, students, prospective students, social partners of the educational institution. But students do not want really much to publish their achievements in a university e-environment. Students use social networks, but it seems that only in their private life. Students are not willing to put status upgrades into a university e-environment. University environment should focus on motivating students not to hesitate to present own achievements. Students should know that information in a university e-environment is valuable and trustworthy for a potential employer. Because the presentation of achievements improves students' status, the circles of acquaintances and interactions are expanding. Possession of means and methods of professional activity results presentation today is an important competence for a representative of any professional field. Offering students at the initial stages assignments aimed at presentation of educational activity results in an e-learning environment, the preconditions for the formation of such competence are created.

It can be seen what students want to find at a university website when they are at the phase of choosing where to study. Students are looking for interesting content of the university website with a user-friendly interface and current information. Students also want to see cooperation with schools, nursery schools, educational centres, companies, etc.

Students can effectively use the benefits of ICT tools and an e-learning environment, thereby achieving the purpose of improving the quality of education, formation and development of professional competencies, formation of competitiveness in the labour market. These benefits are fully compliant with the emerging knowledge society and the ideas of lifelong learning.

**Acknowledgment** The research leading to these results has received, within the framework of the IRNet project, funding from the People Programme (Marie Curie Actions) of the European Union's Seventh Framework Programme FP7/2007-2013/ under REA grant agreement No: PIRSES-GA-2013-612536.

## References

1. Barberan JM, del C-O, Gutierrez JM, León FMC (2013) Detection of learning needs in the teaching staff regarding the use of a virtual campus at La Laguna University. Procedia Soc Behav Sci 93:1333–1336. doi:10.1016/j.sbspro.2013.10.039
2. Benta D, Bologa G, Dzitac I (2014) E-learning platforms in higher education. Case study. Procedia Comput Sci 31:1170–1176. doi:10.1016/j.procs.2014.05.373
3. Bílek M, Slabý A, Turčáni M, Cyrus P (2007) Web as effective environment for science and technology education. In: László Szentirmai (ed) Joining Forces in Engineering Education Towards Excellence: Proceedings; SEFI and IGIP Joint Annual Conference. University of Miskolc, pp 75–76
4. Capay M (2014) Online learning systems in various forms of studies. In: DIVAI 2014 10th International Scientific Conference on Distance Learning in Applied Informatics, Sturovo, Slovakia, pp 285–293
5. Cápay M, Mesárošová M, Balogh Z (2011) Analysis of students' behaviour in E-learning system. In: 2011 Proceedings of the 22nd EAEEIE Annual Conference (EAEEIE), Maribor, Slovenia
6. Cápay M, Tomanová J (2010) Enhancing the quality of administration, teaching and testing of computer science using learning management system. WSEAS Trans Inf Sci Appl 7: 1126–1136

7. Costa C, Alvelos H, Teixeira L (2012) The use of Moodle e-learning platform: a study in a Portuguese University. Procedia Technol 5:334–343. doi:10.1016/j.protcy.2012.09.037
8. Dabbagh N, Kitsantas A (2012) Personal learning environments, social media, and self-regulated learning: a natural formula for connecting formal and informal learning. Internet High Educ 15:3–8. doi:10.1016/j.iheduc.2011.06.002
9. Deng L, Tavares NJ (2013) From Moodle to Facebook: exploring students' motivation and experiences in online communities. Comput Educ 68:167–176. doi:10.1016/j.compedu.2013.04.028
10. Drlik M, Skalka J (2011) Virtual faculty development using top-down implementation strategy and adapted EES model. Procedia Soc Behav Sci 28:616–621. doi:10.1016/j.sbspro.2011.11.117
11. Drlík M, Švec P, Skalka J, Kapusta J (2008) E-learning portal integration to the information system of Constantine the Philosopher University in Nitra, Slovakia. In: EUNIS 2008 Vision IT: visions for IT in higher education. University of Aarhus, Aarhus, 2008. ISBN 978-87-91234-58-3, CD-ROM
12. Frydrychova Klimova B, Poulova P (2014) Hybrid learning. Theory and practice. Lect Notes Comput Sci (including Subser Lect Notes Artif Intell Lect Notes Bioinformatics). doi:10.1007/978-3-319-08961-4
13. Gorard S, Smith E (2006) Beyond the 'learning society': what have we learnt from widening participation research? Int J Lifelong Educ 25:575–594. doi:10.1080/02601370600989269
14. Gutiérrez-Esteban P, Alonso-Díaz L, Smyrnova-Trybulska E, Capay M, Ogrodzka-Mazur E, Gonçalves Pinto P, Noskova T, Gajdzica A, Pavlova T, Yakovleva O (2015) Intercultural and digital competence in teacher training from an international perspective: Poland, Portugal, Slovakia, Spain and Russia (La competencia digital e intercultural en la formación del profesorado desde una perspectiva internacional). Revista Latinoamericana De TecnologíA Educativa - RELATEC 14(1):145–157. doi:10.17398/1695-288X.14.1.145
15. Harandi SR (2015) Effects of e-learning on students' motivation. Procedia Soc Behav Sci 181:423–430. doi:10.1016/j.sbspro.2015.04.905
16. Howe N, Strauss W (2009) Millennials Rising: the Next Great Generation. Vintage, New York
17. Klocoková D (2011) Integration of heuristics elements in the web-based learning environment: experimental evaluation and usage analysis. Procedia Soc Behav Sci 15:1010–1014. doi:10.1016/j.sbspro.2011.03.230
18. Kommers P, Smyrnova-Trybulska E, Morze N, Tatiana N, Tatiana P, Yakovleva O (2014) First outcomes of WP2 research carried out within the framework of the IRNet Project – International Research Network. In: DIVAI 2014 10th International Scientific Conference on Distance Learning in Applied Informatics. Sturovo, Slovakia, pp 158–166
19. Laal M, Aliramaei A, Laal A (2014) Lifelong learning and art. Procedia Soc Behav Sci 116:4047–4051. doi:10.1016/j.sbspro.2014.01.888
20. Labrosse I, Ventimiglia A, Campbell D, Haustein S, Côté G, Archambault É (2013) Scientific Output and Collaboration of European Universities. Available via Science-Metrix. http://www.science-metrix.com/en/publications/reports/scientific-output-and-collaboration-of-european-universities-2013. Accessed 18 Feb 2016
21. Federal'nyy zakon ob obrazovanii "Ob obrazovanii v Rossiyskoy Federatsii" ot 29.12.2012 № 273-FZ (red. ot 31 dekabrya 2014 g.) [Federal Law "About Education in the Russian Federation" from 29.12.2012 № 273-FL (ed. 31.12.2014)].
22. Lemos S, Pedro N (2012) Students' expectation and satisfaction in post-graduate online courses. http://www.slideshare.net/SusanaLemos/students-expectation-and-satisfaction-in-postgraduate-online-courses. Accessed 18 Feb 2016
23. Morze N, Smyrnova-Trybulska E, Umryk M (2015) Designing an e-university environment based on the needs of net-generation students. Int J Cont Eng Educ Life Long Learn 25:466–486
24. Morze N, Spivak S, Smyrnova-Trybulska E (2014) Personalized educational environment – as one of the trends of modern education. In: Kostolanyova K, Kapounova J (eds)

Information and Communication Technology in Education, University of Ostrava, Rožnov pod Radhoštěm, pp 158–166
25. Morze N, Varchenko L, Smyrnova-Trybulska E (2015) Ways of formation of effective students' collaboration skills based upon the usage of WBT. Int J Web Based Communities 11:25. doi:10.1504/IJWBC.2015.067085
26. Mozhaeva G, Feshchenko A, Kulikov I (2014) E-learning in the evaluation of students and teachers: LMS or social networks? Procedia Soc Behav Sci 152:127–130. doi:10.1016/j.sbspro.2014.09.168
27. Nehme M (2010) E-Learning and Students' Motivation. In: 20 Legal Education Review 223-239. Available at SSRN: http://www.austlii.edu.au/au/journals/LegEdRev/2010/11.html. Accessed 15 Feb 2016
28. Noskova T, Pavlova T (2012) New priorities of the educational activities in the educational environment of the modern university. Sci Tech J SPBSPU 2:329–335
29. Noskova T, Pavlova T (2015) E-learning environment: a new quality of the resource base. http://www.irma-international.org/chapter/e-learning-environment/142465/. Accessed 18 Feb 2016
30. Noskova T, Pavlova T, Yakovleva O, Smyrnova-Trybulska E, Svec P, Tomanova J, Drlik M, Capay M (2015) IT tools in a university – e-learning environment: students' opinion survey. In: Smyrnova-Trybulska E (ed) IT tools – Good Practice of Effective Use in Education. Studio-Noa, Katowice-Cieszyn, pp 27–41
31. Noskova T, Yakovleva O (2015) Educational interactions quality in e-learning environment. http://www.igi-global.com/viewtitlesample.aspx?id=142438&ptid=132792&t=educational interactions quality in e-learning environment. Accessed 18 Feb 2016
32. Paechter M, Maier B, Macher D (2010) Students' expectations of, and experiences in e-learning: their relation to learning achievements and course satisfaction. Comput Educ 54:222–229. doi:10.1016/j.compedu.2009.08.005
33. Patarakin Y, Shilova O (2015) Concept of learning design for collaborative network activity. Procedia Soc Behav Sci 214:1083–1090. doi:10.1016/j.sbspro.2015.11.709
34. Pavlova T, Yakovleva O, Morze N, Drlík M Quality features of university information environment in its external indicators. Int J Contin Eng Educ Life Long Learn 26(2): 196–216
35. Rahimi E, van den Berg J, Veen W (2015) Facilitating student-driven constructing of learning environments using Web 2.0 personal learning environments. Comput Educ 81:235–246. doi:10.1016/j.compedu.2014.10.012
36. Skubała P, Trybulska ES (2014) E-learning as an effective modern method of building a sustainable society. Int J Contin Eng Educ Life Long Learn 24:62. doi:10.1504/IJCEELL.2014.059335
37. Stranovska E, Fraterova Z, Munkova D, Muglova D (2012) Politeness factors in requests formulated in the 'category width' cognitive style. Stud Psychol (Bratisl) 54:111–124
38. Talebian S, Mohammadi HM, Rezvanfar A (2014) Information and communication technology (ICT) in higher education: advantages, disadvantages, conveniences and limitations of applying e-learning to agricultural students in Iran. Procedia Soc Behav Sci 152:300–305. doi:10.1016/j.sbspro.2014.09.199
39. Uzun L (2012) The Internet and computer-mediated artefacts for foreign language learning and practice, and intercultural communication: MOODLE, Second Life, and others. Procedia Soc Behav Sci 46:3296–3300. doi:10.1016/j.sbspro.2012.06.054

# Design and Evaluation of the Makahiki Open Source Serious Game Framework for Sustainability Education

**Yongwen Xu, Philip M. Johnson, George E. Lee, Carleton A. Moore and Robert S. Brewer**

**Abstract** Sustainability education and conservation have become an international imperative due to the rising cost of energy, increasing scarcity of natural resource and irresponsible environmental practices. This paper presents Makahiki, an open source serious game framework for sustainability, which implements an extensible framework for different organizations to develop sustainability games. It provides a variety of built-in games and content focused on sustainability; game mechanics such as leaderboards, points, and badges; a variety of common services such as authentication, real-time game analytics and ability to deploy to the cloud, as well as a responsive user interface for both computer and mobile devices. The successful implementation of six sustainability educational games in different organizations provided evidence regarding the ability to customize the Makahiki framework successfully to different environments in both organizational and infrastructure aspects. A serious game stakeholder experience based access method (SGSEAM) was used to formally evaluate Makahiki in order to understand the strengths and weaknesses of Makahiki as a useful serious game framework for sustainability.

Y. Xu (✉) · P.M. Johnson · G.E. Lee · C.A. Moore · R.S. Brewer
Collaborative Software Development Lab, Department of Information and Computer Science, University of Hawaii at Manoa, Honolulu, HI, USA
e-mail: yxu@hawaii.edu

P.M. Johnson
e-mail: johnson@hawaii.edu

G.E. Lee
e-mail: gelee@hawaii.edu

C.A. Moore
e-mail: cmoore@hawaii.edu

R.S. Brewer
e-mail: rbrewer@hawaii.edu

 
T. Issa et al. (eds.), *Sustainability, Green IT and Education Strategies in the Twenty-first Century*, Green Energy and Technology, DOI 10.1007/978-3-319-57070-9_18

## 1 Introduction

The rising cost, increasing scarcity, and environmental impact of fossil fuels as an energy source makes a transition to cleaner, renewable energy sources an international imperative. Moving away from petroleum is a technological, political, and social paradigm shift, requiring citizens to think differently about energy policies, methods of generation, and their own consumption than they have in the past. Unfortunately, unlike other civic and community issues, energy has been almost completely absent from the educational system. To give a sense for this invisibility, public schools in the United States generally teach about the structure and importance of our political system (via classes like "social studies"), nutrition and health (through "health"), and even sports (through "physical education"). But there is no tradition of teaching "energy" as a core subject area for an educated citizen, even though energy appears to be one of the most important emergent issues of the twenty-first century.

On the other hand, energy and water competitions or challenges have been introduced to college dormitories and residential homes as ways to facilitate and incentivize resource reduction to achieve sustainability goals. A survey by Hodge [12] found that there are 163 college residence hall energy competitions taking place or being planned for the 2010–2011 academic year in North America to engaging students in sustainability issues. Hodge found that the average reduction in electricity use during these competitions is 9%. In 2014, 109 schools participated in the Campus Conservation Nationals (CCN) [15], a nationwide electricity and water use reduction competition for colleges and universities. Overall, the competition resulted in 4.5% average electricity competition reduction at the building level, with the top 10 schools achieving reductions ranging from 11% to 24%.

Another emergent issue is the explosive spread of game techniques, not only in its traditional form of entertainment, but across the entire cultural spectrum. The adoption of game techniques to non-traditional areas such as finance, sales, and education had become such a phenomenon that the Gartner Group included "Gamification" [6] on its Hype List.

This paper describes Makahiki, an open source serious game framework for sustainability education and conservation, in which we attempted to create synergy between these two emergent issues. The result of over three years of research and iterative development, Makahiki explored one section of the design space where virtual world game mechanics were employed to affect real world energy behaviors. The goal of the Makahiki project was to provide a framework for organizations to learn to not just affect energy behaviors during the course of the game, but to produce long lasting, sustained change in energy behaviors and outlooks by participants.

We initially used Makahiki to create an energy challenge called the Quest for the Kukui Cup (hereafter, the Kukui Cup) for approximately 1,000 first year students living in four residence hall towers at the University of Hawaii in 2011. During the three weeks of the competition, over 400 of the eligible students played the game, for a total of 850 game play hours. The game mechanics were designed to create a self-reinforcing *virtuous circle* between the real world and virtual world activities. The challenge was well received and had been repeated in academic years 2012 and 2014 at the University of Hawaii. In addition, Makahiki, as a serious game framework, had been used in several organizations including Hawaii Pacific University and the East West Center, an international education institute in Hawaii, to implement their own version of the Kukui Cup serious game.

In order to assess a serious game framework such as Makahiki, it is important to assess the serious games the framework produces. One fundamental question in evaluating a serious game is the extent to which the game achieves its "serious" purpose. This is quite different from the evaluation of traditional entertainment games. SGSEAM (Serious Game Stakeholder Experience Assessment Method) is an assessment method for assessing the strengths and weaknesses of a serious game framework, with respect to the experiences of different stakeholders in a serious game lifecycle [23].

In this paper, we detail the innovative features of Makahiki framework, along with our experiences using them to implement several serious games related to sustainability education and conservation in different organizations. The application of SGSEAM to Makahiki is also described to provide insights into the strengths and weaknesses of Makahiki as a serious game framework for sustainability.

## 2 Related Work

Our research draws on previous work done in the areas of energy behavior research, energy competitions, gamification, and serious games.

To reduce energy consumption, providing energy feedback is a critical foundation. Darby's survey of energy consumption studied from the past three decades found that consumption in identical homes could differ in energy use by a factor of two or more depending on the behavior of the inhabitants [5]. Another survey of energy feedback conducted by Faruqui et al. found that residents that actively used the in-home displays with near-real-time feedback averaged a 7% reduction in energy usage [8]. Darby also pointed out that feedback alone is not always enough: other factors such as training could lead to higher rates of energy conservation [4].

Energy competitions or challenges have been introduced to college dormitories and residential homes as ways to facilitate and incentivize energy reduction. Petersen et al. described their experiences deploying a real-time feedback system in an Oberlin College dorm energy competition in 2005 that includes 22 dormitories over a 2-week period [18]. Web pages were used to provide feedback to students. They found a 32% reduction in electricity use across all dormitories. The Building Dashboard [14], developed by Lucid Design Group, was used to support Oberlin's dorm energy competition, as well as the Campus Conservation Nationals, a nationwide electricity and water use reduction competition on college campuses [15]. The Building Dashboard enabled viewing, comparing and sharing building energy and water use information on the web in compelling visual interface, but the cost of the system creates the barrier for wider adoptions. In addition, the building dashboard solutions focus on providing energy information as a passive media. There is little interaction between participants and the system.

Games on the other hand, had been shown with great potential as successful interactive media that provide engaging interfaces in various serious contexts [17, 20]. Priebatsch attempted to build a game layer on top of the world with his location-based service startup [19].

Reeves et al. described the design of Power House, an energy game that connects home smart meters to an online multiple player game with the goal to improve home energy behavior [21]. In the game, the real world energy data were transformed into a "more palatable and relevant form of feedback", and players were incentivized by the in-game rewards to complete more energy-friendly real-world behaviors.

ROI Research and Recyclebank launched the Green Your Home Challenge as a case study of employing gamification techniques online to encourage residential green behavioral changes offline [10]. With Google Analytics, the results shown a 71% increase in unique visitors. Ninety-seven percent of participants surveyed said that the challenge increased their knowledge about how to help the environment.

The blending of real and virtual worlds had been explored in broader contexts. McGonigal designed the award winning serious Alternative Reality Game (ARG) "World Without Oil" [7] and later "Evoke" [25] with the goal to empower people to come up with creative solutions to our most urgent real-world problems. ARGs had also been used to support learning. Connolly et al. discussed the development of an educational ARG to motivate secondary school students across Europe to learn foreign languages [3]. The results of the pilot run of the game in 2009 indicated that 92% of students felt the game motivated students to learn a second language. One of problems the team identified is the limitation of Moodle platform the game is based on.

The report of the ARGOSI project provided insights to the use of ARGs in game based learning and the challenges in the field of higher education [24]. The pilot was run at the University of Bolton with the aim to provide an engaging alternative to traditional methods of introducing students to university life. The overall up-take of the game was fairly low with only 23 active players out of 173 total. The project identified a number of questions surrounding educational ARGs, such as motivation, relationship to curriculum, marketing and timing. The report suggested that a complete ARG model may not be appropriate for wholesale learning, but there is certainly potential in using game elements.

Game frameworks (also known as game engines) [22] are "comprised of a collection of different tools, utilities, and interfaces that hide the low-level details of the various tasks that make up a game". One of the benefits of using a game framework is that, if correctly designed, it will provide useful and reusable "building blocks" with which to develop a variety of games. Similarly, serious game frameworks also provide building blocks that enable the serious game developer to focus more time and thought on content and results instead of on the technical details and infrastructure for creating the serious game.

Building Dashboard is commercial platform that "enables energy reduction competition and empowers building occupants to become active participants in energy management" [14]. It is similar to Makahiki that they are both frameworks for supporting sustainability competitions, but the cost of Building Dashboard as a commercial system creates a barrier to wider adoption. In addition, the Building Dashboard solution focused on providing energy information as a passive media. Besides a scoreboard, there was little interaction between participants and the system. There were no game elements other than providing a scoreboard to display the ranking of the competing teams. Moreover, there was no individual points or ranking. Unlike Makahiki, Building Dashboard did not have the concept of individual registered player account and personalization, thus it did not have the capability to provide the evidence of individual player engagement.

Freitas et al. [9] pointed out that there were few frameworks to support the evaluation of serious games. They introduced a four dimensional framework for evaluating educational games and simulations. The framework consists of: the context, the pedagogy, the representation, and the learner (or player). Their approach focused on evaluation of a single serious game, as opposed to a serious game *framework*.

There also existed some general purpose assessment tools such as QUIS (Questionnaire for User Interaction Satisfaction) [11] and GEQ (Game Engagement Questionnaire) [2] that are used for measuring game engagement and user interactions. QUIS is a usability assessment tool developed in the HCI lab at the University of Maryland, College Park. It was designed to assess user's subjective satisfaction regarding the human/computer interface of software systems. Currently licensing is required to access the QUIS questionnaires. GEQ was used to assess the engagement in a game. The questionnaire provides a "psychometrically" strong measure of levels of engagement of players during the game. Examples of GEQ questionnaire items include "I lost track of time", "If someone talks to me, I don't hear them", and "I play without thinking about how to play" etc. While the GEQ could measure the engagement level of positive game experience, the original intent of the research is to "examine risk and protective factors for negative game impact". Both QUIS and GEQ focused on assessing the engagement and usability of the game system from the player's perspective. They did not address the specific requirement of the assessment of a serious game framework.

SGSEAM [23] identified the most important stakeholders (in addiction of the players in a game) of a serious game framework and provided a method for gaining insight into the strengths and shortcomings of the framework with respect to each stakeholders' needs. SGSEAM identifies the major stakeholders whose experiences affect a serious game framework as a software infrastructure as: players, system admins, game designers, game managers and game developers. For each stakeholder, multiple in-vivo and in-vitro assessment approaches can be used, depending on the resources available. The more approaches applied, the higher one's confidence in the accuracy of the assessment results.

## 3 Makahiki System Design

Makahiki is an open source "serious game framework for sustainability". It provides a framework for creating serious games for the purpose of education and behavioral change regarding energy, water, food, and waste generation and use. Makahiki intends to create synergy between the need to create knowledge and engagement regarding energy and the ability of so-called "serious game" techniques and energy feedback to create participation and engagement.

The initial version of Makahiki (Version 1) [13] was originally developed as a custom built software solution for the Kukui Cup Energy Competition at the University of Hawaii. The first version of Makahiki provided such functionalities to support energy challenges at the University of Hawaii as: (1) a synergistic mixture of real-world and virtual world activities; (2) real-time feedback on energy consumption by residence hall teams; (3) incentives in the form of prizes and raffle games.

The current version of Makahiki (Version 2) built upon the prior version with the purpose of providing a serious game Framework for sustainability that enables different organizations to easily create customized serious games in the context of sustainability education and behavioral change. In additional to the features from version 1, the current Makahiki framework includes the following features:

- The ability to tailor system functionality to support the requirements of different organizations.
- The ability to support sustainable resource challenges such as water, food, and waste in addition to energy.
- The ability to extend the framework with new mini-games and modules to support different requirements.
- The use of HTML5/CSS3 "responsive" design techniques for support of laptop, tablet, and smart phone interfaces.
- Real-time game analytics to help assess the impact of game mechanics during challenges.
- Support for PaaS (Platform as a Service) facilities such as Heroku. This enables organizations to create and deploy challenges without obtaining physical hardware and its requisite IT support.

The design of Makahiki follows closely the serious game lifecycle using a framework as illustrated in Fig. 1:

In Makahiki, the system admins install or create a new default game instance by instantiating from the Makahiki framework; then the game designers configure or customize the instance according to their organization's specific needs. if needed, the game developers extend the instance by adding new game widgets. The system admins deploy the finalized instance to the infrastructure of the choice of the organization, either locally or to the cloud. Once the game instance is deployed and running, the game managers can manage the game and the players play the game. In Makahiki, the game analytics are available real-time during the game and after the game.

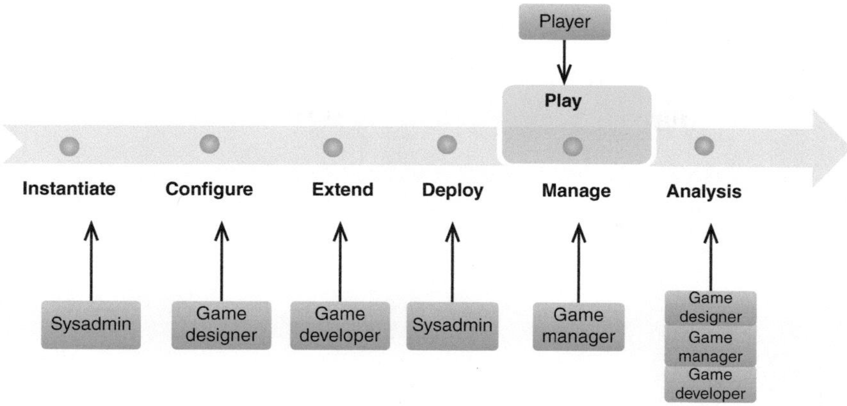


**Fig. 1** Serious game lifecycle using a framework

Makahiki consists of a configurable game engine that can be customized to the needs of different organizations. It includes a library of pre-built game "widgets" that implement a variety of game mechanics. Using the widgets, an organization can create a custom energy challenge in which players can compete individually and/or in teams to earn the most points by reducing their energy consumption as well as by learning about energy concepts in general.

Figure 2 illustrates a home page of the system implemented by using the Makahiki framework.

## *3.1 Architecture*

Figure 3 illustrates the overall architecture of Makahiki system that we developed.

The core component of Makahiki is a configurable game engine that can be customized to the needs of different organizations. It includes two libraries of games and game mechanics. These libraries consist of a set of pre-built "widgets". By selecting and configuring these game and game mechanics widgets, an organization can create a customized serious game.

Makahiki interfaces with the outside environment in three different ways. First, the top side of the architecture diagram shows that Makahiki has two primary user interfaces: one for the players of the serious game, who directly interact with the game and game mechanics widgets; the other for the administrators of the system, who configure the system and monitor the real-time game analytics.

Second, the right side of the diagram illustrates that Makahiki must obtain real-world environmental data as the challenge progresses in order to provide feedback to users about the impact of their actions. In some cases, environmental data can

**Fig. 2** Makahiki home page

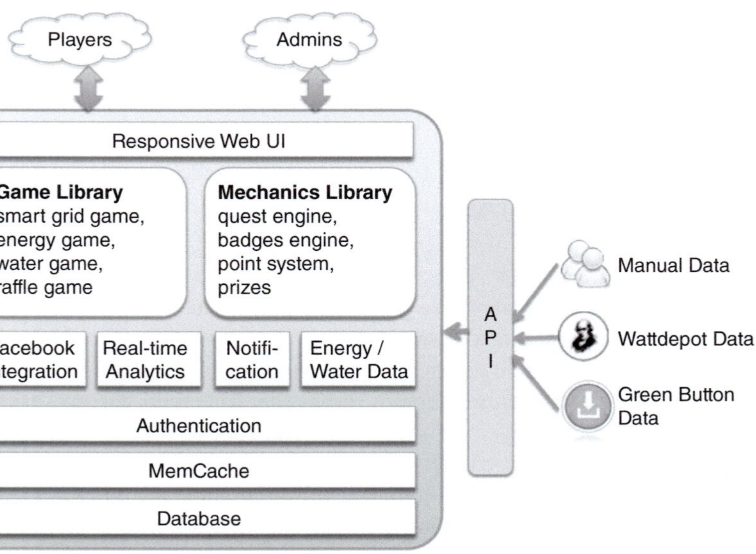


**Fig. 3** Architecture of Makahiki

be input automatically into the system through a combination of "smart" meters and additional services, such as WattDepot [1]. If that is not possible, then manual meters can be read by administrators on a regular (typically daily) basis and input into Makahiki using an administrator interface.

Third, the bottom side illustrates that Makahiki stores its data in a database repository (currently PostgreSQL). To reduce database access and improve performance, Makahiki provides support for caching (currently memcached).

## *3.2 A Library of Configurable Games and Mechanics*

Makahiki builds in a set of configurable games and mechanics that can be turned on or off, or customized by the game designers to the needs of different organizations.

### 3.2.1 Energy and Water Game

A fundamental requirement for enabling more active participation in sustainability behavior is feedback regarding their resource such as energy and water usage. The Energy and Water game in Makahiki are implemented as the Daily Resource Goal Game. The Daily Energy Goal Game widget provides a way for players to see the outcome of the energy reduction behavior, and to make it a game by earning points from their behavior. By reducing their teams' daily energy consumption from a baseline by a set percentage, the players in the team will all earn the configured amount of points. Figure 4 illustrates this widget.

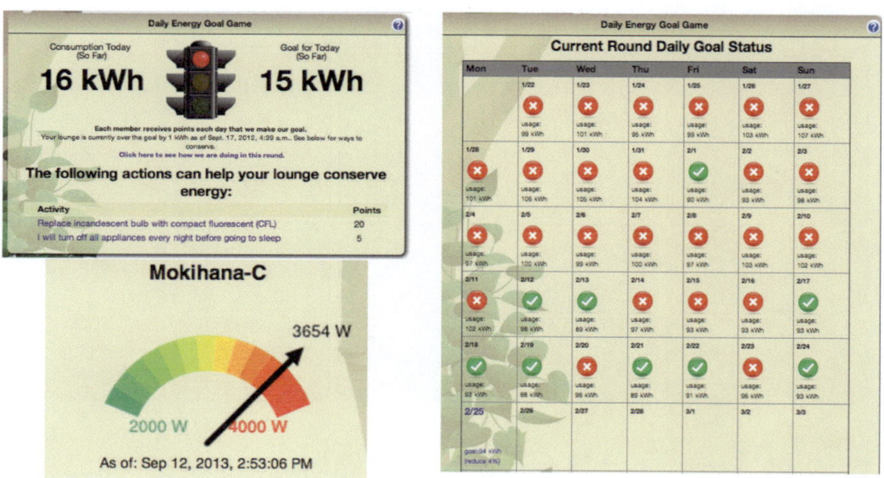


**Fig. 4** Daily Energy Goal Game widget

This interface uses a stoplight metaphor to show at a glance whether or not the team is making the goal. In this case, the stoplight is green, indicating they are currently below the goal. We have found additional perspectives of the energy feedback to also be useful. One useful perspective to a team is a real time power meter visualization that shows the current power usage of a team. This visualization displays the real time power consumption which updates at a specified interval such as 10 s. Another useful perspective to a team is a historical, calendar-based visualization that shows the results of the energy goal game for each day of the current round.

### 3.2.2 Smart Grid Game

Smart Grid Game (SGG) is Makahiki's approach to support "gamified" delivery of educational experiences. Educational actions are organized into a grid of squares (hence the name "Smart Grid") and organized by category columns and levels. Players use its grid interface to discover "actions" they can perform. Successful completion of an action earns the player a variable number of points depending upon the difficulty of the action, and can potentially "unlock" additional actions and higher levels in the SGG. Figure 5 shows a typical Smart Grid Game interface for players.

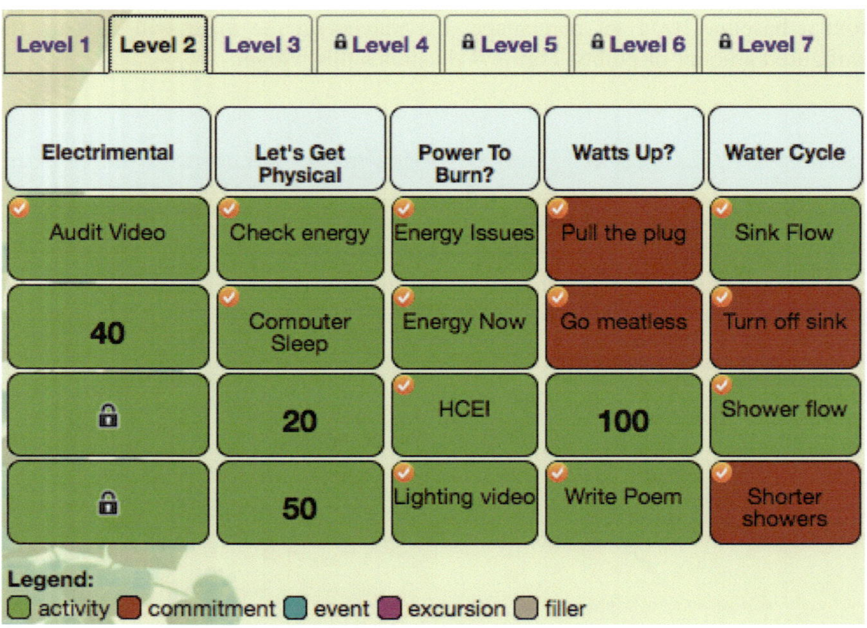


**Fig. 5** Smart Grid Game widget

To make your SGG more interesting to players, and more pedagogically sophisticated, Makahiki supports the definition of "path" through the educational content or actions. In most cases, when a new player sees the SGG for the first time, there should only be a few actions available to them, possibly only one. All of the rest should be locked. Makahiki provide a set of predicates that can be used to define the path. The predicates determine if an action or level is locked or unlocked for a player, which in term depends on the outcome of another action or multiple other actions.

### 3.2.3 Raffle Game

The Raffle Game widget provides a way to incentivize participation from all individuals, even those who are not in the running for a top prize. For every 25 points a player earns, they receive one virtual raffle ticket. Players can dynamically allocate their tickets to any raffle prizes they are interested in at any time, up to the end of the raffle. Figure 6 shows an example of the Raffle Game.

Raffle tickets are independent from a player's score, and allocating a raffle ticket does not affect their rank. The system provides random selection of the winner of each raffle item at the end of a round.

**Fig. 6** Raffle Game widget

#### 3.2.4 Social and Referral Bonuses Game Mechanics

The Social and Referral Bonus widgets are the game mechanics that help encourage participation by providing additional points to players who participate in activities with other players, and facilitate the entry of new players into an energy challenge.

The social bonus is a configurable option when an action is created in the Smart Grid Game. Players earn extra points if they perform the action with another player. When a player submits a response for an action with a social bonus, the player can provide the email address of the person who jointly completed the action. Once the other player completes the action, the social bonus is awarded.

Players are led through a setup process when logging into Makahiki for the first time. One of the steps in this process is the referral bonus. If a player was referred by another player in the system, he can use this step to input their email address. Once the new player earns a certain number of points in the competition, both players are awarded a referral bonus of a configurable number of points.

### *3.3 Real-Time Analytics*

Makahiki is designed to support energy challenges involving hundreds or thousands of users lasting weeks or months. In these circumstances, effective use of the technology requires the ability to understand the state of the game, such as: Who is using it? What are they doing? What is the player response to activities, commitments, excursions, and events? Such state information is important for planning purposes, such as assessing the transportation needs for an upcoming excursion by seeing how many players signed up. It can also be used for making in-game changes to game design, such as changing the point values associated with activities to encourage or discourage participation. It can also help identify breakdowns in game play, such as significant numbers of unallocated raffle tickets indicating that users do not understand the nature of that game mechanic. To address these needs and others, Makahiki includes a variety of widgets that work together to provide high level overview of game play state to the administrators of a challenge. Figure 7 shows examples of some game analytic widgets.

### *3.4 Responsive Mobile Support*

We believe that mobile support is essential for this kind of sustainability challenge, especially for the new generation players. Makahiki implemented responsive web design technology to support multiple devices to enhance the player experience. Figure 8 shows the responsive interface in Makahiki that supports both desktop view and mobile view with the same code base.

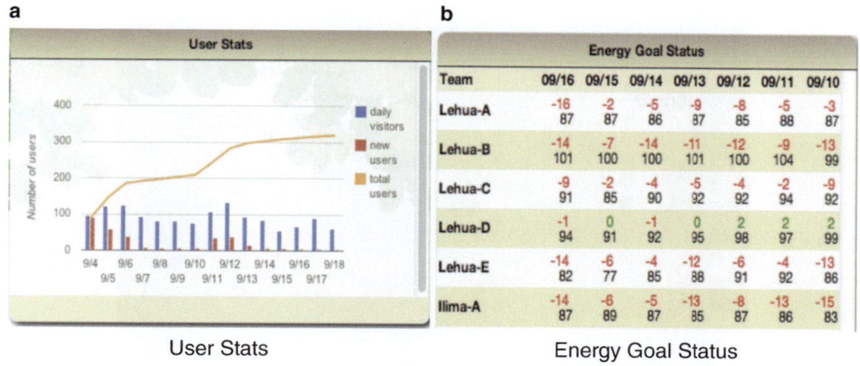


| Team | 09/16 | 09/15 | 09/14 | 09/13 | 09/12 | 09/11 | 09/10 |
|---|---|---|---|---|---|---|---|
| Lehua-A | -16 87 | -2 87 | -5 86 | -9 87 | -8 85 | -5 88 | -3 87 |
| Lehua-B | -14 101 | -7 100 | -14 100 | -11 101 | -12 100 | -9 104 | -13 99 |
| Lehua-C | -9 91 | -2 85 | -4 90 | -5 92 | -4 92 | -2 94 | -9 92 |
| Lehua-D | -1 94 | 0 91 | -1 92 | 0 95 | 2 98 | 2 97 | 2 99 |
| Lehua-E | -14 82 | -6 77 | -4 85 | -12 88 | -6 91 | -4 92 | -13 86 |
| Ilima-A | -14 87 | -6 89 | -5 87 | -13 85 | -8 87 | -13 86 | -15 83 |

**Fig. 7** Game analytic widgets: user stats and energy goal status

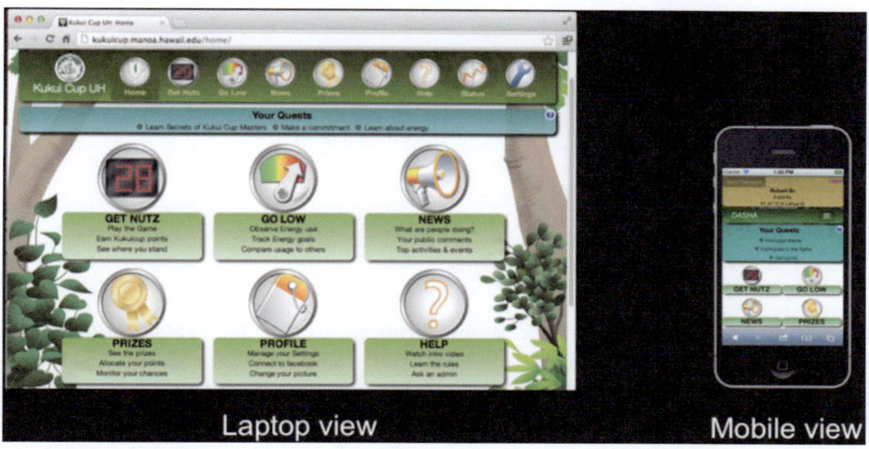


**Fig. 8** Responsive design supports both desktop and mobile

## 3.5 *Cloud Deployment Support*

Another Makahiki feature is the ability to deploy to a cloud platform such as Heroku. Cloud computing has the advantage of simplifying IT administration by eliminating the need to acquire and maintain hardware and operating system software. This also can dramatically decrease the cost of deployment. Figure 9 shows a screen shot of the Dashboard showing the 2012 East West center Kukui challenge deployed in Heroku, one cloud platform provider, and the monthly cost for this deployment.

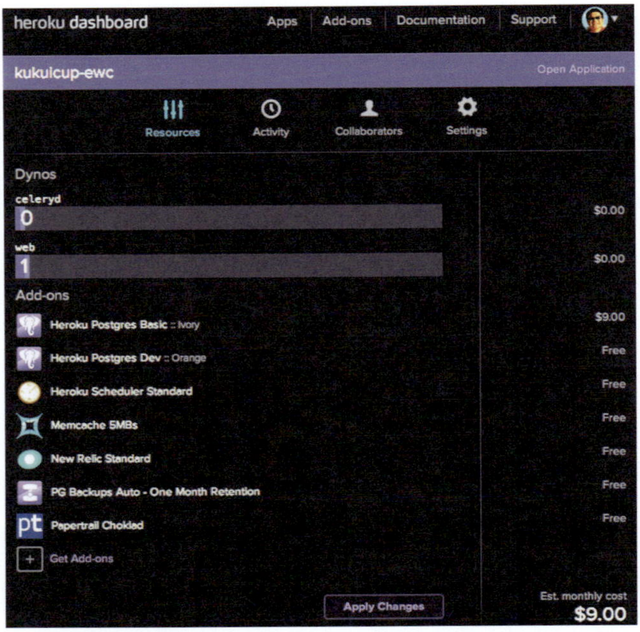


**Fig. 9** Heroku cloud deployment

# 4 Experiences with Makahiki

We had used Makahiki to create seven different energy and water serious game instances, all called "Kukui Cup" challenges. Three Kukui Cup Energy challenges were held at the University of Hawaii at Manoa (UHM) in 2011, 2012 and 2014 for over 1,000 first year students each year living in the residence halls. Hawaii Pacific University (HPU) held a Kukui Cup Energy challenge in 2012 and 2013 for about 200 students each year. An international organization called the East-West Center (EWC) held a Kukui Cup Energy and Water challenge for approximately 600 international residents living in their residence halls in 2012. A Hawaii private school called Holy Nativity School (HNS) held a pilot Kukui Cup challenge for their elementary school students in 2013. Table 1 outlines the organizational differences between these game instances:

Through the game interface, students learned about sustainability both online and in reality to conserve energy and water consumption and other sustainability related behaviors. The in-game survey results from the multiple UHM instances indicate that the games had positive impacts on their energy literacy, self-reported sustainability awareness and self-reported behavior changes. Both the UHM and HPU instances achieved about 30% player participation rate, which may lead to impacts to the community around the participants. The engagement metrics based

**Table 1** Different configuration between the game instances

| Instances | Participants | Teams | Duration | Rounds | Game element | | | |
|---|---|---|---|---|---|---|---|---|
| | | | | | Energy | Water | Prize | Quest |
| UHM2011 | 1070 | 20 | 3 weeks | 3 | ✓ | | ✓ | ✓ |
| UHM2012 | 1080 | 20 | 9 months | 4 | ✓ | | ✓ | ✓ |
| UHM2014 | 1093 | 20 | 2 weeks | 2 | ✓ | | ✓ | ✓ |
| HPU2012 | 189 | 6 | 3 weeks | 3 | ✓ | | ✓ | ✓ |
| HPU2013 | 197 | 6 | 3 weeks | 3 | ✓ | | ✓ | ✓ |
| EWC2012 | 130 | 2 | 2 weeks | 1 | ✓ | ✓ | | |
| HNS2013 | 10 | 2 | 7 months | 1 | | | ✓ | ✓ |

on the game data from 2011 and 2012 UHM instances also indicates high levels of player engagement during the 3-week period.

The successful creation of these serious game instances provides evidence that Makahiki can be successfully tailored to the needs of different organizations. Among these instances, UHM and HPU used different metering infrastructure for energy data collection, EWC collected their resource data manually, and HNS did not have energy data at all. While UHM and HPU challenges involved only energy consumption data, the EWC challenge involved both energy and water consumption data. The IT infrastructure at UHM and HPU provided authentication services using CAS and LDAP, while EWC used both CAS and the built-in Django internal authentication, and HNS used only the internal authentication. Moreover, the user interface was customized to "brand" each challenge with the logo, thematic elements, and the education content of the sponsoring organizations.

Lastly, the hosting infrastructure among these instances are different in two categories: local hosting for HPU and the 2011 UHM instances, and cloud hosting for EWC, HNS and 2012, 2014 UHM instances. These two types of hosting requirements created unique challenges for Makahiki. There are different costs associated with these hosting solutions, as described in Table 2.

Although there are some incurred costs in the case of hosting the Makahiki games in the Heroku cloud infrastructure, there is no need for maintaining local infrastructure, such as keeping the server running, upgrades, patching as well as locating physical space for the server. There is no need to allocate IT support staff resource to do system admin. In addition, the dynamic resource allocation capability of the cloud infrastructure provides a flexible and efficient utilization of infrastructure resources in running the competition, especially in the case of short competition duration and varying user demands during the game.

The successful use of Makahiki in the Heroku cloud platform provided a good alternative for organizations such as EWC and HNS who have less technology capacity. It also provides greater flexibility in infrastructure scaling up and down depending on the usage demand. In the context of serious games, which are often running for a period of time and have various usage demand throughout the game period, using the cloud infrastructure is a good alternative.

**Table 2** Hosting configuration differences between Makahiki instances

| Instances | Hosting | Other cloud services | Duration | Cloud service cost |
|---|---|---|---|---|
| UHM2011 | Local Mac OS server | N/A | 3 weeks | N/A |
| UHM2012 | Heroku cloud | Amazon S3, Memcache | 9 months | $908 |
| UHM2014 | Heroku cloud | Amazon S3, Memcache | 2 weeks | $150 |
| HPU2012 | Local CentOS server | N/A | 3 weeks | N/A |
| HPU2013 | Local CentOS server | N/A | 3 weeks | N/A |
| EWC2012 | Heroku cloud | Amazon S3, Memcache | 2 weeks | $9 |
| HNS2013 | Heroku cloud | Amazon S3, Memcache | 7 months | $146 |

# 5 Evaluation of Makahiki

The evaluation of Makahiki was carried out using the SGSEAM method described [23]. It includes both qualitative and quantitative sources of data regarding the system in the aspects of different stakeholders in a serious game lifecycle, as described in Fig. 1.

The SGSEAM assessment for Makahiki was implemented in two categories of assessment approaches. The in-vivo approaches of pre-post effectiveness study, in-game surveys, and post-hoc interviews were applied to the real-world UHM, HPU and EWC Makahiki instances. The other in-vitro approaches were applied to the UHM ICS691 serious game course as an in-lab experiment. Table 3 provides the overview of assessment for each SGSEAM stakeholders of the Makahiki framework.

Makahiki provides custom quantitative instrumentation that enables us to track when, where, and for how long each player accessed each page of the site. Unlike generic web server logs, we could track per-player application-specific behaviors. We also gathered qualitative data through a survey that players could complete as part of a Smart Grid Game activity during the final week of the competition. The survey asked participants to provide short answers to questions regarding the way the competition and website was designed.

To assess the effectiveness of the framework for designing games that improve player literacy in sustainability, we conducted two energy literacy surveys during the 2011 Kukui Cup Challenge at the University of Hawaii at Manoa. One survey was administrated before the challenge (pre-game) and one after (post-game). Twenty-four players completed both surveys. Out of the total 19 energy literacy questions, the average number of questions answered correctly is 7.54 before the challenge, and 8.96 after the challenge. This result indicates an 18% improvement on the energy literacy. We also surveyed non-players as a control condition, and found that their literacy did not change, indicating that the improvement in player literacy was indeed due to the game.

To assess the effectiveness of the framework for designing games that produce positive change in sustainability behaviors, we recorded and analyzed energy

**Table 3** SGSEAM assessments for Makahiki

| Stakeholder | Assessment approach | Participants |
|---|---|---|
| Players | Pre-Post effectiveness study | UHM Kukui Cup players |
| | Self-reported effectiveness survey | |
| | Self-reported usability survey | |
| | Engagement metrics | |
| System admins | In-lab installation study | UHM ICS691 students |
| | Post-hoc system admin interview | HPU Kukui Cup sysadmin |
| Game designers | In-lab game design study | ICS691 students |
| | Post-hoc game designer interview | HPU & EWC Kukui Cup game designers |
| Game managers | Post-hoc game manager interview | HPU & EWC Kukui Cup game managers |
| Developers | In-lab game development study | ICS691 students |

consumption data before, during and after the challenge. Before the challenge, an energy usage baseline was established. During the challenge, compared to the baseline, 12 out of the total 20 teams reduced their energy consumption, with the highest reduction of 16.1%. However, three teams actually increased their energy consumption, with the highest increase of 11.7%. Overall, the average reduction of the 20 teams was low, approximately 2%.

To access the self-report player perception of the game, our survey asked how would they describe the Kukui Cup. The players were asked to check all that applies to their agreement to the following descriptions of the Kukui Cup: "Educational", "Fun", "Addictive", "So-so", "Difficult", "Boring", "Not useful", and "Other" where they will type in their free text responses. There were 43 responses from the 2011 survey. The numbers of responses and their percentages are listed in Table 4. The "Other" responses are: "AWSOME- NESS" (1), "engaging" (1), "fun competition" (1), "Great way to bond with others" (1), "impressive" (1).

Another survey question was asked about the players' self reported behavior change during the challenge in the 2012 UHM Kukui Cup. The question is "Did you change your behavior during the competition based on the commitment(s) you made? if so, how?" It is a free response question. The results are listed in Table 5.

We also assessed player engagement of the game. We calculated a variety of engagement metrics based on the analytics data collected by the Makahiki framework. The participation rate of this challenge is 37%, which is good compared to other sustainability challenges. Over the course of the challenge, an average player spent about 27.7 min per day on the website. One player spent 8.5 h on one day. There were an average of 266 activity submissions and 208 social interactions between players per day. The average number of website errors per day was 0.6.

**Table 4** Self-reported perception of the 2011 UHM Kukui Cup (n = 43)

| Question: How would you describe the Kukui Cup? | Number of responses | Percentage |
|---|---|---|
| Educational | 41 | 95% |
| Fun | 39 | 91% |
| Addictive | 19 | 44% |
| So-so | 9 | 21% |
| Other | 5 | 12% |
| Difficult | 3 | 7% |
| Boring | 1 | 2% |
| Not useful | 0 | 0 |

**Table 5** Self-reported behavior changes in 2012 UHM KC (n = 45)

| Question: Did you change your behaviour during the competition based on the commitment(s) you Made? | Number of responses | Percentage |
|---|---|---|
| Yes | 39 | 87% |
| Already a habit | 4 | 9% |
| No | 2 | 4% |

The in-lab installation study and the post-hoc interviews were carried out to assess the experiences for the stakeholders of the serious game lifecycle besides players. They are system admins, game designers, game managers and developers.

We performed interviews with the system admin, game designer and game manager of the HPU Kukui Cup challenge to assess their experiences with the Makahiki framework. We analyzed the problems they encountered during the serious game lifecycle. The data showed that the configuration to use HPU's LDAP server and email server were the most difficult tasks, and there were some confusion in the usability of the game design admin interface. In general, the real-word experiences were very positive.

We performed the in-lab study using the UHM ICS691 serious game course to assess the experiences in game installation, game design and game development. The participants were the students in ICS691 Spring 2013 in the information and computer sciences department at the University of Hawaii at Manoa. The students were tasked with installing the Makahiki system into their local computers as well as deploying to the Heroku cloud environment, designing a game using Makahiki and eventually develop an extension to the framework. A Google Form was used for each task to ask the students to record the time they spent completing each step and the problems they encountered. The students also provided feedback about their experiences in the form of blog posts. There were a total of eight students who voluntarily participated in the experiments. The participants were either senior undergraduates or graduate students majoring in Computer Science. The in-lab installation study identified database installation, documentation and the usability of the install script as the weak points in the process of installing Makahiki. The in-lab game design study revealed the most time consuming task is the "Smart Grid Game Design" with difficulty and lack of documentation on the predicate system used to define dependencies between game activities. The in-lab game development study revealed problems with developer efficiency in Makahiki such as lack of documentation for the development libraries. Five out of eight students stated that they had spent more than the assignment estimated hours and had to move on without completing the assignment.

The application of SGSEAM to Makahiki identified several strengths and weaknesses of Makahiki as a serious game framework for sustainability. SGSEAM revealed that the major strengths of Makahiki are the ability to create an engaging and effective serious game for an organization, and the ability to provide an easy-to-use interface for game designers and managers. SGSEAM revealed the major weaknesses of Makahiki are difficulty in developing new enhancements to extend the framework from a game developer's perspective, difficulty to integrate external services such as LDAP and email server from a system admin's perspective, and the lack of WYSIWYG content authoring tool from a game designer's perspective.

## 6 Conclusion and Future Direction

The Makahiki research presented an innovative information technology infrastructure that can support effective and efficient development of serious games for sustainability education and conservation. Makahiki enables different organizations to easily create low cost, customized serious games in the context of sustainability education and behavioral change. Its tailorability and game analytics also provided a useful platform for research on gamification, sustainability education, and behavior change. Makahiki is an open source system hosted on GitHub [16] with less restricted MIT open source license. The application of SGSEAM to evaluate Makahiki had effectively identified several limitations and weaknesses of Makahiki as a serious game framework for sustainability, thus provided focused improvement to the framework which will benefit future sustainability games created by Makahiki.

Our research suggested that several enhancements to the Makahiki framework would be useful. One is real-time player awareness. It is not possible currently in Makahiki to know who is currently "on line" and playing the game. Creating this awareness opens up new social gaming opportunities (performing meaning tasks together), new opportunities for communication (chat windows), and potentially entirely new games (play "against" other online players).

Makahiki currently ships with over 100 possible "actions" already developed for the Smart Grid Game. However, the content is intimately tied to the pre-made Smart Grid Game implementation. An enhancement is to provide a "content management system" for "actions", separating "content" from the "presentation", thus new contents can be added and more games can be developed using this content library.

We are also exploring a future direction involving the development of a consortium of organizations in order to scale the use of the Makahiki framework in new settings. Moving outside of the context of either Hawaii or college-aged players will necessitate development of significant new forms of content, as well as new game mechanics.

**Acknowledgement** This research was supported in part by grant IIS-1017126 from the National Science Foundation; the HEI Charitable Foundation; Hawaii Electric Company; and the State of Hawaii Department of Business, Economic Development, and Tourism.

## References

1. Brewer R, Johnson P (2010) WattDepot: an open source software ecosystem for enterprise-scale energy data collection, storage, analysis, and visualization. In: Proceedings of the First International Conference on Smart Grid Communications, October 2010, Gaithersburg, MD
2. Brockmyer J et al (2009) The development of the game engagement questionnaire: a measure of engagement in video game-playing. J Exp Soc Psychol 45(4):624–634
3. Connolly T et al (2009) Arguing for multilingual motivation in web 2.0: a games-based learning platform for language learning. In: European Conference on Games Based Learning, October 2009, Graz, Austria
4. Darby S (2000) Making it obvious: designing feedback into energy consumption. In: Proceedings of 2nd International Conference on Energy Efficiency in Household Appliances and Lighting, September 2000, Naples, Italy
5. Darby S (2006) The effectiveness of feedback on energy consumption. Technical report, Environmental Change Institute, University of Oxford, Oxford
6. Deterding S et al (2011) From game design elements to gamefulness: Defining "gamification". In: Proceedings of the 15th international academic MindTrek conference, September 2011, Tampere, Finland
7. Electric Shadows (2010) World without oil. http://www.worldwithoutoil.org/metaabout.htm
8. Faruqui A et al (2010) The impact of informational feedback on energy consumption – a survey of the experimental evidence. Energy 35(4):1598–1608
9. Freitas F et al (2006) How can exploratory learning with games and simulations within the curriculum be most effectively evaluated? Comput Educ 46(3):249–264
10. Haiges H (2011) Gaming for good. Technical report. ROI Research Inc, Montreal, QC, Canada
11. Harper B. et al (1993) Improving user satisfaction: the questionnaire for user interaction satisfaction version 5.5. In: Proceedings of the 1st Annual Mid-Atlantic Human Factors Conference, February 1993, Virginia Beach, VA, pp. 224–228
12. Hodge C (2010) Dorm energy competitions: passing fad or powerful behavior modification tool? In: Presentation at the 2010 Behavior Energy and Climate Change Conference, November 2010, Sacramento, CA
13. Lee G (2012) Makahiki: an extensible open-source platform for creating energy competitions. Master's thesis. University of Hawaii, Honolulu, Hawaii
14. Lucid Design Group (2008) Building dashboard. Available via Lucid Design Group, Inc. http://www.luciddesigngroup.com/. Accessed Dec 2008
15. Lucid Design Group (2011) Campus conservation nationals. Lucid Design Group, Inc. http://www.competetoreduce.org/. Accessed Nov 2011

16. Makahiki (2015) An open source serious game engine for sustainability. http://github.com/csdl/makahiki. Accessed Oct 2015
17. McGonigal J (2011) Reality is broken: why games make us better and how they can change the world. Penguin Press, New York
18. Petersen E et al (2007) Dormitory residents reduce electricity consumption when exposed to real-time visual feedback and incentives. Int J Sustain High Educ 8(1):16–33
19. Priebatsch S (2010) The game layer on top of the world. Available via TED. http://www.ted.com/talks/seth_priebatsch_the_game_layer_on_top_of_the_world.html. Accessed Dec 2010
20. Reeves B et al (2009) Total engagement: using games and virtual worlds to change the way people work and businesses compete. Harvard Business Press, Boston
21. Reeves B et al (2011) Leveraging the engagement of games to change energy behavior. In: Proceedings of the CHI 2011 Workshop on Gamification, May 2011. Vancouver, BC, Canada
22. Sherrod A (2006) Ultimate 3D game engine design & architecture. Charles River Media, Inc, Rockland
23. Xu Y. et al (2013) SGSEAM: assessing serious game frameworks from a stakeholder experience perspective. In: ACM International Conference on Gamification 2013, October 2013, ON, Canada
24. Whitton N (2009) Alternate reality games for orientation, socialisation and induction (ARGOSI). JISC Technical Report. Manchester Metropolitan University, Manchester, England
25. World Bank (2010) Urgent evoke. Available via World Bank Institute. http://www.urgentevoke.com/. Accessed Dec 2010

# System-Events Toolbox
## Designing a System of Events that Relies on Local Resources to Create Urban Places for Social Cohesion

**Davide Fassi and Roberta Motter**

**Abstract** This chapter is a reflection on the use of public spaces in towns and the development of a system-events toolbox to make them places of social cohesion. It is the result of a 3-year action research developed together with POLIMI DESIS Lab of the Department of Design to develop design solutions to open up the public spaces of the campus to its surrounding neighborhood. The results have been tested in an event format called "il Sabato della Bovisa" held at the "Durando Campus" of the Politecnico di Milano. It presents a series of initiatives to understand, observe, and benefit designed by members of a public university to trigger a process of knowledge exchange between local residents and students in a climate of conviviality.

## 1 Introduction

"Public spaces mirror the complexities of urban society" [7] and in our time they become fluid spaces [2] of exchange and participative action, not only as contexts but as scenarios able to embody new meanings, and contributing to the creation of community identity [3]. The public space, forgotten for years by industrialized society, is reclaiming its importance due to a renewed interest of urban citizens and institutions. They are becoming places of social innovation, offering a context where creative communities act [11] to bring original solutions to everyday problems, which the current economic system is no longer able to provide.

The bottom-up initiatives of active groups of citizens are often combined with top-down actions of institutions [13] that, together, trigger a virtuous process of social engagement. This creates opportunities for social transformation and

D. Fassi (✉)
Design Department, Politecnico di Milano, Milano, Italy
e-mail: davide.fassi@polimi.it

R. Motter
Milano, Italy
e-mail: motter.roberta@gmail.com

 
T. Issa et al. (eds.), *Sustainability, Green IT and Education Strategies in the Twenty-first Century*, Green Energy and Technology, DOI 10.1007/978-3-319-57070-9_19

sustainable growth that modify the current pattern, replacing the old individualistic values with a new sense of community, sharing, exchange of knowledge and information, and mutual support. The reclaiming of public spaces allows people to come together in an inclusive way – opening roads, town squares, and the city itself for the free enjoyment of all. This happens when people become fully aware of their rights and responsibilities and open horizons to endless opportunities. It shows how citizens are resilient towards external changes due to the contemporary way of life. According to Manzini [9] we need to look at "resilience" with a positive understanding by moving from a mainly defensive one "(resilience as a necessity imposed by the risky times in which we live) to a more positive one: resilience as a deeper expression of the human character and, at the same time, as ground for a possible reconciliation between human beings and nature, between human beings and the irreducible complexity of our world". Consequently, when design is dealing with resilience it uses several tools to reach an effective outcome. In this chapter we will focus mainly on the following two tools to show how design for social innovation can be realized through rapid, small experiments and be prototyped in the real context of local events to create scenarios for long-term solutions:

- co-design workshops: engaging local residents to obtain specific knowledge by collecting information on how to solve wicked problems and exploring fuzzy opportunities [16];
- prototyping events: Participatory Action Research (PAR) [10] where ideas are immediately tested at a one-day event involving people as users of design toolkits. The toolkits are made to be used directly by the end-users, thus empowering them to develop certain actions or set specific goals. This kind of rapid small design experiments allowed quick conclusions to be reached and continued towards more stable and structured solutions [5];

## 2 Hidden Public Spaces and Prototyping Events

In the following paragraphs we are going to show how these design tools and processes (co-design and prototyping events) suit the need for transformation of a changing and fluid urban context. The two tools are combined to create a "system-events toolbox" thanks to strategic design for services methodology [12].

This is composed of a set of toolkits combined together into a system that provides the opportunity to open up a public place (in this chapter the focus will be on university campuses) through a series of actions and activities. It relies on local resources, university staff (students, professors researcher) and on neighborhood communities, such as associations, NGOs, citizen groups, public and private institutions, who want to actively collaborate in the initiative, where the space becomes the background and context, open and accessible to everyone. The toolbox defines a system of events: an accelerator of synergies with the aim of generating cohesion, consolidating relationships among different groups and communities, creating social innovation and a new sense of belonging.

The system-events toolbox can be used in different contexts by local public institutions to activate a public space in original and innovative ways, such as schools, libraries, parks and public gardens, town squares, and city streets. The main purpose of the toolbox is to be an enabling tool for transforming a public space into an accessible public place and to trigger social innovation processes by initiating collaborative actions to engage citizens in active participation and start conversations with the local communities.

In this chapter we will see how the toolbox has already been tested in a public space related to historical memory and no longer used: the Durando Campus of the Politecnico di Milano (Italy) which was established at the beginning of the 1990s in the suburban area of Bovisa in the north of Milan. Situated on 60,000 square metres, this is the area where the old Ceretti & Tanfani company produced funicular railways, and where most of the local residents were employed until the 1960s. The conversion of the industrial area by Politecnico di Milano severed an emotional attachment to the place for residents who considered the company part of the neighbourhood. Politecnico di Milano is a public university and those spaces belong to its properties are public too.

In 2010, the Department of Design with POLIMI Desis Lab[1] began a process of opening the campus to the district of Bovisa through a series of initiatives to allow people to discover the campus and build a new dialogue between the two communities which, until now, had had few opportunities for interaction: the internal community comprises students, faculty and staff, and the external one, the inhabitants of the Bovisa neighbourhood. This process included a combination of research activities and postgraduate courses. The researchers used the tool of scenario building to design periodic and structured occasions for opening places on campus through a series of actions to be set up as public events. These opportunities allowed residents into a continuous appropriation process and at the same time gave the possibility to generate the dynamics of social innovation.

As a result of the research and didactic activities (explained in the next paragraphs) a toolbox project has been developed specifically for the Durando Campus: "il Sabato della Bovisa" (Saturdays in Bovisa). This explores opportunities, both within the university and the local Bovisa district, to create a system able to reconnect the two communities using the campus's urban space as a potential incubator for social experiments that create new opportunities for dialogue. "Il Sabato della Bovisa"[2] is a monthly social event, which proposes a series of initiatives to understand, observe and enjoy a public university space open to local

[1]The POLIMI Desis Lab is a research lab belonging to the Desis Network www.desis-network.org together with more than 40 labs based in the university with the design discipline dealing with social innovation and sustainability. The network was founded by Prof. Ezio Manzini in 2008.

[2]https://www.facebook.com/ilsabatodellabovisa/

citizens, thus triggering a process of knowledge exchange between residents and students in a friendly and relaxed atmosphere. These include: guided tours; book-sharing and public readings in collaboration with the local library; sports activities organized by local associations; fashion design and product laboratories that can connect traditional know-how with innovation linked to university research. There are also opportunities to meet people, share skills and forge new relationships that may have an impact not only in terms of personal connection, but also by creating a place to enhance the development of new micro-economies in the local context.

## 3 The Process

The process that led to the toolbox creation covered a period of twelve months through five main steps. The design experiment has evolved during the course of the project and is still on-going. Traditional research was combined with an "in-the-field immersion" at the local level, followed by a progressive abstraction in order to create a format that could be applied to different contexts.

In the first step a desk research was conducted to outline the contemporary complex framework derived from the historical memory of the neighbourhood embedded in everyday life. That is made up of multiple variables with the main feature being its multidisciplinarity, across different fields of research and action leading to a hybridization of characteristics with three principal directions of investigation:

- the public place and the neighborhood as a small scale context. i.e. creative communities acting locally;
- the relationship between the university and the broader society; its projections on the urban landscape and through the specific case of the toolbox prototyping in the Bovisa district;
- the design of the experiment and its achievement through integration with strategic design and design for services.

The designer's point of view is developed by focusing on public spaces in the city and their daily use, presenting scenarios for enhancing local resources that create multiple levels of opportunities.

The public place is described by its two main features: the social aspect and the formal aspect. The social aspect is the true essence of society that affects aesthetics and form of the urban site. As pointed out by Manzini [8] a contemporary designer must be able to interpret the city as "social laboratories where new ideas and new solutions are being invented and experimented within all fields of daily life", where culture leads to new, forms of citizens' participation and a renewed and profound sense of community.

In the second step the desk research was combined with the immersion on the Bovisa context taking advantage of the university elective masterclass Temporary

Urban Solutions (2012)[3]. During this short and intensive course (7 full and consecutive days), more than 50 international students were involved in developing toolkits to engage the local residents and the university community in social activities inside the Bovisa Campus and test them directly during the last day of the course in a prototyping event open to the public. Instructors used PAR as a main didactic approach. The aim of the course was enabling local residents to reclaim what is, in all respects, a public space. The final event named "C'è Spazio per Tutti" (There's room for all) was held in November 2012 with more than 500 visitors involved in the test of the actions designed by the postgrad students.

This course and the final prototyping event were an opportunity to collect feedback from several stakeholders: 80 comments by the users of the event (the local residents and university community), 64 feedback from the design students and from the professors/tutors involved in the course.

The third step in the design process was the analysis and synthesis of the feedback gathered which underlined that the local community enjoyed to take part to the events, feel part of a community and re-discover the campus public space through social activities. The activities proposed by the students have been related to the preferences of the participants (collected during the in-the-field research) according to their age and profile to understand better which kind of activity can be proposed once again in a more structured framework.

The synthesis of the collected ideas were then used in the fourth step: the creation and evaluation of scenarios. Through tools such as mind map, the scenario building and the evaluation matrix (Fig. 1), four different scenarios came out: opening the campus 24/7, a guided and interactive tour at the campus spaces, open lectures and the concept of a recurrent event to prototype several and different actions.

To identify the scenario to be further developed in design stage, a matrix was used (Fig. 2). Scores range from 0 to 4, where 0 is the lowest rating, while 5 the highest score. The evaluation was made according to the following criteria:

- Participation: active user involvement
- Attractiveness: assuming the level of interest that could stir the solution proposed
- Community: the degree of belonging to a community created by the scenario
- Feasibility: both in terms of organization, costs and technologies needed

According to the evalution given by the research team, the fourth scenario was chosen and it was named "Il Sabato della Bovisa". It was shaped as a system-events

[3]Temporary Urban Solutions is an elective postgraduate course held yearly at the School of Design at the Politecnico di Milano since 2011 by prof. Davide Fassi. It is a studio-based course mainly run as an intensive workshop with a mix of international students with several backgrounds in the field of design (communication, product, interior, service, fashion etc). Students are asked to work together in small teams of 4-5 people and to come out with an action (merging the hardware and the software sides of design) to be tested at the end of the course with real users. Briefs are given by the instructor who developed them in advance with his research team and local actors.

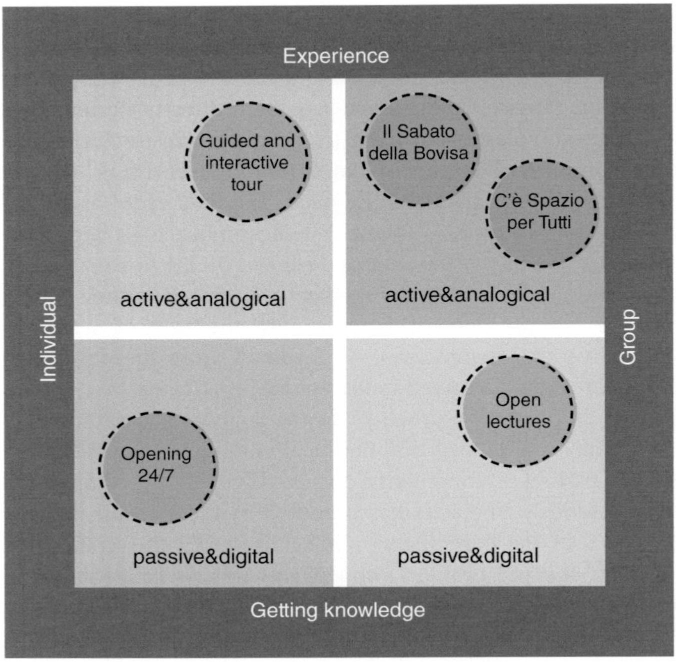


Fig. 1 Evaluation matrix on scenarios

| | Participation | Appeal | Community | Feasibility | |
|---|---|---|---|---|---|
| IL SABATO DELLA BOVISA | ●●●● | ●●●○ | ●●●● | ●●●○ | 14 |
| GUIDED AND INTERACTIVE TOUR | ●●●○ | ●●○○ | ●●○○ | ●●●● | 11 |
| OPENING 24/7 | ●○○○ | ●●○○ | ●○○○ | ●●●○ | 7 |
| OPEN LECTURES | ●●○○ | ●●●○ | ●●○○ | ●●●● | 11 |

Fig. 2 Evaluation matrix on scenarios: assessment

toolbox (a collection of tools and toolkit) and tested in the fifth step of the research: the prototype stage (described in the paragraph 4).

The toolbox provides all the elements needed for setting up actions (events) in a place with the task of answering the new necessities for sociability and conviviality, culture and community. Since public spaces are elements of the built environment that may foster sense of community by facilitating chance encounters between neighbors [14], this toolbox could help to increase the quality of a public space and consequently the connection (emotional and interactive) between the citizens and the space [6].

Designing a "box of possibilities" creates new relationships and gives new meaning to spaces; innovative relations are interwoven with the physical area, and new language and new forms of expression are developed that go beyond formal constraints.

## 4 Definition of the Toolbox

The system-events toolbox has been defined as a useful series of tools for shaping events in the public space. It relies on local resources and it is significant to the inhabitants of a neighborhood and participating communities, assessing the needs of various target groups (elderly, NEET, immigrants etc.). This format enables the immediate proposal of events in a public space and facilitates the understanding of the whole event system's structure. Related to each contingent situation, the toolbox is composed of several elements designed to be adapted and further developed following the emerging trends of the events themselves:

- the manifesto with primary and secondary objectives
- the program with an adaptable activity-box
- the materials for implementation
- the team of volunteers

The core of the system, the program that changes over time, will be discussed later, while the materials of the toolbox are organized in traditional forms of communication, such as postcards, posters, press releases and press kits, integrated with digital media, such as a Facebook page, and a web invitation linked with mailing lists of participants, associations and other collaborators local to the project. To facilitate the management of the system several tools have been created such as a "to do list", a Google map for promotion of the area of interest, and a list of requests for required organizational permissions.

The team of volunteers who participate in the dissemination and promotion of the project, as well as in the preparation and management of some activities, can be made up of people from the neighborhood, local associations, institutional staff, and design students. The administration of tasks is facilitated through meetings and use of the Google group.

The system-events toolbox provides a general framework and structure that can be applied in several contexts where communities would love to be more connected to the public spaces they live, especially in multicultural circumstances and environments where active participation and local resources work together. In order to be replicated, the toolbox needs to be adapted by providing features relevant to the specific context of interest.

## *4.1 The Core of the System*

The instrument used to define the event schedule is a flexible program structured to assume various configurations and to be adapted to situations as they arise. Each activity involved in setting the schedule is developed to be dedicated to a precise target during different events. For instance, a vegetable garden can organize an activity dedicated to children for the first event, while for the second event it can be focused on the elderly, but both activities are linked through the same activity-box (the garden) and topic (the culture of food).

The toolbox was tested in the university campus of Politecnico di Milano starting in spring 2013 by involving as team for volunteers, a group of international students, and by being connected to some local associations who were aware about the research (see Paragraph 3.2) and would have loved to join the prototyping phase. The research team settled five events, one per month, form March 2013 to July 2013 in the open green space of the campus. Each events proposed new activities, designed by the team of volunteers together with the involved stakeholders with the help of the provide toolbox. The first event was made up of four main activities, which proposed different activities dedicated to the different target groups involved. Within the large area of the campus the individual activities have appropriate sites with two central sites: the "Ovale" considered by the university community as the central square of the campus and the community vegetable garden, the other focal point for community gatherings:

- "Sveglia all'Ovale" (Wake up in the Oval) (Fig. 3): Qi Gong lesson offered by a local association of the Dergano-Bovisa neighborhood.
- "Ricuciamo Bovisa" (Reconnecting Bovisa): knitting and sewing activities for adults and seniors of the neighborhood in collaboration with bachelor students of Fashion Design at the Politecnico di Milano and the campus canteen as mail location.
- "Coltivando, the convivial garden at the Politecnico di Milano": the garden was a result of "C'è spazio per tutti 2011" event and it was further developed as a long term project. During "Il sabato della Bovisa" it was used to offer various activities for children and adults related to gardening and food culture.
- "Bo-Legge": book-sharing in collaboration with the local library. It was both a flash-event and the way to establish a proper book crossing corner as a permanent point inside the campus; it even offers public readings and workshops for children, young adults, adults and seniors.
- The first event held in March 2013 (see also Paragraph 4) tested the effectiveness of the toolbox and confirmed how the activities proposed by the university

**Fig. 3** "Sveglia all'Ovale" at the "il Sabato della Bovisa" on April 2013

community, groups of active citizens and by other local actors could be different in each event. The success of the first event trigger the attention by the neighobourhood and there were many requests to join the following initiatives, including a swapping activity, a multi-ethnic choir, yoga and tai chi, activities related to film and music, as well as urban and collective regeneration projects.

## *4.2 Informing, Engaging and Managing*

The creation of the system-events toolbox was complemented by the need to involve and spread the initiative both at the institutional and local level in the area of interest (i.e the Bovisa neoghbourhood in Milan). A fundamental step was to identify possible partnerships between local associations, volunteers, groups of citizens, and commercial activities that could actively take part in the organization in the communication and circulation of the project and also in proposing activities for the event.

Schools, associations and local markets, as well as potentially being active partners were also channels for conveying information, promoting and disseminating the events, along with social networks (Facebook, Twitter, Flickr), and promotion via radio, newspapers, blogs, etc. In the case of "il Sabato della Bovisa", the project established the collaboration among several research areas and functional department within the Politecnico di Milano. The initiative has become part of the Polimi Desis Lab, and supported by Polisocial, the programme of engagement and social responsibility at the Politecnico di Milan. Support has been extended, therefore, to the university instrumental laboratories (the exhibition lab, the fashion lab and the photo and movie lab) and some faculty members, e.g. professors of fashion design, who were taking an active part in the organization.

The municipality of Milan has also offered its patronage to the initiative. At the local level, several informal groups are collaborating with the neighbourhood: associations (i.e. "de.de.p. – democratic and participated design", "il Vaso di Pandora" – offering yoga and tai-chi courses), regional services such as the district library, and urban renewal projects (i.e. a self-funded one: "ZUP (Zuppa Urban Project)", a regional funded one "ColtivAzioni Sociali" both on use the food as a mean of connecting people in the urban context). As mentioned above, it has become increasingly important in recent events to forge an even stronger link between university resources and local communities. The key to social interaction in public spaces is the quality of the public space provided [17] and the activities settled in the university campus aimed at getting this result.

### *4.3 Resources and Investment*

The non-profit nature of the project led to a voluntary commitment on the part of the organizational team, the group of local volunteers and the design students, associations and local authorities, and this constitutes the main resource for the success of the project.

Other resources already invested in the project are mainly allocated to the production of information and communication materials such as posters, flyers and postcards to be distributed throughout the neighborhood, the cost of which is definitely low budget and covered by university funds. Further investment could be proposed that would introduce activities related to increased integration of different cultures (in the Bovisa neighborhood there is a strong Chinese community and North African one) and expand promotional activity by using more communication channels and other activities, without losing the non-profit nature of the project. The low investment in economic terms generates an important impact in which the value exchanged creates new value, for example, value of a social nature.

## 5 Il Sabato Della Bovisa at the Politecnico Di Milano: The 2013 and 2014 Editions

The first edition of "il Sabato della Bovisa" was held as a monthly event from March to July 2013, followed by the second edition in March 2014. In 2013, the toolbox was tested directly in context and at the local scale with the first three events: the inauguration on March 16th, the second event on April 13th, and the third event on May 11th, and then implemented in June and July 2013. The event was organized by a main project design manager[4], who has worked on defining the activities on the basis of different targets to be involved and seeking the cooperation of local associations who wish to participate proposing activities during the event days. The event was promoted through university social media channels, design blogs, local and national radio, primary, middle and secondary schools in the area. The communication of events in the neighborhood has also been promoted by some volunteers (mainly students) through the distribution of flyers and by doing a flash-mob at the local weekly market. The volunteers were a valuable support also during the events, dealing with the arrangement of the areas of the campus and the management of specific activities. At the institutional level, the research leader[5] has been involved in promoting the initiative within the university creating the interest of some university sponsors like “Polisocial” (the university social responsibility program) and “Desis Network” (Design for social innovation and sustainability international network). He has continued a dialogue with the different organizational areas of the Politecnico di Milano, as the “Communications and External Relations” office, the “Management Infrastructure and Services” office, the Design Department and the School of Design.

The opportunity to test more times the toolbox framework allowed the team to gather feedback and implement new ideas. The design cycle on the discovery, immersion, study, design, and feedback collection [15] was overlaid with new reflections and rediscoveries, ranging from the generation of knowledge to critical reflection. These coordinated cycles were combined to optimize the understanding of the proposed solution and led to an incremental improvement during the subsequent events.

For the first season, the overall judgement on the system-events toolbox prototyping on the Bovisa district was extremely positive with an incremental increase in attendance. The total number of visitors for the first three events was around 130–150 adults each, including several target groups from the elderly to children and families[6]. Participants were mainly people of Bovisa district, families with children interested in playing together with other kids and elderly people living nearby the campus eager to spend time in community activities. Even the Politecnico students, using the campus for educational purposes, during the event enjoyed to take part in some of the activities proposed.

During the event days some feedback was directly collected on basic aspects of communication and organization, and on specific activities. A second request for feedback was made through the Facebook page inviting people to give comments, suggestions and proposals. This work is continuing through the identification of an appropriate communication channel that will disseminate and promote the

[4]Roberta Motter, co-author, postgradute student in Product Service System (Politecnico di Milano) at that time, was developing her final master thesis on opening the campus to the neighbourhood.

[5]Davide Fassi, associate professor at the Design department (Politecnico di Milano)

[6]Video of the opening and of the following event can be seen here: https://www.youtube.com/watch?v=SnJP6LCyqsk

initiative broadly to the different targets involved in the Bovisa district and, more widely, in the city of Milan. This trial event started a process that is gradually establishing a dialogue between the organizations that launched "il Sabato della Bovisa" and the local community in a fertile ground for experimentation. It is a process that is able to place the university within the district network, thus emphasizing the renewed interest of participation in a highly accessible initiative.

Several activities have been directly proposed by associations and citizens groups during the events and they helped the research team to decide which could be considered and developed later. The possibility to create micro-communities around certain activities, such as "Ricuciamo Bovisa", or "Bo.Legge" convinced the university administration to include them as an ongoing activity by involving knitting and sewing groups, or by setting up a permanent book exchange facility.

The following two years of experimentation (2014 and 2015) led to the consolidation of the flexible structure that is reinforcing the neighborhood with new and increasingly important bonds, i.e. the local associations. This increasing network of actors is becoming the voice of and force behind the exchange between the internal and external communities and is thus the new core of the toolbox.

The event in 2014 included the creation of seven activities to be presented to people in the neighborhood. Thirty-six students, divided into teams, collaborated with seven local associations who helped the groups develop topics related to the needs of the district. Indeed, the goal was very well-interpreted: experimenting temporary solutions in the university's public space to create a public place that would engage the Bovisa community and reinforce the bonds between students and citizens in a two-way relationship that can create mutual benefits. The sustainability of the event was combined with the sustainability of individual activities through a variety of interpretations: new community-minded behavior of recycling and reuse; social gardening; respect for food and nature; sharing stories to create common experiences of the neighborhood community; children's multicultural street games to support integration and knowledge. Following the already ongoing work of the associations, the design phase was developed in 5 days of intense work in which students developed the seven activities, through research, development, prototyping and action, resulting in some cases in the relevant toolkit. Some action has been designed in a more technologically sophisticated way, mixing physical and digital support, matching meaningful temporary interactions with the preservation of the event result (e.g. Web series with "Plug TV", a local social web tv) and combining different design approaches for unique and interesting results.

The event was held on March 15th, 2014, with nearly 700 people taking part – local residents faculty members, residents of the suburbs and of greater Milan. Two of the proposed activities were a continuation of the previous edition, but redesigned for the occasion: "Bo-Legge" and "Swapping". Bo-legge is the university "book-exchange space" created in the first event, which was improved through interviews, fast prototyping and increased collaboration with the library of Dergano-Bovisa. The event was the occasion to relaunch the book exchange and the opportunity for the local residents to rediscover how easily they can exchange books on the university campus in a free and open way. The "Swapping", named for the occasion "Swap Party 2.0", was designed as an upgrade of an existing system. The inclusion of storytelling to create a stronger connection among the people who participated in the "Swap Party" extended the swap activity. Each participant (more than 50 people) was asked to write down on a label some memories related to the clothes they were going to exchange. By doing this, the people were not just swapping clothes, but were becoming aware of the memories of their neighbors, thus creating a stronger community feeling. Two activities were targeted specifically at children: "Abbattimpronta – Game Up" is a reinterpretation of a board game by the architect Chiara Frassi: following a movable path to learn how to play a more active role in environmental issues and create a greener world. The second one, "Tu Play" created by "de.de.p – democratic and participated design" association is a collection of eight street games from eight different cultures to promote and facilitate multicultural interaction through playing. Activities related to urban gardening, as evidenced by many cases, including the Coltivando community garden of Politecnico di Milano, are very effective in creating a collective feeling of collaboration and respect in the city's districts. Guerrilla gardening has been developed through "Giardini Extravaganti" with the ZUP – Zuppa Urban Project association – an action to set up a green pathway between the abandoned garden of Via Carnevali (a street close to the campus) and the Bovisa Campus. Together with creative recycling to create urban green spots, Coltivando sponsored "Zunnel" (Fig. 4), a co-design activity between students and Coltivando

**Fig. 4** "Zunnel – The pumpkins tunnel" at the "il Sabato della Bovisa" on March 2014

gardeners to design a shelter under which to grow pumpkins, where people can share convivial moments. The activity was designed by shifting the focus to the process of construction, rather than to the final outcome, involving users in the development of a common activity sharing a community context. The combination of two design approaches, participatory action and narrative environment, led to the design of the Micro Set for Web Series with the support of Plug Social TV. The participants were invited to share stories about their life in the Bovisa community. In this case the aim of the design action was to investigate, experiment and prototype stories as a driver for increasing community identity and involvement.

The support of the associations was essential for all the proposed activities to move the design to issues and projects that could bring real benefit to the community of Bovisa. The large amount of participation is closely linked to the research developed in the first year: the design of a flexible tool, the system-events toolbox, and the strong identity of "il Sabato della Bovisa" are completely connected to the local context and recognized as such by the community.

## 6 Scaling Up: Il Sabato Della Bovisasca

The 2015 event was doubled. "Il Sabato della Bovisa" belongs to the "campUS – Incubation and settings for social practices" research programme and it led the research team to figure out a scaling-up opportunity to allow the event to be held outside the campus in the neighbourhood of "il Sabato della Bovisasca".

"campUS" is a research-funded project[7] developed by the Design department (together with the Architecture and Management Engineering departments) at the Politecnico di Milano. It was created as an implementation of several small design actions (coming from the earlier "Il Sabato della Bovisa" event) developed within the Polimi DESIS lab in the previous year, to establish a more structured system to support local initiatives.

"campUS" works for a positive relationship between the space and skills available on universities campuses and the local context in which they occur. It is divided into two areas of intervention, "campUS in", actions inside the campus and "campUS out" actions outside the Campus (in the local district and beyond):

- "campUS" in: through research-action, activation of the campus spaces as incubators of social practices where defining, testing and prototyping social actions (services, spaces, communication systems) with a method of co-design and

[7]This research is funded by Polisocial – The social responsibility programme of the Politecnico di Milano, which makes an annual call for research projects with social purpose, and is funded by the income from national tax declarations. In Italy, every citizen can donate five out of every thousand euros of their tax declaration to public institutions. Politecnico uses these donations for the Polisocial programme.

participatory planning. The results aim at building a package of tools for the dissemination of good design practices, cohesion and social innovation for specific regional communities in specified areas of the city; the aggregation of a number of figures to support the production of content and development of a communication platform of the district as a system of narrative social practices, catalyst actions and partnerships. "campUS" in emerged from the success "Il Sabato della Bovisa", 2013 and 2014 by benefitting from the strong network created with the local associations.
- "campUS" out: definition of a landscape of permanent actions in the neighbourhood that have the potential to lead to social enterprises, through an exchange with virtuous prototyping actions ("campUS" in). The research gives design support, helps adoption and diffusion of instruments of identity and community-building in the neighbourhood and aims at identifying an innovative business model for the long-term management of these initiatives.

This research consists of a theoretical, meta-design dimension and an applied one to experience the dynamics of effective involvement, to test tools and to prototype models of innovative social practices. The university campus and districts of Bovisa-Dergano represent the real case study, where actions and interventions in the public space may actually involve citizens and other social actors, allowing them to explore original forms of relationship among stakeholders. The research also aims at creating "a new community built in collaboration with the existing community and the development of a lived public space with shared memories" [1].

"campUS" has four main work-packages to be developed in close connection with the local neighbourhood: the convivial garden, social tv, the travelling pavilion, and a business model.

"Il Sabato della Bovisa" and "il Sabato della Bovisasca" were the two events held in March 2015 according to the campUS in and campUS out scales of intervention. The system design toolbox was used on both these occasions, being adapted to each specific event. "Il sabato della Bovisasca" followed the same format as the previous year's event on campus but was held in Bovisasca, a small neighbourhood 2 km from the campus, in a 5000 square meter green public area. The city of Milan gave this area for free to a local association, named "9x9", to use it as a community garden. The event was an opportunity to help them to set up and test some ways of running the garden with the involvement of the neighborhood.

Two main projects were developed: "Piantastorie" and "Beyond the boundary".

- "Beyond the boundary" was a project to develop a threshold to circumscribe the community garden's area. The 5000 square meters area needed a 500 m length fence by municipality law. The "9x9" association were asking to design a low or no-budget solution for it. The construction of the fence itself aimed to involve the future gardeners, stimulating their participation in the handicraft process: in this way the collective action starts from the very beginning, in order to create a bond among the co-growers. Plastic crates from the local fruit market (a leftover to be recycled) have been used to create the fence by joining

them with connectors and explaining the set-up through visual instructions to be followed by the participants during the event.
- "Piantastorie" is a red wooden bench where people were invited to tell stories, connected to key local places, using a toolkit, to create short video interviews. The kit included a picture of one of the local places and a seed to be planted in the garden to leave a memory of the interview.

Both the activities were run for a full day with the involvement of more than 120 local citizens. People started to build the fence and they continued the construction in the following weeks. The stories collected through Piantastorie became a web-series to be broadcasted on "PlugTv".[8]

## 7 System-Events Toolbox as Design Opportunity

The objective of a project like the system-events toolbox is to activate urban places through co-designed actions that lead to social cohesion, integration and social innovation. The dynamics that can be generated through a series of events relies on there being numerous local resources that have a high potential for involvement: not only the reclaiming of a public space by its inhabitants, but deriving new meanings by creating unexpected forms of language and belonging.

This project has been developed through a strategic design process which combines traditional research with an in-context immersion and a consequent abstraction for creating a model. The experimental nature enhances the opportunities for social innovation, where the system-events toolbox, organized in boxes with different levels of access that act in synergy to create opportunities, is thought to be as open as it possibly can be in order to be able to change during the course of the work. The toolbox is designed to be used mainly by public institutions, which, with an appropriate contextualization, can follow the guidelines and use the material, following a real example of opening up a public space. In this sense, the inclusion of the project in the Desis Network facilitates the international spread of the toolbox solution. The design of a system-events toolbox composed of activities is that the tool that best suits the situation is able to create fixed activities as well as variables, together with the ambitious goal of building micro-communities that share multidisciplinary and cross-generational interests, knowledge and collective values.

In this particular case, the toolbox applied to the Durando campus of the Politecnico di Milano as "il Sabato della Bovisa" was: an opportunity to stimulate the processes of public and cultural renewal: prototyping space for students' projects, a place for associations and local projects to be seen, an opportunity for university staff to promote and disseminate research projects, and the context for active citizens to experience, discuss and propose ideas. The university is not only

[8]Piantastorie web-series could be seen here https://www.facebook.com/piantastorie/ and PlugTv is available here https://www.facebook.com/plugsocialtv/ (Italian language only).

the context, it becomes a promoter of active participation in finding its integration in the urban district and regaining renewed possession of its fundamental social function, opening up to the surrounding area in an active and practical way. It is important to point out how the nature of a space changes with the change of use. The development of the system-events toolbox, as demonstrated by the prototyping in the university campus spaces, is a successful method for rethinking public places, and merging processes of cohesion and sharing, in particular for those public spaces with a surplus of unused space.

The project is therefore moving in the direction of implementation within a different context regarding both destination and dimension, in order to test the performance of the existing toolbox and to make improvements to the toolbox design.

## 8 Conclusion

In these 3 years of research, the trend regarding the re-appropriation of public space has grown rapidly both in metropolitan areas and in small towns increasing awareness of the environment and enhancing the power of social actions: in Milan for example, the Lambrate district or the NoLo (North Loreto) district are now redesigning their identity and regional development in order to promote the culture of the place, and to regenerate and enhance the area. There are also several cases of shared and convivial gardens, neighbourhood parties, swapping communities and other social activities supported by local associations, public and even private institutions that are creating new "communities of scope" with the willingness to share time and resources and are giving new meanings to the space in which they interact.

Using the temporariness to create real scenarios [4] allows stakeholders to design short-term solutions so to figure out long-term strategies for a public space (and not only). This is the case of this System design toolbox: it started as a way to support a one single local event and it ended to be a strategy to replicate such event to create meaningful experiences in a campus connected with a neighbourhood. The design experiment has continued for the following years and, as previously explained, it is still in place. The flexible and unique toolbox has allowed strong identity to be created, "Il Sabato della Bovisa", and therefore boosted the potential of the system-events toolbox. The involvement every year of more than 60 young international designers of the university course working in collaboration with local associations allowed new collaborative actions to be tested and to open up the campus and create a more consistent network of actors.

Further steps will be taken in the application of the system-events toolbox to a different context by public institutions. For instance to a school or public library in a multi-ethnic district, meanwhile implementing the toolbox and starting the creation of a network of interconnected activities in local communities.

To start the application of the system-events toolbox in a different context we have to start creating a network of actors who will be involved in the initiative in the area of interest, beginning with the public institution involved and then expanding into the neighborhood to connect the various actors and authorities within the system.

Design, as facilitator and/or activator, has supported the creation of these meanings and of a common goal by involving several local actors and by designing solutions to improve the quality of a public space. The example showed in this chapter aims to underlined how designers' support is needed to make these collaborative services accessible, efficient and attractive to the users. Vice versa, the citizens (associations, informal groups etc.) role to the design process emerged to be as important as the designers one to create awareness towards the quality of the public space and to guarantee a long last success of the output.

## References

1. Austin T Culture-led city regeneration: design methodologies. Essay, Central Saint Martins College of Arts and Design, London.
2. Baumann Z (2000) Liquid modernity. Polity, Cambridge
3. Borlini B, Memo F (2008) Il quartiere nella città contemporanea. Bruno Mondadori, Milan
4. Fassi D (2012) Temporary urban solutions. Maggioli Editore, Sant'Arcangelo di Romagna, Rimini
5. Fassi D, Meroni A, Simeone G (2013) Design for Social Innovation as a form of Design Activism: an action format. In: "Social Frontiers: The next edge of social innovation research" online conference proceedings, NESTA, London, http://www.scribd.com/doc/191848489/Design-for-social-innovation-as-a-form-of-designingactivism-An-action-format. Accessed 18 May 2014
6. Francis J, Giles-Corti B, Wood L, Knuiman M (2012) Creating sense of community: the role of public space. J Environ Psychol 32:401–409. doi:10.1016/j.jenvp.2012.07.002
7. Madanipour A (2010) Whose public space? International case studies in urban design and development. Routledge, Abingdon
8. Manzini E (2010) The social construction of public space. In: Golicnik Marušic B, Nikšic M, Corier L (eds) Human Cities. Celebrating public space. Stiching Kunst Boek, Bruxelles
9. Manzini E (2015) Design, when everybody designs: an introduction to design for social innovation. MIT Press, Boston
10. McIntyre A (2007) Participatory action research. SAGE, London
11. Meroni A (2007) Creative communities. People inventing sustainable ways of living. Edizioni Poli.Design, Milan
12. Meroni A, Sangiorgi D (2011) Design for services. Gower Publishing, Farnham
13. Mulgan G, Tucker S, Ali R, Sanders B (2007) Social innovation: what it is, why it matters and how it can be accelerated. SaidBusiness School, Oxford
14. Talen E (2000) Measuring the public realm: a preliminary assessment of the link between public space and sense of community. J Archit Plan Res 17:344–360
15. Tim B (2009) Change by design. How design thinking transforms organizations and inspires innovation. HarperCollins, New York
16. Visser FS, Stappers PJ, Van der Lugt R, Sanders EB (2005) Contextmapping: experiences from practice. CoDesign 1:119–149
17. Zhang W, Lawson G (2009) Meeting and greeting: activities in public outdoor spaces outside high-density urban residential communities. Urban Des Int 14:207–214. doi:10.1057/udi.2009.19

# Teaching Assembly for Disassembly: An Under-Graduate Module Experience

**Eleftheria Alexandri**

**Abstract** This chapter is about the methodology of teaching Assembly for Disassembly to fourth year undergraduate architecture students within the module of sustainable design. When designing a sustainable building one should take into consideration the fact that the building is going to be demolished at some point; thus components should be assembled in such a way so that the building's elements do not become urban waste, but disassembled in a practical way, so that its materials and components can be reused or at least recycled in other buildings or products or be decomposed. Yet, the structure should be safe, hygienic, comfortable and energy efficient during its lifetime. With this technology, a building's sustainability spans also to its "grave". In this chapter the methodology for teaching students how to design for the disassembly of buildings is presented. Apart from the respective lectures, students have been asked to design structures, following, apart from bioclimatic and the zero-$CO_2$ emissions building design, an "assembly for disassembly" design. They have also been encouraged to use, apart from conventional building materials, materials that could be collected from municipal waste, provided they have a small environmental footprint. Frequent tutorials have been made during this project. The pedagogical methodology and the students' designs are presented for three academic years. From this experience, conclusions are drawn on the most efficient pedagogical approach of teaching Assembly for Disassembly to architecture students, hoping to open up a dialogue on how this sustainability construction technique should be taught.

## 1 Introduction

When teaching sustainability to architects, students should be able to practice sustainable architecture both in a holistic and in a serialist way by the end of the module, so that they are able, according to Race [20], to both "comprehend" the issues, so that

---

E. Alexandri (✉)
University of Patras, Patras, Greece
e-mail: el_alexandri@yahoo.gr

 
T. Issa et al. (eds.), *Sustainability, Green IT and Education Strategies in the Twenty-first Century*, Green Energy and Technology, DOI 10.1007/978-3-319-57070-9_20

they can make decisions, and also to use certain tools (algorithms, sun charts etc.). In sustainable design, apart from energy and water savings, the use of ecological materials, passive heating/cooling systems and the implementation of Renewable Energy Sources, students have to be introduced to the Life Cycle Analysis of buildings. Assembly for DISAssembly (ADISA) is one of the most prominent principles to designing sustainable buildings, which will not form building waste, but, if demolished properly, their materials and components can be reused, or at least recycled or decomposed. ADISA, as a sustainable construction technique, is taught during the spring semester module "Sustainable Design II", preceded by the winter semester module "Sustainable Design I" to the fourth-year students of the 5-year course in Architecture at the University of Patras, Greece. Since the academic year 2012–2013 further emphasis has been placed on ADISA. The aim is to teach students that apart from the ecology of building materials, bioclimatic design, low energy and water consumption as well as the low waste production of buildings, their designs cannot be holistically sustainable, unless their disassembly has been put into consideration since the beginning of their design.

As ADISA is not so often part of the curricula of sustainable design modules, the pedagogical approach that has been followed for teaching ADISA to undergraduate architecture students for three academic years and its results are presented in this chapter, hoping to open up a dialogue on the methodology of teaching ADISA to undergraduate architecture and engineering students.

The methodology for teaching ADISA as part of the module on sustainability in the University of Patras for the academic years 2012–2013, 2013–2014 and 2014–2015 is presented in detail and its results are recorded and evaluated. Lecture diaries and notes are examined. The evaluation of the results from the students' participation in the ADISA projects is used as a crucial index on whether the pedagogical approaches used every year have been fruitful or not. Informal, direct oral feedback from students has been received by the lecturer, but is not examined in this chapter, as it is not a trustworthy methodology for evaluating teaching practice [29]. A simple, structured questionnaire [5] has been handed in to students during the academic year 2013–2014, regarding the module as a whole and not specifically ADISA, which is also used in the evaluation methodology. The lecturer's reflections and self-critique on the outcomes are also used for evaluating the teaching approach.

The objective of this research is to point out the optimum pedagogical approaches when teaching ADISA to architecture students, so that ADISA becomes an inextricable part of the building design in the future, as, e.g., its statics are today. As the examined sample is a limited one (three academic years in the same university, taught by the same lecturer), the evaluation of the pedagogical approaches that have been followed cannot be generalized imprudently; further research should be made and especially surveys from independent bodies on how ADISA is taught in different institutions and what the results have been, so as to form the optimum methodology. Nonetheless, the discussion in this chapter can be used as a start-up for putting forward and investigating the effective ways of teaching ADISA either as part of a module on sustainable design or as part of a module on construction technology.

## *1.1 What is ADISA?*

Assembly for Disassembly (ADISA) or Design for Deconstruction (DfD) or Design for Disassembly (DfD) are terms applied when expressing the aim of reusing and recycling materials, as a means of sustainable design [19]. Some architects have even suggested the Design for Disassembly of buildings with the use of data matrix bar coding on each building element, so that it can be easily assembled and disassembled by non-qualified workers [12]. Nevertheless, ADISA or DfD and the reuse of materials in constructions is not necessarily a "high-tech" technique, but a building technique which is as ancient as buildings themselves.

As elements had not been industrially produced before the nineteenth century, their production was labor intensive, making them more precious. Materials, such as stones, marbles, ceramics etc., were not composite and had a long life span. It was common practice throughout all cultures, from vernacular to monumental architecture to disassemble and reuse building materials and other elements in both buildings and streets. The reuse of stones from older buildings (called "spolia") was a common practice since antiquity in the area of Greece [16]. The reuse of old stones, marble debris and ceramics, as well as whole elements such as marble columns and capitals of pagan Athenian temples and other structures has continued in Byzantine masonries, some of which can still be found in existing churches (Fig. 1a). The symmetric marbles of the temple of Artemis of Ephesus, with their magnificent colours, have been used as

**Fig. 1** (a) Stones, capitals and columns of ancient temples and buildings used in the eleventh century AD Christian church "Panagia Kapnikarea" in Athens and (b) the marble statue of a "kouros" from approximately 610 BC, used in the substrate of the Sacred Way (Iera Odos) in 479/8 BC, as was revealed during the 1990s excavations[1] (exhibited in the Archaeological Museum of Kerameikos, Athens)

[1]On the statue's left back there is a trail from the wheels of chariots rolling on top of it for centuries.

marbles on the walls of the church of Saint Sofia in Istanbul (this act also being a symbolic act of the dominance of the new religion – Christianity – over the old one – Twelve-god Paganism).

Among the stone or marble covered streets of Athens since antiquity [25] debris of ceramic amphorae, tiles and glass have been used either as surface materials or as repair materials in the substrate [16]. Even old statues have been used as substrate material, in the example of the Sacred Way (Iera Odos) in 479/8 BC (ibid) (Fig. 1b).

Apart from public and monumental architecture, the disassembly and reassembly of stones, tiles, bricks, composite elements like windows, doors, fireplaces etc. has been a common practice in vernacular architecture around the globe up to recently. With the Industrial Revolution and mostly with the consumerism of building materials, this technique has been abandoned since the previous century in the developed world. Composite materials, with hazardous and toxic components have been widely used, without any reusing considerations. Merely as consuming commodities, their destiny, after the building's death has been to end up in waste fields.

This rationale has led to large amounts of construction waste and primary materials production and consumption. Construction and demolition waste account for 970 million tons per year in the European Union (EUROSTAT 2010 cited in [4]). Before Greece's economic crisis construction and demolition waste made up to 31% of the municipal waste in the country [9]. Under this enormous waste pressure, the Waste Framework Directive 2008/98/EC [10], and its respective Greek Law (ΚΥΑ 36259/1757/Ε103/2010) [17] have been established on minimizing construction and demolition waste.

All the above are presented and explained to students as an introduction, so that they understand that ADISA is not something impossible to design or construct; also that it is not just an academic or ecological concern, but a demand they will have to face during their professional lives, if we want to deliver a sustainable world not only to the future generations but also to the present ones. The legislative framework's objectives and demands are introduced to them, about how building designs should be prepared so that a large amount of building materials and components could be reused and recycled.

## 2 Teaching Methodology

Teaching ADISA in the module "Sustainable Design II" (2012–2013) and in the module "Detailed Elements of Sustainable Design" (2013–2015) is based on lectures, class exercises, tutorials and assignments. Classes usually start with a lecture, followed by class exercises. Assignments are also discussed during the lecture. Tutorials are made on the assignments, from the first weeks of the module, discussing students' primary sketches and ideas up to one or two weeks before handing in the assignments. The module is marked through assignments and class exercises; no exams are given in this module neither in its preceding winter semester.

## 2.1 Lecture Description

During the first academic year (2012–2013) two lectures have been made on ADISA, which have expanded to three during the next two academic years. Lectures focus on an introduction to the Life Cycle Analysis of buildings as well as on the principles of ADISA, as have been put forward by Berge [3]; use of standardized, simple materials, separate layers for each of the building's strata with the possibility for disassembly within each layer. The rationale of "reduce-reuse-recycle" is reminded in every step of the design and ADISA is thoroughly explained; it is made clear that when designing, from blocks of buildings to structural details, one should bear in mind that building materials should directly be reused, so as to lower building waste. Thus, the building's separate layers should be technically separated (e.g. piping separated from walls); each layer should be easily disassembled with minimal mechanical effort; standardized, mono-material components should be used, so as to encourage reusing. In this way, ADISA constructions can be easily separated into identical components and materials during demolition, ready to be useful parts in another structure and not pile up in municipal waste. During the second and third year, as lectures evolve, issues of joints and connections [19] are further discussed during the class and not only during tutorials. Apart from that, structural techniques such as Structural Insulated Panels (SIPs), STEKO system, straw bale structures, clay, adobe structures etc., which are not commonly used today in Greece, but could easily be used in ADISA designs, are further discussed.

The main ADISA strategies are further analysed to students; the fact that toxic and hazardous materials should be avoided in ADISA buildings (as well as any in ecological building), also due to the handling risks they might impose, is strongly emphasised. A list of toxic and hazardous materials, explaining the substances that make them toxic or hazardous and to which ecosystems, is presented.

It is also suggested that composite materials which cannot be easily reused or composite materials with a short life span should be avoided. Emphasis is given on the standardisation of the design and the materials used. Good and bad examples of such designs are provided during lectures. The "joke" used during the presentation of standardisation, so as to alert the students, is the comparison of ADISA components with those of LEGO structures, most of the students made when they were kids. The need to minimize the different materials that make up a building component is also further explained and justified. Finally, the type of joints is discussed, separating structural techniques and materials into cast and joints with bolts, screws or nails, encouraging students to use the latter, provided connections are easily accessible.

So as to make standardised dimensions and modular design easily understood to students, the example of a brick is used, which is a common structural material in Greek structures. The example of the specific brick that is looked into is the so called "eco block", a brick with low U-value (compared to other bricks) and quite large dimensions. This kind of brick is not easily cut (needs an electric cutter and

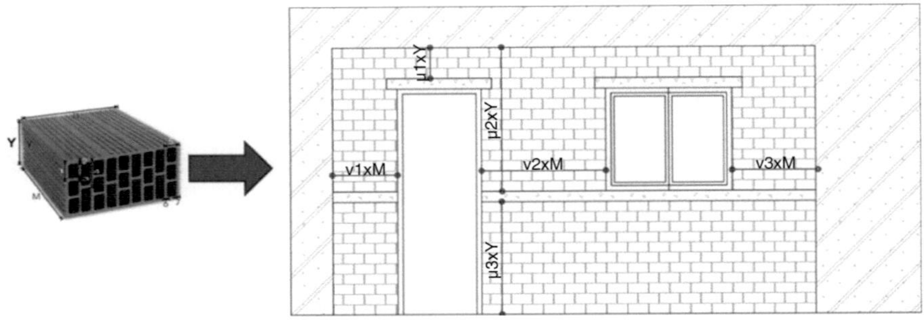


**Fig. 2** Example of the use of the brick "eco block" on a wall, without the unnecessary need of cutting down the brick (from the respective lecture's presentation)

as bricks are brittle, it breaks down easily, causing a lot of material loss); students are encouraged to decide about their walls' dimensions and openings distances as integral multiple of the block's width and height (Fig. 2), minimizing construction waste and allowing more bricks to be reused after the building's demolition.

The ecology, embodied energy, local production, reusability, recyclability and durability of building materials as well as the availability of primary resources are also presented thoroughly, covering all building layers. The notion of the Life Cycle Analysis of buildings is put forward.

Students are reminded that in order to be able to make sustainable constructions and design for disassembly, one should be well aware of:

- the weak points of each material, so that they are protected,
- their strengths, so that they are used at their maximum (within their safety span),
- their maintenance needs.

Finally, most commonly and less commonly used structural materials and their properties are examined, whether and how they can adapt to these design strategies. Emphasis is also given on properties with which the students are already familiar, such as compressive and tensile strength, heat conductivity and heat capacity. The environmental problems of composite materials with toxic components, adhesives or in situ cast materials with toxic additives are presented. Building materials are evaluated as ADISA-capable materials, also taking into consideration their static and thermal properties, environmental-friendliness, life-span and maintenance needs. Tables which evaluate all these properties are presented to students for some materials, although they are not asked to implement such analysis. From these tables conclusions are drawn that no material can be used as panacea for all designs or elements and that their choice should be made with careful reflection on the role each material has to play in the structure and its relationship to other building elements.

As lectures also focus on the importance of reducing building and municipal waste, the rationale of turning municipal waste into building materials is discussed. It is shown that waste such as glass or plastic bottles, cans, tires, timber pallets, newspapers etc. can be reused in constructions as parts of walls, insulation materials or for casting light (e.g. the reuse of glass bottles on a wall, made by the painter Hunterwasser, as shown in Rand [21] is presented to students). The work of the architect Michael Reynolds is also put forward [22] as well as some of the shelters suggested by Kahn [15]. A documentary is also shown to students, the "Garbage Warrior". In this documentary, students can see Reynolds' work on bio-climatic design, the reuse of municipal waste in buildings and how all these have been applied to provide temporary shelters in earthquake-struck areas [14]. For pedagogical reasons, the whole documentary is not shown, but only the parts that focus on the materials and structures from municipal waste, rainwater harvesting and waste management, as well as some issues of bioclimatic design, already discussed during the preceding semester. During the documentary show, comments are made by the lecturer and further information is given to students where they can watch the whole documentary, if they wish to.

As the certification of building materials has already been discussed during other lectures, it is made clear to students that there is no such methodology developed yet for the certification of reclaimed materials. Some examples are given on the thermal properties that have been measured for such municipal waste [7].

Students are also provided with examples of modern buildings with ADISA strategies; examples of Renzo Piano DfD designs, industrial buildings as well as 1950s UK residencies made up from airplane parts are shown to them. Apart from that, the Gaia House [23], the CK Choi Building, the Marie Short House, the OPEN_1 House and R128 House [13] are discussed with them. Respective notes are also uploaded on the module's electronic platform (e-class).

## 2.2 Class Exercises Description

During all three academic years, lectures are also followed by class exercises, as has been suggested by Gibbs and Habeshaw [11] and Webb and Willis [28]. Class exercises focus mostly on ADISA structural details and decisions on the materials that are used and how they are combined together. Students are asked to design structural ADISA details of joining walls and roofs, with materials of their choice, both from reclaimed municipal waste and from certified building materials. Another exercise asks them to make an ADISA green roof structural detail, while the topic of another class exercise is which materials are going to be used and how in an ADISA wall with low U-value and high thermal capacity. All structural details are delivered in 1:20 or 1:10 scale. Another class exercise (given to students only during the first academic year) asks them to discuss whether municipal waste could be used for the construction of safe and healthy buildings, especially recently that the country is under great economic depression. This class exercise is not given to students during the next academic years, as it does not provide them with further technical skills; instead, a respective discussion has been made during the lectures.

Students are given 20 minutes to half an hour to answer to class exercises. While students work on their exercises, the lecturer goes around, discussing the issues that occur, solving queries that arise. In this way, some of the topics that have been presented during the lecture, which some students might not have grasped, are further discussed and explained. This laboratory approach to the course has been more efficient, during the academic years that the course is not compulsory and classes have been smaller; the lecturer is then able to carry further students' questions and queries and discuss some of them with the rest of the class.

Class exercises have proved one of the most efficient ways so that students leave the class having acquired some knowledge. As Bostock [6] points out, students in university usually do not try to understand their subject, but adopt surface approaches, so that they get through the assessments. Class exercises are used as a vehicle, so that students start thinking about what they have learnt and thus they get further from factual knowledge, when they put all the pieces together. In this way an effort is made to guide them to more "conceptual" and "procedural" knowledge, according to Bloom's Taxonomy [1].

Class exercises are corrected and feedback is given back to students during the next class so that they not only learn while taught, but also from their mistakes. Exercises are marked for the participation and not for their content, as students have just been introduced to the respective concepts and a negative mark might be discouraging and unfair. As Race [20] underpins "trial and error" are very important for learning; by class exercises students are able to learn from their mistakes, without the disappointment of poor grades.

## *2.3 Assignment Description*

Great emphasis is given on assignments; as one learns through practice and assessments should be oriented towards learning [11, 28], more stress is given on students' practice on mastering how to design Assembly for Disassembly through the respective assignments. All tutorials evolve around the module's assignments.

During the first academic year (2012–2013) ADISA is introduced in the compulsory spring semester module "Sustainable Design II", which follows the compulsory winter semester module "Sustainable Design I". Students are asked to perform a group assignment (two to four people), which makes up to 40% of the module's final mark; they are asked to design an ADISA small shelter for a homeless person, which could be easily placed, either as a stand-alone shelter or as a group of shelters in the city of Patras public spaces (squares, parks), providing zero-$CO_2$ emissions, water savings, zero-waste, healthy, ergonomic living space for the homeless. They are also asked that the shelter or group of shelters are placed in such a way so that they do not form ghettoes nor encourage depravation in the public space. The structure has to be earthquake-proof and easily assembled and disassembled. Students have been encouraged to provide designs that could be constructed by unqualified people (e.g. by the homeless themselves) and be easily deconstructed, with reusable components, so that they are not left behind when the homeless have found a proper home, as had happened with the prefabricated

**Table 1** Participants in the module and the assignment for the three academic years examined

| Academic year | Number of enrolled students | Number of students who have handed-in the assignment | Percentage of students who have handed in the assignment |
|---|---|---|---|
| 2012–2013 | 129 | 95 | 74% |
| 2013–2014 | 23 | 15 | 65% |
| 2014–2015 | 20 | 7 | 35% |

containers, homeless shelters of the 1999 earthquake in the greater area of Athens. Budget limitations have not been placed. The number of students who have enrolled in this module is 129, 95 of which have handed in the assignment (Table 1).

The next two years (2013–2014 and 2014–2015), ADISA has been taught during the module "Detailed Elements of Sustainable Design" (again a spring semester module, only this time not a compulsory one), presided by the winter semester compulsory module "Sustainable Design". During the academic year 2013–2014, 23 students have enrolled in the module, 15 of which have participated in the ADISA assignment (Table 1). The group assignment is focused on the design of a small canteen in the surroundings of a university building and the redesign of the surroundings of the building. The small canteen, apart from being bioclimatic, with environmentally friendly materials and ADISA, should also be a zero energy building (including the energy for cooking), have low water consumption and its organic waste should be locally managed. The assignment, which also includes some microclimatic measurements and the design of the surroundings of the university building, makes up for 50% of the module's final mark.

During the last academic year (2014–2015), students who participate in the optional module "Detailed Elements of Sustainable Design" have been 7 (while 20 have officially registered with the module). They have been asked to design for their ADISA assignment 2 conference centers and 10 small hostels for the participants in both centers. One center will be used in winter and will be placed in a mountainous location near Patras and the other one will be used in summer, placed by the sea, also near Patras. Apart from ADISA, bioclimatic design, sustainable materials, low water consumption, local organic waste management, local food production and zero energy buildings, students have been asked to use components of the one conference center to the other, moving them out to each center every 6 months. The assignment also makes up for 50% of the module's final mark.

## 2.4 Tutorial Description

Great emphasis is given on the assignment's tutorials. All members of the group are asked to be present during the tutorials, where their sketches, drawings and ideas are reviewed, discussed and corrected. Apart from ADISA, the choice of materials, bioclimatic design, low or zero energy consumption, low water consumption and the small environmental footprint of the structure are discussed. About six official tutorials occur for this assignment (and more unofficial ones for students who have asked further supervision), providing guidance through every step of the students' designs.

During the first year emphasis is not given on construction details during lectures, so as to encourage students to think out of the box and come up with their own ideas about the materials they chose and whether they use construction materials or municipal waste for their assignment. Although there has been a plethora of information on how to design for the homeless, with both municipal waste and structure materials, from both universities and practitioners (e.g. [2, 8, 18] etc.), this information has not been provided to students, so that they mostly focus on ADISA and make up their own structural details or research. Structural details are further examined during the assignment tutorials.

From the first year experience, it has been found out that students, who have not participated so eagerly in the tutorials, have not produced satisfactory ADISA structural details. As during the construction technology module, which is taught in the third year of their studies, no ADISA considerations have been put forward, students have to rethink what they have already been taught; this may cause them some confusion. Thus, during the next two years (2013–2014 and 2014–2015), more emphasis has been given on analyzing some structural details during the lectures, as has been mentioned above, so that students start comprehending the rationale of ADISA more easily. Students are also encouraged during tutorials to visit the webpage http://www.lifecyclebuilding.org, so as to study both the designs and the structural details of the projects that have been rewarded in the Life Cycle Building Challenge Competition.

The main effort during tutorials is to understand students' conceptions, respect them and guide them to the module's rationale. Students are asked to explain the rationale of their designs and how the assignment's requirements are met through their designs. When some requirements are not met, further discussion and suggestions are made, sketches are drawn and given to students, so that their designs become more sustainable and their structures are designed for disassembly. Notes are kept by the lecturer at the end of the tutorial, with comments on what should be made with the specific design, so that a continuation of improvement is achieved during the next tutorial.

There is also a mid-term presentation (crit) of the assignments (approximately 1–2 months before the final submission), where students are asked to either make a small presentation or to pin-up their work and present it in front of the class and usually also in front of another lecturer; in this way students are encouraged to practice their presentational skills and also to get feedback from their colleagues. During the first academic year, that more students have participated in the module, a timetable of presentations has been uploaded in the module's electronic platform (e-class), so that students would not feel as "sitting around, wasting their time and getting nervous", as had been reported by students during previous academic years. This has not been necessary for the next two academic years, when up to 20 people have participated in the module.

Students are encouraged to comment on their colleagues' projects (in a well-disposed way) and further comments, if necessary, are made by the lecturers.

This is a milestone point for the students' assignment; they have to develop their project further and to be able to justify their rationale in front of their colleagues and another lecturer. Students are given up to 5 minutes to present their work. After that, the lecturer asks first their colleagues for their comments (to encourage student participation) and then the other lecturer for his/her comments (up to 15 minutes). When some malicious comments are made, the lecturer intervenes, so that no further stress is placed on the students who present their project. As the success of the crit strongly depends on the atmosphere created [24], the effort is to establish a positive dialogue among the students themselves and between students and lecturers, so that there is a fruitful feedback to students, regarding their further steps for developing their projects.

## 3 Results

Generally, students welcome class exercises, as a way of further understanding what they have just been taught and also gaining a bonus mark for their participation. From an unofficial questionnaire that has been distributed to them at the beginning of the spring semester of the academic year 2013–2014, asking them how they would like the teaching of the spring module to occur, 100% of them (sample of 20 students) have answered that they would like to carry on with class exercises during each class.

Regarding class exercises on ADISA structure details, some students manage to come up with efficient ADISA structural details, while for others, influenced by their existing knowledge on concrete – brick structural details from their previous years of studies, it has been harder to think in a different way just by participating to lectures. As lectures and class exercises evolve (academic years 2013–2015) and students are given a feedback on their previous ADISA class exercises designs, it is easier for them to come up with ADISA structural details; less corrections and comments are needed during the following class exercises.

Regarding the class exercise on the discussion on the use of municipal waste as structural materials (during the first academic year), there is a number of students who think that transforming municipal waste into building materials would provide shelters and generally, less expensive buildings during an economic crisis; furthermore that this could be a solution to the management of municipal waste. Nevertheless, there is a group of students who are skeptical that this would be used only for poor people's shelters and housing, leading to more social depravation and social injustice. It is interesting to mention that the latter group of students does not incorporate municipal waste in their designs, but prefer to use ordinary building materials for their ADISA shelter.

Tutorials have been crucial to further understand ADISA; at the beginning it has been quite hard for students to think about structures where the finishing layer might not be plaster or that piping and electrical wires would not be incorporated inside the walls. Examples of primitive architecture, and the work of Kahn [15] and Triantafyllou [26] have been put forward, as well as the lecturer's ADISA construction details (these proposed towards the last tutorials), so as to help them perceive materials that are used in structures in a different way. The space's ergonomics, as a technique of minimizing the use of materials, as well as the consideration of sun and wind (which had also been a priority in their designs during the previous semester) and water savings are also put forward during tutorials. Aesthetics and the structure's relationship with its surroundings are also discussed.

The presentation (crit), in a well-disposed atmosphere, has also helped the students to put their ideas together and be able to justify them. During the presentation, a lot of comments have been made on how materials and elements are combined together and also on the aesthetics of the proposals, as well as on their incorporation with their surroundings, on which the next tutorials after the presentation focus. Apart from that, some students whose assignments have not advanced that much up to the presentation day, seem to be inspired by the presentations of their colleagues; some of the ideas that have been discussed and encouraged during the presentation, have been implemented in the designs of groups who have not been so eager to participate in the tutorials, especially during the first academic year. During the last two years, as there has been a close relationship of the lecturer and the students and among the students themselves, the results from the crit have not been so dramatic; a constant improvement is observed in most cases through tutorials. The crit does not make such great difference as in larger classes.

Regarding the evaluation of the students' performance through an exam paper is beyond this chapter's scope; none of the sustainable design modules has been assessed through exam papers, but only through assignments and class exercises. During the unofficial 2013–2014 survey, all students have agreed that they think they learn more through assignments than through exam papers.

## 3.1 Assignments of the Academic Year 2012–2013

In the assignments that have been handed in (95 students have submitted their assignments), 62% of the main materials that are used are structural materials, while 38% reclaimed municipal waste. Most of the municipal waste is combined with structural materials (e.g. timber or steel frames), while 14% are more radical designs and use only municipal waste as their primary structural source, incorporating also fishing cords and nails to joint reclaimed municipal waste so as to form the shelter. In all designs, no gluing takes place between the different layers of materials; only nails and screws. Some designs have also used paper or clay to join together plastic or glass bottles and fewer cement, while other designs have allowed larger amounts of infiltration through the bottles, so as to make disassembly easier (Fig. 3a, b). No plastering is incorporated in any of the designs. The different layers of the structure are clearly separated (structural frame from walls, insulation from structural elements, or walls that do not need a separate insulation layer are used, separation of drainage, piping and electrical wires from walls etc.), while some designs have used standard units to form the shelter.

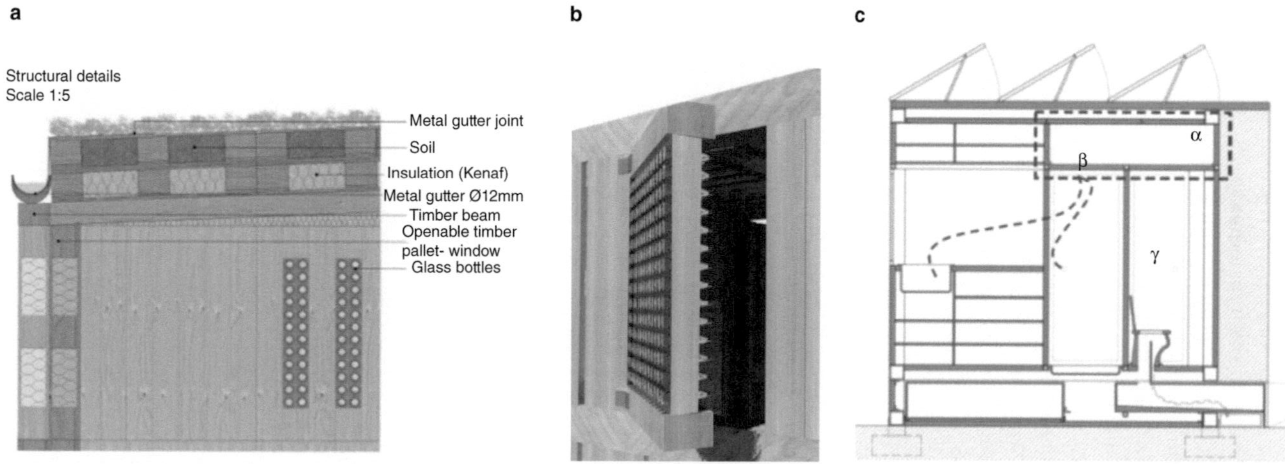


**Fig. 3** [2]Structural details (a) of roof and walls, made of timber pallets, openings made of glass bottles and water piping by students V. Antonopoulos, A. Ioannou and V. Charalampopoulou (b) of a window, made of glass bottles and timber frame, which is nailed to the timber wall by students F. Dimoglou, I. Theodosopoulou and S. Pavlidou and (c) rain water harvesting for sanitary water, compost toilet for water and waste saving, photovoltaics for electricity production and solar panels for Sanitary Hot Water by students G. Kourakos and T. Svoronos

[2]Blueprints' descriptions have been translated in English by the author, apart from those in Fig. 4.

The frequency with which the construction materials or the municipal waste are used is shown in Table 2. As can be observed in the following table, timber pallets are the most frequently used municipal waste (25% when compared to other municipal waste, 9% when compared to all main materials used). Students use them as flooring, wall or roof materials, insulating them with either paper waste or insulation of plant origin (kenaf, cotton, cork or straw).

A lot of the designs have used timber pallets as a standardized material (Fig. 5b), the dimensions of which make up the structure (both floors and walls), making it possible to reuse these items either as pallets or as structural or furniture materials, after the shelter's demolition. Glass bottles are frequently used (19% when compared to municipal waste and 7% when compared to all main materials used), many times within timber pallets as openings and sometimes joint together with paper pulp, cement or cob.

**Table 2** Main materials used in the ADISA assignments and the percentage of their use (a) per all main materials and (b) per construction materials or municipal waste

| | Materials used | Percentage of use of materials per total | Percentage of use of materials per construction materials or municipal waste, respectively |
|---|---|---|---|
| Construction materials | Cob / Rammed earth | 4% | 7% |
| | Paper | 3% | 5% |
| | Metal frame | 9% | 15% |
| | Cork | 2% | 3% |
| | Timber | 14% | 23% |
| | Thermal insulation (kenaf, cotton) | 6% | 9% |
| | OSB or SIP panels | 3% | 5% |
| | Straw and straw bales | 2% | 3% |
| | Metal parts (reused) | 2% | 3% |
| | Glass with aluminum or timber frame | 14% | 22% |
| | Paper | 1% | 1% |
| | Concrete | 1% | 1% |
| | Sand | 1% | 1% |
| Municipal waste | Plastic bottles | 2% | 6% |
| | Glass bottles | 7% | 19% |
| | Timber pallets | 9% | 25% |
| | Plastic membranes | 6% | 15% |
| | Fabric | 6% | 15% |
| | Paper tubes, cartons and newspapers | 4% | 12% |
| | Metal tubes | 3% | 8% |

Regarding structural materials, most students have preferred to use timber in their structure either as frames and / or as walls (23% when compared to other structural materials and 14% when compared to all main materials used), while double-glazed, pure glass with aluminum or timber frame replaces the reclaimed bottles for solar gains and natural daylight (22% when compared to other structural materials and 14% when compared to all main materials used). Some students also use cob / rammed earth as their main construction material (7% and 4%, respectively). Although cob as well as cement may not be the most prominent ADISA materials, students who have used the first one, argue that it can be easily demolished, decomposed and recast, thus reused.

Students who use rammed earth, as well as students who use reclaimed materials are very sensitive about the costs of such a structure, arguing that their design offers a low budget shelter, created by the homeless themselves, while other students have decided that the Municipality of Patras or the Greek Authorities will provide the homeless with the appropriate budget to create a quite high-tech shelter, with photovoltaics and all the necessary facilities.

The majority of shelters (69%) are designed so as to be combined with other shelters and be connected to facilities buildings (existing municipal buildings or new ADISA buildings), where toilet and kitchen facilities are supplied, while 31% of the designs refer to stand-alone, autonomous shelters. Students who prefer the first solution follow the rationale that less materials should be used in the shelter and that waste should be treated centrally, while students, who prefer the second solution, put forward the idea that the new-homeless people should be able to survive independent from a central authority that might not finally work out. In many of the non-autonomous shelters, no sewing systems are necessary and in some no electricity facilities either; night light is provided through photovoltaic table lamps. In the autonomous shelters, piping is either detached from the walls and is visible to the interior, while in other designs, more bold about the economic costs, timber frames cover piping and electrical wires. Many of the autonomous designs use rainwater harvesting to cover sanitary water needs, compost toilets for water savings, photovoltaic panels and solar panels for the production of electricity and sanitary hot water, respectively (Fig. 3c).

The design of the shelter varies from very simplistic, movable boxes (Fig. 4) to autonomous and fully equipped, ergonomic, little homes (Fig. 5). Many students

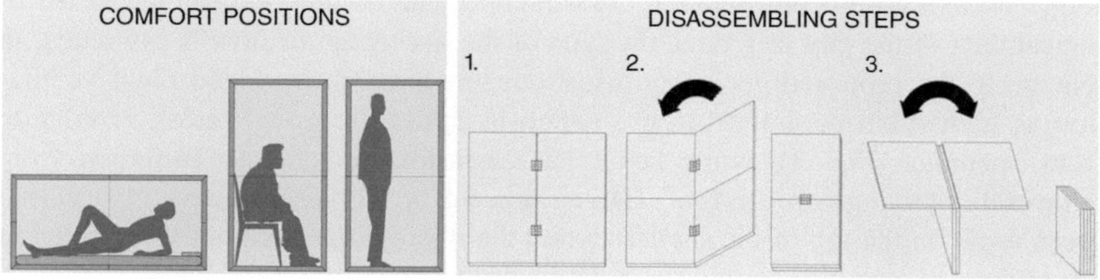


**Fig. 4** A transformable and movable minimal shelter by students V. Paraskevopoulou and Z. Charalampous

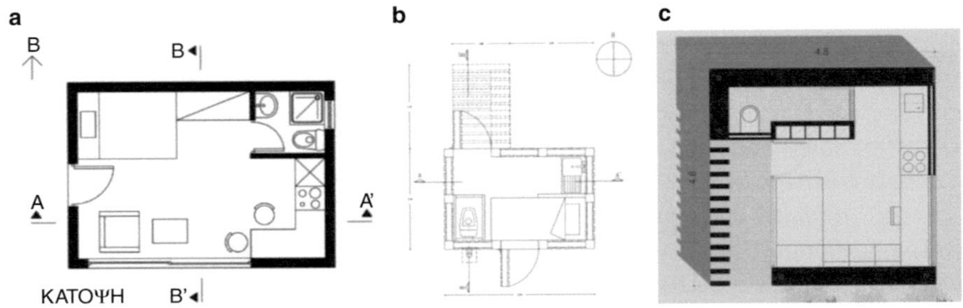


**Fig. 5** Stand-alone, fixed units by students (a) A. Alafi-Andrikopoulou, N. Ioannidou, K. Papadioti and G. Tsantoulis (b) F. Dimoglou, I. Theodosopoulou and S. Pavlidou and (c) A.-P. Mourelatos and S.-P. Pandazopoulos

**Fig. 6** Movability and transformations of a shelter by students E. Vlachaki, A.-P. Dogani and M. Lafazani

(43% of the assignments) have put forward the idea that apart from ADISA, these shelters should be easily disassembled so as to be moved to another place, either in the simplistic form shown in Fig. 4, or in more sophisticated structures with openable and movable members, as the one presented in Fig. 6.

Most of the designs have oriented shelters to the south, so as to achieve solar gains either through direct gain or other passive solar systems, mostly Trombe walls. Eastern, western and northern openings are kept to a minimum in most of the designs, used for ventilation and day-light, if necessary. External shading devices, either movable or permanent, from fabric or timber are placed on the southern, eastern and western openings. Solar winter penetration and summer shading have been calculated for the geographical latitude of Patras (38°N) (Fig. 7). Taking into consideration the wind speed and direction from the meteorological data of the city of Patras, the type of the occurring air flow is estimated, in relation to the proposed geometry. Most designs have incorporated night ventilation as their main cooling strategy, which is achieved through cross ventilation from openings (Fig. 7), while some have considered vertical ventilation from removable openings on the roof. Eleven percent of the designs have also placed green roofs on the top of the shelter so that the microclimate around it is altered in a positive way and also so that food can be produced within the shelter.

The aesthetics of the shelter and its placement in the urban environment have also been very crucial points. Some students come up with more classical

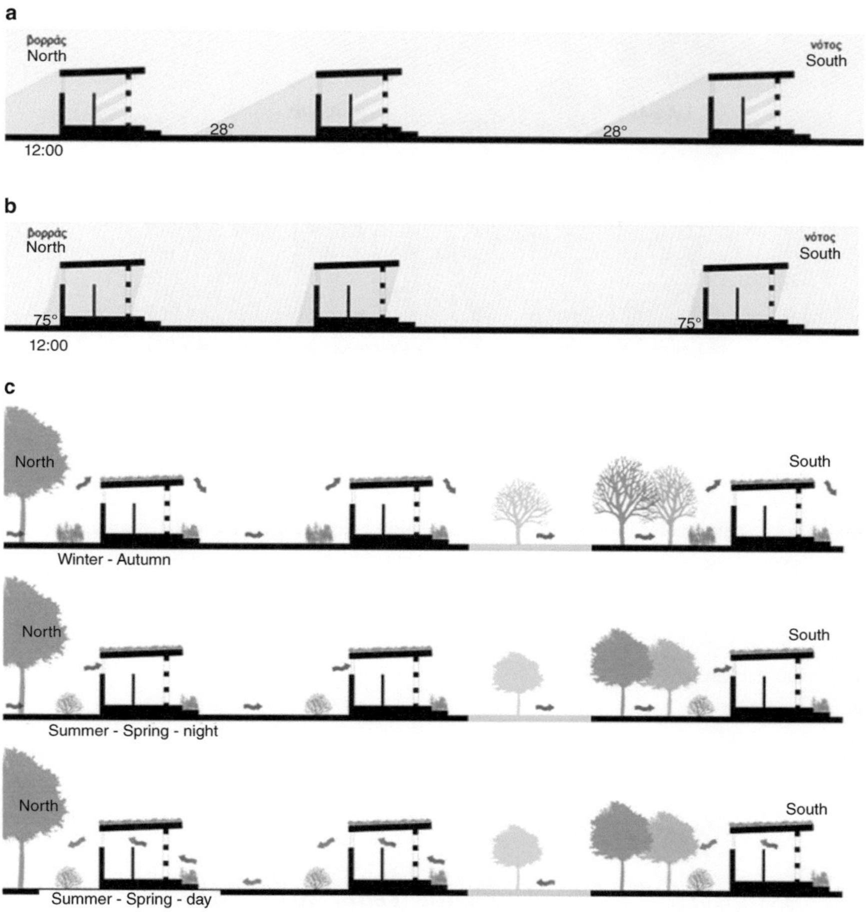


**Fig. 7** Noon solar penetration at the southern façade of the shelter (a) in winter – solar gains and (b) in summer – external shading and (c) ventilation strategy and type of air flow during different seasons in Patras by students V. Antonopoulos, A. Ioannou and V. Charalampopoulou

relationships between shelters and the open space, allowing for large areas between the shelters, where air can flow and people can move (Fig. 7), while others have considered their shelters to be a living sculpture (Fig. 8), adding up to the aesthetics of the urban space.

In many designs, photovoltaic panels are used for the production of electricity (45%), as shown in Table 3. Solar panels are also used for the production of sanitary hot water or for space heating (31%). Geothermal pipes for space heating and cooling is used only by 3% of the assignments, while rainwater harvesting, water recycling and compost toilets are used by the majority of students (54%, 51% and 46%, respectively), especially in the stand-alone units.

**Fig. 8** ADISA shelters placed so as to form a living sculpture in an urban park in the city of Patras by students E. Bagianou and K. Papathanasopoulou

**Table 3** Technology for electricity, sanitary hot water, space heating / cooling (apart from passive solar systems and ventilation), water collection and savings used in the assignments

| Technology | Percentage of its use in the assignments |
|---|---|
| PV panels for electricity | 45% |
| Solar panels for SHW and space heating | 31% |
| Rainwater harvesting | 54% |
| Water recycle | 51% |
| Compost toilets | 46% |
| Geothermal pipes | 3% |

## 3.2 Assignments of the Academic Years 2013–2014 and 2014–2015

As there have been much fewer students who participate in the class (Table 1)[3], when the module becomes non-compulsory, the statistical analysis of their choices does not reveal any tendencies, from such small samples. Thanks to the small number of students, the relationship between the lecturer and the students has been closer, throughout the whole pedagogical approach (lectures, class exercises, tutorials). Thus all students have followed the same guidelines regarding the orientation of their buildings, the detailed study of solar patterns, ventilation strategies, the necessity of RES applications and ADISA structural details.

[3]As there is a large choice of optional modules in the curriculum, the number of participants is the expected one. There is also a number of students who do not participate in the modules, but just enroll, so that they do not miss the academic year.

For the design and the integration of an ADISA, zero-energy consumption, low-water consumption canteen in the surroundings of university buildings (academic year 2013–2014), students have been asked to select the location of their design. During a class exercise of a previous lecture on outdoors thermal comfort, where students have been filling in questionnaires and measuring micro-climatic parameters, they have been encouraged to ask people who work at the university's canteen about the thermal conditions of their environment and what they would like to improve. When the assignment is announced, students are reminded about that class exercise and to implement these findings in their designs.

Regarding the location, most students have used building surroundings (e.g. Fig. 9a), while others the roof tops (Fig. 9b). They use either reclaimed municipal waste or structural materials. It is interesting to mention that 20% of the students use reclaimed municipal waste (while it has been 38% the previous year). Some students have used recycled metal frames (Fig. 9a), while some other have used certified sustainably grown timber frames (Fig. 9b), all with ADISA structural details (Figs. 11 and 12) and bioclimatic design strategies (Fig. 10). Again, plastering and gluing are avoided, screws and nails are used in the joints and materials with a low environmental footprint are chosen. Despite the fact that at the beginning ADISA structural details have seemed bizarre to students, it has been easier

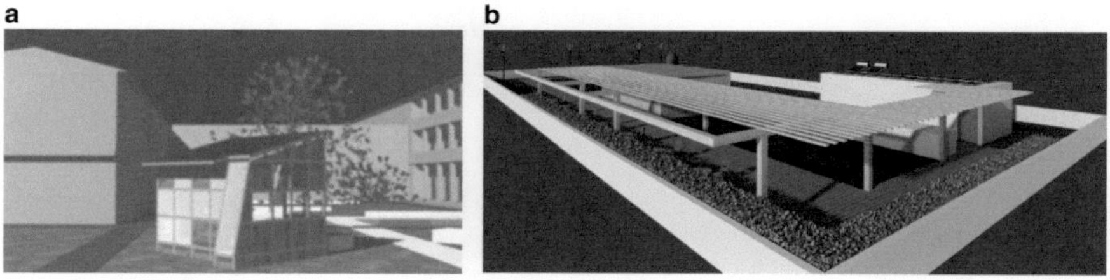

**Fig. 9** (a) ADISA canteen with metal frame, heraklith insulation and timber finishing, with PV panels, solar oven, compost toilets, rainwater harvesting and its surroundings, by students G. Giasia, A. Zafiratou and C. Krika and (b) ADISA canteen with timber frame, roof and walls, wind turbines, photovoltaics, solar collectors, external solar oven, compost toilets, rainwater harvesting and its surroundings on the roof top of a building, by students E. Daliani, A. Papathanasiou and S. Tsakos (academic year 2013–2014)

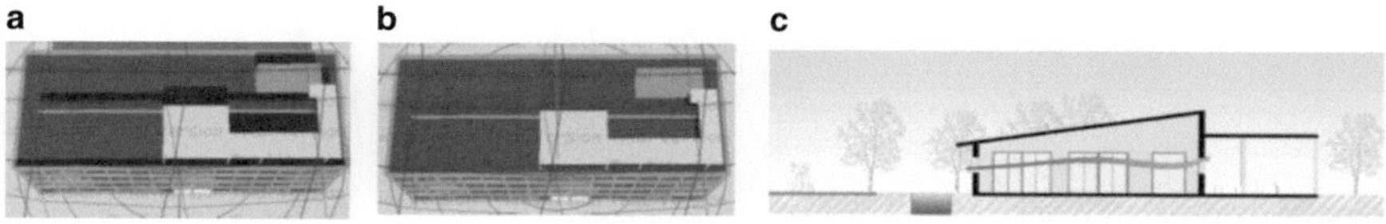

**Fig. 10** Shading patterns for (a) 21st December at 12:00, (b) 21st June at 12:00 by the students E. Daliani, A. Papathanasiou and S. Tsakos and (c) cross ventilation strategy by students M. Arifi, A. Zikou, C. Pagoni and A. Tsitsimpi (academic year 2013–2014)

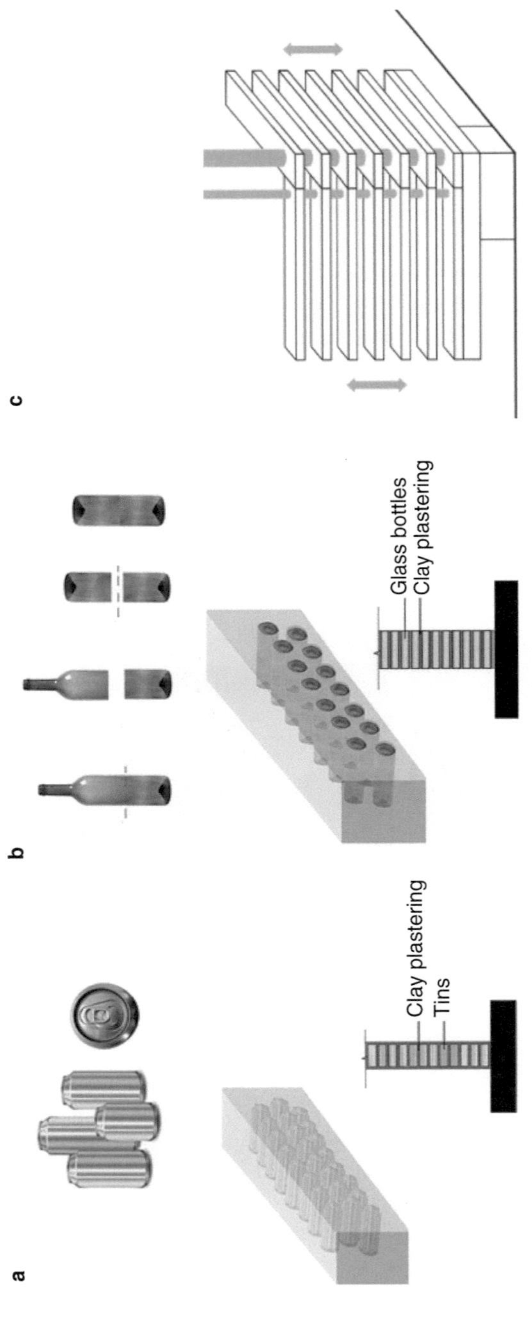


**Fig. 11** Wall details from reclaimed municipal waste (a) tins and (b) glass bottles by students M. Arifi, A. Zikou, C. Pagoni and A. Tsitsimpi and (c) detail of timber louvers joint together with timber rods by students E. Daliani, A. Papathanasiou and S. Tsakos (academic year 2013–2014)

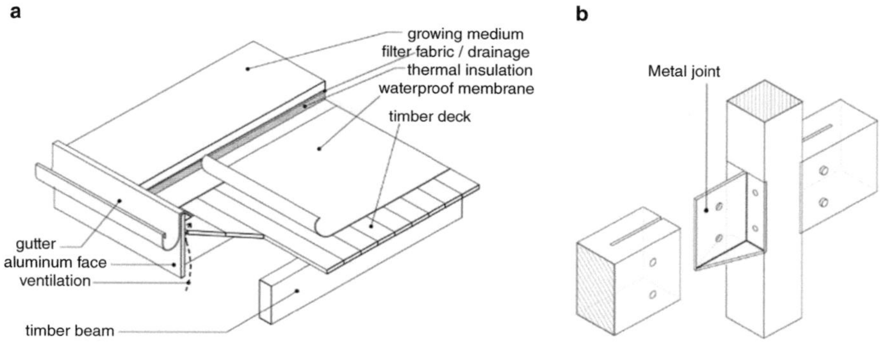


**Fig. 12** (a) "Brown" green roof details and (b) timber beams and columns connection detail by students T. Giannea, K. Lamprou and P. Tounta (academic year 2013–2014)

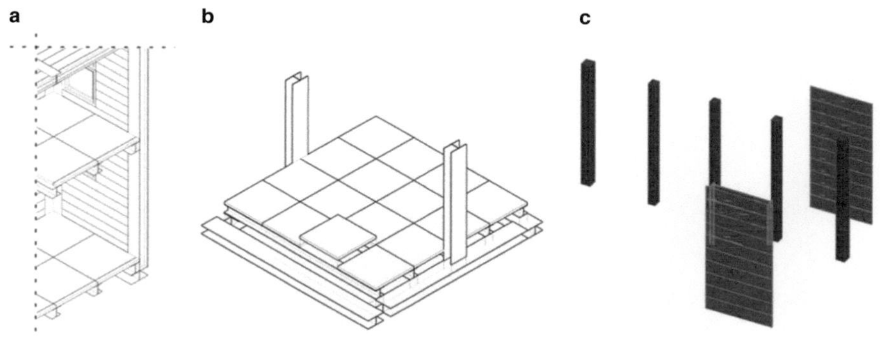


**Fig. 13** (a and b) Details of the metal frame / modular design by students Z. Vidaki and P. Koulakoglou and (c) details of the timber wall joint with timber columns modular design by student M. Zacharopoulou (academic year 2014–2015)

for this year's students to design ADISA structures, thanks to the fact that classes have been functioning as laboratories, because of the smaller number of students.

Similar results can be observed for the academic year 2014–2015, as well, when the number of students has been similar to that of the previous academic year. So as to focus on standardised dimensions, the parameter of transferring parts of the buildings from one conference centre to the other has been added in the 2014–2015 assignment. Students find out from the first tutorials that they have to design modularly, so as to achieve the transferability of specific components (Fig. 13). Some students have based their designs on metal frames, while others on timber (again, recycled and certified sustainably grown, respectively). Most students have used a vertical and horizontal grid for all movable members of their structures, moving all (apart from flooring) parts of hostels and a large percentage of the main conference building. Students who have placed vegetation on the

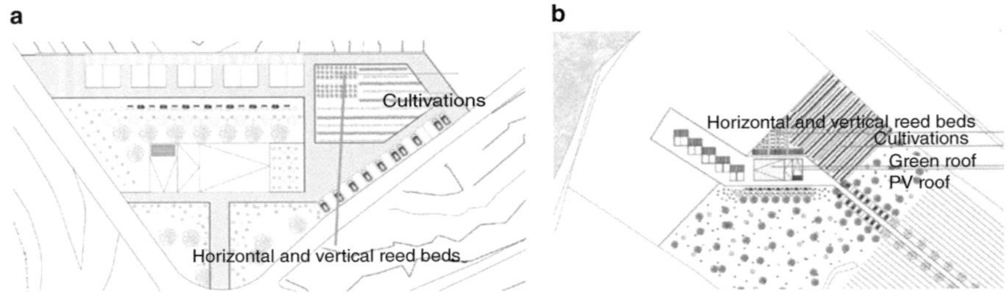


**Fig. 14** (a) Winter and (b) summer conference centre and small hostels by students G. Kyriakopoulou and I. Kavalari, allowing for winter solar penetration in design (a), but protecting from summer sun rays in design (b). In both cases SIPs are used in walls, roofs are green, horizontal and vertical reeds are used for water treatment, PVs for electricity production and food is cultivated in each site (academic year 2014–2015)

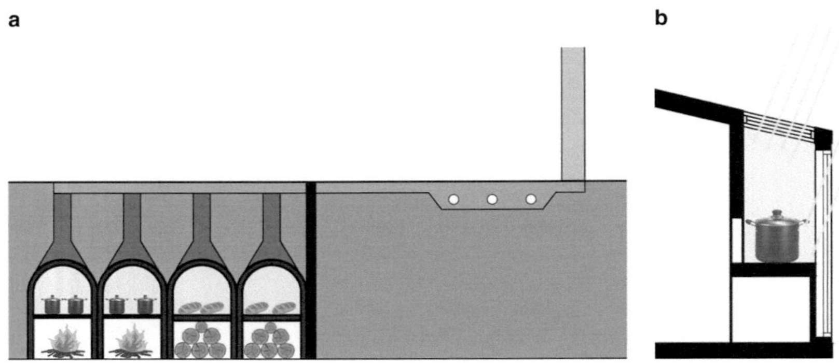


**Fig. 15** (a) Biomass oven in the winter building, where the hot air from the ovens is directed towards the main spaces of the conference building for heating and (b) solar oven in the summer building, thermally detached with insulation from the rest of the building, by student M. Zacharopoulou (academic year 2014–2015)

roofs, also keep the beams and columns and transfer only walls and openings from one location and assemble them to the other. Some students have used the combination of straw bale with clay walls (for the parts of the buildings that are not transferred), while other students have preferred prefabricated SIPs.

Regarding bioclimatic strategies, students have designed winter and summer conference centres with bioclimatic strategies of welcoming solar energy / avoiding wind and avoiding solar energy / taking advantage of local winds, respectively (Figs. 14 and 16c). Some students have chosen to provide energy for cooking with biomass in winter, when solar radiation is smaller, reusing the hot air for heating the building, while in summer, they have incorporated insulated solar ovens in the main building (Fig. 15). Rain water harvesting has also been used as a cooling technique, as well as cooling towers and dormer windows (Fig. 16).

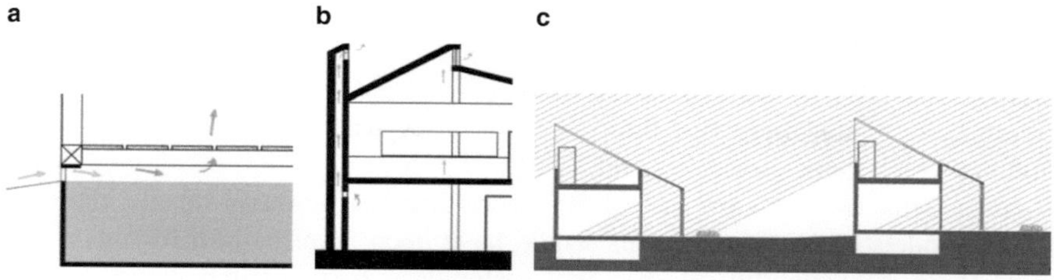


**Fig. 16** Passive cooling strategy from (a) the rainwater harvesting pond in the basement of hostel and conference buildings and (b) from cooling towers and dormer windows for the summer buildings and (c) passive heating strategy through a sun space, south-oriented openings and the appropriate placement of buildings for the winter dormitories, by student M. Zacharopoulou (academic year 2014–2015)

# 4 Lessons Learnt

Through this process, it has been found out that the most important pedagogical tools for learning ADISA, apart from lectures, has been the assignment and the respective tutorials, combined with class exercises. Students who have participated in all tutorials, where their designs are thoroughly discussed and advice is given on how to improve the implemetation of their ideas, have been able to hand in high quality work. Students who have been less eager to participate in tutorials have come up with more simplistic solutions, some of which have also been presented in this chapter. The mid-term presentation (crit), acting also as a way of communicating ideas, seems to help students in large classes, especially those who have not worked their project so intensively up to then, to get engaged in the assignment and get inspired by the ideas their fellow-students have come up with. Lectures, notes, documentary show have also been helpful, but not as much as asking students to think and to practice themselves through class exercises and the assignment. The student – lecturer communication through the tutorials for the achievement of "thinking out of the box" ADISA designs is also a key element in the educational process. This has been more effective with smaller groups of students. Nonetheless, "thinking out of the box" on structural details finally has occurred more efficiently through the first academic year, when students have not been shown during lectures ADISA designs. But as this has been achieved only with students who have been enthusiastic about the class, as has been mentioned above, structural details have been added to the lectures of the subsequent years. The result of this has been that students have mostly copied or imitated the details they have been taught during the class or have seen in respective web pages, rather than try to think their own structural details.

Teaching ADISA as a laboratory module (active interaction of the lecturer with students especially during class exercises and frequent and efficient tutorials) is the ideal way to engage students in creative and renovative thinking. When the module becomes non- compulsory and less students participate in it (academic years 2013–2015), it is possible to develop it as a laboratory module with smaller groups, which has been more efficient in students' understanding and practicing ADISA. It would be better that the module is taught as a laboratory one (more lecturers, for more productive tutorials), especially when taught as a compulsory module, which more students attend.

As students are not used to look into very detailed structural details, the creation of models of structural details would have been more helpful for them, provided there is availability of structural materials – or municipal waste – and a respective laboratory, where they could assemble and disassemble in practice the structural details they have designed. In this way they would comprehend ADISA design more thoroughly. Visiting ADISA construction sites would also be extremely helpful and educational.

Through this research, it could be concluded that ADISA should be part of the construction technology module, so that students start thinking about disassembly when they are first introduced to structural details. This would give the optimum pedagogical results for learning ADISA, as it has been found out that already existing knowledge of conventional structures often blocks the ADISA way of thinking and designing. Whether ADISA is taught in a sustainable design module or a construction technology one, it has been found out that class exercises, well-intended crits and assignments are very crucial for the learning process. Students seem to appreciate class exercises and prefer to be assessed through assignments than exams. Exam papers have not been put forward in this methodology; their effectiveness could be evaluated by some future work on teaching ADISA.

Finally, a survey from an independent body on the students' opinions would be very helpful, so as to see from the students' point of view how the subject of ADISA should be taught within the module of Sustainable Design. This survey would be even more promising, if students are asked one to two years after they have passed the module what they remember on ADISA and how useful they think it is. Comparison of the effectiveness of different pedagogical approaches for teaching ADISA in different institutions would be extremely useful for improving pedagogical techniques and finding the optimum teaching approach, which has been this chapter's preliminary request, nevertheless limited to one institution.

## 5 Conclusion

Through the intensive tutorials on the assignment, lectures, class exercises and presentation (crit), the submitted assignments have been, in most cases, of high quality, especially those produced by students who participated throughout the whole process (lectures, class exercises, all tutorials and presentation / crit). Students have been given the freedom to choose materials, either typical structural materials or municipal waste as well as the type of the design, from permanent to movable structures, from simplistic to sophisticated designs.

Although it has been hard at the beginning for students to comprehend ADISA and to rethink about the structural details they had learned during their third year of studies, through the collaboration of the lecturer and the students, as well as through their collective work, the majority manages to incorporate ADISA in their sustainable design, in both a comprehensional and operational level. After this experience, students are able to both set the sustainability goals of their design, take the appropriate decisions and also to design the details that lead to the desirable result. One of the students of the academic year 2012–2013, A.-P. Mourelatos, has won a national students competition for the visitor centre of the Stavros Niarchos Park (the design selected by Renzo Piano), designing a "recyclable" building [27], which has been constructed as a DfD visitors' centre.

As our educational system tends to lead to more standardised ways of thinking and designing, one of the hardest parts of this pedagogical approach has been to make students think out of the box and apply Assembly for Disassembly either with conventional building materials or with reclaimed municipal waste, making their own decisions, designs and structural details. As new environmental problems arise every day and the architects and engineers of the future should be able to challenge them quickly, apart from ADISA, students should be asked more frequently to think out of the box through our educational program. Tutorials and the close communication between students and the teacher are very helpful for this.

As the displacement of large portions of populations unfortunately is in the agenda of recent and future global developments, ADISA shelters could be a small relief for populations on the move. The 2012–2013 designs of some of the students, especially those where municipal waste have been used as primary materials could be further developed and used as a temporary solution for sheltering these people. ADISA and material reclaim is not only a way of sustainable design and construction, but also, as Reynolds has taught us [14], a way for directly facing the need of shelter in times of crisis.

## References


1. Anderson LW, Krathwohl D (eds) (2001) A taxonomy for learning, teaching, and assessing: a revision of Bloom's taxonomy of educational objectives. Longman, New York
2. Azari A (2009) Almost home. RMIT University. Available from: http://www.designresearch.rmit.edu.au/dri/programs/design-challenge/entries/2011/almost-home. Accessed 28 Mar 2013
3. Berge B (2009) The ecology of building materials. Architectural Press, Oxford
4. Bio Intelligence Service (2011) Service Contract on Management of Construction and Demolition Waste – Sr1, A Project under the Framework Contract Env.G.4/Fra/2008/0112. Final Report Task 2, Bio Intelligence Service S.A.S., Paris, France
5. Bostock S (2012a) Staff guide to the module evaluation scheme, student evaluation of modules (SEM), Electronic Version. Glyndwr University, Wrexham
6. Bostock S (2012b) Notes on Learning and Teaching, Electronic Version. Glyndwr University, Wrexham

7. Chasapis D, Androutsopoulos A, Aidonis A (2009) Use of materials with low or zero cost as insulation layer in storing energy areas. In: 9th National Conference on Clean Energy Sources, Pafos, Cyprus, 26–28 Mar 2009, pp 159–166
8. D'Apolito T (2012) Shell House. Technology engineering design project, technology engineering education. tennessee-teea.org/shell%20/house.ppt. Accessed 28 Mar 2013
9. Dimoudi A (2006) Ecological structural materials. Democritus University of Thrace, Environmental Engineering Department, Xanthi, Greece
10. European Union Directive (2008) Directive 2008/98/EC on waste and repealing certain directives. *Official J Eur Union*, L312/3-L312/30
11. Gibbs G, Habeshaw T (1992) Preparing to teach. Technical & Educational Services Ltd, Bristol
12. Gines M, Beorkrem C (2010) The transformation of architecture: design for dis-assembly. In: ARCC/EAAE International Conference on Architectural Research, 23–26 Jun 2010, Washington DC, USA
13. Guy B, Ciarimboli N (2006) Design for disassembly in the built environment; A guide to closed-loop design and building. http://www.lifecyclebuilding.org/docs/dfdseattle.pdf. Accessed 28 Mar 2013
14. Hodge O (2007) Garbage Warrior. Documentary, Open Eye Media Ltd, ICA Films
15. Kahn L (Ed) (1973) Shelter. Shelter Publications Inc, Bolinas
16. Kostaki L (2009) Road network of Athens. In: Korres M (Ed) Attica's streets; Ancient roads of Attica. Melissa Editions, Athens, pp 96–111
17. KYA 36259/1757/E103/2010 Measures, conditions and program for the alternative management of waste from excavations, structures and demolitions. *Official J Greece*, ΦΕΚ 1312 B/ 24-8-2010: 17600–17609
18. Meinhold B (2013) A-Kamp47 vertical urban campground provides subversive shelter for the homeless in France. http://inhabitat.com/a-kamp47-vertical-urban-campground-provides-subversive-shelter-for-the-homeless-in-france. Accessed 26 Oct 2014
19. Nordby AS, Berge B, Hestnes AG (2007) Salvageability of building materials. In: Portugal SB07, Sustainable Construction, Materials and Practices, Lisbon, Portugal, 12–14 Sep 2007, pp 593–599
20. Race P (2001) The lecturer's toolkit. Kogan Page, Bath
21. Rand H (1993) Hundertwasser. Taschen, Köln
22. Reynolds M (1993) Earthship, Vol 1&2. Solar Survival Press, Taos
23. Roaf S et al (2007) Ecohouse. Elsevier – Architectural Press, Oxford
24. Sara R, Parnell R (2013) Fear and Learning in Architectural Crit. Field 5(1):101–125. http://www.field-journal.org/uploads/file/2013%20volume%205/field%205(1)%20sara%20&%20parnell.pdf. Accessed 17 May 2015
25. Travlos I (1960/2005) City-planning evolution of Athens from the prehistoric era up to the beginning of the 19th century. Kapon Editions, Athens
26. Triantafyllou G (2010) Archetypes. Byzantine and Cristian Museum Ed, Athens
27. SNF (2013) Visitor Center; Design. The National Library of Greece, The Greek National Opera, The Stavros Niarchos Park, Stavros Niarchos Foundation Cultural Center. http://www.snfcc.org/visitors-center/design. Accessed 6 Apr 2014
28. Webb A, Willis L (2010) Enhancing feedback for engineering students. The Higher Education Academy, Engineering Subject Centre, Loughborough University, UK
29. Wyman M, Bostock S, Pike C (2009) Guidance in good practice in module evaluation. Document approved by the Learning and Teaching Committee 14 October 2009 with a minor edit approved afterwards by the Chair, Keele University, UK

# Designing Participatory Learning Systems

**Henriikka Vartiainen**

**Abstract** The knowledge society has opened up the possibility for people to design and construct the contexts of their own learning in entirely new ways and to cross the boundaries of existing communities and prevailing knowledge by creating novel and extended networks of learning. These new opportunities challenge educators to advance participatory pedagogy, and co-develop dynamic learning systems that connect students' learning ecologies with multiple contexts and communities, their social practices, information resources, and tools. To meet these challenges, this chapter presents perspectives on participatory learning and introduces a design-oriented pedagogy (DOP) for connecting learning across spaces and communities. The presented research-based insights may help educators in different institutions to promote inquiry activities that enable students to participate in the co-development of the skills and ways of knowing needed in twenty-first-century society.

## 1 Introduction

The twenty-first century calls on us to solve some of the most challenging problems ever created and faced by humankind, such as climate change, renewable energy, world hunger, and refugee crises. When addressing these pressing issues of our time and unforeseen problems of the future, there is an ever-increasing need for robust lifelong learning, collaboration, creativity, and systems-level thinking [14, 4]. These complex challenges require us to cross boundaries of existing communities and prevailing knowledge by creating novel and often far-reaching networking linkages to experts, communities, and networks representing heterogeneous knowledge and competence [18].

H. Vartiainen (✉)
University of Eastern Finland (UEF), School of Applied Educational Science
and Teacher Education, Savonlinna, Finland
e-mail: henriikka.vartiainen@uef.fi



T. Issa et al. (eds.), *Sustainability, Green IT and Education Strategies in the Twenty-first Century*, Green Energy and Technology, DOI 10.1007/978-3-319-57070-9_21

Moreover, the networked age with its new tools, technologies, and information resources has opened up the possibility for people to design and construct the contexts of their own learning in entirely new ways [38]. By systematically building their own personal learning networks that stretch beyond the boundaries of their immediate community, people can now transform the ecosystem of their own learning and development [18]. Furthermore, the socio-technological developments have been increasing visibility and accessibility in many ways, and have opened up the possibility for citizens to play a much more active role in shared decision making, problem solving, and in co-producing new innovations [8]. According to Clinton et al. [6], these rapid advancements in technology and socio-cultural developments have shifted the cultural logics and social practices, and point us toward a more participatory culture, one in which people have an expanded capacity to share and circulate their ideas, and one in which networked communities can shape our collective agendas. Yet, our current education systems are largely based on a re-production model that was developed to meet the labor needs of the Industrial Revolution [31]. The model including one classroom, one teacher, one class, and one subject at a time is being increasingly questioned [17]. Instead of socializing ourselves within existing practices, the knowledge society challenges us to pursue novelty [18]. Consequently, the transformation efforts in education call for moving from an industrial system toward organic, complex, and adaptive learning systems that evolve and connect students' learning ecologies with multiple contexts and communities, their social practices, and tools [17, 23, 25, 31, 8]. One of the major challenges for this time seems to be *how to design vital learning systems that connect extended environments and communities with formal learning, and in a way that enables our students to participate in co-developing the skills and ways of knowing needed in twenty-first-century society.*

To enhance the creation of the new kinds of learning systems required in twenty-first-century societies, this chapter introduces a specific way of enhancing participatory learning in extended learning environments and communities. The discussion on participatory learning is organized around a sociocultural framework with the intention of introducing a design-oriented pedagogy (DOP) [21, 22, 40, 41] to connect the elements of the learning system in vital interaction.

## 2 Design Perspectives for Participatory Learning

Participatory learning originally derives from the sociocultural framework of Lev Vygotsky [45]. A sociocultural approach considers learning as participation in cultural practices [47] in which the learning activities are bound to the context in which they take place [36], and to the people, tools, and artifacts that mediated them [34]. Rather than being seen as an individual mental process, learning from this perspective is considered as a collaborative process that takes place in groups, communities, and networks that have a shared object of activity [11]. It involves a deepening process of participation in which newcomers gradually move from the periphery to becoming fully recognized practitioners in the community [47]. Furthermore, to be a competent participant in many activities involves the mastery of a range of tools that enhance our thoughts and actions [37].

Accordingly, learning is approached as a systemic process, where the elements are the subjects (the actor or actors participating in the activity), the object of their activity, and the tools and artifacts as mediational means that actors use for acting on the object. Examples of such learning systems are, for example, the home, school, workplace, or even pastime communities. These systems can be observed on the basis of their aims as organizations, where they direct their actions, what kinds of tools they offer for support, and who the central actors are. Consequently, the majority of what children learn is as a result of their constant participation in the learning systems of everyday life within particular sociocultural communities. Moreover, learning is also mediated by the children's own interests, what kind of objects for learning they set for themselves, and what kind of support, tools, and resources are available to them, and the ways in which all of these aspects are integrated in the pursuit of these objectives.

However, instead of socializing into the existing practices of a particular sociocultural community, the present is characterized by a rapidly changing society and an unforeseen future. Both local and big global problems are challenging us to constantly renew cultural practices and to cross borders in order to exploit and develop divergent expertise. Furthermore, online spaces are becoming increasingly important places where participants from heterogeneous backgrounds connect with each other, negotiate what it means to engage with others, explore and develop identities through encounters with others, and negotiate shared values and rules of collective activities [8]. Nowadays, participatory learning is increasingly situated in ubiquitous digital environments that have given rise to participatory knowledge cultures in which people work together to collectively classify, organize, construct, and evaluate information [20]. Jenkins et al. [13] define cultures of participation as cultures which share a strong sense of community, have relatively low barriers to participation, informal mentorship structures, and support for creating and sharing one's creations.

In contrast to formal learning, the participation in these new knowledge cultures is typically voluntary and interest-driven. These cultures exploit the practices of collaborative-knowledge work by having a multifaceted object of activity where participants are continually contributing to, and enhancing the shared intellectual resources, so that at both the individual and group level, there is continual movement beyond current understanding and capacity [33]. They can be characterized as self-organizing learning systems, in which the members of various backgrounds share expertise and insights, and feel that there is personal ownership of their contributions. As noted by Facer [8], the common element in such an open source culture is that there is always the possibility for the participants to contribute and take action, build on the work of others, and to remix and reshape the existing in pursuit of personalization or improvement. In this way, these practices not only filter and reproduce the existing ways of knowing, but actively fuel the co-development by constantly producing and sharing new meaning or improved ideas.

Rheingold [30] argues that these new opportunities challenge educators to advance participatory pedagogy, assisted by digital media and networked publics, which focuses on catalyzing, inspiring, nourishing, facilitating, and guiding learning that is essential to individual and collective life in the twenty-first century. Joseph and Charnecki [14] note that we should no longer merely focus on questions pertaining to digital media access, but, increasingly, to inequalities in access regarding opportunities for participating in the practices of cultures supporting the development of these new competencies and skills, such as working effectively and respectfully with diverse teams, exercising flexibility and having a willingness to make compromises to accomplish common goals, and assuming shared responsibility for collaborative efforts while valuing individual contributions.

The participatory culture can be seen as exemplifying how new forms of collaboration and communication may have important transformative potentials for developing participatory learning systems in schools. Participatory learning aims to provide opportunities and resources for learners to engage in social activities, to create shared understanding among diverse stakeholders, and to join in culturally and personally relevant activities that can lead to broader communal and societal outcomes [46, 9]. The principles of this participatory culture can be capsulized into the following premises for designing twenty-first-century learning systems [40, 21, 22]:

- Learning is interest-driven participation in extended and generative communities
- The objects of learning are multifaceted phenomena that challenge systems-level thinking
- The learning process is emergent and self-organizing by nature, mediated by other people, diverse physical, digital, and social tools, technologies, and information resources
- The progress of learning is based on co-development, driven by the continuous improvement of existing knowledge and practices
- The context of learning is constantly evolving through collaborative design

What follows is the need to exemplify how these theoretical insights about participatory learning have been applied as pedagogical practices for nurturing the development of vital learning systems.

## 3 Design-Oriented Pedagogy

To enhance participatory activities situated in schools, out-of-school environments (especially natural and cultural environments), and in technological environments, our research group has developed a DOP. The DOP has been constructed on three pillars: participatory learning in extended and generative communities as a vital concept for learning; diverse technological resources and infrastructure as social and personal tools; and co-development, both as a pedagogical model, and as a

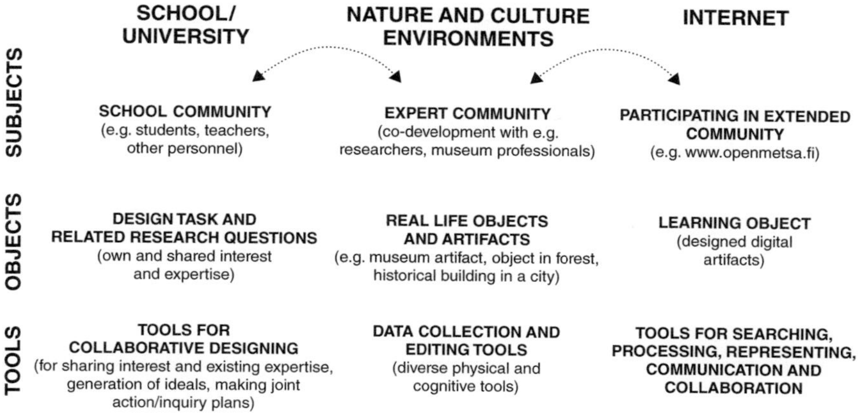


**Fig. 1** Design-oriented learning process, described as a system [40]

powerful social innovation for solving the multidisciplinary and complex problems people face in their everyday lives [21, 22, 40, 41]. The DOP shares several similarities with inquiry-based pedagogies such as knowledge building [32], progressive inquiry [10], project-based learning [16], and learning through collaborative design [35]. The DOP aims to transform learning by paying attention to the ways that diverse people, objects, artifacts, and tools interact with each other, and offers a pedagogical model and process, together with the underlying conceptual system embodied in the design. Figure 1 presents a design-oriented learning system of interconnected elements that derive their full meaning in relation to each other [40].

## 4 Extending the Learning Community

Drawing on sociocultural perspectives, one important dimension of implementing participatory learning is the emphasis on the interactions with others – with peers and with more experienced others, such as teachers and external experts. In DOP, this is supported by promoting the possibility of being able to use the students' own interests as a basis for their participation in activities undertaken with more mature members of the culture [40]. Consequently, learning is deliberately situated in extended environments and generative communities [41], consisting of students, fellow educators, and professional experts from museums, community centers, and other settings, who can mediate and support a student's learning on demand [26].

Participation in an expert community is driven by the students' own interests and research questions, where they work together in teams in pursuit of advancing their own understanding to be shared with the extended community [40]. The aim is to enhance the activities in which students learn and work together with different roles, perspectives, and responsibilities, and apply their own expertise.

Furthermore, when extending the school environment and community, the extended social networks with heterogeneous participants may mediate diverse ways of working together and enhance individual and communal expertise.

## 5 Extending the Objects of Learning

Considering the major challenges we face in the twenty-first century and the situations we encounter in our lives, there is a well-grounded need to exceed the narrow boundaries of traditional school tasks and engage students in the collaborative pursuit of varying complex and multifaceted problems that require individuals with different points of view to collaboratively design and redesign solutions from various perspectives [12, 28]. Following Hakkarainen et al.'s [12] notion of trialogical inquiry, the emphasis is on the co-development of shared objects, which requires the participants to go beyond mere dialogues. The shared object is negotiated and developed in the DOP system through an open-ended design task, which encourages approaching global phenomena from the perspective of local objects and the learners' own ideas and interpretations [21]. Complex design tasks and objects can be addressed from different perspectives [12, 35], and they intentionally bring into play multiple disciplines, multiple ways of working, and different habits of the mind, and community [24]. They also provide students with opportunities to design and perform inquiries [16], and to choose different kinds of perspectives and paths with which to engage in inquiry with the extended learning community [21]. Thus, it aims to offer opportunities for the learner to join and be part of a larger social network, where people learn through their interactions and participation with others, in fluid relationships that are the result of shared interest [38].

## 6 Extending the Learning Resources

Current education systems tend to anchor learning in formal environments, mostly in classrooms and from textbooks [26]. In contrast to traditional 'chalk and talk' classrooms in which knowledge is abstracted from real-life situations, the DOP aims at opening up possibilities to observe the world's complex phenomena in a systemic way, through different interests, perspectives, and conceptual and material artifacts linked to them. In pursuit of complex design tasks, the object under development is understood as a relational problem, and taking different perspectives into account (e.g., biologists, anthropologists, engineers, economists) may help the students explore what kinds of knowing and methods of inquiry count, for whom, in what contexts. Moreover, the emphasis is on activities in which students connect with the world around them through the objects and artifacts they self-organize for their own action and thinking. Rather than just emphasizing "what is already known," the "design" metaphor stresses the creative element in the interpretive activities of learners that goes beyond giving back what is already there [37]. As noted by Mäkitalo et al. [27], we are now held accountable not just for what is in one particular artifact, text, or even in large numbers of information resources; the summarizing of what is known is not enough. Rather, it is our ability to make insightful and productive use of the collective resources in locally relevant ways that is of importance [27]. In other words, the students are supported in not just filtering from a rich information landscape, but in evaluating the value of information in relation to a context, and applying it in the problem-solving process with the intention of creating something new.

The articulation of the shared task and the students' own related research questions can be understood from one point of view as identifying, negotiating, and selecting the real-life artifacts that become part of the students' own learning resources, in relation to their own interests, past experiences, and future intentions [40]. As noted by Kangas et al. [15], these intentions guide the design process, but may transform when the process advances. It includes the process of perceiving the function and meaning of the selected real-life artifacts and related information resources in terms of achieving a particular goal. The artifacts in nature and in cultural environments can be approached from different perspectives with various questions in mind, and can take on different functions when the students select and embed them in their own activities [40]. Furthermore, the human relationship with the real-life objects and artifacts is not considered as constant, but it may develop as they are encountered differently in evolving design processes in which connections are established with other artifacts, tools, and subjects [45].

In DOP, the learners are deliberatively provided with the possibility and means through which to share their ideas, thoughts, and their own designs related to real-life artifacts with the extended community in the form of a learning object (see an example of a learning object: www.openmetsa.fi/dopvideo). The notion of the learning object in design-oriented learning is defined as "designed digital representations from real objects in contexts that are related to the phenomenon in question and to tools that mediate the process of the negotiation of meaning" [41]. The construction of learning objects shares the idea of trialogical inquiry by engaging students' learning in creative work with externalized ideas, and the objectification and materialization of thoughts in respect of creating their own (digital) artifacts in interaction with which the subsequent inquiry takes place [12, 35]. As pointed out by Thomas and Brown [38], promoting the opportunity for the students to share the outcomes of their inquiry activities with an extended collective that is structured around participation (e.g. www.openmetsa.fi) is very different to putting the outcome onto the school wall or into the public domain.

By offering the students the chance to collaborate with their peers, to take part in face-to-face interactions with expert communities, and to be involved in mediated contact online with a more dispersed population, the students are deliberatively offered different forms of participation [13]. Viewed through the lens of participatory culture, it offers the opportunities for and the means of participating in the practices of the social production of knowledge [13] in distributed networks of communities, resources, and tools.

## 7 Tools for Making, Thinking, and Sharing

The DOP also aims to enhance the opportunity to apply diverse physical, cognitive, and social tools and technologies in collecting, developing, and sharing information. The technologies that the students own, provide tools to enhance learning across different contexts, and to collect various empirical data when implementing inquiries in authentic environments. When working with expert communities, the learners are provided with the possibility of being able to use domain-specific tools that characterize such expertise. Additionally, social media provides tools for learners to organize, develop, and share knowledge, and to collaborate within and outside the school community [40, 41].

However, rather than dealing with technology in isolation, the DOP takes a more systemic approach by considering the interrelationship among tools, artifacts, and the communities, and the activities in which they are embedded. The tools derive their full meaning and functional role in relation to the other elements of the learning system in situated social practice: the subject(s) using the tools (e.g., students, experts, and teachers' agency); the object (e.g., shared tasks, students' own research questions); and the artifacts of their actions (e.g., material and conceptual); and the context of using the tools (e.g., designing, making inquiries, sharing the results). Thus, the subject and real-life artifacts are not connected by the tool in a mechanical manner, but are dynamic interactions, and are grounded in particular activities. Different tools are needed during the process of collaborative designing and when implementing the inquiry activities in extended learning environments, and during this evolving process, the same tool may be used in different ways and may serve different purposes [40]. Furthermore, the development of tools such as 3D printers has opened up the possibilities to rapidly design and build new tools and artifacts to meet particular purposes and interests [8]. As Claxton [5] argues, if the main thing we know about the future is that we do not know much about it, then the educators should not only provide young people with the tools of today, but should help them to become confident and competent designers and makers of their own tool environments when solving emergent problems. Thus, novel tools and technologies should not simply represent new ways of resourcing existing learning activities in schools; instead, they can be perceived as tools for developing new social practices of learning and instruction.

## 8 Situated Context

Like activity systems [7], the elements of the DOP system are not static but are continuously interacting with each other, through which they define the emerging learning system as a whole. This emergent form of the system ultimately shifts our focus to the situated context that people, objects, and tools form, and promotes the students' possibilities of shaping it. It proposes a clear transformation from a predetermined learning environment toward the creation of dynamic and extending learning networks [40]. As argued by Liljeström et al. [22], the focus is transformed in emerging learning ecosystems that offer the students the opportunity to self-organize and utilize the afforded community, technology, and information resources to construct their own interpretations of their chosen research tasks and related inquiries. Then, the students may benefit from the existing networks and resources that they are already connected with to explore the extended networks that others have access to, and construct their understanding of how best to design, develop, and use these networks and tools for solving emergent problems [8]. This view overlaps with Barab and Roth's [2] notion of affordance networks. They define this concept as the collection of facts, concepts, tools, methods, practices, agendas, commitments, and even people, with respect to an individual, that are distributed across time and space, and are viewed as necessary for the satisfaction of particular goal sets. According to Barab and Roth [2], education should connect learners to an ecological system that stimulates an appreciation for, and a desire to be a part of contexts through which these networks take on meaning, as well as equipping students so that they can create new and useful affordance networks. From this perspective, learning and participation is about successfully participating as part of an ecosystem, which involves increasing the possibilities for action in the world [2].

The DOP utilizes the notion of self-organizing systems of participatory cultures by emphasizing that the process is not scripted in detail in advance, but has to be negotiated and actively designed by the learners themselves. It matters that various resources and more experienced community members (e.g., teachers, experts) are supporting and available for the use of the learners, but it is essential that the learners should be positioned in a key role when defining the specific network of artifacts, tools, and information resources in terms of their own intentions and negotiated research questions. Yet, the learners are supported by the instructional model and by joint activities with mature members of the community to design and build learning paths that mediate the practices of innovative professional or scientific communities (see [41]).

Schools, museums, and other public institutions have an essential role to play in creating more equitable opportunities for participating and contributing one's own expertise to a process that connects many intelligences and communities outside of the school [13]. As pointed out by Facer [8], the idea about what makes for successful public services has been changing as citizens are seen to play a much more active role as "co-producers." When the students participate in practices to address shared intentions beyond the school, the students become, at that moment, enculturated, participatory, contributing community members, and the students and the extended community's ecosystems may overlap [1]. At the heart of the idea is to allow students to participate in knowledge-creating activities around shared objects and to share their efforts with the wider community for further knowledge building that is a legitimate part of civilization [32].

## 9 Conclusion and Future Directions

Following the principles of design-based research, our research group has conducted several design experiments to develop the pedagogy and the related design principles to their present "state of the art." Liljeström et al. [21, 22] have demonstrated how open-ended learning tasks and collaborative inquiry approaches can mediate the kind of authentic practices that scientists apply. Likewise, Vartiainen and Enkenberg [42, 44] have examined the types of learning systems that emerge when different student groups collaboratively design their own specific network of museum artifacts, tools, and other resources in terms of the shared design task and their own specific research questions. Vartiainen and Enkenberg [43] have also assessed how applicable and acceptable the DOP is from an international perspective. Furthermore, our research team, representing multidisciplinary expertise concerning educational and forest sciences, has constructed the OpenForest portal (http://www.openmetsa.fi/) for people to share, develop, and organize knowledge, and to collaborate within and outside of the education community and institutions. In addition to an open wiki environment, there is significant effort being dedicated to creating accessible tools that allow students, teachers, researchers, and others who are interested to upload and access digital records (e.g. real-time data, scientific knowledge, research findings, and learning objects created by the students) connected with physical locations by means of GPS (global positioning system) [39]. This offers interesting possibilities for future research concerning the ways of connecting the digital, social, and physical landscapes of learning, and to explore how these can enhance each other in the pursuit of networked communities around forest-related phenomena.

One particular and pervasive challenge is how to bring new methods and learning environments into existence within schools, especially when the current school system is often very different from the framework conceptualized by the new educational innovations [3]. The teachers have to work in a re-productive system that has created organizational structures, practices, rituals, and belief systems, which impose significant restrictions for teachers to implement and develop new pedagogical practices [44]. Consequently, it indicates the need to build new kinds of networks of teachers, school leaders, researchers, and external members and experts to work together to enhance the creation of a new kind of participatory learning culture.

Following the principles of participatory culture, our current research on DOP focuses on the ways of using the model of crowdsourcing facilitated by socio-technological tools to draw interested people into the processes of identifying, designing, and co-developing new opportunities and models for twenty-first-century learning. The emphasis is on a shift from the top-down models toward co-development within and between different levels of the educational system (e.g. students, teachers, school leaders, administration), and outside of it (e.g. museums, the natural environment, and related social networks). Connecting the learning processes at these different levels is important for any kind of innovation, but particularly for the ones that are truly systemic in nature. Figure 2 represents

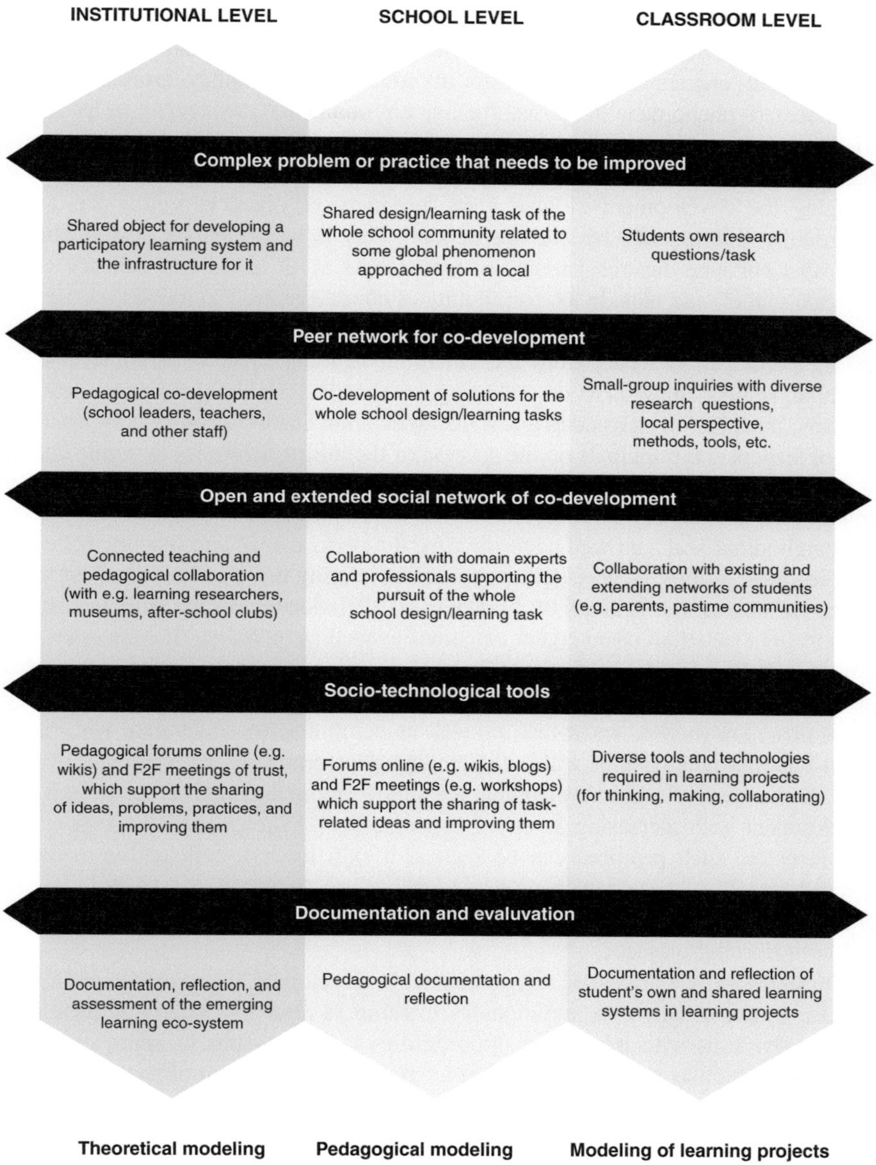


**Fig. 2** Framework for developing a participatory learning culture in schools

the framework modeling the co-development of a participatory learning culture in schools.

The long-term goal of a sustainable new cultural practice may be considered as a shared object of activity that the researchers, together with teachers, and others who are interested, all develop collaboratively [12]. In this context, the DOP pedagogy may be seen as a tool for the pursuit of larger social innovations in schools that engage the community in collaborative inquiry around selected multifaceted phenomena (e.g. well-being, environmental awareness, a particular contemporary problem such as unemployment). It includes the intention of creating classroom-based projects with small-group activities, and at the same time crowd-sourcing the development of new ideas, products, or practices within the whole community that extends beyond the school. This is supported by socio-technical tools that enhance the transparency of the process at all levels and enhance collaboration within, and outside the community of education and institutions.

To reflect on and improve existing practices, co-development is supported by collecting participant-led data from the emerging learning systems at all levels. For example, students may focus on collecting multimedia data about their own learning projects, teachers on the pedagogical documentation of the learning activities, and school leaders and principals on the success of the infrastructure for systemic change at the institutional level. Such crowdsourcing involving constant evaluation aims to enhance the pursuit of shared objects under development in and across all levels at the longitudinal scale, as well as to improve the performance and practice of individual learners, groups, projects, and organizations during the practical actualization of the co-development. As noted by Lemke et al. [19], documentation activities should support the evaluation of emergent outcomes as well as recording the processes over time by which the outcomes are produced. Yet, valued outcomes include more than just acquired knowledge. Such context-bound learning outcomes may include e.g. know-how, know-who, know-that, as well as being able to collaborate with others when building relationships within the emerging learning system. Moreover, evaluation should always support planning for future improvements of the design and development with increasing levels of expertise [19]. When combined with education research, such practices can be seen as a shift toward participative modes of research in which those who have agreed to become involved in a study have produced some or even all of the data [29] as a part of their professional development and own learning projects.

Furthermore, we are especially interested in developing international collaboration for future research on participatory learning in networked communities. This could provide us with interesting opportunities through which to approach global phenomena such as sustainable development as a shared design object for learning and crowdsourcing in an international network of students, teachers, researchers, experts, and interested others. At the same time, it would provide researchers with interesting opportunities to examine how and in what ways the participants from different backgrounds use and share their own interests, their own and afforded tools and technologies, and local physical and social environments as resources for learning driven by joint co-development. The promotion of inquiry activities that enable students to participate in the co-development process with local and international communities in addressing some of the most pressing issues of our time might be particularly important for learning in a world of constant change in which the use of diverse knowledge resources, tools, and network connections are essential aspects when solving complex, emergent problems, and creating situation-based solutions. It also emphasizes the importance of collaborative design in twenty-first-century learning [40], and ultimately, of building our students' capacity to design and co-develop a sustainable world that they and we might all want to inhabit [8].

**Acknowledgements** I would like to express my sincere gratitude to our research group, especially to Emeritus Professor Jorma Enkenberg, Professor Sinikka Pöllänen, Anu Liljeström, and Dr. Petteri Vanninen. This study is supported by the Finnish Cultural Foundation and the "Forest as a Learning Environment: South Savo as a Key to Communal Forest Knowledge" Project (UEF, project no. A70133).

## References

1. Barab S, Chekes-Julkowski M, Swenson R, Garrett S, Shaw R, Young M (1999) Principles of self-organization: learning as participation in autocatakinetic systems. J Learn Sci 8(3–4): 349–390
2. Barab S, Roth W-M (2006) Curriculum-based ecosystems: supporting knowing from an ecological perspective. Educ Res 35(5):3–13
3. Bielaczyc K (2013) Informing design research: learning from teachers' designs of social infrastructure. J Learn Sci 22(2):258–311
4. Binkley M, Erstad O, Herman J, Raizen S, Ripley M, Rumble M (2011) Defining 21st century skills. In: Griffin P, McGaw B, Care E (eds) Assessment and teaching of 21st century skills. Springer, New York, pp 17–66
5. Claxton G (2002) Education for the learning age: A sociocultural approach to learning to learn. In: Wells G, Claxton G (eds) Learning for life in the 21st century. Blackwell, Oxford, pp 21–34
6. Clinton K, Jenkins H, McWilliams J (2013) New literacies in an age of participatory culture. In: Jenkins H, Kelley W (eds) Reading in a participatory culture: remixing Moby-Dick in the English classroom. Teachers College Press, New York, pp 3–23
7. Engeström Y (1987) Learning by expanding: an activity-theoretical approach to developmental research. Orienta-Konsultit, Helsinki
8. Facer K (2011) Learning futures. Education, technology and social change. Routledge, London
9. Fischer G (2013) Learning, social creativity, and cultures of participation. In: Sannino A, Ellis V (eds) Learning and collective creativity: activity-theoretical and sociocultural studies. Taylor & Francis, New York
10. Hakkarainen, K (1998) Epistemology of inquiry and computer-supported collaborative learning. (Unpublished doctoral dissertation) University of Toronto, Toronto
11. Hakkarainen K (2010) Communities of learning in the classroom. In: Littleton K, Wood C, Kleine Staarman J (eds) International handbook of psychology in education. Emerald, Bingley, pp 177–225. https://www.academia.edu/350358/Draft_of_Hakkarainen_K_2010_Communities_of_learning_in_the_classroom. Accessed 28 Jan 2014
12. Hakkarainen K, Paavola S, Kangas K, Seitamaa-Hakkarainen P (2013) Socio-cultural perspectives on collaborative learning: towards collaborative knowledge creation. In: Hmelo-Silver C, Chinn C, Chan C, O'Donnell A (eds) International handbook of collaborative learning. Routledge, New York, pp 57–73
13. Jenkins H, Clinton C, Purushotma R, Robison AJ, Weigel M (2008) Confronting the challenges of participatory culture: media education for the 21st century. John D. and Catherine T. MacArthur Foundation, Chicago, http://www.nwp.org/cs/public/print/resource/2713. Accessed 28 Jan 2014

14. Joseph B, Charnecki K (2013) Leveraging digital media to create a participatory learning culture among incarcerated youth. In: Delwiche A, Henderson D (eds) The participatory cultures handbook. Routledge, New York, pp 220–231
15. Kangas K, Seitamaa-Hakkarainen P, Hakkarainen K (2011) Design expert's participation in elementary students' collaborative design process. Int J Technology Des Educ 23(2): 161–178
16. Krajcik J, Blumenfeld P (2006) Project-based learning. In: Sawyer K (ed) The Cambridge handbook of the learning sciences. Cambridge University Press, Cambridge, pp 317–333
17. Kumpulainen K, Mikkola A, Jaatinen AM (2013) The chronotopes of technology-mediated creative learning practices in an elementary school community. Learn Media Technology 39(1):53–74
18. Lehtinen E, Hakkarainen K, Palonen T (2014) Understanding learning for the professions: how theories of learning explain coping with rapid change. In: Billett S, Gruber H, Harteis C (eds) International handbook of research in professional practice-based learning. Springer, Dordrecht, pp 199–224
19. Lemke J, Lecusay R, Cole M, Michalchik V (2015) Documenting and assessing learning in informal and media-rich environments. The MIT Press, Cambridge
20. Lévy P (2013) The creative conversation of collective intelligence. In: Delwiche A, Henderson D (eds) The participatory cultures handbook. Routledge, New York, pp 99–108
21. Liljeström A, Enkenberg J, Pöllänen S (2013) Making learning whole: an instructional approach for mediating the practices of authentic science inquiries. Cult Stud Sci Educ 8(1): 51–86
22. Liljeström A, Enkenberg J, Pöllänen S (2014) The case of design-oriented pedagogy: what students' digital video stories say about emerging learning ecosystems. Educ Information Technologies 9(3):583–601
23. Loi D, Dillon P (2006). Adaptive educational environments as creative spaces. Camb J Educ, 36(3), 363–381
24. Lombardi M (2007). Authentic learning for the 21st century: an overview. https://net.educause.edu/ir/library/pdf/ELI3009.pdf. Accessed 15 Nov 2007
25. Lto M, Gutiérrez K, Livingstone S, Penuel B, Rhodes J, Salen K, Schor J, Sefton-Green J, Watkins SC (2013) Connected learning: an agenda for research and design. Digital Media and Learning Research Hub, Irvine
26. National Education Technology Plan. (2010). Transforming American education. Learning powered by technology. www.ed.cov/technology/netp-2010. Accessed 5 Aug 2013
27. Mäkitalo Å, Jakobsson A, Säljö R (2009) Learning to reason in the context of socioscientific problems. Exploring the demands on students in 'new' classroom activities. In: Kumpulainen K, Hmelo-Silver C, Cesar M (eds) Investigating classroom interaction. Methodologies in action. Sense Publishers, Rotterdam, pp 7–26
28. Perkins D (1986) Knowledge as design. Erlbaum Associates, Hillsdale
29. Prosser, J, Loxley, A (2008). Introducing visual methods. National Centre for Research Methods Review Paper No. 10. http://eprints.ncrm.ac.uk/420/. Accessed 28 Jan 2014
30. Rheingold H (2013) Participatory pedagogy for a literacy of literacies. In: Delwiche A Henderson D (ed) The participatory cultures handbook. Routledge, New York, pp 215–219
31. Robinson K, Aronica L (2015) Creative schools: the grassroots revolution that's transforming education. Viking, New York
32. Scardamalia M, Bereiter C (2006) Knowledge building: theory, pedagogy, and technology. In: Sawyer K (eds) Cambridge handbook of the learning sciences. Cambridge University Press, New York, pp 97–118
33. Scardamalia M, Bransford J, Kozma R, Quellmalz E (2011) New assessments and environments for knowledge building. In: Griffin P, McGaw B, Care E (eds) Assessment and teaching of 21st century skills. Springer, New York, pp 231–300
34. Schoultz J, Säljö R, Wyndhamn J (2001) Heavenly talk: discourse, artifacts, and children's understanding of elementary astronomy. Hum Dev 44(2–3):103–118

35. Seitamaa-Hakkarainen P, Viilo M, Hakkarainen K (2010) Learning by collaborative design: technology-enhanced knowledge practices. Int J Technology Des Educ 20(2):109–136
36. Sfard A (1998) On two metaphors for learning and the danger of choosing just one. Educ Res 27(2):4–13
37. Säljö R (2010) Digital tools and challenges to institutional traditions of learning: technologies, social memory and the performative nature of learning. J Comp Assisted Learn 26(1):53–64
38. Thomas D, Brown J (2011) A new culture of learning: cultivating the imagination for a world of constant change. CreateSpace, Lexington
39. Vanninen P, Liljeström A, Vartiainen H, Enkenberg J, Pöllänen S (2013) Forests in teacher education: OpenForest portal as a novel resource for learning. In: Issa T, Sharef NM, Issa T, Isaías P (ed), Proceedings of the International Conference on Sustainability, Technology and Education (STE2013). IADIS Press. http://www.iadisportal.org/ste-2013-proceedings. Accessed 22 May 2016
40. Vartiainen, H (2014) Principles of design-oriented pedagogy for learning from and with museum objects. Publications of the University of Eastern Finland. Dissertations in Education, Humanities, and Theology, no. 60. University of Eastern Finland, Joensuu
41. Vartiainen H, Liljeström A, Enkenberg J (2012) Design-oriented pedagogy for technology-enhanced learning to cross over the borders between formal and informal environments. J Univers Comput Sci 18(15):2097–2119
42. Vartiainen H, Enkenberg J (2013a) Learning from and with museum objects: design perspectives, environment, and emerging learning systems. Educ Technology Res Dev 61 (5):841–862
43. Vartiainen H, Enkenberg J (2013b) Reflections of design-oriented pedagogy for sustainable learning: an international perspective. J Teach Educ Sustain 15(1):43–53
44. Vartiainen H, Enkenberg J (2014) Participant-led photography as a mediating tool in object-oriented learning in a museum. Visit Stud 17(1):66–88
45. Vygotsky LS (1978) Thought and language. MIT Press, Cambridge
46. Wells G (2010) Schooling: the contested bridge between individual and society. Pedagogies 5(1):37–48
47. Wenger E (1998) Communities of practice: learning, meaning, and identity. Cambridge University Press, Cambridge

# The Concern about Environmental Responsibility in Public Service Media Management

**Tania Fernández-Lombao, Andrea Valencia-Bermúdez and Francisco Campos-Freire**

**Abstract** Corporate Social Responsibility is the commitment to ethical and moral principles and good corporative practices. Indeed, it has become an improved management tool for private and public organizations in the last 50 years. Social, environmental, ethical responsibilities are social demands to which public and private companies should respond with the same priority. Environmental policies are a relevant issue, as it has been evidenced at the Paris Summit (2015) on climate change (COP21). The environmental commitment and responsibility is at the epicentre of management, since it belongs to the major challenge of present-day society. Verifying that commitment – which is the proposal of this paper, related to public broadcasters – is checking how those organizations are fulfilling the objectives of the third millennium. The results show that some public media in Europe report on environmental issues, but there is a lack of detailed strategies to reduce their impact.

## 1 Introduction

The increasing concern about Corporate Social Responsibility is more a symptom of necessity than a diagnosis of objectives achieved. Undoubtedly, the two concepts are part of the PSM's management agenda, since that is also a concern of societies in which audiovisual corporations – both public and private – are working. In the case of public companies this is even more important, since to those internal exigencies

T. Fernández-Lombao (✉) · A. Valencia-Bermúdez · F. Campos-Freire
University of Santiago de Compostela, Santiago, Spain
e-mail: t.lombao@gmail.com

A. Valencia-Bermúdez
e-mail: andrea.v.bermudez@gmail.com

F. Campos-Freire
Prometeo of the SENESCYT in the Particular Technical University of Loja and the Pontifical Catholic of Ibarra, Ibarra, Ecuador
e-mail: francisco.campos.freire@gmail.com



T. Issa et al. (eds.), *Sustainability, Green IT and Education Strategies in the Twenty-first Century*, Green Energy and Technology, DOI 10.1007/978-3-319-57070-9_22

are added the triple estimation of intrinsic, exchange, and use value [1]. That double responsibility – social and environmental – compromise providers of public audiovisual services, since it offers them a new tool for re-legitimation of their remit and economic sustainability behind their publics and stakeholders.

Until just a decade ago, CSR – as a self-regulatory system – was only conceived as a strategy and management tool for private organizations, but not suitable for public institutions, because it had been assumed that regulatory frameworks of public service media covered all their responsibilities. That conception has become, however, obsolete, since the institutional legal framework is too limited and failing to address technological challenges on time.

That is why many companies started to integrate hetero-regulation frameworks, assuming social and environmental responsibilities as a voluntary strategy of interaction with stakeholders, social legitimation and reputation improvement.

Despite the popularity and advances in the integration of social and environmental responsibility into companies' management, there is yet not unanimity on the definition and scope of both concepts. Neither is there full consistency between the assumption of their moral philosophy and the effectiveness of their implementation in all organizations.

When switching from theory to practice, some organizations are only pursuing reputation through social marketing, and others through CSR management, as a sequence of bureaucratic processes more dominated by evidences' verification than by the social significance of their actions. The development of CSR could be divided into three historical periods: first one, the assumption of that social commitment; second one, the implementation, systematization and verification; and third one, the significance of its social impact.

The evolving concepts of sustainability and social responsibility may be explained in terms of their historical trajectory. Sustainability is the satisfaction of needs of current generations without compromising the ability of future generations to meet their own needs [2]. According to Clarke [3], it is important to distinguish between sustainable growth, sustainable use, and sustainable development. The first one is contradictory, since there cannot exist indefinite growth. The second concept can be applied only to renewable resources. And the third one is used to indicate the improvement of human's quality of life within the limits of ecosystems' capacity [4]. Sustainability converges with social responsibility since its overall performance within the organization includes policies, decisions, and actions to achieve economic, environmental and social outcomes [5].

As regards the European Commission [6, 7, 20], CSR refers to "companies voluntarily going beyond what the law requires to achieve social and environmental objectives during the course of their daily business activities. For the World Business Council for Sustainable Development" [8, 9], CSR is the commitment by business to behave ethically and contribute to economic development while improving the quality of life of the workforce, the families, and the local community and society at large.

Corporate Social Responsibility encompasses "not only what companies do with their profits, but also how they make them. It goes beyond philanthropy and compliance, and addresses how companies manage their economic, social, and environmental impacts, as well as their relationships in all key spheres of influence" [10]. It is also presented from a business approach, including opened and transparent practices, ethical behaviour, respect to stakeholders and commitment in the contribution of economic, social, and environmental value [11].

The global financial crisis in 2008 highlighted a need to restore ethical and integral values of governance, in order to enhance companies' credibility. What has become evident in this crisis is the lack of regulation, ethics, solidarity, moral principles, and accountability within a complex scenario with structural economic and social problems.

The concept of governance [12–15], as a multi-level management system designed to respond to complexity, an open-government model for society and stakeholders, not only focused on internal and immediate profitability, but also on external medium and long-term utility. This approach is the one that fits with the commitment and current practice of Corporate Social Responsibility.

The corporate response to society's general concerns is part of what some refer to as the third generation of corporate codes of ethics [16], considering that it was first the consistency of ethics with business, and then it responsibility against stakeholders. Now these renewed postulates propose – in response to the perception that ethics has been removed from the collective agenda, being the global crisis one of the consequences – joining economy and society back together to generate shared social value [17].

The commitment to ethics and corporate social responsibility should not be understood as an imposition or an administrative procedure, but as potential guiding and driver of company's internal and external values [18]. This clarification is important since this double interpretation is highlighted in some organizations. That kind of actions are not performed because they are understood as a requirement that can overextend the possibilities of management; and, also, they are seen as bureaucratic actions based only on the achievement of indicators' report for the use of their public relations policies. Both attitudes are wrong and have an impact on management's moral responsibility, which occurs when ethical expectations are violated, compromising business value [3].

Public service broadcasting is under threat across Europe, while the awareness of its contribution to democratic, social, cultural, and an identity values still remain. The traditional model of public 20th century-broadcasters faces the challenge of readapting itself as a new public audiovisual service, which is still necessary in the digital era. Undoubtedly, its particular challenges are reforming its governance systems and the sustainability of its funding models; satisfying the requirements of accountability; adapting to the digital ecosystem; fostering creativity and innovation; and improving quality of contents, pluralism, diversity and contribution to democratic values.

Social and environmental responsibilities are included in these new governance practices. They explicitly assume the commitment to social and environmental responsibility, although its management does not go much beyond standard practices of accountability. Analysis of reports and websites show that there are formal policies, but there is a lack – except in some rare cases – of innovative proposals, which would help them to convert those actions in reputation and social impact improvements of those organizations.

This chapter aims to analyse the way in which public service broadcasters are facing the new challenges of sustainability, transparency and accountability. This issue, until recently seen as a model of private-sector success, has been incorporated into public media policies as a way to renew and regain their legitimacy. The chapter will study how the environment becomes an increasingly significant part of corporate social responsibility policies. Results show that, although some public broadcasters report on environmental issues, there is a lack of a clear and defined strategy.

## 2 Methodology

The research analyses how the environment becomes increasingly relevant in CSR policies implemented by European pubic service media. For that purpose, only corporations with CSR reports are studied (BBC, RTE, ORF, YLE, ZDF, RTP, RTVE, FTV, and RTBF). Also, Spanish public broadcasters are analysed because of its distinct public audiovisual arena. The chapter is divided into two different parts. Firstly, it is carried out a content analysis of the last Corporate Social Responsibility statements, which are available on their corporate websites: BBC Corporate Social Responsibility Performance Review 2013, from the BBC – United Kingdom; Corporate Responsibility 2009, from RTÉ – Ireland; Nachhaltigkeit Im ORF 2011, from ORF – Austria; Vastuurapotti 2012, from YLE – Finland; Mit Perspektive für alle. Gesellschaftliches Engagement und unternehmerische Verantwortung des ZDF 2011–2012, from ZDF – Germany; Relatório de Sustentabilidade 2012, from RTP – Portugal; Informe del Pacto Mundial 2013, from RTVE – Spain; Rapport Annuel 2012, from FT – France; and Rapport Annuel 2013 from RTBF – Belgium. Secondly, it is conducted an observation of the environmental treatment on websites of the above-mentioned public broadcasters.

The analysed items are those proposed by the Sustainability Reporting Guidelines (Global Reporting Initiative [19]), which establish environmental indicators with a media sector supplement. This paper recommends reporting on: use of paper and water, impact of inks and printing materials, recycling policies, greenhouse gas emissions and measures to be environmentally sound.

This international initiative has made its guides the most widely implemented by companies around the world when reporting on corporate performance. In 2013, it published the G4 guidelines for sustainability reporting, although it has not yet made available any specific complement for media companies. For that reason, it is decided to use the criteria proposed in the GRI Media Sector Supplement. On the other hand, it is determined whether public broadcasters report on environmental policies in their websites. To this end, it will be taken contents displayed as a part of a crosscutting strategy instead of specific actions related to this topic.

## 3 Environmental Performance of CSR Reports of Public Service Media in Europe

The first part of the research, based on the content analysis of the last CSR reports of the nine above-mentioned broadcasters, allow us to identify areas to which public service media pay the most attention. Symbols "✓" and "X" in Table 1 indicate whether or not the organization develop environmental responsibility actions into that category. A more detailed evaluation of environmental performance is carried out from sections 3.1 to 3.6.

### *3.1 Use of Paper*

Four of the analysed organizations have developed actions in this field. The Austrian ORF, engaged in a process of digitization, reduced the use of paper from 180.98 tonnes in 2009 to 173.08 in 2011. On the other hand, the German ZDF recycled 20.3 tonnes of paper between 2011 and 2012, while the French FT also diminish the use of paper and the printer replacement, but it does not provide specific data.

T he Portuguese RTP employed 27.6 tonnes of paper in 2013 – a 3% less than the previous year – even though every worker uses more sheets – eight – as a result of staff reductions. To counter-balance these negative effects, the use of recycled paper was increased by 4%.

**Table 1** Indicators of environmental performance in Public Service Media in Europe (prepared by the authors, 2016)

| Indicator of environmental performance | | | | | | | | | |
|---|---|---|---|---|---|---|---|---|---|
| Environmental performance | BBC | RTE | ORF | YLE | RTP | ZDF | RTVE | FT | RTBF |
| Use of paper | X | X | ✓ | X | ✓ | ✓ | X | X | X |
| Use of water | ✓ | X | X | X | ✓ | ✓ | X | X | X |
| Impact of ink and printing materials | X | X | X | X | X | X | ✓ | X | X |
| Recycling policies | ✓ | ✓ | ✓ | ✓ | ✓ | ✓ | X | X | X |
| Greenhouse gas emissions | ✓ | ✓ | X | ✓ | ✓ | ✓ | ✓ | X | ✓ |
| Measures to be environmentally | ✓ | ✓ | ✓ | ✓ | ✓ | ✓ | ✓ | ✓ | ✓ |
| Sound | | | | | | | | | |

## 3.2 Use of Water

Regarding the use of water, the English BBC reduced by 7% its expenditure during the period from 2012 to 2013. The ZDF used less water in 2012 – 4.4 million litres – but figures are not specified. The RTP spent 27% less than in 2012, intending to continue along these lines with the installation of water savers in the taps.

## 3.3 Impact of Ink and Printing Materials

The Spanish RTVE – through its SCR memory – is the only organization that provides information on this matter, since it makes available its project of printers' optimization, on which 389 out of 2,099 units were removed in 2012.

## 3.4 Recycling Policies

Five organizations provide information on their policies in this field through their SCR memories. The BBC points the improvement of recycling ratios – 81% in 2012–2013. Furthermore, the RTÉ recycled 68% of its waste in 2008, and the RTP supports resources management, but no figures are supplied.

The Austrian ORF opts for economic efficiency and has acquired long-living technological products that, even so, require more alternating current and air-conditioning. Also, the ORF takes into account the availability of replacement pieces in all its purchases. The organization indicates an increase in non-recyclable waste from 132.74 tonnes in 2009 to 168.6 in 2011. Meanwhile, the ZDF cleaned up hard drives of equipment – based on the ISO 900I – revising an annual average of 500 computers and 120 laptops.

## 3.5 Greenhouse Gas Emissions

All analysed corporations include information on greenhouse gas emissions in their memories, with the exception of Austria, Finland and France. The BBC shows its commitment through control measurements with the "Albert" system, for the sustainability in television production. On the figures, it points a 4% reduction in energy consumption and a 6% reduction in CO2 emissions during the period 2012–2013. In Ireland, the RTÉ indicated a 19% decrease in the use of natural gas and an 11% decline in power consumption during 2008.

The SCR report of RTP details the different types of emissions and their corresponding quantities. In this regard, RTP points out that greenhouse gas emissions were reduced by 17% in 2013. Also, the memory proposes to continue the same trend, reducing the number of journeys and the fleet of vehicles. The ZDF reduced power consumption to 902.289 kilowatt-hours and lowered CO2 emissions to 24 tonnes a year. The organization bets on electric vehicles and proves that the entire fleet complies with the Euro 5 emissions standard.

The only measure presented by the Spanish RTVE is the installation of low energy air-conditioning devices. In Belgium, the RTBF notes the contract of green electricity supply and the pilot implementation of compressed natural gas vehicles.

### *3.6 Measures to be Environmentally Sound*

Most of the data displayed in SCR memories is related to this topic, since its generic title allows the inclusion of all kind of environmentally friendly measures.

With a view of minimising the impact in environmental terms, the BBC moved its headquarters to a new sustainable place and replaced physical travel with videoconferencing. In 2008, the Irish public broadcasters joined the program "A little step", developed by the Transport Office in Dublin to encourage staff walking to work. Also, the RTÉ joined a programme to promote the use of bicycles by its staff, with financial assistance to cover the cost of these vehicles. Furthermore, it abolished the use of polystyrene.

The ORF optimized hours of air-conditioning use, gradually replaced lamps, reduced air and water in cooling systems, and installed solar panels in its headquarters, which allowed it to save up to 20%. The RTP reduced by 7% power consumption and 12% fuel consumption.

The German ZDF is the most active corporation on environmental measures. In its last SCR report, it notes the replacement of obsolete components and devices; the launching of the Ecoprofit project for the optimized use of resources; the development of an environmental guide; the installation of solar cells on the headquarters' roof; the assembly of a new cooling system, which represents a saving of 25,000 million a year; the use of train whenever possible; and the use of LED lights and certified timber. The result of these measures has been the saving of 120,000 euros and the yearly 1.3% reduction in energy consumption. Also, the ZDF adds that 19% of purchases are 100% certified organic.

As regards the fleet, ZDF reports that vehicles should comply with Euro 5 emissions standard, establishing limits to fuel consumption and CO2 emissions. The corporation promotes electric mobility within the company. In fact, it has bought a new sustainable vehicle in 2014 with an efficiency of 90%. Also, they call for the use of green cleaning products.

In the same vein, the Spanish RTVE installed long life lamps in its headquarters; demolished asbestos-affected installations; procured more sustainable technologies and tools; installed more efficient energy – with inverter systems – and launched the Corporate Videoconference Service, with the objective of reducing personnel movement – in 2012, 613 meetings were convened by videoconferencing.

The French FT reports on the implementation of the following environmental measures: use of environmentally friendly furniture; employment of natural detergents and low-energy LED lights; remodelling of meeting rooms equipped with videoconference technologies; long-term purchases and promotion of the Ecoprod Collective to promote responsible film productions.

The Belgian RTBF supports the use of electric bikes in the organization's buildings and the acquisition of long-term office equipment. Also, it created 10 *ecogroups* to come up with new environmental solutions. The Finnish YLE distinguished five elements in this area: reducing company premises by 40%; applying environmental criteria for vehicle purchase and travel policies; reducing flights; encouraging videoconferences and the use of public transport; and following the recommendations of the WWF Green Office in environmental matters.

## 4 Environmental Responsibility in Websites of European Public Service Media

The analysis of corporate PSM's websites shows three distinct realities: (1) BBC from the UK, FT from France, and YLE from Finland include fixed areas for environmental responsibility; (2) RTÉ from Ireland and RTP from Portugal only publish reports on that issue ("The Green broadcaster" and "Relatorios de Sustentabilidade", respectively); (3) RTVE from Spain, ZDF from Germany, RTBF from Belgium, and ORF from Austria do not have a section on environmental responsibility. While environmental responsibility is still in the process of becoming a management's priority, below is a description of the main characteristics of the four pioneer broadcasters.

The British BBC highlights three main approaches to sustainability: (1) Drive innovation in their core business in order to lead the broadcasting industry in sustainable production; (2) Continually work to reduce the impacts of our operations, through targets which address their environmental footprint and save the BBC money; (3) Inspire their staff to exemplify sustainable behaviour in all that they do.

The BBC approach to environmental responsibility is called "The Difference", including five defined areas: (1) get the basics right, reducing the impacts of the infrastructure they provide and operate, for example by investing in energy efficiency for their buildings and technology; (2) improve the way of working by reducing travel, by increasing video conference, and by using bespoke tools such as albert the carbon calculator for TV production to encourage good practice; (3) require suppliers to improve the sustainability of their products and services; (4) lead the industry, embedding sustainability in the TV production process; (5) engage their staff through training courses and awareness campaigns.

BBC's sustainability team works with a group of external advisors, who they meet twice, annually to analyse the BBC's progress and provide advice and guidance. In 2008 the Royal Charter set a framework of objectives for a period of 8 years. Among environmental goals, they have targets to reduce energy consumption and $CO2$ emissions from buildings by 20%; to reduce water use by 25% per person;

to recycle more of their waste by 70%; and to tackle emissions from travel by 20%. Also, the corporation is expected to reach 100% of sustainable production.

The BBC also launched a carbon calculator called Albert, a tool that help productions measure their footprint, and to help the industry make the switch to more sustainable ways of working. Albert's use is mandatory for BBC Television and CBeebies. From 2011 onwards, the BBC has formed a partnership with BAFTA, where it is made available for free.

Recently, the corporation has developed a complementary tool – Albert + – which aims at helping production teams make their programmes in more sustainable ways, keeping track of their achievements, and rewarding them with an albert + rating – One, Two, or Three stars, determined by the range and complexity of green initiatives implemented by each team. All working areas may be controlled by Albert+, from training and production offices, to travels and studies.

As regards the French FT, it declares that its immediate priority is to address climate change through three objectives: (1) reduce environmental footprints in operations; (2) reduce the environmental impact of their products; and (3) inform and make people aware of environmental responsibility.

As regards the first goal, it attaches the highest importance to the development of a greenhouse gas management system, using the carbon footprint method. These systems are organized into five main areas: (1) building operation; (2) green policies for ICT – Information and Communication Technologies; (3) vehicle fleet's mobility and management; (4) waste management; (5) and responsible purchases. This goal is accompanied by internal communication's strategies with two key messages: climate change is a global problem that has to be mitigated, and actions for reducing carbon footprint have to be carried out at all levels.

France Télévisions is one of the largest TV-content producers in France and, consequently, it assesses the environmental impact of their products and tries to gradually reduce it. The CSR department entered into an opened dialogue with production managers, technicians, and all departments for encouraging them to adopt environmental-friendly approaches.

In this regard, FT is one of the founding members of Ecoprod, which in 2009 established a resource centre for audiovisual professionals in order to reduce the ecological footprint on their production processes. That initiative created the Carbon' Clap, the first carbon-footprint calculator for audiovisual productions, which may be an analogous to the above-mentioned Albert. Also, a practical guide on how to make eco-productions and to create a charter for audiovisual companies complemented the tool.

For activities related to the third objective, the FT stresses the need for reporting on environmental problems, climate change and energy measures. The aim is to provide different publics with the keys for understanding the main social problems; for example, the launching of opening dialogues with staff, NGOs, and public at large. Also, FT lets publicly know its commitment in 2008 to the Charter of Sustainable Development for Public Establishments and Companies, with the objective of promoting and proposing future advanced solutions for sustainable broadcasters.

Furthermore, the French broadcaster launched in 2014 a carbon strategy in order to gradually reduce it. Within the Horizon 2020, FT is aimed at reducing greenhouse gas emissions by 8% with respect to the data in 2011. This global objective is applied to buildings, mobility, resource management, waste management, green ICTs, and sustainable purchases.

The Finnish YLE launched its environmental policy in 2009, when it established a programme for the promotion of environmentally friendly practices, including the improvement of natural resource efficiency and the recycling of used materials.

In the period 2014–2016, YLE updated its objectives both in terms of environmental footprint and cost savings. On the one hand, it reduced its real estate holdings and started to make more efficient use of facilities in Tampere and Pasila. Also, it modified its travel policy, reducing national flights, increasing videoconference meetings, giving priority to public transport and encouraging teleworking.

As regards energy, the web states that the consumption of electricity was reduced in 2014. Moreover, it was installed LED lighting, both outdoors and indoors. Electricity consumption in 2007 was 34.6 million kW, while in 2013 did not reach 30 million. Also, the cooling system was improved and the water use was reduced 13% in 2014, compared with the previous year.

Traffic is another working area included in the YLE's environmental strategy. In 2014, the fleet had 275 vehicles, 30 of which were hybrids or low-fuel consumption. The objective is to continue the renovation of low-emissions cars until reaching further to about a fifth of fleet. Additionally, YLE provides its staff with electric bicycles for short distances. When it comes to recycling, YLE promotes the product separation and recovery in their facilities. In 2014, it produced about 560,000 kilos of waste, being 7.3% mixed. The goal is to maintain the percentage below 10%. YLE also recycles furniture, appliance, and office supplies. According to data published, the recycling percentage today is 92.7% and continues to grow. From the management point-of-view, YLE created an Environment Committee to monitor strategies where members meet from two to four times a year.

## 5 The Case of Spanish Local Corporations

Spain has a unique model of public service media. It is divided into two main groups: the state corporation (RTVE) and 12 regional organizations, framed within territories known as Autonomous Regions. As regards these 12 audiovisual services, 5 of them are developing CSR actions, while there are significant differences in approaches and objectives.

Galicia, Catalonia and the Basque country launched crosscutting strategies including the main elements of CSR: environment, human resources, local communities, staff, audiences, and management. However, responsibility strategies in corporations from Balearic Islands and Castile-La Mancha are mainly focused on the social field, developing lines of work based, in most of the cases, on partnerships with non-profit making organizations.

The 12 analysed Spanish broadcasters launched web sections named transparency portals, except from *Canal Extremadura* and *Radiotelevisión Canaria*. An in-depth analysis shows that these sections include information on policy frameworks and economic management, although there are differences between them.

The *Corporació Catalana de Mitjans Audiovisuals* from Catalonia has a transparency unit divided into six subsections: corporate information, activity, management, recruitment, employment information, control, and assessment. It is worth stressing that CSR is not included.

The CRTVG from Galicia and the EITB from the Basque country do include CSR in their transparency portals. The first one also includes institutional information, corporate activity, and users interaction in the same section. The second one adds five elements: planning and results, regulation, group presentation and organization, recruitment and content self-regulation.

The Murcian broadcaster also has a corporate section with transparency objectives, in which include information on contractor's profile, normative, annual statements; budgets, minor contracts and contracts, chart and staff; human resources, audiovisual sector, and funding in 2015. On the other hand, the *Corporación Aragonesa de Radio y Televisión* groups contents together into four major blocs: institutional, organisational, and political information; data on planning, management, control, funding, and budgets; information on contracts, agreements, and assignments; and, finally, contact, right of access, and other information.

The transparency unit from the Asturian Broadcaster provide users with documents on legislation, structure, organization chart, framework terms of reference, contract program, management, recruitment, employment information, and control and assessment.

The Balearic broadcaster has also a transparency portal with six sections: recruitment, collaboration agreements, economic and budget information, and relation of real estate properties, remunerations, and right of access. The corporation from Castile-La Mancha shows four elements: institutional and organizations information, economic and budgetary data, public service – in which CSR is included – and right to access information.

The transparency section from the Andalusian RTVA includes from institutional and organizational to documents on procedures, services charter, citizen participation, and relevant legal information. Also, it includes data on high positions, planning and assessment, agreements, and subsidies. In the same vein, *Telemadrid* presents a section divided into three subsections: institutional and organizational information, relevant legal information, and economic and budgetary data.

Among these five above-mentioned corporations, only three of them have global strategies. Also, it should be noted the way environment becomes a priority issue in performance reports. The following table shows the information contained in CSR reports from the CCMA – 2013, EITB – 2013, 2014, and CRTVG – 2014 (Table 2).

**Table 2** Indicators of environmental performance in Spanish local broadcasters (prepared by the authors, 2016)

| Indicator of environmental performance | | | |
|---|---|---|---|
| Environmental performance | CCMA | EITB | CRTVG |
| Use of paper | ✓ | ✓ | X |
| Use of water | X | X | ✓ |
| Impact of ink and printing materials | ✓ | ✓ | X |
| Recycling policies | ✓ | ✓ | ✓ |
| Greenhouse gas emissions | X | ✓ | ✓ |
| Measures to be environmentally sound | ✓ | ✓ | X |

## *5.1 Use of Paper*

Corporations from Catalonia and the Basque country merely publish information on paper consumption in 2013 and 2014, respectively. In the case of CCMA, the amount reaches 22,920 kilos, while the EITB reaches 19,577 – lower than the figure of 20,130 in 2013. Further, it notes that 18,780 kilos of paper waste were produced – 16,880 in 2013 – from which 95.9% were recycled. The Basque EITB also mentions the launching of Zero Paper, an app for conducting processes without paper.

## *5.2 Use of Water*

CRTVG is the only one that provides information on this issue, while it is not very precise. It reports on the application of water reduction systems, but there are no specific figures.

## *5.3 Impact of Ink and Printing Materials*

The Catalan CCMA notes that the corporation used 300 kilos of toners in 2013, while the EITB employed 46.2 in 2014, 5 less than the previous year.

## *5.4 Recycling Policies*

The CCMA reports that it sent 872 elements for recycling in 2014, and donated material to foundations, education centres, and NGOs – 2.252 elements and 56 receivers in 2012. Also, it expresses its commitment to recycle the rest of material (2012: 18,960 kilos of wood, 379 of cans, 837 of batteries, 20 of plastic, 129 of

CD and DVD, 11,156 of electronic waste, and 2,550 of video tapes. The CRTVG focuses its activity on waste separation and disposal of obsolete equipment.

The EITB presents a list of results on its recycling policy between 2013 and 2014: (1) electronic devices: 630 kilos in 2013 and 510 in 2014; (2) fluorescent: 143 kilos in 2013 and 269 in 2014; (3) batteries: 153 kilos in 2013 and 289 in 2014; (4) sanitary waste: 9 kilos in 2013 and 13 in 2014; (5) paper and cardboard: 16,880 kilos in 2013 and 18,880 in 2014; (6) plastic: 800 in 2013 and 3,200 in 2014; (7) toner: 52 kilos in 2013 and 46 in 2014; (8) wood: 13,660 kilos in 2013 and 9,920 in 2014.

### 5.5 *Greenhouse Gas Emissions*

The EITB reached a 14.2% reduction of greenhouse gas emissions from 2012. Also, it emitted 4,103 CO2 tons in 2014 – 4,340 in 2013 – and reached 10,026.952 kW/h of electrical consumption in 2013, higher than previous year – 9,962.544 kW/h. Besides, EITB created an Audit Department on Energy Efficiency through which lights were changed and replaced by a new lighting installation; air conditioning was improved; energy measurement systems were renewed; and automatic switching off computers at night was installed. On the other hand, the CRTVG refers to the renovation of fifteen cars of its fleet in order to comply with the European rule on polluting gases.

### 5.6 *Measures to be Environmentally Sound*

This section includes all kind of actions aimed ultimately at reducing the impact on environment. In this regard, the action plan for was management and the manual of good environmental from the CCMA should be highlighted.

The EITB report includes the Mobility Plan and the renewal of the cars fleet for environmental reasons. Moreover, it notes that all public contracts should be based on environmental criteria. On the other hand, all environmental policies are certified by AENOR – Spanish Association for Standardization and Certification – rule ISO I400I.The main measure introduced by CRTVG was the renovation of the cars fleet, purchasing 120 LED lights for its facilities.

## 6 Towards a New Way of Reporting

The research is based on the parameters proposed by the Global Reporting Initiative in its G3.1 Sustainability Reporting Guidelines; specifically, the Media Sector Supplement. In the environmental section, it warns on the need to pay attention to: materials, energy, water, biodiversity, emissions, effluents and waste;

products and services, compliance, transport, suppliers' environmental assessment, and grievance mechanisms on the environment. This is evidence of a tightening for companies of environmental performance's criteria.

When it comes to materials, the guide recommends indicating the weight, volume and percentage of recycled products, as well as reducing consumption and energy requirements for products and services. As regards water use, items highlighted include total water collection through various sources and percentage of recycled and reused water.

GRI intends to monitor biodiversity trough data from operating facilities; description of the most significant impacts; protected and restored habitats; and number of affected species because of operations. Also, it calls for data on direct and indirect greenhouse gas emissions, as well as their intensity and measures for reducing them.

Regarding wastes, core indicators include measuring total emissions by quality and destiny; total weight of waste and effluents; number and volume of discharges; weight of transported weight; identification, size, statues of biodiversity in water bodies; and habitats affected in any way by water spills resulting from corporate activities.

When talking about the products and services aspect, the GRI want companies to report on initiatives to mitigate environmental impacts and percentage of products sold and their packaging materials. Reporting on compliance should include monetary value of significant fines and total number of non-monetary sanctions for non-compliance with environmental laws and regulations. Then, it calls for data on significant environmental impacts of transporting products and other goods and materials.

The overall section is referred to total environmental protection expenditures and investments by types, while suppliers' environmental assessment is focused on the percentage of suppliers examining environmental issues and negative impacts, both real and potential, on the supply chain. Grievance mechanisms refer to managed and resolved complaints.

## 7 Conclusions

While the European Union hardens environmental legislation, international initiatives as the Global Reporting Initiative looks into the requirements that corporations have to fulfil if they are to be responsible and environmentally friendly. Although this work is based on public service companies, whose commitment itself is directly related to their public mission, they cannot escape from corporate social responsibility rules, since they need to regain credibility and reputation in the eyes of a citizenry that has lost faith in their public media.

Results obtained reach a main conclusion: public broadcasters in Europe and Spanish Autonomous regions with global CSR strategies are not strictly following the rigid parameters proposed by the Global Reporting Initiative. Indeed, the content analysis shows that the GRI indicators in which the nine broadcasters are most active are: (1) measures to be environmentally sound; (2) greenhouse gas emissions; (3) recycling policies. On the contrary, much less attention is given to the impact of ink and printing materials. That means that public broadcasters are developing many actions, but there is a lack of detailed strategies. And also, those actions are included in content blocks with general and imprecise titles. In most of the cases, performance reports do not go beyond providing figures, leaving aside the real impact and procedure.

These results are not enough to evaluate environmental policies, since any public broadcaster report on a consistent strategy for reducing impacts. They are concrete short-term measures to provide one-off solutions. This is why it was opened a second line of research on environmental policies. A public broadcaster should have specific strategies if they are to set up targets; to evaluate, correct and implement new actions based on the results. That part of the study showed that the implementation of environmental policies is not widespread for the time being. The same applies for regional corporations in Spain, since only three out of 12 are working in this sense.

This article leads to the conclusion that, while there is some awareness of environmental matters on the part of European public broadcasters, there is a lack of a global strategy. Also, these nine organizations, which have implemented their strategies for corporate social responsibility as a self-regulation tool, represent less than half of the EU28.

**Acknowledgments** The work is part of the R+D research project CSO2015-66543-P, belonging to the National Programme for Encouraging Excellent Scientific and Technical Research, national subprogram of knowledge creation from the Spanish Ministry of Economy and Competitiveness, entitled "Indicators related to broadcasters' governance, funding, accountability, innovation, quality and public service applicable to Spain in the digital context". Also, it is part of the Prometeo Program, which belongs to the National Secretary of High Education, Science, Technology, and Innovation (SENESCYT) of the Ecuadorian government, and is developed at the Technical University of Loja and the Pontifical Catholic University of Ecuador – Ibarra headquarters. Authors of this paper are also members of the International Research Network on Communication Management – XESCOM.

## References

1. EBU (2015) Public service media contribution to society produced the Media Intelligence Service (MIS) of the European Broadcasting Union. www.ebu.ch/mis. Accessed Nov 2015
2. Bruntland Commission (1987) Our common future. Oxford University Press, Oxford
3. Clarke T (2014) Ética, valores y gobierno corporativo. La toma decesiones comerciales es un ejercicio moral. Available via BBVA OpenMind. https://www.bbvaopenmind.com/articulo/etica-valores-y-gobierno-corporativo/?fullscreen=true. Accessed May 2015
4. IUCN, UNEP, WWF (1991) Caring for the earth. A strategy for sustainable living. https://portals.iucn.org/library/efiles/edocs/CFE-003.pdf. Accessed May 2015
5. AccountAbility (2005) Stakeholder engagement standard. AccountAbility, London
6. European Commission (2003) Corporate social responsibility: a business contribution to sustainable development. European Union, Brussels

7. European Commission (2004) Corporate social responsibility: national public policies in the European Union. Unión Europea, Brussels. http://ec.europa.eu/social/BlobServlet?docId=1577&langId=en
8. World Business Council for Sustainable Development (2002) Corporate social responsibility: the WBCSD's journey. WBCSD, Geneva
9. World Business Council for Sustainable Development (2011) Global corporate citizenship initiative: redefining the future of growth: the new sustainability champions. International Labour Office, Geneva
10. Kennedy School of Government Corporate Responsibility Initiative (2004) Leadership, accountability and partnership: critical trends and issues in corporate social responsibility. Harvard University, Cambrigde
11. SustainAbility (2011) Global economic recovery and sustainability survey. SustainAbility, London
12. Moore MH (1995) Creating public value. Strategic management in government. Harvard University Press, New York
13. Muñoz E (2005) Gobernanza, ciencia, tecnología y política: trayectoria y evolución. Revista Arbor, CSIC 715:287–300
14. Prats J (2005) De la burocracia al management. Del management a la gobernanza. Las transformaciones de las Administraciones Públicas de nuestro tiempo. INAP, Madrid
15. Le Galés P (2007) "Gobernanza y cohesión social". In: P HalL (dir.): Congreso Regiones Capitales, Madrid, Comunidad de Madrid, pp 327–347
16. Olsen JP (2008) EU Governance: where do we go from here?. Working Paper. University of Oslo. https://www.sv.uio.no/arena/english/research/publications/arena-publications/workingpapers/working-papers2008/wp08_10.pdf. Accessed Feb 2015
17. Stohl C, Stohl M, Popova L (2009) A new generation of corporate codes of ethics. J Bus Ethics 90(4):607–622
18. Porter ME, Kramer MR (January–February 2011) Creating Shared Value. Harvard Business Review 89(1–2):62–77
19. Global Reporting Initiative (2012) Guía para la Elaboración de Memorias de Sostenibilidad y Suplemento Sectorial para Medios. https://www.globalreporting.org/resourcelibrary/Spanish-MSS.pdf. Accessed June 2015
20. European Commission (2009) Review of the EU Sustainable Development Strategy (EU SDS). http://ec.europa.eu/environment/eussd/

# Index


**A**
Analyze, 196, 202, 220, 258, 375, 445, 446
Architecture, 78, 84, 85, 87–88, 91, 509–511, 544
Assembly for disassembly, 543–567
Australia, 2, 17–31, 37, 42, 45, 46, 53, 54, 252, 341–342, 376, 377, 498
Awareness, 2, 5, 18, 20, 29, 31, 103–143, 149–181, 185–211, 215–240, 283–304, 307–335, 339–350, 353–389, 391–424, 427–469, 483, 521, 587, 592

**B**
Bhutan, 243–278
Brazil, 155, 391–424
Building materials, 544, 545, 546, 547, 548, 549, 553, 567
Business strategies, 18–19, 20, 45, 47, 50, 245, 253, 360, 396, 438

**C**
China, 77–97, 155, 251, 252, 366, 445
Cloud, 2, 5, 159, 515, 516, 517, 518, 520
Cochran Q-test, 488
Cohesion, 527, 539, 540, 541
Colombia, 353–389
Corporate Social Responsibility, 19, 39, 41, 105, 111, 113–117, 142, 159–160, 244, 248, 253, 396, 399, 434–435, 585, 586, 587, 588, 593, 594, 595, 598, 599
Curriculum, 4–6, 154, 180, 218, 239, 288, 303, 308, 310, 335, 560

**D**
Data mining, 433
Design for Deconstruction, 545, 549
Design oriented pedagogy, 572–573, 574, 575, 576, 577
Developed nations, 4, 312, 332, 340, 341, 342
Developing countries, 4, 104, 105, 111, 112, 119, 150, 154, 163, 180, 192, 244, 251, 252, 253, 274, 310, 335, 340, 355, 358, 359, 360, 363, 364, 366–368, 370, 371, 391, 398–399, 430, 437–441
Diverse cultures, 54, 75

**E**
Eduation programs, 216, 475–476, 567
Educating architecture students, 544
Education, 2–3, 31, 36–37, 41, 42, 54, 75, 78, 81, 108, 117–118, 150, 166, 191, 192, 198, 199, 201–202, 215–216, 218, 220, 222, 238, 239, 284, 285–288, 300, 303, 308, 310, 312, 318, 340, 342, 346, 349–350, 359, 422, 474, 475, 476, 477, 478, 479, 481, 482, 483, 486, 487, 488, 498, 499, 578
e-environment, 473–499
e-learning, 474, 476, 477, 479, 480, 481–485, 488, 498, 499
Environment, 2, 3, 4, 18, 27, 29, 30, 40, 41, 42, 47, 80, 81, 82, 84, 104, 107, 112, 113, 114, 117, 118, 150, 153, 187, 188, 191, 248, 249, 252, 274, 277, 278, 286, 288, 308, 311, 313, 354, 363, 364, 385, 393, 395, 397, 420, 429–430, 434, 438, 465, 479, 481–485, 498, 499, 541, 558, 578




T. Issa et al. (eds.), *Sustainability, Green IT and Education Strategies in the Twenty-first Century*, Green Energy and Technology, DOI 10.1007/978-3-319-57070-9

Environment issues, 21, 79, 82, 106, 142, 157, 163, 180, 187, 210, 284, 303, 304, 354, 361, 363, 365, 366, 369, 385, 395, 401, 429, 537, 588, 598
Environmental, 2, 3, 4, 19–20, 36–37, 40, 41, 46, 78, 79, 80, 81, 82, 83, 84, 86, 87, 91, 92, 104, 105, 106, 107, 111, 112–113, 115, 116, 117, 139, 141, 142, 150, 151, 152, 153, 154, 156, 157, 161, 163, 177–178, 188, 190, 210, 217, 237, 245, 247, 248, 254, 274, 284, 285, 287–288, 303, 304, 311, 313, 333, 342, 349, 354, 356, 360, 361, 362, 363, 365, 366, 369, 370, 371, 388, 394, 395, 396, 397, 398, 399, 400, 401, 417, 428, 429, 431–432, 433–434, 435, 437, 438, 439, 442, 443, 444, 465, 466, 467, 468, 469, 509, 548, 586, 587, 588, 589–592, 596, 597, 599
Environmental responsibility, 117, 237, 268, 274, 278, 585–599
Ethical strategies, 37, 38, 40, 45, 49, 51, 53, 54, 56
Europe, 42, 46, 53, 54, 56, 80, 247, 399, 473–499, 587, 589–594, 598
Evaluation, 2, 7–10, 43, 45, 47, 57, 174, 478, 503–522, 529, 530, 544, 554, 580
Exploratory study, 400, 403, 421, 422, 423, 424

## F

Failure, 18, 29, 30, 138, 177, 178, 181, 186, 187–188, 191, 237, 312, 333, 335, 349, 366, 417, 418, 465, 468, 469
France, 43, 46, 53, 54, 56, 57, 252, 588, 590, 592, 593
Future, 3, 4, 13, 31, 36, 38, 75, 85–86, 87, 90, 91, 104, 105, 107, 112, 151, 163, 181, 186, 187, 197, 204, 211, 216, 217, 221, 229, 244, 247, 248, 252, 288, 303, 304, 310, 311, 313, 334–350, 355, 356, 359, 363–364, 387, 392, 393, 394, 396, 397, 401–402, 422, 423, 428, 430, 442, 443, 467, 468, 521–522, 567, 575, 578–581

## G

Game framework, 503–522
Gamification, 504, 505, 506, 521
Green IT, 2, 3, 4, 5, 6, 12, 13, 14, 261, 288, 346
Green Supply Chain Management, 429, 431–433, 434, 436, 441, 467, 469

## H

Higher education, 2, 4–6, 17–31, 35–58, 149–181, 215–240, 283–304, 308, 309, 310, 311, 314, 316, 329, 333, 334, 335, 342, 345, 350, 391–424

## I

Impact, 2, 3, 12, 13, 30, 31, 37, 39, 41, 48, 75, 86, 104, 109, 110, 113, 114–115, 124, 150, 154, 157, 161, 180, 187, 188, 191, 192, 193, 218, 245, 251, 363, 365, 371, 430, 431, 469, 474, 479, 480, 495, 504, 516, 587, 588, 590, 593, 596, 598
India, 215–240
Interactive learning, 71
IRNet project, 474, 498
Iterative improvement, 75

## L

Learning benefits, 484
Learning experiences, 2, 10, 45, 54, 63, 66, 73, 75, 154
Learning outcomes, 5, 66, 68, 70, 71, 73, 75, 480, 580
Learning styles, 66, 67, 68, 72, 73, 75
Learning system, 474, 569–581
Learning technology, 474, 481

## M

Mindset, 2, 3, 4, 6, 25, 30, 36, 38, 45, 53, 55, 57, 263, 346, 379

## N

Negative, 2, 18, 19, 20, 25, 27, 28, 29, 30, 31, 40, 45, 52, 87, 90, 104, 180, 188, 250, 309, 359, 371, 392, 403, 404, 414–419, 431, 444, 465, 488, 598
Net generation, 63–75
Nigerian organizations, 427–469

## O

Opportunities, 40, 87, 104, 113, 125, 135, 136, 141, 185–211, 246, 254, 255, 257, 258, 265, 278, 287, 307–335, 340, 342, 347, 349, 364, 365, 379–382, 383, 474, 485, 521, 527, 528, 540, 572, 580

## P

Pakistan, 307–335
Participatory Action Research, 526
Participatory learning, 569–581

Positive, 11, 18, 21, 25, 26, 28, 30, 31, 52, 90, 113, 152, 202, 341, 364, 392, 403, 414–419, 420, 421, 423, 436, 456, 463, 479, 488
Prototyping, 526, 528, 529, 535, 536
Public broadcasters, 588, 591, 598–599
Public media, 588, 598
Public organizations, 246, 247, 254, 257, 258, 277
Public places, 527, 528, 536, 541

**Q**
Qualitative, 2, 42, 43, 45, 86, 196, 257, 258, 277, 402, 404, 429, 446, 463, 479, 483, 518

**R**
Reuse of building materials, 545, 547, 549
Risks, 135, 138–139, 141, 151, 161, 185–211, 215–240, 244, 246, 254, 265, 266, 267–268, 270, 273, 278, 293, 294, 300, 308, 314, 316, 318, 328, 330, 331, 335, 340, 346, 347–348, 349, 355, 379–380, 383–385, 392, 414, 417, 420, 421, 422, 423, 547

**S**
Saudi Arabia, 339–350
Serious game, 503–522
Service design, 161, 315, 528, 542
Social innovation, 525, 526, 527, 535, 540, 580
Social Performance, 112
Students' competencie, 478, 484
Sustainability, 1–14, 17–31, 36, 37, 38, 40, 77–78, 79, 80–84, 86, 87, 94–95, 96, 97, 243–278, 427–469, 503–504, 505, 508, 516, 519, 521, 536, 543, 586, 588, 592
Sustainability awareness, 20, 31, 103–143, 149–181, 185–211, 215–240, 283–304, 307–335, 339–350, 353–389, 391–424
Sustainability in Thailand, 119–124
Sustainability pedagogy, 544
Sustainable construction, 544, 548
Sustainable development, 5, 29, 31, 36, 37, 38, 39, 40, 45, 46, 48, 49, 50, 51, 55, 57, 78, 79, 81, 84, 87, 97, 104, 105, 107, 108–111, 117, 118, 124, 142, 150, 151, 155, 156, 180, 244, 245, 246, 247, 248, 249, 252, 274, 284, 285, 286, 288, 300, 304, 310, 312, 313, 345, 354, 355, 356–357, 358, 359, 360, 362, 363, 364, 371, 388, 393, 394, 395, 396, 398, 400, 420, 422, 423, 424, 429, 437–438, 580, 586
Sustainable Growth, 354, 357–358, 388, 429, 526
Sustainable student engagement, 63–75
System events Toolbox, 525–542

**T**
Thai users, 104, 129, 131, 132, 133, 136, 138
Thailand, 103–143
Triple Bottom Line, 3, 19, 55, 111–113, 152, 153, 160, 188–190, 244, 248, 253, 278, 309, 313, 360, 401, 434, 435, 463
Tukey test, 488
21st century skills, 6, 7, 13, 42, 154, 482, 538, 570, 572

**U**
University, 2, 5, 10, 12, 13, 86, 474, 475, 476, 477, 478, 481, 482, 483, 486, 487, 492, 495, 497, 498, 532, 535, 536, 550, 561

MIX
Papier aus verantwortungsvollen Quellen
Paper from responsible sources
FSC® C105338

If you have any concerns about our products,
you can contact us on
**ProductSafety@springernature.com**

In case Publisher is established outside the EU,
the EU authorized representative is:
**Springer Nature Customer Service Center GmbH**
**Europaplatz 3, 69115 Heidelberg, Germany**

Printed by Libri Plureos GmbH
in Hamburg, Germany